SPECTRAL TRANSFORM AND SOLITONS

STUDIES IN MATHEMATICS AND ITS APPLICATIONS

VOLUME 13

Editors:
J.L. LIONS, *Paris*
G. PAPANICOLAOU, *New York*
R.T. ROCKAFELLAR, *Seattle*
H. FUJITA, *Tokio*

NORTH-HOLLAND PUBLISHING COMPANY
AMSTERDAM · NEW YORK · OXFORD

SPECTRAL TRANSFORM AND SOLITONS:
TOOLS TO SOLVE AND INVESTIGATE NONLINEAR EVOLUTION EQUATIONS

VOLUME ONE

Francesco CALOGERO
and
Antonio DEGASPERIS

*Dipartimento di Fisica, Università di Roma,
Istituto Nazionale di Fisica Nucleare, Sezione di Roma*

1982

NORTH-HOLLAND PUBLISHING COMPANY
AMSTERDAM · NEW YORK · OXFORD

© *North-Holland Publishing Company, 1982*

All rights reserved. No part of this publication may be reproduced, stored in a retrieval system or transmitted, in any form or by any means, electronic, mechanical, photocopying, recording or otherwise, without the prior permission of the copyright owner.

ISBN 0444 863680

Publishers:

NORTH-HOLLAND PUBLISHING COMPANY
AMSTERDAM · NEW YORK · OXFORD

Sole distributors for the U.S.A. and Canada:
ELSEVIER SCIENCE PUBLISHING COMPANY, INC.
52 VANDERBILT AVENUE,
NEW YORK, N.Y. 10017

Library of Congress Cataloging in Publication Data

Calogero, F.
 Spectral transform and solitons.

 (Studies in mathematics and its applications; v. 13)
  Bibliography: p.
  Includes index.
   1. Evolution equations, Nonlinear. 2. Solitons.
 3. Spectral theory (Mathematics) 4. Transformations
 (Mathematics) I. Degasperis, Antonio. II. Title.
 III. Series.
 QC20.7.E88C26   515.3'55   81-22599
 ISBN 0-444-86368-0 (v. 1)   AACR2

PRINTED IN THE NETHERLANDS

CONTENTS

Contents	v
Contents of Volume II	x
Preface	xi
Foreword	xiii
Chapter 0. Introduction	1
0.N. Notes to chapter 0	3
Chapter 1. The Main Idea and Results: An Overview	7
1.1. Solution of linear evolution equations by Fourier transform	7
1.2. A class of solvable nonlinear evolution equations	14
1.3. The spectral transform	15
1.3.1. Direct spectral problem	16
1.3.2. Inverse spectral problem	18
1.3.3. The spectral transform	20
1.4. Solution of nonlinear evolution equations via the spectral transform	22
1.5. Relation to the Fourier transform technique to solve linear evolution equations	26
1.6. Qualitative behaviour of the solutions: solitons and background	27
1.6.1. Solitons	28

	1.6.2. Background	32
	1.6.3. Generic solution	33
1.7.	Additional properties of the solutions	34
	1.7.1. Bäcklund transformations	35
	1.7.2. Nonlinear superposition principle	41
	1.7.3. Conservation laws	42
1.8.	A list of solvable equations	48
1.N.	Notes to chapter 1	63

Chapter 2. The Schroedinger Spectral Problem on the Line — 68

2.1.	Direct spectral problem	68
	2.1.1. Transformation properties	79
2.2.	Inverse spectral problem	81
2.3.	The spectral transform	87
2.4.	Formulae relating two functions to the corresponding spectral transforms	89
	2.4.1. Wronskian integral relations	89
	2.4.1.1. Additional wronskian integral relations	105
	2.4.2. Spectral integral relations	108
	2.4.2.1. Additional spectral integral relations	115
	2.4.3. Connection between wronskian and spectral integral relations	117
2.N.	Notes to chapter 2	118

Chapter 3. Nonlinear Evolution Equations Solvable by the (Schroedinger) Spectral Transform — 120

3.1.	KdV and higher KdV's	120
3.2.	Analysis of the solutions	131
	3.2.1. Solitons and solitrons	132
	3.2.2. The weak field limit	164
	3.2.3. Generic solution	165
	3.2.4. Special solutions of the KdV equation	167
	3.2.4.1. Rational solutions	167
	3.2.4.2. Asymptotically diverging solutions	171
	3.2.4.3. Similarity solutions and ODE's of Painlevé type	171
3.N.	Notes to chapter 3	175

Chapter 4. Bäcklund Transformations and Related Results — 179

    4.1. Bäcklund transformations — 181

    4.2. Commutativity of Bäcklund transformations and nonlinear superposition principle — 191

    4.3. Resolvent formula — 199

    4.4. Nonlinear operator identities — 201

    4.5. Generalized Bäcklund transformations and resolvent formula — 202

    4.N. Notes to chapter 4 — 205

Chapter 5. Conservation Laws — 208

    5.N. Notes to chapter 5 — 224

Chapter 6. Extensions — 226

    6.1. More variables — 227

    6.2. Coefficients depending linearly on x — 234

    6.3. Solutions of the KdV equation that are asymptotically linear in x — 255

        6.3.1. The Schroedinger spectral problem with an additional linear potential — 258

        6.3.2. Solution of a nonlinear evolution equation including as a special case the cylindrical KdV equation — 268

        6.3.3. Conservation laws — 270

        6.3.4. The cylindrical KdV equation — 276

    6.4. Solutions of the KdV equation with one real double pole — 281

    6.5. Evolution equations associated with the spectral problem based on the ODE $-\psi_{xx}(x)+u(x)\psi(x)=k^2[\rho(x)]^2\psi(x)$ — 287

    6.N. Notes to chapter 6 — 296

Appendices

    A.1. On the number of discrete eigenvalues of the Schroedinger spectral problem on the whole line — 299

    A.2. Orthogonality and completeness relations for the Schroedinger spectral problem on the whole line — 306

A.3.	Asymptotic behaviour (in k) of the transmission and reflection coefficients	309
A.4.	Dispersion relations for the transmission coefficient	311
A.5.	The inverse spectral Schroedinger problem on the whole line	313
	A.5.N. Notes to appendix A.5	330
A.6.	Wronskian integral relations: proofs	330
A.7.	Spectral integral relations: proofs	338
A.8.	A formula for the variation of the coefficients of the asymptotic expansion of the phase of the transmission coefficient	341
A.9.	Properties of the operators $\Lambda, \tilde{\Lambda}, L, \tilde{L}$, and other formulae	343
A.10.	The two-soliton solution of the KdV and higher KdV equations	365
A.11.	Miura and Gardner transformations and related results	370
A.12.	Bäcklund transformations, Darboux transformations and Bargmann strip	378
A.13.	Asymptotic expansion of $C(k) = 2ik[1 - 1/T(k)]$	386
A.14.	Conserved quantities for generalized KdV equations	388
A.15.	Reflection and transmission coefficients at $k=0$	390
	A.15.N. Notes to appendix A.15	394
A.16.	The spectral transform outside of the class of *bona fide* potentials	394
A.17.	Applications of the wronskian and spectral integral relations to the Schroedinger scattering problem on the whole line	398
A.18.	On the class of equations $\eta(L)u_t = \alpha(L)u_x$	412
A.19.	Examples of functions with explicitly known spectral transform	418
A.20.	A general approach based on the algebra of differential operators; connections with, and amongst the spectral transform method, the Lax approach and the AKNS technique	444
	A.20.N. Notes to appendix A.20	461
A.21.	Local conservation laws: proofs	462

A.22. "Variable phase approach" to the Schroedinger scattering
     problem on the whole line   470
A.23. KdV and higher KdV equations as hamiltonian flows: an
     outline   477
     A.23.N. Notes to appendix A.23   486

References   488

Subject Index   511

List of Symbols   514

CONTENTS OF VOLUME II

Chapter 7. Introduction to Spectral Problems Based on Systems of Linear ODEs

Chapter 8. The Class of Nonlinear Evolution Equations Solvable via the Zakharov–Shabat Spectral Transform

Chapter 9. The Class of Nonlinear Evolution Equations Solvable via the Matrix Schroedinger Spectral Transform

Chapter 10. Other Spectral Problems and Associated Nonlinear Evolution Equations

Appendices

PREFACE

The idea of this book has developed from a set of lecture notes, produced by one of us (FC) in the spring 1979 at the Institute of Theoretical Physics in Groningen [C1979]. It is a pleasure in this connection to thank Professor David Atkinson, who went through them thoroughly, improving the English and sometimes the substance as well. Also to be thanked is Mrs. Marianne Morsinsk, for her accurate typing of these lecture notes.

The present version of this volume has involved a drastic rewriting of the lecture notes and the addition of new material. We were actually planning to cover much more ground, including a treatment of the classes of equations solvable via the (so-called) Zakharov–Shabat spectral problem, via the spectral problem associated with the matrix Schroedinger equation (with which our own research work has been largely associated) and via other spectral problems; but the growth of the material under our hands has forced us to defer the inclusion of these topics to a second volume, that we hope to publish without undue delay.

This volume has been partly written during the academic year 1979–80, while both authors were in England, FC at the Department of Applied Mathematics of Queen Mary College (QMC) in London and AD at the Department of Mathematics of the University of Manchester Institute of Science and Technology (UMIST). It is a pleasure for us to thank the Faculty of Sciences of Rome University for granting to both of us a one-year leave, the Science Research Council of Great Britain for the financial support that has made our visits to England possible, Professors R.K. Bullough of UMIST and I.C. Percival of QMC for arranging our visits and for providing, together with their colleagues at UMIST and QMC, a pleasant and stimulating environment. We would also like to thank the Laboratory of Mathematical Physics at the University of Montpellier, and in particular professor P.C. Sabatier, for the friendly hospitality on several occasions, which fostered our work on this book. For the same reason one of us (AD) also wishes to thank the International School for Advanced Studies in Trieste; and he also gratefully acknowledges the NATO research grant No. 057.81 that has provided partial support to his visits to Manchester and Montpellier in 1981. Finally, the other one (FC) would of course like to thank the University of Groningen for the appointment as (first) Zernike Visiting Professor in the spring 1979 (see above).

We also wish to thank C. Pöppe, J.K. Drohm, L. Kok and E. Abraham, for computer-producing the various graphs that appear in this volume (their contributions are separately acknowledged at the appropriate places below), and Mrs. P. Queeley in Manchester and Miss S. Maiolo in Rome for typing the text.

This book would not have been written had we not been able to interact closely, over the last few years, with our colleagues and friends, from the United States, the Soviet Union, Europe, Japan, China, Latin America, India, who make up the "soliton" community. We dedicate our book to them all, in the hope that our collaboration and friendship be allowed to flourish in the years that come.

FOREWORD

As it is apparent from the table of Contents, this book has a multi-layered structure. This results from a subdivision into chapters, sections, subsections and appendices, that should be useful both to the student who works through the whole book and to the researcher interested only to retrieve the treatment of one specific topic. Moreover, in chapters one to five, a vertical line to the left of the text identifies those parts that can be omitted in a first reading (or read through in a cursory way). The idea is, for the self-taught student, to read and reread through the body of the book following these guidelines, and to work through the Appendices whenever he feels a specific need for it. Of course, no reader is expected (much less advised) to follow these guidelines with absolute consistency; nor have we strictly conformed to them in our presentation (equations from the part meant for a second reading are occasionally, if seldom, referred to in the part meant for the first reading). Thus the main usefulness of this typographical device is to provide an evident indication of which parts are less essential for a basic understanding.

As for the appendices, they are of different kinds. Some complement the main text (for instance, by providing explicit proofs); some elaborate in more detail topics that are only mentioned or outlined in the main text; some treat a specific topic, largely independently from the main text. The title of each appendix should provide a fairly clear indication of its contents; we have tried to make their presentations as self-consistent as possible.

Each chapter (and a few appendices) is concluded by a section devoted to Notes. They follow closely the main text, giving a historical, and especially bibliographical, commentary; they are meant to provide guidelines for additional reading (few references are mentioned in the text).

Within each chapter, section, subsection and appendix, all equations are numbered progressively; within the same unit, they are referred to by their number; outside their unit, they are identified by their complete address, including, in addition to their number, the indication of chapter, section and, if need be, subsection, or, if they appear in an appendix, of the code identifying it (the letter A followed by a number). Thus equation (13) of subsection 1.3.2 is identified as (13) within that subsection and as (1.3.2.-13) elsewhere; equation (34) of appendix A.1 is identified as (34) within that appendix and as (A.1.-34) elsewhere. To facilitate retrieving equations, the indication of the chapter, section and, if need be, subsection, or of the appendix number, is reported on the top of each page.

References are identified by an acronym; they are listed at the end, in strict alphabetical order (of the acronyms). We have not attempted to be complete in reporting references; nor would have this been useful (or, for that matter, feasible), due to the enormous growth of the number of publications in this field that has occurred over the last few years. We have tried to provide a complete list of all the references closely connected to the type of approach followed in this book; and to report also all the review articles and books that have appeared in the last few years and that are devoted, in full or in part, to this topic. Finally we have, very selectively, mentioned a number of papers because of their historical significance and/or to provide insights into other approaches or related developments. In this last respect we are aware that omissions due to oversight have certainly occurred, and we apologize for this to the readers as well as to the authors who may have been unjustifiably excluded (indications of omissions will be appreciated and might give us a chance to remedy our faults in volume II).

This volume is concluded by a subject index and by a list of mathematical symbols.

This book is primarily aimed at students who are approaching this subject for the first time, although we hope that it also proves useful as a reference text for the expert researcher. Thus, its treatment is biased towards the presentation of the main ideas in the simplest context; more general approaches are reported later. This entails some repetitiveness, presumably justified by the didactic advantages of this method. In any case, the layered structure of the book should make it possible to read through it without being diverted from the mainstream of the argument by those details (including proofs) that can be assimilated at a subsequent stage.

This book should be understandable to any reader who possesses the standard notions of calculus, including an elementary treatment of ordinary and partial differential equations. In the context of ODEs, the elementary notions of spectral theory for Sturm–Liouville singular eigenvalue problems should be known; in the context of PDEs, the elementary notions relevant to the analysis of linear evolution equations. These notions are reviewed in the book, but too tersely to be assimilated by a reader who is totally ignorant of them.

This book should be usable for self-instruction, or as a textbook (possibly also at the undergraduate level). It provides the material appropriate for a one-semester course (25–30 hours); but individual lecturers may want to omit the detailed treatment of some parts and instead cover some of the topics that have been left out, for instance the connections with perturbation theory, or problems with periodic boundary conditions, or the vast field of applications. Part of the material has been tested in the classroom; by FC in the academic years 1977–80 in Rome and in several short series of lectures given in Lausanne, Heidelberg, Groningen and at QMC in London; by AD in short series of lectures given at a summer school in Istanbul and at UMIST in Manchester. Both of us also reviewed parts of this material in various summer schools, meetings and symposia, and several of these contributions have appeared in print ([C1977], [C1978b], [C1979], [C1979a], [CD1977], [CD1980], [D1977], [D1978], [D1979], [D1979a], [C1980d], [D1980]). Moreover, most of the results reported in this book have been already published in original papers, that are referred to below wherever appropriate. The knowledgeable reader will however also notice a number of novel results; although it should be emphasized that our main goal has not been their display, but rather the presentation of a unified and transparent picture.

CHAPTER ZERO

INTRODUCTION

The soliton was discovered (and named) in 1965 by Zabusky and Kruskal [ZK1965], who were experimenting with the numerical solution by computer of the Korteweg–deVries (KdV) equation. This nonlinear partial differential equation had been introduced at the end of the last century to describe wave motion in shallow canals [KdV1895]. Zabusky and Kruskal studied it because of its relevance to plasma physics, as well as to the Fermi–Pasta–Ulam puzzle [FPU1955] (for a fascinating account of the motivations that led to the "birth of the soliton", see [K1978]). The first scientific description of the soliton as a natural phenomenon goes however back to the first half of the nineteenth century, and was reported by J. Scott-Russell in the following prose: [SR1845]

> "I was observing the motion of a boat which was rapidly drawn along a narrow channel by a pair of horses, when the boat suddenly stopped—not so the mass of water in the channel which it had put in motion; it accumulated round the prow of the vessel in a state of violent agitation, then suddenly leaving it behind, rolled forward with great velocity, assuming the form of a large solitary elevation, a rounded, smooth and well defined heap of water, which continued its course along the channel apparently without change of form or diminution of speed. I followed it on horseback, and overtook it still rolling on at a rate of some eight or nine miles an hour, preserving its original figure some thirty feet long and a foot to a foot and a half in height. Its height gradually diminished, and after a chase of one or two miles I lost it in the windings of the channel. Such, in the month of August 1834, was my first chance interview with that singular and beautiful phenomenon...".

But the real breakthrough occurred in 1967, when the idea of the spectral transform technique was introduced by Gardner, Greene, Kruskal and Miura as a means to solve the Cauchy problem for the KdV equation

[GGKM1967]. Soon afterwards Lax put the method into a framework, that provided a clear indication of its generality and greatly influenced future developments [L1968]; and a few years later Zakharov and Shabat, by a nontrivial extension of the approaches of GGKM and Lax, were able to solve the Cauchy problem for another important nonlinear evolution equation, the so-called nonlinear Schroedinger equation [ZS1971]. The way was thereby opened for the search and discovery of several other nonlinear evolution equations, or rather classes of such equations, solvable by these techniques; a process that continues unabated to the present days. The subject has moreover branched out into other areas of mathematics (algebraic and differential geometry, functional and numerical analysis), and its applications are percolating through the whole of physics (from nonlinear optics to hydrodynamics, from plasma to elementary particle physics, from lattice dynamics to electrical networks, to superconductivity and to cosmology), and are indeed appearing also in other scientific disciplines (epidemiology, neurodynamics, etc.). This is of course related to the central rôle played by nonlinear evolution equations in mathematical physics, and more generally in applied mathematics; and to the fact that the spectral transform approach constitutes in some sense (which will be made more precise below) an extension to a nonlinear context of the Fourier transform technique, whose all-pervading rôle for solving and investigating linear phenomena is of course well known.

The broadness of scope, both in pure and in applied mathematics, that has been outlined here, as well as the dynamical stage of development of this field of enquiry, excludes the possibility of providing a systematic and complete coverage of the theory and/or its applications. This book focusses on one approach, and leaves out any specific treatment of applications. The guiding thread is the analogy of the spectral transform technique for solving (certain classes of) nonlinear evolution equations, to the Fourier transform method for solving linear partial differential equations.

The material covered in this book is displayed in sufficient detail by the Tables of Contents, not to require an additional review here. Let us rather mention the main topics we have omitted. These include the relationship to hamiltonian dynamics (except for a terse outline in appendix A.23); discretized problems (finite-difference equations, mappings, dynamical systems with a finite number of degrees of freedom); problems on a finite interval, and in particular problems with periodic boundary conditions; the "inverse" problem of ascertaining, given a nonlinear evolution equation, whether it belongs to some class of equations tractable by spectral transform techniques, and if so, what is then the appropriate spectral problem that provides the basis to introduce the spectral transform (this is still largely an open problem, whose investigation is now actively pursued mainly in the framework of differential geometry); the study of equations "close" to those solvable by these techniques, or equivalently the use, as the point of departure of a perturbative approach, of some nonlinear evolution equation solvable via the spectral transform (rather than a "brutally" linearized equation); the many other approaches that overlap and complement that treated in this book; and of course the open ended range of applications.

One final word about our standards of mathematical rigour, which have not been set very high. Our justification for neglecting the pursuit of rigour is, firstly, pedagogical, namely a reiteration of our intent to opt for maximal simplicity; secondly, it hinges on the relative novelty of the topic treated here. Indeed, although a more rigorous presentation than that given below could be reported, much still remains to be done in that direction. (The analogy with the Fourier transform case is instructive: there are many interesting results associated with it that were established soon after it was publicised [Fou1822] or even before, such as Parseval's theorem [Par1805]; yet only now, more than a century later, is the theory of the Fourier transform attaining a reasonably rigorous shape; indeed in order to reach such a stage it has been necessary to wait for the introduction of distributions or generalized functions [Sch1950], [GSV1958]).

Obviously this disclaimer to mathematical rigour should not be construed as a justification for mistakes or sloppiness; we have tried to stay clear from both these sins (or at least, to avoid mistakes and to minimize sloppiness). But we will generally be vague about the class of functions we are working with, except for the most essential specifications; it being generally understood that all the nice properties (boundedness, smoothness, analyticity, asymptotic convergence, integrability; you name it) may be invoked if need be. A punctilious analysis of the more permissive conditions under which each of the results reported below can be proved, let alone any attempt to extend these restrictions, lies clearly beyond the scope of this presentation.

0.N. Notes to chapter 0

In addition to those referred to in the text, the main contributions to the early history of "soliton" theory include: [Mi1968], [MGK1968], [SG1969], [G1971], [KMGZ1970], [GGKM1974], [Hi1971], [ZF1971], [ZDK1968], [T1971].

It is perhaps of interest to note that the keyword "soliton" became an entry of the Subject Index of Physics Abstracts in 1973 (January-June issue).

A list of "solvable" equations is given in section 1.8. Among the more important papers that have enlarged the class of solvable equations are [ZS1974] and [AKNS1974]; the main papers that have extended the class of solvable equations using the approach described in this book are [C1975], [CD1976a], [CD1977a], [CD1978a], [CD1978c], [CD1978h], [CD1981].

For developments towards other branches of mathematics see for instance: [DMN1976], [H1976], [H1977], [Man1978].

The literature on applications is enormous. The following review papers and books provide useful guidance: [SCM1973], [Bu1977], [Bu1977a], [Ce1977], [BS1978], [Bu1978], [FML1978], [LS1978], [GE1979], [TB1979], [W1979], [BC1980].

The following list of books and survey papers, that are entirely devoted to these topics or include some treatment of them, should prove useful especially to readers interested in approaches different from that followed in this book: [SCM1973], [Ar1974], [N1974], [W1974], [KZ1975], [Mo1975], [DMN1976], [H1976], [Mat1976], [Mi1976], [Bu1977], [Ce1977], [CS1977], [H1977], [Mar1977], [Ab1978], [Ba1978], [C1978], [CC1978], [Cra1978], [DDJ1978], [FML1978], [LS1978], [Mak1978], [Man1978], [Sab1978], [ChD1979], [MZ1979], [R1979], [W1979], [ZM1979], [BB1980], [BC1980], [BPS1980], [Lam1980], [ZMNP1980], [AS1981], [E1981], [EVH1981], [DEGM1982].

The following list of papers that are closely related to the approach followed in this book should instead prove particularly useful to the reader who wishes to pursue some detail insufficiently covered in the book, or to test whether some development, suggested by the results given in the book, has already been accomplished: [C1976], [C1975], [C1976a], [C1975a], [CD1976c], [CD1976a], [CD1976b], [CD1977a], [CD1976d], [CD1976e], [E1977], [E1977a], [CaC1977], [CD1977b], [CD1978a], [CD1978b], [CD1978c], [CD1978d], [CD1978e], [CD1978f], [CD1978g], [CD1978h], [CD1978i], [C1978b], [COP1979], [BLR1978], [BLR1978a], [CD1980], [C1978a], [CL1977], [Le1978], [LR1978], [LR1979], [LR1979a], [KM1979], [San1978], [San1979], [Pil1980], [GK1980a], [GK1980b], [C1979a], [C1979b], [C1980d], [CD1980a], [CD1981], [BLR1978b], [BLR1979], [D1979], [D1979a], [BLR1980], [BMRL1980], [BR1980a], [BR1980b], [BR1980c], [D1980], [DOP1980], [LB1980], [LOPR1980], [LRB1980], [MS1980], [MST1980], [San1980], [BR1981a], [BR1981b], [BRL1981], [Le1981], [LPS1981], [LSP1981], [R1981], [San1981], [BLOPR1982], [BLR1982], [BLR1982a], [BLR1982b], [BR1981c], [D1982], [LR1982], [LRS1982]. This is a list of the papers on the topics treated in this book (including volume II), that have been produced by us and by our colleagues in Rome over the last few years, arranged in an order that compromises between logic and chronology; some papers that refer to discretized systems are included, as well as a few papers by other authors, due to their close topicality.

The main original papers on the relationship with hamiltonian dynamics are [G1971] and [ZF1971]. The subsequent literature is vast, see e.g. [FN1974], [GD1975], [GD1976], [GD1977], [L1977], [B1978], [GD1978a], [GD1978b], [Man1978], [GD1979], [GDo1979], [A1979], [A1979a], [BePe1980], [Ma1980]; and see appendix A.23 (including the Notes).

For discretized equations see the review papers [Ab1977], [Ab1977a], [Ab1978], [Le1978].

The recent literature on dynamical systems is very large. The main papers that have originated this revival are [T1967], [C1971], [H1974], [F1974a], [F1974b], [Ma1974], [Mo1975a]. Review papers are: [T1975], [C1977a], [C1979b], [C1980b], [T1980], [OP1981]. See also [CC1978], [ChD1979] and [T1981]. For a terse discussion of the remarkable connection between certain nonlinear evolution equations solvable by the spectral transform technique and certain integrable dynamical systems (with a finite number of degrees of freedom) see subsection 3.2.4.1 (including Notes).

The treatment in this book is confined to problems on the whole line, generally with vanishing boundary conditions at infinity. The mathematical techniques relevant to analogous problems, but in a finite interval, bear, in the case with periodic boundary conditions, the same sort of relationship to the problems treated here, as the Fourier series to the Fourier integral. In fact, however, such problems involve generally a higher dose of mathematics than is used in this book, for instance familiarity with elliptic functions and some algebraic geometry. The relevant literature is fairly ample; see for instance [L1975], [MKM1975], [DMN1976], [MKT1976], [Kri1977].

The research on prolongation structures (the geometrico-differential technique to investigate whether a given nonlinear evolution equation is integrable) is surveyed in [EW1977]. See also [H1976], [H1977] and [PRS1979]. An elementary recent introduction requiring hardly any geometrico-differential background is provided by [Kau1980].

The construction of a "perturbation theory" based on integrable nonlinear evolution equations is an important task that is far from completed. Steps in this direction are described in [N1977], [KML1977], [MLS1977], [KM1978], [KN1978b], [EVH1981], [KA1981].

It is impossible to outline here the many other approaches that now exist to solve and investigate nonlinear evolution equations; but the list of books and survey papers given above should be sufficient to provide adequate guidance. The same for applications.

The implicit notion that there exist different degrees of mathematical rigour may appear disturbing to some readers: a theoretical physicist friend of us, having once claimed to have proved something "rigorously", was promptly asked by a mathematician whether it would not have sufficed to prove it "normally"; the implication being of course that a proof is a proof and needs no adjective to qualify it. Not so: for even within "pure" mathematics the standards of rigour vary immensely. As a telling instance, see the popularized analysis [NN1958, pp. 37–39 and 104–109] of the hidden assumptions—that ought to be brought out in order to make the proof "rigorous"—contained, in the context of formalized mathematical logic, in the proof of even the most elementary theorem of arithmetic, such as that negating the existence of a largest prime number.

CHAPTER ONE

THE MAIN IDEA AND RESULTS: AN OVERVIEW

In this chapter we introduce, in the very simplest context, the basic idea of the technique to solve (certain classes of) nonlinear evolution equations via the spectral transform and we outline its main implications. The results surveyed here are then taken up and treated in more detail in subsequent chapters.

1.1. Solution of linear evolution equations by Fourier transform

Consider the linear partial differential equation

(1) $$u_t(x,t) = -i\omega\left(-i\frac{\partial}{\partial x}\right)u(x,t),$$

where $\omega(z)$ is, say, a polynomial. The study of many natural phenomena can be reduced to the investigation of the solution $u(x,t)$ of (1) characterized by the initial condition

(2) $$u(x,0) = u_0(x), \quad -\infty < x < \infty,$$

where $u_0(x)$ is regular (for all real values of x) and vanishes asymptotically, say

(3) $$\lim_{x\to\pm\infty}\left[|x|^{1+\epsilon}u_0(x)\right] = 0, \quad \epsilon > 0.$$

Note that, if the odd part of $\omega(z)$ is real and the even part is imaginary (i.e. if $[\omega(z^*)]^* = -\omega(-z)$), (1) is a real equation; then $u(x,t)$ is also real for $t > 0$ if $u_0(x)$ is real. Moreover if $\omega(z)$ is real ($[\omega(z^*)]^* = \omega(z)$), then (1) is purely dispersive (i.e., non dissipative); for instance it is then easily seen (see

7

(4) and (7) below) that the integral over all values of x of $|u(x,t)|^2$ is time-independent.

The *initial-value problem* or *Cauchy problem* characterized by a boundary condition of type (2) and (3) is the typical kind of problem we will be investigating throughout this book; although our main focus below will be on *nonlinear* evolution equations rather than on *linear* evolution equations like (1).

A central role in the solution of (1) is played by the (direct and inverse) Fourier transform formulae for (the dependence on the variable x of) $u(x,t)$:

(4) $\quad u(x,t) = (2\pi)^{-1} \int_{-\infty}^{+\infty} dk \exp(ikx) \, \hat{u}(k,t),$

(5) $\quad \hat{u}(k,t) = \int_{-\infty}^{+\infty} dx \exp(-ikx) \, u(x,t).$

This is due to the fact that, if $u(x,t)$ evolves according to the *partial* differential equation (1), the Fourier transform $\hat{u}(k,t)$ evolves according to the *ordinary* differential equation

(6) $\quad \hat{u}_t(k,t) = -i\omega(k)\hat{u}(k,t),$

that can be immediately integrated to yield

(7) $\quad \hat{u}(k,t) = \hat{u}(k,0) \exp[-i\omega(k)t].$

Thus the solution of (1) and (2) (with (3)) is accomplished in three steps. Firstly, at the initial time $t=0$, the Fourier transform

(8) $\quad \hat{u}(k,0) = \hat{u}_0(k) = \int_{-\infty}^{+\infty} dx \exp(-ikx) \, u_0(x)$

is evaluated (see (5) and (2)); then, the Fourier transform $\hat{u}(k,t)$ is obtained from (7); and finally, at time t, the function $u(x,t)$ is recovered from $\hat{u}(k,t)$ using the Fourier transform formula (4).

The technique of solution that we have illustrated here is conveniently summarized by the schematic diagram displayed at the top of the next page, where the broken line indicates the (difficult) problem that is generally directly related to applications, while the 3 continuous lines indicate the 3 (easier) steps whose sequence yields the solution.

We submit that the main reason why the Fourier transform is such an important tool in mathematical physics and in applied mathematics is

1.1 Solution by Fourier transform

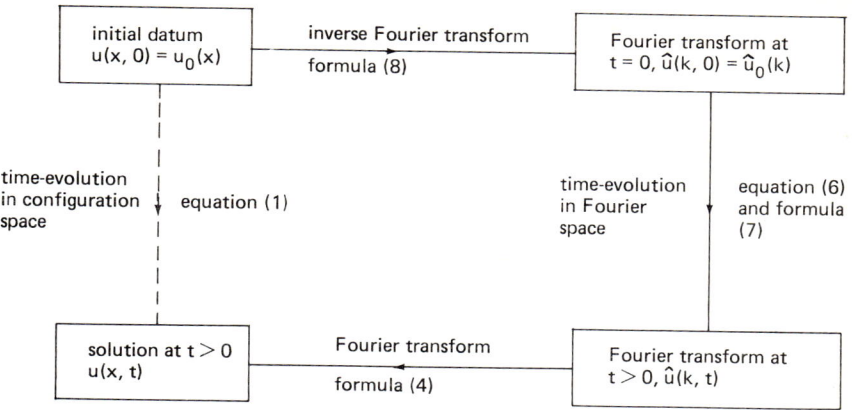

because it provides, as we have indicated above, the appropriate technique to solve the problem characterized by (1) and (2); for indeed this mathematical problem, as well as its generalizations that are also analogously solvable by Fourier methods, constitute the proto-typical (often, of course, approximate) schematization of many natural phenomena.

In fact what is really important is not the possibility to exhibit explicitly the solution of (1) and (2) (indeed only in rare instances can the integrals in (4) and (5) be analytically calculated), but rather the insight into the behaviour of $u(x,t)$ implied by this technique of solution. Clearly the time evolution is completely determined by the "dispersion function" $\omega(k)$; and the (standard) analysis of the long-time behaviour of $u(x,t)$, as given by (7) and (4), establishes that a solution $u(x,t)$, characterized by initial data $u_0(x)$ that are localized around a position x_0 and moreover have a Fourier transform $\hat{u}_0(k)$ that is localized around a value k_0, behaves generally as a "wave packet" moving with the *group velocity*

$$(9) \qquad v_g = \left.\frac{d\omega(k)}{dk}\right|_{k=k_0}$$

and dispersing asymptotically (the peak amplitude of its envelope is localized around $x = x_0 + v_g t$ and decreases proportionally to $t^{-1/2}$ as $t \to \infty$). Note that this behaviour, although of course well understood, is by no means evident from the structure of the partial differential equation (1). Contrast this with the extremely simple time evolution (7) of the Fourier transform $\hat{u}(k,t)$. It is thus seen that *the dynamics is much simpler in k-space than in x-space*. The implications of this message, that synthetizes the lesson to be drawn from the solvability of (1) via Fourier transform, are pervasive;

not only do they suggest the proper language to theorize about many natural phenomena (all those whose mathematical description can be, one way or another, patterned on (1)), but indeed they dictate how best to experiment with them (for instance the motivation for experimenting, in optics, with monochromatic beams, can be traced precisely to this origin; and the same applies, in a way, even to the drive to build particle accelerators of higher and higher energy!). It is reasonable to expect that an analogous situation develop for those *nonlinear* evolution equations that are solvable by a technique—based on the spectral transform—that as we will see presents a close similarity, but also some significant novelties, relative to the Fourier transform approach discussed thus far.

Let us also mention here some other properties of the solutions of (1) that will also be shown to have analogues in the nonlinear cases treated in the following.

Firstly, note that, if $u^{(1)}(x,t)$ satisfies (1) and $u^{(2)}(x,t)$ is related to $u^{(1)}(x,t)$ by the formula

$$(10) \quad g\left(-i\frac{\partial}{\partial x}\right)u^{(2)}(x,t)+h\left(-i\frac{\partial}{\partial x}\right)u^{(1)}(x,t)=0,$$

then $u^{(2)}(x,t)$ also satisfies (1). Indeed, the relation in Fourier space corresponding to (1) reads of course

$$(11a) \quad g(k)\hat{u}^{(2)}(k,t)+h(k)\hat{u}^{(1)}(k,t)=0,$$

or, equivalently,

$$(11b) \quad \hat{u}^{(2)}(k,t)=-[h(k)/g(k)]\hat{u}^{(1)}(k,t),$$

clearly implying that, if $u^{(1)}(k,t)$ satisfies (6), so does $u^{(2)}(k,t)$. Although (10), for given $u^{(1)}(x,t)$, reads as an ordinary differential equation for $u^{(2)}(x,t)$ only if $g(z)$ and $h(z)$ are polynomials, or rational functions, the functions $g(z)$ and $h(z)$ here are largely arbitrary, except for those conditions which guarantee that $u^{(2)}(x,t)$ is regular and vanishes asymptotically according to (3), if also $u^{(1)}(x,t)$ does (for instance, $g(z)$ should not have any real zero).

Secondly, we note that, if $u^{(1)}(x,t)$ and $u^{(2)}(x,t)$ are solutions of (1), this equation is also satisfied by any linear combination of them (with constant coefficients),

$$(12) \quad u(x,t)=c_1 u^{(1)}(x,t)+c_2 u^{(2)}(x,t).$$

This is of course the *superposition principle*, that corresponds to the linear structure of (1).

Thirdly, we remark that the solution of (1) and (2) can be formally written through the *resolvent formula*

$$(13) \qquad u(x,t) = \exp\left[-it\omega\left(-i\frac{\partial}{\partial x}\right)\right] u_0(x),$$

corresponding to a straightforward integration of (1) (from 0 to t, using (2)). In the particular case $\omega(z) = -z$, when equation (1) reads

$$(14) \qquad u_t(x,t) = u_x(x,t)$$

and has the general solution

$$(15) \qquad u(x,t) = f(x+t),$$

(13) corresponds therefore to the well-known *operator identity*

$$(16) \qquad f(x+a) = \exp\left(a\frac{d}{dx}\right) f(x),$$

where we have written a in place of t to emphasize that the validity of this formula has nothing to do with the time-evolution problem: indeed (16) expresses a property of the "translation operator" $\exp(a\,d/dx)$, and it holds for any function $f(x)$ (strictly speaking, for any function $f(z)$ holomorphic in the disc $|z-x| \leq |a|$).

While for the sake of simplicity we have introduced no explicit time-dependence in (1), it should be emphasized that the technique of solution described in this section works (with trivial changes) also if this partial differential equation is explicitly time-dependent, i.e.

$$(17) \qquad u_t(x,t) = -i\omega\left(-i\frac{\partial}{\partial x},t\right) u(x,t).$$

Then of course in place of (6) one has

$$(18) \qquad \hat{u}_t(k,t) = -i\omega(k,t)\hat{u}(k,t),$$

and (7) is replaced by

$$(19) \qquad \hat{u}(k,t) = \hat{u}(k,0) \exp\left[-i\int_0^t dt'\, \omega(k,t')\right].$$

If instead the partial differential equation (1) contains an explicit x-dependence, the applicability of the Fourier transform approach becomes less simple. A case in which it continues to work (in the sense of reducing the solution to quadratures, and to the solution of a, generally nonlinear, *ordinary* differential equation; see below) is when the x-dependence is linear, so that in place of (1) one has

$$(20) \quad u_t(x,t) = -i\left[\omega\left(-i\frac{\partial}{\partial x}, t\right) + x\omega'\left(-i\frac{\partial}{\partial x}, t\right)\right] u(x,t).$$

This equation is now characterized by the two functions $\omega(z,t)$ and $\omega'(z,t)$ (for the sake of generality we are also introducing a t-dependence). Its counterpart in Fourier space reads

$$(21) \quad \hat{u}_t(k,t) = [-i\omega(k,t) + \omega'_k(k,t)] \hat{u}(k,t) + \omega'(k,t) \hat{u}_k(k,t),$$

and it is explicitly solvable (by the method of characteristics), yielding

$$(22) \quad \hat{u}(k,t) = \hat{u}_0[k_0(t,k)] \exp\left\{\int_0^t dt' [-i\omega(\chi,t') + \omega'_k(\chi,t')]\right\}$$

where

$$(23) \quad \chi \equiv \chi[t', k_0(t,k)],$$

the function $\chi(t, k_0)$ being defined by the (ordinary) differential equation

$$(24) \quad \chi_t(t, k_0) = -\omega'[\chi(t, k_0), t]$$

and by the boundary condition

$$(25) \quad \chi(0, k_0) = k_0;$$

while the function $k_0(t, k)$ is defined by χ through the (implicit) formula

$$(26) \quad \chi(t, k_0) = k.$$

Thus the solution of the (Cauchy) problem characterized by (20) and (2) is given by (4) and (22), $\hat{u}_0(k)$ being given by (8). Of course we are here assuming that all integrals converge, and that the integrations by parts necessary to derive (21) from (20) are permissible. That this is not always the case is clear from (22), since this formula, together with (23–26), need not imply that $\hat{u}(k,t)$, for $t>0$, vanish asymptotically in k, even though

$\hat{u}(k,0) \equiv \hat{u}_0(k)$ does. We shall discuss this problem in more detail in section 6.2 and appendix A.16, since this difficulty similarly arises in the context of nonlinear evolution equations.

Our motivation for reviewing in this section some well-known facts concerning the solution of *linear* evolution equations by Fourier transform is because of the close similarity of this approach to the method of solution, via the spectral transform, of (certain classes of) *nonlinear* evolution equations, that constitutes our main interest. The correspondence applies also to the extensions that we have just mentioned (equations with t-dependent and linearly x-dependent coefficients). Let us end this section by mentioning two directions in which instead such a close correspondence does not yet seem to exist.

First and most important, is the extension of the approach to more (space) variables. This can be done rather trivially in the linear case by introducing the multidimensional Fourier transform; a comparably straightforward extension to problems with more space variables does not exist in the nonlinear case (although there are some ways to introduce extra variables, as we shall see later, in section 6.1).

Returning to the simplest case of one space and one time variable, there is another kind of extension that can be done very simply in the linear case but still has no simple counterpart in the nonlinear context: the inclusion of certain classes of integro-differential equations. Consider for instance, in place of (1), the evolution equation

$$(27) \quad u_t(x,t) = \int_{-\infty}^{+\infty} dy\, K(x-y,t) u(y,t).$$

Then the treatment described above is again applicable, with (6) (or rather (18)) replaced by

$$(28) \quad \hat{u}_t(k,t) = \hat{K}(k,t) \hat{u}(k,t),$$

where of course $\hat{K}(k,t)$ is the Fourier transform of $K(x,t)$:

$$(29a) \quad K(x,t) = (2\pi)^{-1} \int_{-\infty}^{+\infty} dk\, \hat{K}(k,t) \exp(ikx),$$

$$(29b) \quad \hat{K}(k,t) = \int_{-\infty}^{+\infty} dx\, K(x,t) \exp(-ikx).$$

No spectral transform technique is as yet available for solving simple nonlinear evolution equations of integral type; although recently there appears to have been progress also in that direction (see section 1.8).

1.2. A class of solvable nonlinear evolution equations

A class of *nonlinear* evolution equations solvable via the spectral transform technique can be written in compact form as follows:

(1) $\quad u_t(x,t) = \alpha(L) u_x(x,t).$

Here $\alpha(z)$ is, say, a polynomial, and L is the integro-differential operator defined by the following formula that specifies its action on a generic function $f(x)$:

(2) $\quad Lf(x) = f_{xx}(x) - 4u(x,t)f(x) + 2u_x(x,t) \int_x^{+\infty} dy\, f(y).$

Here, and generally below, we restrict consideration to functions such that all integrals we write are convergent. Of course f may also depend on other variables, for instance on t.

Note that the operator L depends on u; this causes the r.h.s. of (1) to be nonlinear in u. Since L is integro-differential, it might appear that (1) is generally an integro-differential equation. But this is not the case: as long as $\alpha(z)$ is a polynomial in z, (1) is a (nonlinear) *partial differential* evolution equation. This is due to an important and nontrivial property of the operator L, that shall be proven below (see appendix A.9) and that can be synthetically formulated as follows:

(3) $\quad L^n u_x(x,t) = g_x^{(n)}, \quad n = 0, 1, 2, \ldots,$

where $g^{(n)}$ is a polynomial (of degree $n+1$) in u and its x-derivatives (up to the derivative of order $2n$). For instance:

(3a) $\quad Lu_x = u_{xxx} - 6uu_x = (u_{xx} - 3u^2)_x,$

(3b) $\quad L^2 u_x = u_{xxxxx} - 10 u u_{xxx} - 20 u_x u_{xx} + 30 u^2 u_x$
$\quad\quad = (u_{xxxx} - 10 u u_{xx} - 5 u_x^2 + 10 u^3)_x.$

Thus the simplest nonlinear evolution equation contained in the class (1) is the Korteweg–de Vries (KdV) equation

(4) $\quad u_t + u_{xxx} - 6 u u_x = 0,$

corresponding to $\alpha(z) = -z$.

As in the case discussed in the preceding section, our main interest will be in the Cauchy problem associated to (1), characterised by a given initial condition

(5) $\quad u(x,0)=u_0(x), \quad -\infty<x<+\infty,$

with $u_0(x)$ regular for all real values of x and vanishing asymptotically (see (1.1.-2,3)).

We have written out the class of evolution equations (1) at this early stage to provide a motivation for subsequent developments. We shall return to this class several times in the following. The technique of solution of (1) hinges on the spectral transform of (the x-dependence of) $u(x,t)$, that plays an analogous role to that of the Fourier transform for the solution of (1.1.-1). Thus the following section is devoted to a terse outline of the spectral transform.

But before closing this section we would like to point out that, in the limit in which $u(x,t)$, and its derivatives, are so small that it is justified to neglect all powers of them, the *nonlinear* evolution equation (1) goes over into the *linear* equation

(6) $\quad u_t(x,t)=\alpha\left(\dfrac{\partial^2}{\partial x^2}\right)u_x(x,t)$

or equivalently (see (1.1.-1))

(6a) $\quad u_t(x,t)=-i\omega\left(-i\dfrac{\partial}{\partial x}\right)u(x,t)$

with

(6b) $\quad \omega(z)=-z\alpha(-z^2).$

Note that, for real $\alpha(z)$, $\omega(z)$ turns automatically out to be real and odd, yielding therefore not only, via (6a), a real equation (as of course it must), but indeed a purely dispersing (i.e. nondissipative) one.

1.3. The spectral transform

This section is divided into three subsections. In the first the Schroedinger spectral problem is tersely described, and the spectral transform is introduced. In the second, the solution of the corresponding *inverse* spectral problem is outlined. In the third, the spectral transform is discussed.

The spectral problem that we discuss in this section is that appropriate to the solution, via the spectral transform technique, of the class of nonlinear evolution equations introduced in the preceding section. This spectral problem is discussed in more detail in the following chapter 2. Other spectral problems are appropriate for the solution of other classes of nonlinear evolution equations; some are treated in volume II.

Let us emphasize that the topic treated in this section (and in chapter 2) has no reference to the time-evolution problem; just as the study of the Fourier transform has nothing to do with the eventual use of it to solve (linear) evolution equations.

1.3.1. Direct spectral problem

The spectral problem that we discuss here is familiar to students of quantum mechanics, being based on the stationary Schroedinger equation

$$(1) \qquad -\psi_{xx}(x,k) + u(x)\psi(x,k) = k^2 \psi(x,k), \quad -\infty < x < +\infty.$$

This is a second-order linear ordinary differential equation. In the context of this section $u(x)$ is a given real function, that we assume to be regular for all (real) values of x and to vanish (sufficiently fast) asymptotically, say

$$(2) \qquad \lim_{x \to \pm \infty} \left[|x|^{1+\varepsilon} u(x) \right] = 0, \quad \varepsilon > 0.$$

In the quantum-mechanical context the stationary Schroedinger equation (1) obtains from a time-dependent Schroedinger equation. It is perhaps useful to emphasize that this has nothing to do with the time evolution we are discussing in (other sections of) this book. Indeed it is probably advisable to forget altogether the physical interpretation of (1) in terms of one-dimensional quantal scattering or bound states, but rather to consider the spectral problem defined by (1) as a typical Sturm–Liouville eigenvalue problem; in fact, a *singular* Sturm–Liouville problem, since one is considering the differential equation (1) on the whole real line, namely on an infinite interval. In the following we shall take this point of view, although we keep some of the terminology (such as "reflection" and "transmission" coefficients; see below) that clearly has originated from the quantal scattering problem.

The spectrum of the Sturm–Liouville problem characterized by (1) has two components: a continuum, including all positive values of the eigenvalue k^2, and a number of discrete negative eigenvalues, $k^2 = -p_n^2$, $p_n > 0$, $n = 1, 2, \ldots, N$.

It is convenient, in order to characterize the continuum part of the spectrum (corresponding to real values of k, so that $k^2 > 0$), to introduce the solution of (1) characterized by the asymptotic boundary conditions

(3a) $\quad \psi(x,k) \to T(k)\exp(-ikx), \quad x \to -\infty,$
(3b) $\quad \psi(x,k) \to \exp(-ikx) + R(k)\exp(ikx), \quad x \to +\infty.$

This asymptotic behaviour is clearly consistent with (2), and it identifies uniquely the eigenfunction $\psi(x,k)$, as well as the *transmission coefficient* $T(k)$ and the *reflection coefficient* $R(k)$.

Let N be the number of (discrete) negative eigenvalues,

(4) $\quad k^2 = -p_n^2, \quad p_n > 0, \quad n = 1, 2, \ldots, N.$

To each of these eigenvalues there corresponds a solution $f_n(x)$ of (1), uniquely identified by the asymptotic boundary condition

(5) $\quad \lim_{x \to +\infty} [\exp(p_n x) f_n(x)] = 1, \quad n = 1, 2, \ldots, N.$

This solution vanishes also as $x \to -\infty$, proportionally to $\exp(p_n x)$, and is therefore normalizable; indeed it is precisely this condition that identifies the (discrete negative) eigenvalue $-p_n^2$. It is convenient to define the *normalisation coefficient* ρ_n by the formula

(6) $\quad \rho_n = \left[\int_{-\infty}^{+\infty} dx\, f_n^2(x) \right]^{-1}, \quad n = 1, 2, \ldots, N.$

We assume hereafter that the number of discrete eigenvalues N is finite; a sufficient condition is that (2) hold for some $\varepsilon > 1$ (see Appendix 1). Of course N might vanish, namely there might be no discrete eigenvalue; a sufficient condition for this is that $u(x)$ be nowhere negative, $u(x) \geq 0$ for $-\infty < x < +\infty$; a sufficient condition to exclude this is that $u(x)$ be nowhere positive, $u(x) \leq 0$ for $-\infty < x < +\infty$. Indeed N may be viewed as a (global) measure of the negativeness of the function $u(x)$, as implied by the upper and lower bounds on its value given in Appendix 1.

The *spectral transform* S of the function $u(x)$ is, by definition, the collection of data

(7) $\quad S[u] = \{ R(k), -\infty < k < +\infty; p_n, \rho_n, n = 1, 2, \ldots, N \}.$

The motivation for such a definition is, that there is a one-to-one correspondence between functions $u(x)$ (in an appropriate functional class, as indicated above), and the spectral transform (7). The analysis of this subsection indicates how S is determined (clearly uniquely) by u; this is the *direct* spectral problem. In the following subsection the *inverse* spectral problem is discussed, namely the determination of u from a given S.

1.3.2. Inverse spectral problem

Given the spectral transform

(1) $\quad S=\{R(k), -\infty<k<+\infty; p_n, \rho_n, n=1,2,\ldots,N\},$

the following procedure yields the corresponding function $u(x)$.

Define first of all the function

(2) $\quad M(x)=(2\pi)^{-1}\int_{-\infty}^{+\infty}dk\exp(ikx)R(k)+\sum_{n=1}^{N}\rho_n\exp(-p_n x).$

Note that, except for the contribution from the discrete spectrum, $M(x)$ is precisely the inverse Fourier transform of $R(k)$.

Consider next the *Gel'fand–Levitan–Marchenko* (GLM) integral equation

(3) $\quad K(x,y)+M(x+y)+\int_{x}^{+\infty}dz\,K(x,z)M(z+y)=0, \quad y>x.$

This is a Fredholm integral equation, and it determines uniquely the function $K(x, y)$. Note that the integral equation (3) refers to the dependence of $K(x, y)$ on its second argument, y; as for the dependence of $K(x, y)$ on its first argument, x, this occurs, as it were, parametrically, being caused by the appearance of x both in the argument of the inhomogeneous term $M(x+y)$ and as lower limit of integration.

Once $K(x, y)$ is determined, the function $u(x)$ follows from the simple formulae

(4) $\quad w(x)=2K(x,x+0),$

(5a) $\quad w(x)=\int_{x}^{+\infty}dy\,u(y),$

(5b) $\quad u(x)=-w_x(x).$

Of course, the function $u(x)$ yielded by this procedure is real and regular for all x, and vanishes asymptotically, only provided the spectral transform S satisfies certain properties; for instance, the reflection coefficient $R(k)$ must obviously be Fourier transformable. This is further discussed in the following chapter and in appendix A.16.

We end this subsection treating some examples.

Consider the very special case of a spectral transform having a vanishing contribution from the continuum, and containing a single discrete eigenvalue:

(6) $\quad S=\{R(k)=0, -\infty<k<+\infty; N=1, p_1=p, \rho_1=\rho\}$.

As we shall see, this case is going to play a very important role. In the present context, it provides a simple explicit illustration of the way the inverse spectral problem works. For in this case, $M(x)$ becomes simply an exponential,

(7) $\quad M(x)=\rho\exp(-px)$,

and therefore the GLM equation (3) becomes separable and is easily solved, yielding

(8) $\quad K(x,y)=-p\exp[p(\xi-y)]/\cosh[p(x-\xi)]$,

(9) $\quad w(x)=-4p/\{1+\exp[2p(x-\xi)]\}$,

(10) $\quad u(x)=-2p^2/\cosh^2[p(x-\xi)]$.

In these formulae

(11) $\quad \xi=(2p)^{-1}\ln(\rho/2p)$.

Actually this procedure can be explicitly carried out also in the more general case of a spectral transform, having again no contribution from the continuum, but with N discrete eigenvalues:

(12) $\quad S=\{R(k)=0, -\infty<k<+\infty; p_n, \rho_n, n=1,2,\ldots,N\}$.

Indeed in such a case the GLM Fredholm equation (3) is still separable, although now of rank N. The function $u(x)$ corresponding to (12) can be written in the compact form

(13) $\quad u(x)=-2\dfrac{d^2}{dx^2}\{\ln\det[I+C(x)]\}$,

where I is the unit matrix of order N and $C(x)$ is the symmetrical matrix of order N with elements

(14) $\quad C_{mn}(x)=(\rho_m\rho_n)^{1/2}(p_m+p_n)^{-1}\exp[-(p_m+p_n)x]$.

Consider finally the case when there are no discrete eigenvalues and moreover $R(k)$ is very small (in modulus), so that $u(x)$ is also very small (see below). Note the consistency of this last assumption with the absence of discrete eigenvalues; indeed if $u(x)$ is very small, there can be at most one discrete eigenvalue, as implied by the upper bounds on N given in appendix A.1.

Assume then that the smallness of R, and therefore of M (see (2)), justifies the neglect of the last term in the l.h.s. of (3), yielding

(15) $\quad K(x, y) \approx -M(x+y)$

and therefore

(16) $\quad u(x) \approx (2\pi)^{-1} \int_{-\infty}^{+\infty} dk \exp(ikx) ikR(\tfrac{1}{2}k).$

It is therefore seen that, in this approximation, there is a simple relationship between the reflection coefficient $R(k)$ and the Fourier transform $\hat{u}(k)$ of $u(x)$ (see (1.1.-4)), namely

(17a) $\quad \hat{u}(k) \approx ikR(\tfrac{1}{2}k).$

Indeed this formula, or rather the completely equivalent version

(17b) $\quad R(k) \approx (2ik)^{-1} \hat{u}(2k)$

corresponds, in the context of the quantum-mechanical scattering problem, just to the familiar "Born approximation" formula. As it is clear from (17b), this approximation generally breaks down in the neighbourhood of $k \approx 0$; this will be further discussed in chapter 3.

1.3.3. The spectral transform

The spectral transform has been introduced at the end of subsection 1.3.1. The results of that subsection, and of subsection 1.3.2, imply that there is a one-to-one correspondence between the function $u(x)$ and its spectral transform $S[u]$:

(1) $\quad u(x) \Leftrightarrow S[u].$

There are moreover constructive procedures to go from a function u to its spectral transform S, and from a spectral transform S to the corresponding function u. Both these procedures involve the solution of *linear* problems; the Schroedinger differential equation (1.3.1.-1) for the direct spectral problem ($u \Rightarrow S$); the GLM Fredholm integral equation (1.3.2.-3) for the inverse spectral problem ($S \Rightarrow u$). The relation (1), between a function and its spectral transform S, is on the other hand clearly *nonlinear*; except in the approximate case of small $u(x)$, when, as discussed at the end of the preceding subsection (and see also below), the spectral transform coincides essentially with the Fourier transform.

The analogy with the Fourier transform (see (1.1.-4)) is moreover apparent from the exact formulae

(2) $$u(x) = \pi^{-1} \int_{-\infty}^{+\infty} dk \left[f^{(+)}(x,k) \right]^2 2ikR(k) - 4 \sum_{n=1}^{N} \rho_n P_n [f_n(x)]^2,$$

(3) $$w(x) = -\pi^{-1} \int_{-\infty}^{+\infty} dk\, f^{(+)}(x,k) \exp(ikx) R(k)$$
$$- 2 \sum_{n=1}^{N} \rho_n f_n(x) \exp(-p_n x),$$

where $f^{(+)}(x,k)$ is the solution of the Schroedinger equation (1.3.1-1) characterised by the asymptotic boundary condition

(4) $$\lim_{x \to +\infty} \left[\exp(-ikx) f^{(+)}(x,k) \right] = 1,$$

while $f_n(x)$ is analogously defined, but for $k = ip_n$, $p_n > 0$ (see 1.3.1.-5)). Note that (3) provides an expression for the integral of $u(x)$,

(5) $$w(x) = \int_{x}^{+\infty} dy\, u(y),$$

rather than for $u(x)$ itself. The equivalence of (3) to (2), as well as the validity of these representations of $u(x)$ and $w(x)$, are nontrivial results; they are derived and discussed in the following chapter 2.

We also display here three formulae that may be viewed as the analogs of the inverse Fourier transform (1.1-5):

(6) $$R(k) = (2ik)^{-2} \int_{-\infty}^{+\infty} dx \left[\psi(x,k) \right]^2 u_x(x),$$

(7) $$R(k) = (2ik)^{-1} \int_{-\infty}^{+\infty} dx\, \psi(x,k) \exp(-ikx) u(x),$$

(8) $$R(k) = (2ik)^{-2} \int_{-\infty}^{+\infty} dx\, \psi(x,k) \exp(-ikx) \left[u_x(x) + u(x) w(x) \right].$$

Here the function $\psi(x,k)$ is the solution of the Schroedinger equation

(1.3.1-1) characterised by the boundary conditions (1.3.1.-3); again the equivalence of these three expressions of $R(k)$ is nontrivial. Also these results are discussed and proved in the following chapter 2. Note that, in writing (8), we have used the definition (5).

The nonlinear character of the relation between a function $u(x)$ and its spectral transform is well displayed by these formulae; it appears from the fact that the functions $f^{(+)}(x, k)$ and $f_n(x)$ in (2) and (3) and the function $\psi(x, k)$ in (6), (7) and (8), depend on $u(x)$. In the approximation in which this dependence is ignorable because $u(x)$ is negligibly small, so that (as implied by (1.3.1.-1), (4) and (1.3.1.-3))

(9) $\quad f^{(+)}(x, k) \approx \exp(ikx),$

(10) $\quad \psi(x, k) \approx \exp(-ikx),$

and (see (1.3.2.-17))

(11) $\quad R(\tfrac{1}{2}k) \approx (ik)^{-1} \hat{u}(k),$

then (2) and (3) go over into the Fourier transform formula (1.1.-4) (provided the contribution of the discrete part of the spectrum is neglected; we have commented in the previous subsection on the consistency of this assumption), while (6), (7) and (8) go over into the inverse Fourier transform formula (1.1.-5).

1.4. Solution of nonlinear evolution equations via the spectral transform

In the preceding section we have introduced the spectral transform S of a function $u(x)$ (regular for all real values of x and vanishing at infinity). Imagine now that u depends also on another variable, call it t ("time"): $u \equiv u(x, t)$. Then of course, in one-to-one correspondence to $u(x, t)$, there is a spectral transform that is also time-dependent:

(1) $\quad u(x, t) \Leftrightarrow S(t).$

Thus, if $u(x, t)$ evolves in time, so does $S(t)$.

The all-important, and highly nontrivial, discovery that has opened up this field of scientific enquiry is, that there exist a class of *interesting* time-evolutions of $u(x, t)$ to which there corresponds *simple* time-evolutions of $S(t)$. These time-evolutions of $u(x, t)$ can therefore be investigated by following the evolution in the spectral space, namely by following the evolution of $S(t)$ rather than, directly, the evolution of $u(x, t)$; taking then advantage, to gain information on the (*interesting*) evolution of $u(x, t)$, of the possibility to go from u to S (say, at the initial time) and from S to u (say, at any later time), via the direct and inverse spectral problems described in the preceding section.

This is, of course, a closely analogous procedure to the technique of solution via Fourier transform of the class of linear partial differential equations discussed in section 1.1. This analogy is further discussed below, in this section and in the following one, and in several other places as well, since it constitutes the main idea on which this book is based.

As the reader should have guessed, the simplest class of *nonlinear* evolution equations that are solvable in this way are those that were introduced in section 1.2, namely the class of *nonlinear partial differential equations*

$$(2) \qquad u_t(x, t) = \alpha(L) u_x(x, t),$$

where $\alpha(z)$ is a polynomial and L is the integro-differential operator defined by the formula

$$(3) \qquad Lf(x) = f_{xx}(x) - 4u(x, t) f(x) + 2u_x(x, t) \int_x^{+\infty} dy f(y),$$

that specifies its action on a generic function $f(x)$ (vanishing at infinity). (The reader is advised at this stage to review the basic properties of this class of evolution equations, as described in section 1.2).

The crucial property of the class of evolution equations (2) is that, as will be proved in chapter 3, the corresponding time-evolution of the spectral transform of $u(x, t)$ is given by *linear ordinary differential equations*, namely

$$(4) \qquad R_t(k, t) = 2ik\alpha(-4k^2) R(k, t),$$
$$(5) \qquad \dot{p}_n(t) = 0, \quad n = 1, 2, \ldots, N,$$
$$(6) \qquad \dot{\rho}_n(t) = -2p_n\alpha(4p_n^2)\rho_n(t), \quad n = 1, 2, \ldots, N$$

(a subscripted t, or a dot on top, are equivalent, both denoting differentiation with respect to t). These equations are explicitly solved according to the formulae

$$(7) \qquad R(k, t) = R(k, 0) \exp[2ik\alpha(-4k^2)t],$$
$$(8) \qquad p_n(t) = p_n(0) = p_n, \quad n = 1, 2, \ldots, N,$$
$$(9) \qquad \rho_n(t) = \rho_n(0) \exp[-2p_n\alpha(4p_n^2)t], \quad n = 1, 2, \ldots, N.$$

Let us pause a moment and note that (5), or equivalently (8), implies that the (discrete) eigenvalues of the Schroedinger differential operator

$$(10) \qquad -\frac{d^2}{dx^2} + u(x,t)$$

do not change, when $u(x,t)$ evolves in time according to (2). This property of the Schroedinger operator (10), to experience an *isospectral evolution* when $u(x,t)$ evolves according to (any one of the evolution equations of the class) (2), plays a crucial rôle in other approaches to these results, more operator-theoretically oriented than the (rather elementary) point of view basically adopted throughout this book.

Let us now consider the solution of the Cauchy problem for the class of nonlinear evolution equations (2). It clearly proceeds through three steps. Firstly, at the initial time $t=0$, from the given datum

$$(11) \qquad u(x,0) = u_0(x)$$

the spectral transform

$$(12) \qquad S(0) = \{ R(k,0), -\infty < k < +\infty; p_n, \rho_n(0), n=1,2,\ldots,N \}$$

is evaluated (solving the *direct* spectral problem; see subsection 1.3.1). Then, the spectral transform at time t,

$$(13) \qquad S(t) = \{ R(k,t), -\infty < k < +\infty; p_n, \rho_n(t), n=1,2,\ldots,N \}$$

is obtained, from the explicit expressions (7), (8) and (9) (note that the isospectrality property, (5) or (8), also implies that the number of discrete eigenvalues N does not change as time evolves). Finally, at time t, the function $u(x,t)$ is recovered from its spectral transform $S(t)$ (solving the *inverse* spectral problem; see subsection 1.3.2). This technique of solution may be summarised by the schematic diagram, closely analogous to that displayed in section 1.1, that appears at the top of the next page. There the broken line indicates the (difficult and interesting) problem of evaluating the time-evolution phenomenon described by (2); the three continuous lines indicate the steps whose sequence yields the solution. Note that these three steps are easier than the direct solution of (2); in particular, they require only the solution of *linear* problems.

Of course, only in special cases (some of which are discussed below) the operations described above can be actually carried out, yielding in analytic

1.4 Solution via the spectral transform

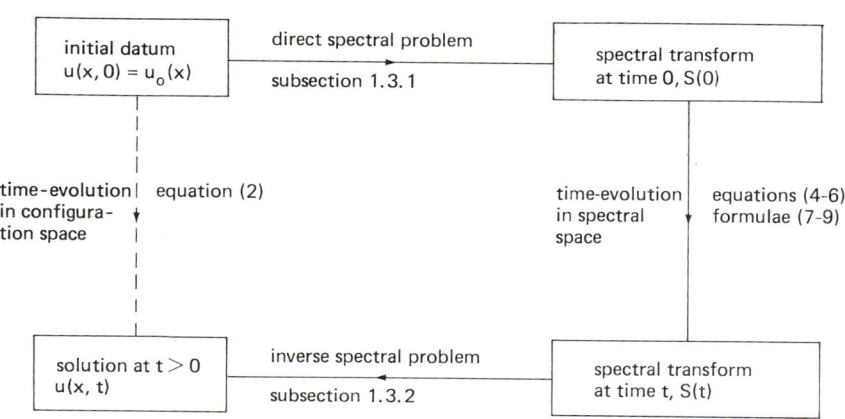

explicit form the solution $u(x, t)$ of (2). The message of general validity that obtains from the technique of solution we have just described is, that for this class of evolution equations *the time evolution is much simpler in the spectral space than in configuration space*. This message, as we will see, has many theoretical implications; it is eventually going to impact also on the experimental techniques used to investigate phenomena that are described, through appropriate, possibly approximate, schematizations, by equations belonging to the class (2) (or to other classes solvable by analogous techniques, see below). The similarity of this situation to that discussed in section 1.1, in the context of the linear evolution equations solvable by Fourier transform, should of course be emphasized.

It is moreover possible to utilize the technique of solution we have just described to evince a qualitative understanding of the behaviour of the solutions of (2), especially of their long-time behaviour; again in analogy to the situation prevailing in the context of the linear evolution equations solvable by Fourier transform, as tersely discussed in section 1.1. This we do in section 1.6, after having treated, in the following section, the relationship between the solution via the spectral transform of the nonlinear evolution equations (2) and the solution via Fourier transform of the linear evolution equations that obtain from (2) if all nonlinear contributions are neglected.

Let us end this section noting that the class of nonlinear evolution equations (2) is not the most general one that can be solved with the help of the spectral transform of section 1.3. Somewhat more general versions are discussed in chapters 3 and 6. Other, still more general, classes of nonlinear evolution equations, solvable via more general spectral problems, are discussed in volume II.

1.5. Relation to the Fourier transform technique to solve linear evolution equations

In this section we clarify in what sense the spectral transform technique to solve (certain classes of) nonlinear evolution equations constitutes a natural extension of the Fourier transform method to solve linear evolution equations. The analogy between these techniques has been already noted. Here we show that, in the limit of "brutal" linearization corresponding to the neglect of all nonlinear contributions, just as the nonlinear evolution equation becomes a linear equation, so the technique of solution via the spectral transform goes over into the technique of solution via the Fourier transform.

Indeed all the elements leading to this conclusion have been provided in the preceding sections. All we need here is to collect the relevant formulae, which are reported below even if this entails some repetition.

The class of nonlinear evolution equations whose solvability via the spectral transform has been described above read

$$(1) \qquad u_t(x,t) = \alpha(L) u_x(x,t),$$

with the integro-differential operator L defined by (1.2.-2). In the approximation in which all nonlinear effects are neglected it goes over into the linear partial differential equation

$$(2) \qquad u_t(x,t) = -i\omega\left(-i\frac{\partial}{\partial x}\right) u(x,t)$$

with

$$(3) \qquad \omega(z) = -z\alpha(-z^2)$$

(see end of subsection 1.2).

The solution of (2), as discussed in section 1.1, results from the fact that the Fourier transform $\hat{u}(k,t)$ of $u(x,t)$ evolves in time according to the simple formula

$$(4) \qquad \hat{u}(k,t) = \hat{u}(k,0) \exp[-i\omega(k)t].$$

On the other hand, as $u(x,t)$ evolves according to (1), the corresponding reflection coefficient $R(k,t)$ evolves according to the formula

$$(5) \qquad R(k,t) = R(k,0) \exp[2ik\alpha(-4k^2)t],$$

as discussed in the preceding section 1.4. This formula, using (3), can be rewritten in the form

$$(6) \qquad R(\tfrac{1}{2}k, t) = R(\tfrac{1}{2}k, 0) \exp[-i\omega(k)t],$$

which indicates that $R(\tfrac{1}{2}k, t)$ evolves, when $u(x, t)$ obeys (1), exactly as $\hat{u}(k, t)$ does, when $u(x, t)$ obeys the linearized version (2) of (1). And this is precisely consistent with the remark reported at the end of subsections 1.3.2 and 1.3.3; in the approximation in which all nonlinear effects are neglected, the reflection coefficient $R(\tfrac{1}{2}k)$ is simply proportional to the Fourier transform $\hat{u}(k)$:

$$(7) \qquad R(\tfrac{1}{2}k, t) = (ik)^{-1} \hat{u}(k, t).$$

This clinches the argument; albeit with two qualifications. Firstly, that we have ignored here the contribution of the discrete eigenvalues; the justification for doing this, in the limit of small u, has already been mentioned above (subsection 1.3.2). Secondly, it is clear from (7) that the above argument breaks down in the neighbourhood of $k=0$, namely for the more slowly varying component of $u(x, t)$. This will be further discussed in subsection 3.2.2.

1.6. Qualitative behaviour of the solutions: solitons and background

In section 1.4 a class of *nonlinear* partial differential equations of evolution type has been identified, that is solvable by the spectral transform technique, in close analogy to the *linear* partial differential equations of evolution type whose solution via Fourier transform has been discussed in section 1.1. As emphasized there, the availability of a constructive technique of solution is important not so much to obtain explicitly the solution—a task that is only rarely feasible—but rather to provide a qualitative understanding of its behaviour. Indeed the main lesson in the *linear* case, solvable via the Fourier transform, is, as we have already emphasized, that the time evolution is much simpler in "k-space" than in "x-space"; this observation constitutes the main guiding principle to investigate these equations, and thereby to understand the behaviour of the natural phenomena whose time evolution is, in the framework of mathematical physics or applied mathematics, described by them; indeed the influence of this guiding principle largely determines the appropriate approach to the experimental study of these natural phenomena.

The situation is completely analogous for the class of nonlinear evolution equations identified in the preceding section. Here again the time evolution is much simpler in "k-space" than in "x-space"; however the structure of "k-space" is now richer, due to the presence, in general, of two different components, namely the continuous ($k^2>0$, k real) and discrete ($k^2 = -p_n^2$, p_n real, $n=1,2,\ldots,N$) parts of the spectrum (see section 1.3). Note that the results of section 1.4 imply that the time evolution of each of these components occurs separately, with no mixing.

It is therefore convenient, in order to arrive at a qualitative understanding of the behaviour of the solutions of the class of nonlinear evolution equations (1.2.-1), to begin by analysing separately the solutions that correspond to the two components, discrete and continuous, of the spectrum. We therefore consider first the (rather special) case characterised by the absence of any contribution from the continuum; then the case characterised by the absence of any discrete eigenvalue; and finally we tersely describe the behaviour of a generic solution, whose spectral transform contains both contributions, from the discrete spectrum and from the continuum. As in the case of the linear evolution equations discussed tersely in section 1.1, the qualitative analysis focusses largely on the asymptotic, long-time, behaviour of the solutions.

1.6.1. Solitons

The simplest solution with no contribution from the continuum corresponds to an initial datum $u_0(x)$ whose spectral transform S_0 has a vanishing reflection coefficient and only one discrete eigenvalue:

(1) $\quad S_0 = \{R(k)=0, -\infty<k<\infty; p, \rho_0\}$,

(2) $\quad u_0(x) = -2p^2/\cosh^2[p(x-\xi_0)]$,

with

(3) $\quad \xi_0 = (2p)^{-1} \ln(\rho_0/2p)$;

note the consistency of these, and the following, formulae, with (1.3.2.-6, 10, 11). The time evolution of this function is then given, very directly, by the results of section 1.4:

(4) $\quad S(t) = \{R(k,t) = 0, -\infty < k < \infty; p, \rho(t) = \rho_0 \exp[-2p\alpha(4p^2)t]\}$,

(5) $\quad u(x,t) = -2p^2/\cosh^2\{p[x-\xi(t)]\}$,

(6a) $\quad \xi(t) = (2p)^{-1}\ln[\rho(t)/2p]$,

(6b) $\quad \xi(t) = \xi_0 + vt$,

(7) $\quad v = -\alpha(4p^2)$.

The solution (5) (with (6b) and (7)) describes a wave of constant shape moving with constant speed (in a graphical picture where $-u(x,t)$ marks the profile of the surface of the water in a canal as a function of the distance x, this wave has indeed "the form of a large solitary elevation, a rounded, smooth and well-defined heap of water", moving "without change of form or diminution of speed"). This is of course the famous *soliton* (indeed the sentences quoted above have been lifted from the Scott-Russell citation reported in the introduction; whose relevance here is highlighted by the fact that the KdV equation was indeed introduced as the appropriate mathematical schematization to describe "long waves advancing in a rectangular canal" [KdV1895]).

The special *single-soliton solution* (5) (with (6b) and (7)) of the nonlinear evolution equation (1.2.-1) contains two (real) parameters, ξ_0 and p. The first characterizes the initial localization of the soliton; its arbitrariness corresponds of course to the translation-invariant character of (1.2.-1); and its relation to the spectral transform is given by (3). The second parameter, p, whose spectral significance is directly related to the value of the discrete eigenvalue (see section 1.3), determines the shape of the soliton (both its height and its width) and moreover its speed; note that the shape is the same for all equations of the class (1.2.-1), while the speed depends on the function $\alpha(z)$, see (7), namely it depends on which specific equation of the class (1.2.-1) one is considering. For instance for the KdV equation (1.2.-4) corresponding to $\alpha(z) = -z$, the speed of the soliton is

(8) $\quad v = 4p^2$;

thus all solitons of the KdV equation move to the right (the fact that this does not correspond to the behaviour of the waves in a canal, that should clearly be parity invariant, need not worry the reader; the KdV equation in the simple form (1.2.-4) is only appropriate to describe waves travelling in one direction; moreover it describes the behaviour of long waves in shallow canals as seen in a reference frame moving with an appropriately chosen constant speed).

Let us now proceed and consider the *N-soliton solution*, namely the function $u(x,t)$, being again characterised by the absence of the continuum contribution in the corresponding spectral transform ($R(k,t)=0$, $-\infty<k<\infty$), but now with N discrete eigenvalues $k^2=-p_n^2$, $n=1,2,\ldots,N$. A closed form expression for this solution (see (1.3.2.-13)) reads

$$(9) \quad u(x,t)=-2\partial^2\{\ln\det[\boldsymbol{I}+\boldsymbol{C}(x,t)]\}/\partial x^2,$$

where $\boldsymbol{I}$ is the unit matrix (of order N) and $\boldsymbol{C}(x,t)$ is the symmetrical matrix of order N defined by

$$(10) \quad C_{mn}(x,t)=[\rho_m(t)\rho_n(t)]^{1/2}(p_m+p_n)^{-1}\exp[-(p_m+p_n)x],$$

where of course the time-dependence of the normalization coefficients, $\rho_n(t)$, is given by (1.4.-9).

To discuss further this solution, we hereafter restrict consideration, for simplicity, to the case of the KdV equation, so that the velocity of the soliton is given by the simple formula (8); but the qualitative essence of the findings discussed below applies equally to all equations of the class (1.2.-1).

For definiteness, let us order the N positive numbers p_n in increasing order,

$$(11) \quad p_1<p_2<\cdots<p_n;$$

and let us introduce the corresponding soliton velocities

$$(12) \quad v_n=4p_n^2, \quad n=1,2,\ldots,N,$$

that are then clearly also ordered in increasing order,

$$(13) \quad v_1<v_2<\cdots<v_N.$$

The solution (9) is not simple, especially if N is large; but it is not too difficult to show that its asymptotic behaviour in t is quite simple:

$$(14) \quad u(x,t)\approx -2\sum_{n=1}^{N} p_n^2/\cosh^2\left[p_n\left(x-\xi_n^{(\pm)}-v_n t\right)\right], \quad t\to\pm\infty,$$

where the $\xi_n^{(\pm)}$'s are $2N$ real constants (related to one another, see below). Thus both in the remote past and future the N-soliton solution (9) divides up into N separated solitons; in the remote past the solitons are ordered, from left to right, in order of decreasing amplitude and they move towards the right with speeds ordered in decreasing magnitude; then the taller and faster solitons, that are more to the left, gradually catch up and eventually "overtake" the fatter and slower solitons (the quotation marks underscore the fact that, whenever two, or more, solitons get together, their individuality is in fact lost; indeed the "overtaking"—as seen, for instance, in a film—may appear as an exchange of identity, with the taller soliton becoming fatter, and vice versa, as they get close together, until they separate again because the one that has become taller speeds up, while the one that has become fatter slows down; see subsection 3.2.1.); and finally in the remote future the ordering of the solitons gets altogether reversed, with the taller and faster heading the escape to the right. The most important implication of (14) is that precisely the same solitons that existed at the beginning are found at the end, the only effect of their "interaction" having been to shift the position of the nth soliton, relative to what it would have been had it been moving in isolation, by the amount

(15) $\quad \Delta_n = \xi_n^{(+)} - \xi_n^{(-)}.$

These shifts are, moreover, determined (while either the N quantities $\xi_n^{(-)}$ or the N constants $\xi_n^{(+)}$ can be chosen arbitrarily; this corresponds to the possibility of choosing arbitrarily the N quantities $\rho_n(0)$ in (1.4.-9)), being given by the formula

(16) $\quad \Delta_n = \sum_{m=1}^{n-1} \Delta(p_n, p_m) - \sum_{m=n+1}^{N} \Delta(p_n, p_m),$

with

(17) $\quad \Delta(p, p') = p^{-1} \ln[(p+p')/|p-p'|].$

Of course, in writing (16), we employ the usual convention that sets to zero a sum if the lower limit exceeds the upper limit.

The formula (16) has a simple "physical" interpretation. Let us first consider the case of two solitons only, $N=2$; it is then seen that the second soliton (the faster and taller one) gets *advanced* precisely by the amount $\Delta(p_2, p_1)$, while the first soliton (the slower and fatter one) gets *delayed* by the amount $\Delta(p_1, p_2)$. Thus the shift (16) experienced by the n-th soliton appears as the sum of the $n-1$ positive shifts derived from its "overtaking" the $n-1$ slower solitons and the $N-n$ negative shifts derived from its being overtaken by the $N-n$ faster solitons. Note that this result makes perfect sense in the case in which each collision occurs separately; it is highly non-trivial in the present context, since it might well happen (depending on the initial conditions) that, at some intermediate time, several solitons all coalesce together.

This behaviour strongly suggests ascribing to each soliton an individuality, even though it does, strictly speaking, show up as a separate entity only in the remote past and future; indeed it was the observation, in numerical experiments with the KdV equation, of precisely such persistence after collisions that motivated Zabusky and Kruskal to introduce the soliton concept [ZK1965]. In the context of the spectral transform technique, the individuality of the soliton has an obvious interpretation; and its permanence is clearly related to the fact that the discrete eigenvalues do not vary as $u(x,t)$ evolves according to (1.2.-1) (thus it will be a feature also of the generic solution of (1.2.-1); see below). The possibility to "identify" the individual solitons for all time also in configuration space is discussed in subsection 3.2.1.

1.6.2. Background

In contrast to the solutions whose spectral transform contains only contributions from the discrete spectrum (see preceding subsection), solutions corresponding to the continuum part of the spectral transform cannot generally be exhibited in explicit form (for some counter-examples see subsection 3.2.4.). Thus the study of these solutions is more difficult.

On the other hand, while the soliton solutions discussed in the preceding subsection reflect in an essential way the nonlinear character of the class of evolution equations (1.2.-1), the solutions to be discussed here include those "weak field" situations in which the drastic linearization of (1.2.-1), consisting in the elimination of all nonlinear terms, provides a reasonable approximation. Indeed, if $u(x,t)$ is very small, its spectral transform contains generally no discrete eigenvalues (or at most only one; see appendix A.1), so that such solutions generally belong to the class being discussed here.

For such solutions, the standard analysis, based on the Fourier transform technique of solution, becomes applicable. Thus their behaviour—in particular, their long-time behaviour—is essentially characterised by the phenomenon of dispersion, and by the group velocity associated through (1.1.-9) with the dispersion function $\omega(k)$, which is itself related by (1.5.-3) to the function $\alpha(z)$ that determines the r.h.s. of (1.2.-1), namely the form of the particular nonlinear evolution equation under consideration.

It might appear that this analysis is only relevant to those solutions $u(x, t)$ that are everywhere sufficiently small to justify the approximation consisting in the drastic elimination of all nonlinear contributions in the r.h.s. of (1.2.-1). But if the spectral transform of $u(x, t)$ has no discrete eigenvalues, namely if $u(x, t)$ contains no solitons, there is no mechanism available to maintain the solution $u(x, t)$ (or any part of it) localized in the long-time limit. Thus, in the asymptotic long-time limit, the solution becomes locally small everywhere. This indicates that the analysis given above is in fact applicable to any solution of the type being considered in this subsection.

We conclude therefore that the long-time behaviour of those solutions of the nonlinear evolution equations (1.2.-1) whose spectral transform contains no discrete spectrum contribution may be qualitatively understood by considering the corresponding linearized equation, namely (1.1.-1) with (1.5.-3). They are therefore characterized by dispersion (with the dispersion function defined by (1.5.-3)); in contrast to the solitons, these solutions do not remain localized. They are often referred to as *background*; sometimes as *radiation*, to underscore the difference from the particle-like nature of the solitons; and sometimes (perhaps a bit disparagingly—motivated by their lack of persistence?) as *hash* [K1974].

1.6.3. Generic solution

The N-soliton solution of the nonlinear evolution equation (1.2.-1) is clearly not generic, since it has the specific form (1.6.1.-9); it is, however, fairly general, since it contains the $2N$ arbitrary constants p_n and $\rho_n(0)$, $n = 1, 2, \ldots, N$, and N itself is arbitrary. That this solution is indeed far from generic is particularly evident if one considers its spectral transform, that is of course characterized by the total absence of the continuum contribution ($R(k, t) = 0$).

The pure background solution, characterized by the complete absence of solitons, is also not quite generic, although it represents a broader class than the pure multi-soliton solutions; the condition on a potential, in the context of the Schroedinger spectral problem, to have no associated discrete eigenvalues (in the language of physics, no bound states) appears indeed less restrictive than the requirement to be completely reflectionless.

The generic solution of the nonlinear evolution equation (1.2.-1) contains a soliton part and a background; the separation into these two components is of course not generally visible in x-space (although it may occur, as it were, automatically, in the long-time limit; see below); this separation is on the other hand quite evident, throughout the time evolution, from the point of view of the spectral transform.

The qualitative behaviour of the generic solution in x-space as time evolves is a combination of those discussed in the preceding two subsections; of course only when the two phenomena get disentangled also in x-space can a simple qualitative picture emerge. This generally happens in the asymptotic long-time limit. Thus in the remote future the solution evolves generally into a finite number of separated well-localised solitons, superimposed on a background that tends locally to zero everywhere according to the standard behaviour of solutions of (non-dissipative) dispersive linear evolution equations.

For instance, for the KdV equation (1.2.-4) ($\alpha(z) = -z$ in (1.2.-1)), the group velocity v_g turns out to be negative ($\omega(k) = -k^3$, $v_g = -3k^2$; see (1.5.-3) and (1.1.-9)), in contrast to the soliton speeds $v_n = 4p_n^2$ (see (1.6.1.-12)), which are all positive; thus in this case a complete separation occurs asymptotically between the background part of the solution, that disperses away towards the left, and the solitons, that travel to the right as described in subsection 1.6.1.

Clearly this behaviour constitutes a distinctive and remarkable feature of this class of nonlinear evolution equations (and of other classes solvable by analogous technique, based on different spectral problems; see below). The physical, or more generally natural, interpretation of these results—for instance the emergence of the solitons as localized individual entities—does of course depend on the particular physical, or more generally natural, phenomenon that the nonlinear evolution equation under consideration is supposed, perhaps approximately, to represent; be it in fluid dynamics or in demography, in solid state physics or in epidemiology, in the investigation of signal transmission through nervous fibres or of models of elementary particles or of plasma disturbances.

1.7. Additional properties of the solutions

The class of nonlinear evolution equations (1.2.-1) has a number of additional remarkable features, all of them originating from the basic property of this class of nonlinear evolution equations, i.e. the fact that the time evolution, although nonlinear in configuration space (x-space), is related (via the spectral transform) to a simple linear evolution in the spectral space (see section 1.4.). In this section we review tersely some of these properties, that shall be then discussed in more detail, and proved, in chapters 4 and 5.

1.7.1. Bäcklund transformations

Consider an evolution equation, say

(1) $\quad u_t(x,t) = F[u, u_x, u_{xx}, \ldots]$.

Let $u^{(0)}(x,t)$ be a generic solution of (1), and let $u^{(1)}(x,t)$ be related to $u^{(0)}(x,t)$ by some explicit relation (involving the differential operator $\partial/\partial x$ as well as the integral operator $\int_x^\infty dy$.) which, for the moment, we indicate by

(2) $\quad G[u^{(0)}, u^{(1)}] = 0$.

If $u^{(1)}(x,t)$ is then also found to satisfy (1), then (2) is, by definition, a *Bäcklund transformation*.

Thus we define here a *Bäcklund transformation* to be any formula relating two functions, say $u^{(0)}(x,t)$ and $u^{(1)}(x,t)$, in such a way as to insure that, if one of them satisfies a (generally nonlinear) evolution equation, then the other does so too. Such transformations were introduced long ago in differential geometry and their name goes back to this origin. Here we tersely outline how, in the context of the spectral transform technique to solve the nonlinear evolution equations (1.2.-1), a class of Bäcklund transformations for the solutions of this equation naturally emerges. Our purpose and scope here is merely to convey the basic idea. A more detailed discussion is then given in chapter 4.

Let us begin by reporting once more here the class of evolution equations (1.2.-1), reading

(3) $\quad u_t = \alpha(L) u_x, \quad u \equiv u(x,t)$,

where $\alpha(z)$ is a polynomial and L is the integrodifferential operator defined by

(4) $\quad Lf(x) = f_{xx} - 4uf + 2u_x \int_x^\infty dy\, f(y)$.

And let us recall that, as $u(x,t)$ evolves according to (1), the reflection coefficient $R(k,t)$ corresponding to $u(x,t)$ via the spectral (or scattering) problem of section 1.3, evolves in time according to the formula

(5a) $\quad R_t(k,t) = 2ik\alpha(-4k^2)R(k,t),$

implying of course

(5b) $\quad R(k,t) = R(k,0)\exp[2ik\alpha(-4k^2)t].$

For the sake of simplicity, in the following we shall limit consideration to the reflection coefficient, assuming that there is a one-to-one correspondence between u and R (which is actually the case only if the spectral transform of u has no discrete eigenvalues; see section 1.3).

We then report a property of the Schroedinger spectral problem of section 1.3, that is proved in the following chapter 2. Let $u^{(0)}(x)$ and $u^{(1)}(x)$ be two functions (both regular for real x and vanishing "sufficiently fast" as $x \to \pm\infty$), related to each other by the (nonlinear integrodifferential) formula

(6) $\quad g(\Lambda)[u^{(0)}(x) - u^{(1)}(x)] + h(\Lambda) \cdot \Gamma \cdot 1 = 0.$

Here $g(z)$ and $h(z)$ are two arbitrary entire functions (say, two polynomials) and the integrodifferential operators Γ and Λ are defined by the following formulae that specify their action on a generic function $f(x)$ (vanishing as $x \to \infty$):

(7) $\quad \Gamma f(x) = [u_x^{(0)}(x) + u_x^{(1)}(x)]f(x)$
$\qquad + [u^{(0)}(x) - u^{(1)}(x)]\int_x^\infty dy[u^{(0)}(y) - u^{(1)}(y)]f(y),$

(8) $\quad \Lambda f(x) = f_{xx}(x) - 2[u^{(0)}(x) + u^{(1)}(x)]f(x) + \Gamma \cdot \int_x^\infty dy f(y).$

Note that Γ and Λ depend (in a symmetrical way) on $u^{(0)}(x)$ and $u^{(1)}(x)$; this causes (6) to be nonlinear in $u^{(0)}$ and $u^{(1)}$. Incidentally, for $u^{(0)}(x) = u^{(1)}(x) = u(x)$, Λ goes over into the operator L, see (4).

Then the reflection coefficients $R^{(0)}(k)$ and $R^{(1)}(k)$, corresponding respectively to $u^{(0)}(x)$ and $u^{(1)}(x)$, are related to each other by the simple linear formula

(9) $\quad g(-4k^2)[R^{(0)}(k) - R^{(1)}(k)] + 2ikh(-4k^2)[R^{(0)}(k) + R^{(1)}(k)] = 0,$

1.7.1 Bäcklund transformations

implying

(10) $\quad R^{(1)}(k) = R^{(0)}(k) \dfrac{\left[g(-4k^2) + 2ikh(-4k^2)\right]}{\left[g(-4k^2) - 2ikh(-4k^2)\right]}.$

Let us emphasize that this result constitutes a property of the spectral problem; it has nothing to do with time-evolution. But if $u^{(0)}$ and $u^{(1)}$ evolve in time (i.e., if they depend on another variable t in addition to x), all the formulae given above remain valid, of course with $R^{(0)}$ and $R^{(1)}$ also depending on t. Indeed also the two arbitrary functions g and h in (6), (9) and (10) could well depend on t; but the following development hinges on the assumption that they be time-independent.

Suppose then that $u^{(0)}(x, t)$ evolves in time according to (3); then the corresponding reflection coefficient, $R^{(0)}(k, t)$, evolves in time according to (5). But then $R^{(1)}(k, t)$, being related to $R^{(0)}(k, t)$ by (10) (with time-independent g and h), also evolves according to (5); hence $u^{(1)}(x, t)$ also evolves according to (3). Conclusion: if $u^{(0)}(x, t)$ satisfies (3) and $u^{(1)}(x, t)$ is related to $u^{(0)}(x, t)$ by (6), then $u^{(1)}(x, t)$ also satisfies (3).

It is thus seen that (6) is a Bäcklund transformation for the (class of) evolution equations (3). Indeed, due to the arbitrariness of the polynomials g and h, this formula yields a large class of Bäcklund transformations. The simplest of these transformations corresponds to g and h being just constants; setting for notational convenience (see chapter 4) $g/h = -2p$, it reads

(11) $\quad u_x^{(0)}(x, t) + u_x^{(1)}(x, t) = \left[u^{(0)}(x, t) - u^{(1)}(x, t)\right]$

$\qquad \cdot \left\{ 2p - \int_x^\infty dy \left[u^{(0)}(y, t) - u^{(1)}(y, t)\right] \right\}.$

An equivalent, but generally more convenient, formula obtains for the integrals $w^{(m)}(x, t)$ of the $u^{(m)}(x, t)$,

(12a) $\quad w^{(m)}(x, t) = \int_x^\infty dy\, u^{(m)}(y, t), \quad m = 0, 1,$

(12b) $\quad w_x^{(m)}(x, t) = -u^{(m)}(x, t), \quad m = 0, 1,$

reading

(13) $\quad w_x^{(0)}(x, t) + w_x^{(1)}(x, t) = \tfrac{1}{2}\left[w^{(0)}(x, t) - w^{(1)}(x, t)\right]$

$\qquad \cdot \left\{ 4p - \left[w^{(0)}(x, t) - w^{(1)}(x, t)\right] \right\}.$

To get (13) from (11) an integration has been performed; the vanishing of $w^{(m)}(x,t)$ as $x \to \infty$,

(14) $\quad w^{(m)}(\infty, t) = 0, \quad m = 0, 1,$

clearly implied by the definition (12a), as well as the asymptotic vanishing of $w_x^{(m)}(x,t)$ implied by (12b), have moreover been invoked to set to zero the integration constant. Note that (13) is, if one assumes $w^{(0)}$ to be known, a Riccati equation for $w^{(1)}$; and vice versa.

The relationship (10) corresponding to the simplest Bäcklund transformation (11) or (13) reads simply

(15) $\quad R^{(1)}(k,t) = -R^{(0)}(k,t)[(k+ip)/(k-ip)].$

An elementary application of the Bäcklund transformation (11) or (13) is to use it in order to discover some (special) solutions of (3). The simpler instance obtains setting

(16a) $\quad w^{(0)}(x,t) = 0$

in (13), since this formula, or equivalently the formula

(16b) $\quad u^{(0)}(x,t) = 0$

implied by it, clearly provides a solution of (3)—indeed, a very trivial solution. The solution of (13) with (16a) (and (14)) is an elementary task, and it yields

(17a) $\quad w(x,t) = -2p\left[1 - \tgh\{p[x-\xi(t)]\}\right]$

namely

(17b) $\quad u(x,t) = -2p^2/\cosh^2\{p[x-\xi(t)]\}.$

In the last two formulae we have omitted for notational simplicity the superscript 1; what we are displaying is merely a solution of (3). It should be noted that this result obtains for positive p (although (17b) is eventually independent of the sign of p); for negative p there is no solution of (13) (with (16a)) consistent with (14). The quantity $\xi(t)$ in the r.h.s. of (17a) and (17b) originates as an integration constant, which means, in the present context, that it does not depend on the variable x ((13) is a differential equation in this variable); this however does not prevent it from depending on t, as we have explicitly indicated.

Obviously (17b) is just the single-soliton solution, see (1.6.1.-5); indeed the time dependence of $\xi(t)$ can be determined by inserting (17b) back into (3), and by using the formula

(18) $\quad Lu_x(x,t) = 4p^2 u_x(x,t),$

that is easily verified from (4) and (17b). There obtains

(19) $\quad \dot{\xi}(t) = -\alpha(4p^2),$

consistently with (1.6.1.-6,7).

This derivation of the single-soliton solution from the Bäcklund transformation (13) is however rather objectionable, for clearly the simplification adopted in this subsection, namely to ignore the discrete part of the spectral transform, is not applicable to the soliton solution. Indeed, while the crucial role in the derivation of the Bäcklund transformation formulae, (6) resp. (11) or (13), has been played by the corresponding formulae for the reflection coefficient, (9) or (10) resp. (15), in the special case that we have now considered (and that has yielded the single-soliton solution (17b)) the latter formulae are merely instances of the identity "zero equal zero". The more detailed treatment of chapter 4 will clarify this point.

Another important property of the Bäcklund transformations that have been introduced above is their *commutativity*. In the (simplified) context to which consideration has been restricted in this subsection this property is rather obvious. Consider indeed two sets of two polynomials, $g^{(m)}(z)$ and $h^{(m)}(z)$, $m=1,2$, and the two Bäcklund transformations (6) generated by them, say BT1 and BT2. Take as a starting point some given function $u^{(0)}(x)$, and associate with it two functions, say $u^{(1)}(x)$ and $u^{(2)}(x)$, obtained respectively from $u^{(0)}(x)$ via these two Bäcklund transformations, BT1 and BT2. Then associate with $u^{(1)}(x)$ a new function, say $u^{(12)}(x)$, obtained applying the Bäcklund transformation BT2; and with $u^{(2)}(x)$ a function $u^{(21)}(x)$ obtained applying the Bäcklund transformation BT1. The property of commutativity is expressed by the equality

(20) $\quad u^{(12)}(x) = u^{(21)}(x).$

It is a highly nontrivial property when viewed, as we have just done, in configuration space; its validity is implied by the corresponding formula in the spectral space,

(21) $R^{(12)}(k) = R^{(21)}(k),$

which is instead trivially valid, since (10) implies that

(22a) $R^{(12)}(k) = R^{(0)}(k) B_1(k) B_2(k),$

(22b) $R^{(21)}(k) = R^{(0)}(k) B_2(k) B_1(k),$

with

(23) $B_m(k) = \dfrac{[g_m(-4k^2) + 2ikh_m(-4k^2)]}{[g_m(-4k^2) - 2ikh_m(-4k^2)]}, \quad m = 1, 2.$

Thus the commutativity of Bäcklund transformations, in the spectral space, is merely the commutativity of ordinary multiplication. Here we see once more the advantage, in terms of simplicity, to work in the spectral space rather than in configuration space.

The property of commutativity that we have just described is conveniently synthetized by the following diagram:

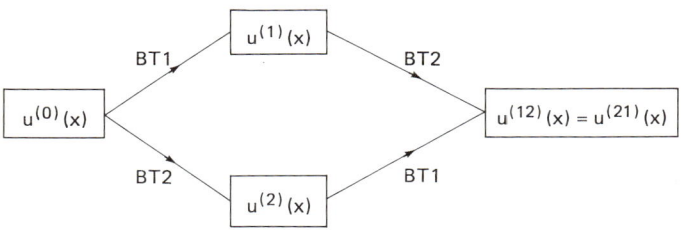

We end this subsection noting that in the limit in which both $u^{(0)}(x)$ and $u^{(1)}(x)$ are so small that all nonlinear terms in (6) can be neglected, this Bäcklund transformation reads

(24) $g(\partial^2/\partial x^2)[u^{(0)}(x) - u^{(1)}(x)] + h(\partial^2/\partial x^2)[u_x^{(0)}(x) + u_x^{(1)}(x)] = 0,$

namely, up to a change of notation, becomes identical to (1.1.-10), displaying once again the fact that all properties derived in the context of the spectral transform approach have counterparts in the Fourier transform context, to which they go over whenever it is justified to ignore all nonlinear effects.

1.7.2. Nonlinear superposition principle

Another remarkable property of the class of evolution equations (1.2.-1) obtains as a straightforward consequence of the commutativity property of Bäcklund transformations described in the preceding subsection, applied to the simplest Bäcklund transformation (1.7.1.-11) or rather (1.7.1.-13) (we work hereafter, in this subsection, with the integrals $w(x,t)$ of the solutions of (1.2.-1), rather than with the functions $u(x,t)$ themselves, see (1.7.1.-12); and, for simplicity of language, we refer directly to these functions w, that are in one-to-one correspondence with their derivatives u, see (1.7.1.-12), as the solutions of (1.2.-1); hoping to cause no misunderstanding).

Let $w^{(0)}(x,t)$ be some solution of (1.2.-1), $w^{(1)}(x,t)$ resp. $w^{(2)}(x,t)$ the solutions of (1.2.-1) related to $w^{(0)}(x,t)$ by a Bäcklund transformation (1.7.1.-13) with the parameter p replaced by p_1 resp. p_2, and $w^{(12)}(x,t) = w^{(21)}(x,t)$ (see (1.7.1.-20)) the solution of (1.2.-1) obtained from $w^{(0)}(x,t)$ via two Bäcklund transformations (1.7.1.-13), with the parameter p replaced once by p_1 and once by p_2:

(1) $\quad w_x^{(0)} + w_x^{(1)} = -\tfrac{1}{2}(w^{(0)} - w^{(1)})(-4p_1 + w^{(0)} - w^{(1)})$,

(2) $\quad w_x^{(0)} + w_x^{(2)} = -\tfrac{1}{2}(w^{(0)} - w^{(2)})(-4p_2 + w^{(0)} - w^{(2)})$,

(3) $\quad w_x^{(1)} + w_x^{(12)} = -\tfrac{1}{2}(w^{(1)} - w^{(12)})(-4p_2 + w^{(1)} - w^{(12)})$,

(4) $\quad w_x^{(2)} + w_x^{(21)} = -\tfrac{1}{2}(w^{(2)} - w^{(21)})(-4p_1 + w^{(2)} - w^{(21)})$.

Now subtract (2) from (1), and (4) from (3) (using (1.7.1.-20)), and subtract again from each other the two resulting equations. Then all differentiated terms in the l.h.s. cancel out, and there results an algebraic relation that can be solved for $w^{(12)} = w^{(21)}$:

(5) $\quad w^{(12)} = w^{(21)}$

$\qquad = w^{(0)} - (p_1 + p_2)(w^{(1)} - w^{(2)}) / [p_1 - p_2 + \tfrac{1}{2}(w^{(1)} - w^{(2)})]$.

Let us emphasize that this formula provides a completely explicit expression of a novel solution of (1.2.-1), in terms of three other solutions of the same equation: an (arbitrary) solution $w^{(0)}$, and the two solutions $w^{(1)}$ and $w^{(2)}$ related to $w^{(0)}$ by the Bäcklund transformations (1) and (2).

As an example of application of the *nonlinear superposition principle* (5), let us again consider the case with

(6) $\quad u^{(0)}(x,t) = w^{(0)}(x,t) = 0$,

that clearly provides a (trivial) solution of (1.2.-1). Then (see (1.7.1.-17))

(7a) $\quad u^{(m)}(x,t) = -2p_m^2/\cosh^2\{p_m[x - \xi_0^{(m)} + \alpha(4p_m^2)t]\}, \quad m = 1, 2,$

(7b) $\quad w^{(m)}(x,t) = -2p_m[1 - \tgh\{p_m[x - \xi_0^{(m)} + \alpha(4p_m^2)t]\}], \quad m = 1, 2,$

where $\xi_0^{(m)}$ has to be either real,

(8a) $\quad \operatorname{Im} \xi_0^{(m)} = 0$

or $\xi_0^{(m)} = \xi_0^{(m)\prime} + i\pi/(2p_m)$ with $\xi_0^{(m)\prime}$ real,

(8b) $\quad \operatorname{Im} \xi_0^{(m)} = \tfrac{1}{2}\pi/p_m,$

in order that $u^{(m)}$ and $w^{(m)}$, $m = 1, 2$, be real. Insertion of (6) and (7b) in (5) yields the two-soliton solution, provided $0 < p_1 < p_2$ and $\xi_0^{(1)}$ satisfies (8a) while $\xi_0^{(2)}$ satisfies (8b) (otherwise the solution produced by (5) is complex or singular).

Having thus obtained the two-soliton solution, it is possible to apply the nonlinear superposition principle (5) to get the three-soliton solution, by inserting in place of $w^{(0)}$ the single-soliton expression (with parameter, say, p_1) and in place of $w^{(1)}$ resp. $w^{(2)}$ the two-soliton expression (with parameters p_1 and p_2 resp. p_1 and p_3); and the process can then be continued (*soliton ladder*). In this manner it is in principle possible to construct all multisoliton solutions by a purely algebraic procedure; and very simple rules can be given (see section 4.2), detailing the restrictions on the soliton parameters p_m and the reality properties of the $\xi_0^{(m)}$'s ((8a) or (8b)), to insure that the solution so arrived at is indeed real and nonsingular, and thus coincides with (1.6.1.-9).

1.7.3. Conservation laws

Another important property of the class of nonlinear evolution equations (1.2.-1) is the existence of an infinite sequence of local conservation laws. Each of these conservation laws yields, for the class of asymptotically ($x \to \pm\infty$) vanishing solutions to which our consideration is confined, a conserved quantity expressed as the integral over all space of an appropriate nonlinear (polynomial) combination of $u(x,t)$ and its derivatives.

In this subsection we review tersely these conserved quantities; these results are further discussed, and proved, in chapter 5. Let us mention that,

1.7.3 Conservation laws

in the framework of the approach developed in this book, the existence of an infinite sequence of conservation laws, remarkable as it is, appears as a secondary feature of the class of evolution equations (1.2.-1). In other approaches to this subject the existence of an infinite number of conservation laws plays a more fundamental role; indeed the discovery of several independent conservation laws for the KdV equation provided the first hint of its peculiarity, and it was actually in the process of investigating this property that the spectral transform technique was invented.

All the results given in the preceding sections hold for, or are trivially extended to, the class of evolution equations

$$(1) \qquad u_t(x,t) = \alpha(L,t) u_x(x,t),$$

where $\alpha(z, t)$ is an (arbitrary) polynomial (with t-dependent coefficients) in z, and L is the integro-differential operator defined by (1.2.-2). This is a more general class than (1.2.-1), since an explicit (and largely arbitrary) t-dependence has been introduced. In the preceding sections this generalization was omitted, although it could have been easily accommodated, since we are interested throughout this chaper in outlining the basic ideas rather than in presenting the results in full generality. Here, however, we refer to the more general class (1), since the existence of an infinite sequence of conserved quantities, in spite of the largely arbitrary explicit time dependence appearing in the r.h.s. of (1), is sufficiently remarkable to deserve special mention already at this stage.

We already noted (see section 1.2) that the class of nonlinear evolution equations (1) has itself the structure of a local conservation law,

$$(2) \qquad u_t(x,t) = \gamma_x^{(0)}(x,t),$$

with $\gamma^{(0)}(x, t)$ a polynomial (now, with t-dependent coefficients) in $u(x, t)$ and its x-derivatives. (The notation employed for the r.h.s. of this equation is motivated by future developments; see below.) This implies the asymptotic vanishing of $\gamma^{(0)}(x, t)$,

$$(3) \qquad \gamma^{(0)}(\pm\infty, t) = 0.$$

Integration of (2) over x from $-\infty$ to $+\infty$ therefore shows that the quantity

$$(4) \qquad C_0 = c_0 = \int_{-\infty}^{+\infty} dx\, u(x,t)$$

is a constant of the motion for the flow defined by (1).

The remarkable fact is, that this quantity is merely the first one of an infinite sequence. There are in fact two equivalent sequences of constants of the motion, whose expression may be given in explicit and compact form. The first reads

$$(5) \qquad C_m = (-1)^m (2m+1)^{-1} \int_{-\infty}^{+\infty} dx\, L^m [x u_x(x,t) + 2 u(x,t)],$$

$$m = 0, 1, 2, \ldots,$$

where the appearance of the integro-differential operator L, defined by (1.2.-1), should be noted. An alternative expression for the quantity (5) is

$$(6) \qquad C_m = (-1)^m (2m+1)^{-1} \int_{-\infty}^{+\infty} dx\, \tilde{L}^m u(x,t), \quad m = 0, 1, 2, \ldots,$$

where $\tilde{L}$ is the integro-differential operator defined by the formula

$$(7) \qquad \tilde{L} f(x) = f_{xx}(x) - 4 u(x,t) f(x) + 2 \int_x^{+\infty} dy\, u(y,t) f_y(y,t),$$

that shows its action on a generic function $f(x)$. As will be seen in the following, also this operator plays an important rôle, comparable to that played by the operator L; indeed, in some sense (that is made precise in appendix A.9), $\tilde{L}$ is just the adjoint of L.

Additional, alternative, expressions of the constants of the motion C_m are given in chapter 5 (see also appendix A.9, in particular (A.9.-57)).

The second sequence of conserved quantities reads

$$(8) \qquad c_m = (-1)^m \int_{-\infty}^{+\infty} dx\, u(x,t) M^{2m} \cdot 1, \quad m = 0, 1, 2, \ldots$$

with the integro-differential operator M defined by the formula

$$(9) \qquad M f(x) = f_x(x) - \int_{-\infty}^x dy\, u(y,t) f(y),$$

that specifies its action on a generic function $f(x)$ (of course here, as well as in (7), f may depend on other variables besides x, for instance on t). Also the quantities (8) possess an alternative expression,

$$(10) \qquad c_m = (-1)^m \int_{-\infty}^{+\infty} dx\, \Lambda^{(0)m} u(x,t), \quad m = 0, 1, 2, \ldots,$$

1.7.3

where $\Lambda^{(0)}$ is the integro-differential operator defined by the formula

(11) $$\Lambda^{(0)}f(x) = f_{xx}(x) - 2u(x,t)f(x) + u_x(x,t)\int_x^{+\infty} dy\, f(y)$$
$$+ u(x,t)\int_x^{+\infty} dy\, u(y,t)\int_y^{+\infty} dz\, f(z).$$

The first three constants of the motion given by (5) read

(12a) $$C_0 = \int_{-\infty}^{+\infty} dx\, u(x,t),$$

(12b) $$C_1 = \int_{-\infty}^{+\infty} dx\, u^2(x,t),$$

(12c) $$C_2 = \int_{-\infty}^{+\infty} dx\, [2u^3(x,t) + u_x^2(x,t)].$$

Note that the explicit x-dependence in the r.h.s. of (5) has disappeared from these formulae; this is indeed the case for all of them, as implied by the alternative formula (6) (with (7)), or by the relationship of the C_m's to the c_m's, see below. It is also obvious that all the C_m's are independent (none of them can be expressed in terms of the others); and their general structure is apparent from the explicit expressions (12) (an elementary dimensional argument implies that each of the monomials appearing in the integrand in the definition of C_m must contain q powers of u and r x-derivatives, with the restriction $2q+r=2m+2$).

The constants c_m are also all independent of each other. There is instead a relationship between the constants of the two series, that is expressed in compact form by the formula

(13) $$\sum_{m=0}^{\infty} c_m z^{2m+1} = \sin\left[\sum_{m=0}^{\infty} C_m z^{2m+1}\right],$$

which is to be understood by expanding in powers of z the r.h.s. and then equating the coefficients of equal powers of z:

(14a) $c_0 = C_0,$

(14b) $c_1 = C_1 - \frac{1}{6}C_0^3,$

(14c) $c_2 = C_2 - \frac{1}{2}C_0^2 C_1 + \frac{1}{120}C_0^5,$

and so on.

Let us emphasize that all these conservation laws hold for the whole class of evolution equations (1); and note that the definition of the conserved quantities does not depend on $\alpha(z, t)$, namely on which particular equation of the class (1) is being considered.

The conservation laws expressed by the time independence of the integrals (5) and (6) can be also rewritten in the differential form of a "continuity equation". In fact, to the "densities" (see (5))

(15) $\quad f^{(m)}(x,t) = L^m[xu_x(x,t) + 2u(x,t)], \quad m = 0, 1, 2, \ldots,$

respectively (see (6))

(16) $\quad g^{(m)}(x,t) = \tilde{L}^m u(x,t), \quad m = 0, 1, 2, \ldots,$

are associated the "currents" $\varphi^{(m)}(x, t)$, respectively $\gamma^{(m)}(x, t)$, satisfying the equations

(17) $\quad f_t^{(m)}(x,t) = \varphi_x^{(m)}(x,t), \quad m = 0, 1, 2, \ldots$

(18) $\quad g_t^{(m)}(x,t) = \gamma_x^{(m)}(x,t), \quad m = 0, 1, 2, \ldots,$

together with the conditions

(19) $\quad \varphi^{(m)}(\pm \infty, t) = 0, \quad \gamma^{(m)}(\pm \infty, t) = 0, \quad m = 0, 1, 2, \ldots,$

that guarantee the time independence of (5) and (6). Note that, while the densities (15) and (16) do not depend on $\alpha(z, t)$ (they are common to all evolution equations of the class (1)), this is not the case for the currents $\varphi^{(m)}$ and $\gamma^{(m)}$. Let us also mention that in the conservation law (18) both the density $g^{(m)}$ and the current $\gamma^{(m)}$ have the remarkable property of being polynomial expressions in u and its x-derivatives (this implies in particular that all the integrations that appear in the r.h.s. of (16) as a consequence of the definition of $\tilde{L}$, see (7), can be performed in closed form). For instance, for the (trivial) case $\alpha(z, t) = 1$, the current $\gamma^{(m)}$ coincides with its corresponding density, $\gamma^{(m)} = g^{(m)}$, while for the KdV equation

(20) $\quad u_t + u_{xxx} - 6u_x u = 0, \quad u \equiv u(x, t),$

(the particular equation of the class (1) corresponding to $\alpha(z, t) = -z$) the

1.7.3 Conservation laws

first two currents are

(21a) $\quad \gamma^{(0)} = -u_{xx} + 3u^2,$

(21b) $\quad \gamma^{(1)} = -u_{xxxx} + 3u_x^2 + 12uu_{xx} - 12u^3.$

An explicit and compact expression for the currents $\gamma^{(m)}$ and $\varphi^{(m)}$ for all $m \geqslant 0$, and for any polynomial $\alpha(z, t)$ characterizing the evolution equation (1), is given in chapter 5.

It should be finally pointed out that there exists an additional conserved quantity, for the evolution equation (1). For instance, for the KdV equation (20) the quantity

(22) $\quad C = \int_{-\infty}^{+\infty} dx \left[xu(x,t) + 3tu^2(x,t) \right]$

is another constant of the motion, independent of all those written above; its time-independence is readily verified differentiating (22) with respect to t and using (20). Note that in the definition (22) the time variable t, as well as the space coordinate x, appear explicitly.

By using the constancy of C, together with that of C_0 and C_1 (see (12)), it is immediately seen that, for the KdV equation (20), *the center of mass*

(23) $\quad X(t) = \int_{-\infty}^{+\infty} dx \, xu(x,t) \bigg/ \int_{-\infty}^{+\infty} dx \, u(x,t)$

undergoes a *uniform motion*:

(24) $\quad X(t) = X_0 + Vt,$

the constants X_0 and V being given by

(25) $\quad X_0 = C/C_0, \qquad V = -3C_1/C_0.$

Let us emphasize that this property of the center of mass $X(t)$, to move with constant speed (as a free particle), holds for a generic solution of the KdV equation (20).

This result in fact generalizes to the whole class (1), the additional conserved quantity being

(26) $\quad C = \int_{-\infty}^{+\infty} dx \left[xu(x,t) + \int_0^t dt' \, \alpha(\tilde{L}, t') u(x,t) \right],$

with $\tilde{L}$ defined by (7). This implies that, for the generic solution $u(x,t)$ of (1), the center of mass $X(t)$, see (23), moves according to the formula

$$(27) \qquad X(t) = X_0 + \sum_{m=0}^{M} (-1)^{m+1}(2m+1)(C_m/C_0) \int_0^t dt'\, \alpha_m(t'),$$

where

$$(28) \qquad X_0 = C/C_0$$

and the functions $\alpha_m(t)$ are the coefficients of the polynomial $\alpha(z,t)$ in (1):

$$(29) \qquad \alpha(z,t) = \sum_{m=0}^{M} \alpha_m(t) z^m.$$

Thus, for all "autonomous" evolution equations of the class (1) (characterized by the lack of an explicit time-dependence: $\alpha(z,t) \equiv \alpha(z)$), the center of mass of the generic solution moves uniformily, see (24), with the (constant) speed

$$(30) \qquad V = \sum_{m=0}^{M} (-1)^{m+1}(2m+1)\alpha_m C_m/C_0.$$

1.8. A list of solvable equations

To complete this introductory chapter we list some nonlinear evolution equations of applicative and/or theoretical relevance. This list is not meant to be complete, but merely representative; it includes equations that are treated in this book (including volume II), and equations that do not fit into the classes that are actually discussed here, but are nevertheless solvable by techniques that are (non trivial) extensions of those treated here. And some examples that are *not* solvable by the spectral transform technique are also mentioned.

The purpose of this section is to provide an idea of the scope of the spectral transform technique to deal with nonlinear evolution equations, not to discuss in any detail the properties of the equations being listed or their applicative relevance.

The *Korteweg–de Vries* (KdV) *equation*

$$(1) \qquad u_t + u_{xxx} - 6u_x u = 0.$$

Introduced in 1895 in the context of fluid dynamics [KdV1895], it has great applicative and theoretical relevance; the former, especially in plasma physics and fluid dynamics; the latter, inasmuch as it is, in some sense, the simplest truly nonlinear novel equation beyond the class of second order partial differential equations for which a fairly general theory exists (consisting in the classification, at least locally, of the three possible behaviours, hyperbolic, parabolic, or elliptic). The technique of solution of this equation has been described in the preceding sections and is treated in more detail in the following chapters.

Let us emphasize that, in writing the KdV equation (1), as well as the equations that follow, we are generally adopting the choice of variables that yields the neater notation (or the more conventional one; see for instance the factor 6 in (1)). The flexibility implied by the possibility to rescale (dependent or independent) variables should however be kept in mind; for instance this allows the introduction of arbitrary constants multiplying the three terms in (1).

The regularized long-wave (RLW) *equation, or PBBM* (Peregrine, Benjamin, Bona and Mahoney) *equation*

$$(2) \qquad u_t + u_x - u_{txx} - 6u_x u = 0.$$

This is a competitor of the KdV equation (1) to describe waves in a rectangular canal, in the limit when the wavelength is large relative to the depth of the canal. The RLW equation (2) differs from the KdV equation (1) in two respects. In the first place, due to the presence of the term u_x; but this difference is marginal, since the introduction of such a term in the KdV equation (1) corresponds to a trivial change of variables, equivalent to measuring the spatial coordinates in a frame of reference moving with unit speed. The second difference, namely the appearance of the term $-u_{txx}$ in (2) in place of u_{xxx} in (1) is, from an analytical point of view, more substantial; it implies indeed that (2) is not an evolution equation at all, at least in the most straightforward and strict sense. On the other hand, to the extent that dispersive and nonlinear effects are negligible in (2), there holds the approximate equation $-u_t \approx u_x$, and this implies $-u_{txx} \approx u_{xxx}$, a relation that eliminates the difference between (2) and (1) (with u_x appropriately added in (1)). This remark is at the root of the physical "equivalence" of the KdV and RLW equations, in the sense that both have been used to describe the same physical phenomenon.

The RLW equation is *not* known to be solvable by the spectral transform technique, and there are reasons to suspect this is an intrinsic property, namely that any search for a spectral transform appropriate to solve (2) is doomed to failure. (It is, however, worth noting that the equation

(2a) $\quad u_t + u_x - u_{txx} + 4u_t u - 2u_x \int_x^{+\infty} dy\, u_t(y,t) = 0,$

whose linear part coincides with the linear part of (2), belongs to the (solvable) class (1.2.-1), with the (nonpolynomial) choice $\alpha(z) = (z-1)^{-1}$. A more detailed comparison of the KdV and RLW equations, providing a justification for the last statement, is made in subsection 3.2.1.

The *transitional Korteweg–de Vries equation*,

(3) $\quad u_t + u_{xxx} - 6f(t) u_x u = 0.$

The name originates from the assumption that $f(t)$ be a monotonically increasing function, with $f(\pm\infty) = \pm 1$; so that (3) described the transition from a wave propagation regime modelled by the KdV equation (1) with a positive coefficient of the nonlinear term to a regime where the evolution equation is still the KdV but with the nonlinear term having the opposite sign. This equation has three constants of the motion whose expression is easily retrieved from appendix A.14. The transitional KdV equation, for any choice of $f(t)$ that has the properties stated above, in *not* solvable by the spectral transform technique; indeed the exceptional case $f(t) = t^{-1/2}$, whereby (3) can be easily transformed into the cylindrical KdV equation (see below), is not of the type we have described above as "transitional". When $f(t)$ is sufficiently smooth compared to the wave-length (3) describes wave propagation on the surface separating two shallow layers of fluid, the function $f(t)$ accounting for a (smooth) variation of the depth of the bottom layer. In particular the zero of $f(t)$ corresponds to the point where the two layers have equal depth, and the propagation across this turning point can be approximately modelled by the equation

(3a) $\quad u_t + u_{xxx} - 6(1 - \varepsilon t) u_x u = 0$

that has been investigated by perturbation and numerical methods in the neighborhood of $t = \varepsilon^{-1}$.

The (so-called) *cylindrical Korteweg–de Vries* (cKdV) *equation*

(4) $\quad u_t + u_{xxx} - 6u_x u + (2t)^{-1} u = 0.$

Note the explicit time-dependence and the singularity at $t = 0$. The Cauchy problem for this equation (from an initial condition given at $t = t_0 \neq 0$) can be solved, within the (physically interesting) class of functions $u(x, t)$ that vanish asymptotically, $u(\pm\infty, t)=0$, by a spectral transform technique based on a novel spectral problem, although one still related to the one-dimensional linear Schroedinger equation (1.3.1.-1). This case is treated in section 6.3. The cKdV equation (4) had been introduced, and investigated numerically, in plasma physics, before its solvability was demonstrated. It obtains by performing an analogous schematization to that leading to the (standard) KdV equation (1), but in a cylindrical rather than a rectangular geometry. (The variable x in (4) does not have, however, the significance of a radial coordinate; indeed (4) is considered, as all other equations discussed in this book, on the whole line, $-\infty<x<+\infty$).

As implied by its solvability via the spectral transform technique, the cKdV equation (4) possesses an infinite sequence of conserved quantities, in the form of integrals over all space of polynomials in u and its x-derivatives (but with explicitly time-dependent coefficients; see section 6.3.4).

The (so-called) *spherical Korteweg–de Vries equation*

(5) $\quad u_t + u_{xxx} - 6u_x u + t^{-1} u = 0.$

This equation is *not* solvable (so far at least) by any spectral transform technique. It arises, and it has therefore been numerically investigated, in the same context as the standard and cylindrical KdV equations, (1) and (4), with the modifications obviously implied by its name. Only three conserved quantities are known for this equation (this is the standard endowement for an evolution equation of this general form; see appendix A.14).

Note that it is not possible to change the numerical factor multiplying the last term in (4) and (5) by rescaling variables, without changing the first term as well.

The modified Korteweg–de Vries (mKdV) equation

(6) $\quad u_t + u_{xxx} \pm 6u_x u^2 = 0.$

Note that this equation differs from the KdV equation (1) only because of its cubic (rather than quadratic) nonlinearity. It has applicative relevance, for instance it has been used to describe acoustic waves in anharmonic lattices and Alfvén waves in a collisionless plasma. The Cauchy problem for this equation is solvable by an appropriate spectral transform, based on a spectral problem that is analogous to that discussed in the preceding sections, except for the replacement of the linear Schroedinger equation by the spinor Pauli equation. A general version of this spectral problem is treated in volume II, together with the class of nonlinear evolution equations solvable through it, including, in addition to the mKdV equation (6), several other equations of applicative and theoretical significance (see below). We refer to this spectral problem as the (generalized) Zakharov–Shabat spectral problem.

As it is typical of all equations solvable via the spectral transform, the mKdV equation (6) possesses an infinite number of conservation laws, its solutions may be related by explicit Bäcklund transformations, and it gives rise to the soliton phenomenology (note however that for real u, solitons may occur in (6) only if a positive sign appears in front of the last term; in contrast to the KdV case (1), this sign cannot be modified, relative to that of the second term, by any real rescaling of variables). The mKdV equation is related to the KdV equation via a transformation known as the Miura transformation; this transformation, that has played an important rôle in the early history of the spectral transform method, is discussed in appendix A.11.

It is generally believed that the generalized KdV equation $u_t + u_{xxx} \pm 6 u_x u^m = 0$ is solvable by spectral transform techniques only if $m=1$ (ordinary KdV; see (1)) or $m=2$ (mKdV; see (6)); and of course by Fourier transform if $m=0$ (linear case). Indeed only in these cases there exist an infinite number of conservation laws (of polynomial type).

As an example of a highly nonlinear evolution equation that differs from the KdV equation (1) only for the nonlinear terms (and includes the mKdV equation (6) as a special case) we report the equation

$$(7) \quad u_t + u_{xxx} - 6 u_x \left\{ u^2 - \left(u + u_{xx} - 2u^3 \right)^2 / \left[a^2 - 4\left(u^2 + u_x^2 - u^4 \right) \right] \right\} = 0,$$

whose solvability is demonstrated in volume II. Clearly (7) reduces to (6) if the constant a^2 is sufficiently large.

Another highly nonlinear equation whose solvability is displayed in volume II reads

$$(8) \quad u_t + u_{xxx} - \tfrac{1}{8} u_x^3 + u_x \left[a \exp(u) + b \exp(-u) + c \right] = 0.$$

This includes, as a limiting case, the equation

$$(9) \quad u_t + u_{xxx} + u_x \left(\alpha u^2 + \beta u + \gamma \right) = 0,$$

that obtains from (8) by replacing u with εu, by setting $a = \alpha \varepsilon^{-2} + \tfrac{1}{2} \beta \varepsilon^{-1} + \gamma$, $b = \alpha \varepsilon^{-2} - \tfrac{1}{2} \beta \varepsilon^{-1}$, $c = -2\alpha \varepsilon^{-2}$, and then letting $\varepsilon \to 0$. The constants α, β

and γ in (9) (as well as the constants a, b and c in (8)) are arbitrary. For $\alpha=\gamma=0$, $\beta=-6$, (9) becomes the KdV equation (1); for $\beta=\gamma=0$, $\alpha=\pm 6$, it becomes the mKdV (6).

Two other highly nonlinear evolution equations that are also tractable by the spectral transform method are the *Harry Dym* (HD) *equation*

(10a) $\quad v_t = v_{xxx} v^3$,

(10b) $\quad u_t + 2(u^{-1/2})_{xxx} = 0, \quad u = v^{-2}$,

and the *Wadati–Konno–Ichikawa–Schimizu* (WKIS) *equation* (for the *complex* field u)

(11) $\quad iu_t + \left[(1+|u|^2)^{-1/2} u\right]_{xx} = 0$.

The corresponding spectral problems are, however, somewhat unconventional. A related way to investigate these equations is by reducing them to other equations by a change of variable (that depends however on the solution itself). For instance we show in section 6.5 how the HD equation (10) can be transformed into the KdV equation (1).

The last few examples we have reported are notable because of their theoretical significance rather than their applicative relevance (so far at least). Let us return now to some other nonlinear evolution equations solvable via the spectral transform technique that are directly related to applications.

The *Benjamin–Ono* (BO) *equation*

(12) $\quad u_t + Hu_{xx} - 6u_x u = 0$

resembles the KdV equation (1), except for the replacement, in the second term, of one space derivative by the Hilbert operator defined by the formula

(12a) $\quad Hf(x) = \pi^{-1} P \int_{-\infty}^{+\infty} dy f(y)/(y-x)$.

(Incidentally, this implies that the dispersion function for the linear part of (12) reads $\omega(k) = -k^2 \operatorname{sign} k$, which can be compared with the corresponding dispersion functions for the linear part of the KdV equation (1), $\omega(k) = -k^3$, and of the RLW equation (2), $\omega(k) = k/(1+k^2)$). The BO equation (12) is of course an integrodifferential, rather than a partial differential, equation; it can however be solved by a method analogous to the spectral transform technique discussed in this book. As a consequence, it possesses all the attributes of the solvable equations discussed above: an infinite number of conservation laws, Bäcklund transformations, the soliton phenomenology.

The BO equation (12) was introduced to describe internal waves in stratified fluids (linear geometry, scale of disturbance small relative to the depth of the fluid). There is actually a more general nonlinear evolution equation, that is appropriate to describe a stratified fluid of finite depth. It reads

(13) $\quad u_t + Tu_{xx} - 6u_x u = 0,$

with the integro-differential operator T defined by the formula

(13a) $\quad Tf(x) = P\int_{-\infty}^{+\infty} dy f(y)\{\cotgh[\pi(y-x)/(2d)]$

$\qquad - \operatorname{sign}(y-x)\}/(2d).$

The (constant) parameter d represents the distance between the bottom and the internal wave layer. It is easily seen, by appropriate rescaling, that in the limit $d \to 0$ (shallow fluid), (13) goes over into the KdV equation (1), while in the limit $d \to \infty$ (deep fluid), (13) goes over into the BO equation (12). Most remarkably, the general equation (13) is itself solvable, and it does possess all the attributes of a solvable equation.

The *Boussinesq* (BSQ) *equation*

(14) $\quad u_{tt} - u_{xx} - u_{xxxx} + 3(u^2)_{xx} = 0.$

Introduced by Boussinesq in 1871 to describe the propagation of long waves in shallow water in rectilinear geometry, it is more general than the KdV equation (1), since it accounts for motion in both directions; indeed the KdV equation (1) (with u_x appropriately added) derives from (14) in the approximation, $u_t \approx -u_x$, that corresponds to restricting consideration to the motion around a wave traveling only in one direction. The BSQ equation (14) has many other applications: long waves in a one-dimensional nonlinear lattice, vibrations in a nonlinear string, ion sound waves in a plasma. Remarkably, the BSQ equation (14) is itself solvable by the spectral transform technique, see volume II. It is therefore graced by all the paraphernalia that come with such solvability: infinite number of conservation laws, Bäcklund transformations, multisoliton solutions.

The *Kadomtsev–Petviashvili* (KP) *equation*

(15) $\quad (u_t + u_{xxx} - 6u_x u)_x \pm u_{yy} = 0.$

This equation is also called the *two-dimensional KdV equation*; it describes slow variations in the y-direction of wave propagation with dispersive and nonlinear effects in the x-direction. Thus, it models the "transverse" perturbation of solutions of the KdV equation (1); and, in particular, the single soliton solution of (1) is stable with respect to this perturbation if (15) is written with the "plus" sign in the last term, and unstable in the other case. This equation can be treated by techniques that are similar to those discussed in this book, though the extension required to introduce the additional spatial coordinate y is far from trivial. The KP equation (15) possesses multisoliton solutions that are however localized only in one direction (not in the other); explicit solutions that are localized in both directions are also known (they depend rationally on the spatial coordinates and are nonsingular only if the "plus" sign prevails in (15)). Other particular solutions of (15) with the "plus" sign can be obtained from solutions of the cKdV equation (4) by using the fact that, if $\bar{u}(x,t)$ is a solution of (4), then

(15a) $\quad u(x,y,t) = \bar{u}(x + y^2/4t, t)$

satisfies (15) (see [J1979]).

The *Burgers equation*

(16) $\quad u_t = u_{xx} + 2uu_x.$

This is the simplest physically interesting nonlinear evolution equation (it may be considered as a one-dimensional reduction of the Navier–Stokes equations). Its solvability, discovered in 1950, before the introduction of the spectral transform, is easily achieved via the Hopf–Cole transformation

(16a) $\quad u = (\ln v)_x,$

that maps the Burgers equation into the linear ("diffusion") equation

(16b) $\quad v_t = v_{xx}.$

The *nonlinear Schroedinger* (NLS) *equation*

(17) $\quad iu_t + u_{xx} \pm 2|u|^2 u = 0.$

This equation appears in many applicative contexts; indeed, it is generally appropriate to describe the time evolution of the envelope of an almost monochromatic wave of moderate amplitude in a weakly nonlinear, dispersive system (note that u is now complex). An example of its emergence in a perturbation approach is discussed in volume II. It is solvable by the spectral transform technique, belonging to the same class as the mKdV equation (6) (the underlying spectral problem is the generalized Zakharov–Shabat spectral problem; see volume II). It possesses all the features of solvable equations: infinite number of conservation laws, Bäcklund transformations, the soliton phenomenology (note however that solitons may appear only if the "plus" sign occurs in front of the last term in (17)). A remarkable property of the NLS equation (17) is that its solutions can be mapped into the solutions of the one-dimensional *Heisenberg ferromagnet* equation

(18) $\quad \vec{S}_t = \vec{S} \wedge \vec{S}_{xx}, \qquad \vec{S} \cdot \vec{S} = 1,$

whose solution (the 3-dimensional unit-vector $\vec{S} \equiv \vec{S}(x, t)$) describes a linear (continuous) spin density. This of course entails the solvability of (18) as well.

The *derivative nonlinear Schroedinger equation*

(19) $\quad iu_t + u_{xx} \pm i(|u|^2 u)_x = 0.$

Also this equation has a number of applications, for instance in plasma physics (propagation of circularly polarized nonlinear Alfvén waves or of radio-frequency waves). This equation is also solvable by an appropriate spectral transform; the spectral problem is a variation of the Zakharov–Shabat problem (the dependence on the eigenvalue is quadratic rather than linear). It has all the properties characteristic of the equations solvable by the spectral transform technique.

The *Hirota equation*

(20) $\quad u_t + iau + ib(u_{xx} - 2\eta|u|^2 u) + cu_x + d(u_{xxx} - 6\eta|u|^2 u_x) = 0.$

We have written this equation in a form intended to emphasize its generality; indeed, this equation is solvable by the spectral transform technique based on the generalized Zakharov–Shabat spectral problem (see volume II) even if a, b, c, and d are (arbitrarily) time-dependent, while η is an (arbitrary) constant. For $a=c=d=0$, $b=-1$, $\eta=\pm 1$, (20) becomes the NLS equation (17); for $a=b=c=0$, $d=1$, $\eta=\pm 1$ and u real, it becomes the mKdV (6).

An interesting extension of the NLS equation (17) is the system of two coupled equations

(21a) $\quad iu_t + u_{xx} - 2\eta|u|^2 u + iuv_x + uv^2 = 0,$

(21b) $\quad v_t - 2\eta(|u|^2)_x = 0,$

that is also solvable by the spectral transform technique; it has been investigated as a model for the interaction between long waves and short envelope waves, respectively described by the fields v and u.

A different model for interacting waves is described by the *3-dimensional 3-wave resonant interaction* (3D3WRI) equation

(22) $\quad \left(\partial_t + \vec{v}_j \cdot \vec{\nabla}\right) u_j + \eta_j u_k^* u_n^* = 0, \quad jkn = 123 \text{ cyclic}.$

Here the three complex fields u_j are functions of the 3-dimensional position vector $\vec{r}$, and of time; the three 3-vectors $\vec{v}_j$ are given constants, the dot indicates the usual scalar product, $\vec{\nabla}$ is the gradient operator and $\eta_j = \pm 1$. Note that this example, in contrast to all those reported above and below, is imbedded in 3-dimensional space (although the first term in each of the 3 coupled equations (22) actually describes wave propagation in one direction only, that of the velocity $\vec{v}_j$). The system of nonlinear evolution equations (22) is important in many applications describing the resonant coupling between different plane waves, that in many contexts provides the first nonlinear correction to a basically linear dispersionless wave-propagation phenomenon. The solution of (22) involves a clever interplay of three spectral problems of the Zakharov–Shabat type, referring to three different space coordinates, each of them collinear to one of the vectors $\vec{v}_j$. Also the 3-wave interaction equations that read as (22) but with only one or two spatial coordinates, are solvable by the spectral transform method.

The *Davey–Stewartson equations*

(23a) $\quad i u_t - u_{xx} + u_{yy} + u|u|^2 - 2uv = 0,$

(23b) $\quad v_{xx} + v_{yy} - \left(|u|^2\right)_{yy} = 0.$

This system of 2 coupled differential equations is relevant to the description of two-dimensional long surface waves on water of finite depth. It is solvable by techniques that are analogous to those developed for the KP equation (15).

The *boomeron equation*

(24a) $\quad u_t = \vec{b} \cdot \vec{v}_x,$

(24b) $\quad \vec{v}_{xt} = u_{xx} \vec{b} + \vec{a} \wedge \vec{v}_x - 2 \vec{v}_x \wedge [\vec{v} \wedge \vec{b}].$

Here u and the 3 components of the 3-vector $\vec{v}$ are the dependent variables, x and t are the independent variables, $\vec{a}$ and $\vec{b}$ are two given 3-vectors, and the scalar and vector products are defined in the standard way ((24) is actually solvable even if $\vec{a}$ and $\vec{b}$ are time-dependent, although for simplicity we assume here they are constant). This system of 4 coupled equations is equivalently described by the single matrix equation of order 2,

(24c) $\quad 2W_{xt} = i[W_x, A] + \{W_{xx}, B\} + [W_x, [W, B]].$

Here the matrices W, A and B are related to the scalar u and to the 3-vectors $\vec{v}$, $\vec{a}$ and $\vec{b}$ by the formulae $W = u + \vec{v} \cdot \vec{\sigma}$, $A = \vec{a} \cdot \vec{\sigma}$, $B = \vec{b} \cdot \vec{\sigma}$, $\vec{\sigma}$ being the 3-vector having as components the three Pauli matrices, and the square and curly brackets indicate commutation and anticommutation ($[A, B] \equiv AB - BA$, $\{A, B\} \equiv AB + BA$). The evolution equation (24) is solvable by the spectral transform technique; the spectral problem is based on the *matrix* Schroedinger equation. The class of equations solvable in this manner is treated in volume II; all these evolution equations possess the properties characteristic of solvable equations, including an infinite number of conservation laws, explicit Bäcklund transformations and multisoliton solutions. The basic property of solitons, to interact with each other, as it were, "elastically", is also preserved; but a novel phenomenon appears, namely the fact that each soliton generally moves with variable speed, i.e. as a particle acted upon by an external force rather than as a free particle. Indeed the interest (and name) of the evolution equation (24) originates from the fact that it provides the simplest instance of this novel phenomenon; generally its solitons boomerang back, in the remote future, to where they came from in the remote past (in the case in which the vectors $\vec{a}$ and $\vec{b}$ are mutually orthogonal another generic behaviour of solitons may occur, namely they may oscillate without ever escaping to infinity, i.e. behave as *trappons* rather than *boomerons*; both behaviours may coexist in a multisoliton solution).

Another equation, that is essentially a subcase of the boomeron equation (24) (in the particularly interesting case with $\vec{a} \cdot \vec{b} = 0$ mentioned above), but has the advantage to describe the evolution of a single scalar field (rather than four coupled fields), is the *zoomeron equation*

(25) $\quad (\partial^2/\partial t^2 - \partial^2/\partial x^2)(Z_{xt}/Z) + 2(Z^2)_{xt} = 0.$

This highly nonlinear equation is a convenient one to display the novel phenomenology associated with boomerons and trappons; moreover its solitons ("*zoomerons*") have an amplitude that changes with time along with their speed. A treatment of this equation is given in volume II.

The *reduced Maxwell–Bloch equations*

(26a) $\quad E_t - v = 0,$

(26b) $\quad v_x - \omega r - Eq = 0,$

(26c) $\quad q_x + Ev = 0,$

(26d) $\quad r_x + \omega v = 0.$

Here E is the electric field, r and v are real combinations of the density matrix related to the atomic polarization, and q is a measure of the atomic polarization; these four functions depend on the variables x and t that are appropriate linear combinations of the physical space and time coordinates, while ω is constant, proportional to the atomic density. All these quantities have been put in dimensionless form by appropriate rescaling. The physical phenomenon (approximately) described by this system of nonlinear evolution equations is the interaction, with a medium of two-level atoms, of a short intense beam of light traveling to the right.

The *Sine–Gordon* (SG) *equation* comes in two forms,

(27a) $\quad u_{xt} \pm \sin u = 0, \quad u \equiv u(x,t),$

(27b) $\quad \varphi_{\xi\xi} - \varphi_{\tau\tau} \pm \sin \varphi = 0, \quad \varphi \equiv \varphi(\xi, \tau),$

the transition from one form to the other corresponding to the change from "light-cone" (x, t) to "laboratory" (ξ, τ) coordinates, i.e.

(27c) $\quad \varphi(\xi, \tau) = u(x, t), \quad x = \tfrac{1}{2}(\xi + \tau), \quad t = \tfrac{1}{2}(\xi - \tau).$

The initial value problem for (27a) is the Goursat problem (the initial datum is given on the characteristic $t = 0$), while for (27b) it is the Cauchy problem (initial datum given at $\tau = 0$). Both these problems are solvable by the spectral transform; in fact (27a) belongs to the same class (associated with the Zakharov–Shabat spectral problem, treated in volume II) as the mKdV equation (6) and the NLS equation (17), or the Hirota equation (20).

The SG equation (27), in one or the other avatar, appears in many applications, from differential geometry (its relation to two-dimensional pseudo-spheres was actually known a century ago) to nonlinear optics, where solitons of (27a) account for the *self-induced transparency* phenomenon (in fact, the reduced Maxwell–Bloch equations (26) become the SG equation (27) when $\omega = 0$, $r = 0$, $E = u_x$, $v = \mp \sin u$, $q = \mp \cos u$), to superconductivity. Moreover it has been extensively investigated as a model of relativistic field theory (both classical and quantal; this is due to its relativistically invariant nature). The SG equation (27) possesses of course all the attributes of a solvable equation (infinite number of conservation laws, Bäcklund transformations, solitons, which however appear as "kinks", namely not u is localized, but its derivative u_x).

The *double Sine–Gordon equation*

(28) $\quad u_{xt} \pm [\sin u + \eta \sin \tfrac{1}{2} u] = 0,$

η being a constant. The relevance of this equation is motivated by its physical applications in nonlinear optics (propagation of ultra-short pulses in a resonant 5-fold degenerate medium) and in low-temperature physics (propagation of spin waves in anisotropic magnetic liquids). There is strong evidence that this equation is not solvable via the spectral transform method.

The (so-called) *phi-four equation*

(29) $\quad \varphi_{tt} - \varphi_{xx} - \varphi + \varphi^3 = 0,$

its name being due to the expression $\mathcal{H} = \tfrac{1}{2}(\varphi_t^2 + \varphi_x^2) - \tfrac{1}{2}\varphi^2 + \tfrac{1}{4}\varphi^4$ of the corresponding Hamiltonian density. This equation has been the subject of intensive investigation in the context of classical and quantised field theory. It has kink-like solutions (similar to the SG kinks) but they are not solitons (they are solitrons, see subsection 3.2.1); in fact the phi-four equation is not solvable by means of the spectral transform.

The *Liouville equation*

(30) $\quad u_{xt} = \exp(\eta u),$

η being an arbitrary constant. The general solution of this equation has been given by Liouville in 1853, and reads

(30a) $\quad u(x,t) = f(x) - g(t) - (2/\eta) \ln \left\{ p \int_{x_0}^{x} dy \exp[\eta f(y)] \right.$

$$\left. + (\eta/2p) \int_{t_0}^{t} ds \exp[-\eta g(s)] \right\},$$

where p is an arbitrary parameter, and $f(x)$ and $g(t)$ are arbitrary functions. Indeed the Liouville equation can be transformed into the wave equation $\varphi_{xt} = 0$, whose general solution is precisely $\varphi(x,t) = f(x) + g(t)$. The Liouville equation can be imbedded in the same class of solvable equations as the NLS, mKdV and SG equations, and it has some relevance as a field-theoretical model.

Another scalar equation that, as the SG and Liouville equations, has the property to be relativistically invariant and that is solvable via the spectral transform technique reads

(31) $u_{xt} + \exp(-u) - \exp(2u) = 0.$

The spectral problem that provides the basis for the spectral transform used to solve (31) involves matrices of order 3 and is discussed in volume II.

There are several other relativistically invariant equations that are now known to be solvable by the spectral transform technique (though in several cases this technique requires a substantial extension of the formulation described in this book), but they generally involve two, or more, coupled fields. For instance the SG equation (27b) is the reduced case, with $v=0$ and $\varphi = 2u$, of the *Pohlmeyer–Lund–Regge-model*

(32a) $u_{\xi\xi} - u_{\tau\tau} \pm \sin u \cos u + (\cos u / \sin^3 u)(v_\xi^2 - v_\tau^2) = 0,$

(32b) $(v_\xi \cotg^2 u)_\xi = (v_\tau \cotg^2 u)_\tau,$

which is also solvable, and the equation (31) is the reduced case, with $v=0$, of the solvable system (see volume II)

(33a) $u_{xt} + \cosh(3v) \exp(-u) - \exp(2u) = 0,$

(33b) $v_{xt} - \sin^3 v \exp(-u) = 0.$

Among other interesting field equations, we mention below (without further comments) a few more that stand out as particularly representative of the present stock of nonlinear partial differential equations that are treatable by the spectral transform method.

The $O(n)$ *sigma-model*

(34) $\vec{v}_{xt} + (\vec{v}_x \cdot \vec{v}_t)\vec{v} = 0;$

here $\vec{v}$ is an n-dimensional real unit vector, $(\vec{v} \cdot \vec{v}) = 1$.

The principal $SU(n)$ *chiral field equation*

(35) $(U^\dagger U_x)_t + (U^\dagger U_t)_x = 0,$

where $U(x,t)$ is a unitary matrix, $U^\dagger U = 1$, of $SU(n)$.

The (reduced) *Einstein equation*

(36) $\quad \left(g^{1/2}\boldsymbol{G}_x\boldsymbol{G}^{-1}\right)_t + \left(g^{1/2}\boldsymbol{G}_t\boldsymbol{G}^{-1}\right)_x = 0,$

$\boldsymbol{G}(x,t)$ being a 2×2 symmetrical matrix, and $g(x,t)$ being its determinant, $g = \det \boldsymbol{G}$.

The *Ernst equation*

(37) $\quad (\operatorname{Re} E)\left(E_{\rho\rho} + \rho^{-1}E_\rho + E_{zz}\right) = E_\rho^2 + E_z^2,$

where E is a complex function of the radial and axial coordinates ρ, z, this equation being another reduced case of the Einstein equation, corresponding to a static axially symmetric vacuum field.

The (reduced) *Yang–Mills equations*

(38) $\quad \left(\boldsymbol{U}^\dagger \boldsymbol{U}_t\right)_t - \left(\boldsymbol{U}^\dagger \boldsymbol{U}_{\bar{z}}\right)_z = 0,$

where $\boldsymbol{U}(x, y, t)$ is a matrix of SU(n), $\det \boldsymbol{U} = 1$, and $\partial_z \equiv \partial_x - i\partial_y$, $\partial_{\bar{z}} = \partial_x + i\partial_y$. Note that this equation involves two spatial coordinates.

The *massive Thirring model*

(39a) $\quad iu_x + v + u|v|^2 = 0,$

(39b) $\quad iv_t + u + v|u|^2 = 0.$

The *Nambu–Jona Lasinio–Vaks–Larkin model*

(40a) $\quad iu_x^{(n)} = v^{(n)} \sum_{m=1}^{N} v^{(m)*}u^{(m)},$

(40b) $\quad iv_t^{(n)} = u^{(n)} \sum_{m=1}^{N} u^{(m)*}v^{(m)}.$

The *Gross–Neveu model*

(41a) $\quad iu_x^{(n)} = v^{(n)} \sum_{m=1}^{N} \left(v^{(m)*}u^{(m)} + u^{(m)*}v^{(m)}\right),$

(41b) $\quad iv_t^{(n)} = u^{(n)} \sum_{m=1}^{N} \left(u^{(m)*}v^{(m)} + v^{(m)*}u^{(m)}\right);$

it is remarkable that these relativistic invariant equations have been recently applied in solid state physics.

We terminate here this list, that has included 37 "solvable" equations and 5 that are, at least so far, not solvable by the spectral transform method. Let us reiterate that this is by no means a complete list; for one thing, novel solvable equations are being discovered at a sustained rate. Moreover, we have confined our list to partial differential or integrodifferential equations, omitting any mention of finite difference equations; yet the techniques based on the spectral transform can be extended to discretized contexts (either the space variable or the time variable or both can be discretized). The same restraint is practised in the rest of this book as well.

Finally, our attention to solvable equations involving more than two variables has been limited to the mere mention of the KP equation (15), the 3D3WRI equation (22), the DS equations (23) and the reduced YM equations (28); also this limitation is essentially maintained throughout this book (section 6.1 being one exception).

1.N. Notes to chapter 1

Some parts of chapter 1 constitute merely an overview of topics that are then treated more fully in subsequent chapters. In these cases the bibliographical references are postponed to the later chapters.

1.1. The solution of linear PDEs via the Fourier transform is standard textbook material; see for instance [W1974]. The progress mentioned at the end of this section refers to the recent demonstration that certain nonlinear *integrodifferential* evolution equations have been recently shown to be solvable; see section 1.8.

1.3.2. An elementary introduction to matrix algebra is given, mainly for notational purposes, at the beginning of volume II; but the reader is supposed to know the basic elementary theory, including for instance the definition of the determinant of a (square) matrix.

1.5. Among the first researchers to emphasize the analogy between the spectral transform technique to solve *nonlinear* evolution equations and the Fourier transform method to solve *linear* partial differential equations were [T1972] and [AKNS1974]. The analogy between the direct and inverse spectral (or scattering) problems and the Fourier transform technique has on the other hand been always well understood by researchers in this field, including the pioneers (see the standard reference texts on the inverse problem, for instance [F1958], [F1959], [F1964], [F1974], chapter 20 of [N1966], [CS1977], [Mar1977]; and the early references quoted there).

1.6.1. The pure multi-soliton solution (1.6.1.-9) is most easily derived (once the spectral transform formalism is available) by the solution of the inverse spectral problem, see subsection 1.3.2. There also exist alternative techniques, among which stands out as particularly effective the so-called "direct method" (for the KdV equation see [Hi1971]; a review of this technique is given in [Hi1980]); this method, mainly due to Hirota, is largely manipulative, but has proved its worth in many instances, by discovering multisoliton solutions for novel equations before other techniques to deal with these equations were developed.

1.8. The *KdV equation*, being the protoype of the nonlinear evolution equations whose integrability has been recently discovered, has been investigated in an enormous number of papers in the last decade. The argument that singles it out as worthy of study from an *a priori* mathematical point of view is given in [K1974a]. For a short readable introduction to the KdV equation, and to its properties, see, for instance, [Mi1978]. The fact that the KdV equation is just the first of a sequence of nonlinear evolution equations, the so-called *higher KdV equations* (set $\alpha(z) = z^n$ in (1.4.-2)),

(1) $\quad u_t = L^n u_x, \quad n = 2, 3, 4, \ldots,$

was pointed out by [L1968]. The static version of (1.4.-2), the so-called Novikov equation, $\alpha(L)u_x = 0$, has been thoroughly investigated in [No1974] (see also [GD1975] and [GD1979]). The wide applicative relevance of the KdV equation is due to its being the natural equation to describe the long-time behavior of any wave-like phenomenon that is non-dissipative, weakly nonlinear and weakly dispersive (see for instance [SG1969], [Mi1974], [Seg1976], [OB1980], [Mil1981]).

The theory of the *RLW equation* has been thoroughly investigated in [BBM1972], and for this reason it is also known as the BBM equation; numerical solutions of this equation were first given in [P1966] (hence the name PBBM, also largely used). It is a matter of dispute [K1974] whether this equation or KdV is more appropriate for applications, and in particular to describe long waves; see also [B1974]. See also Notes to subsection 3.2.1.

For a description of the fluid dynamical phenomena modelled by the *transitional KdV equation*, see [KN1980] and the references quoted there.

The applicative relevance of the *cylindrical and spherical KdVs* (and some numerical results) are discussed in [Max1976], and in the literature quoted there. The solvability of the cKdV equation via the spectral transform method has been first proved in [CD1978h]. For additional references on the cKdV equation see section 6.N.

The *mKdV equation* has been extensively studied, due to its applicative relevance but especially because of its close relation to the KdV equation. Indeed the *Miura transformation* that relates the solutions of these two equations (see appendix A.11) has played an important rôle in bringing about the discovery of the spectral transform technique; see, e.g., [Mi1978] and [AKS1979]. Its solvability (via the Zakharov–Shabat spectral transform; see volume II) was first proved in [W1972].

The generalized KdV equation $u_t + u_{xxx} \pm 6u_x u^m = 0$ is tersely discussed in [SCM1973]; see also [Mi1974] and [Mi1976a], and appendix A.14.

The evolution equations (1.8.-7) and (1.8.-8) have been introduced (and shown to be solvable) in [CD1981].

For the *Harry Dym equation* (1.8.-10) see sections 6.5 and 6.N.

The *WKIS equation* (1.8.-11) is treated in [WKI1979] and [SW1980].

The references for the *BO equation* (1.8.-12) are: [Ben1967], [Ono1975], [Jo1977], [JE1978], [CLP1978], [Cas1978], [BK1979], [Cas1979a], [Cas1979b], [Cas1979c], [M1979a], [Na1979a], [SI1979], [CK1980] and [M1980c]. For a related *modified BO equation*, see [Na1979c], while for the *"higher" BO equations* see [M1979b] and [M1980a].

The references for the *finite depth fluid equation* (1.8.-13) are [CL1979], [M1979c], [SAK1979], [H1980] and [GKu1980]. For the *"higher" finite depth fluid equations* see [M1980b], and for a *modified finite depth fluid equation* see [Na1979b] and [GKu1980].

The solution of the *BSQ equation* (1.8.-14) (introduced in [B1871] and [B1872]) via the spectral transform technique is based on the spectral problem associated with the linear third order equation $\psi_{xxx} + v(x)\psi_x + u(x)\psi = -ik^3\psi$; this problem has been solved, with various degrees of generality, in [K1980d] and [Ca1980]; the references on the solvability of the BSQ equation are: [Z1973], [Hi1973a], [Ch1975], [MK1978], [BP1980], [Ca1982].

References for the *KP equation* (1.8.-15) are: [KP1970], [Dr1974], [MZBIM1977], [GFJ1978], [ZM1979], [Z1980], [MST1980] and [M1981].

For the *Burgers equation* (1.8.-16) see [Bu1974]; the linearising transformation (1.8.-16a) has been independently given in [Ho1950] and [Co1950]. For a symmetry approach to a class of evolution equations including the Burgers equation see [FY1982], and for its formulation in the Hamiltonian formalism see [Taf1981].

For the applicative relevance of the *NLS equation* (1.8.-17) see, for instance, [TY1969], [Seg1976], [LML1980], [N1980] and the references quoted there. The discovery that the NLS equation is solvable via (an appropriate) spectral transform is due to [ZS1971]. For its relation to the *one-dimensional Heisenberg ferromagnet equation* (1.8.-18) see [T1977] and [Lak1977].

The *derivative nonlinear Schroedinger (DNLS)* equation (1.8.-19) has been solved in [KN1978]. A complete treatment of the spectral transform method to solve this equation is given in [GIK1980]. An extended version of the DNLS equation, that also has applicative relevance, has also been shown to be solvable, see [MD1979].

The *Hirota equation* (1.8.-20) has been investigated by the "direct method" in [Hi1973], and so named in [SCM1973]. For its solution via the spectral transform technique, see volume II.

The wave interaction modelled by the equations (1.8.-21) has been investigated via the spectral transform technique in [N1978].

The *3D3WRI equations* (1.8.-22) in the one-dimensional case has been solved in [ZM1973], [ZS1974] and [ZM1975]. The most extensive investigation of this problem has been carried out in a series of papers by Kaup, see for instance [K1976a], [K1977], [K1980a], [K1980b], [K1980c], [K1981d]; the last four papers are based on an advance due to Cornille [Cor1979].

The *Davey–Stewartson equations* (1.8.-23) have been introduced in [DS1974], and their solvability has been demonstrated in [AF1978].

The *boomeron and zoomeron equations* have been first investigated in [CD1976d] and [CD1976e]; for a review on the corresponding soliton phenomenology see [D1977]. Their solution by the spectral transform is discussed in volume II. There exist two films that display the behaviour of the one- and two-soliton solutions of these equations: [E1977], [E1977a].

The *reduced Maxwell–Bloch* (rMB) *equations* (1.8.-26) provide a model to describe the phenomenon of Self-Induced Transparency (SIT), see e.g. [EGCB1973]. They are treated, for instance, in [GCBE1973]; see also [Bu1977], [Bu1977a], [BJKS1979]. The rMB equations approximate the Maxwell–Bloch (MB) equations through the restriction to waves traveling only in one direction. The MB equations themselves are presumably not integrable; they seem to possess solitrons, not solitons (in the language introduced in section 3.2.1) [CE1977].

The literature on the *SG equation* is very extensive, due to its ubiquitous presence in mathematical and physical problems, and also to its providing the first instance of a relativistically invariant solvable equation possessing multisoliton solutions (in contrast with the Liouville equation that has no soliton solutions). For the early physical literature see [Sky1958], [Sky1961] and [BEMS1971]; more recently, its applications have been discussed, for instance, in [Bu1977], [Bu1977a], [Ce1977], [Bi1978], [TB1979]. The early mathematical literature is much older, see for instance [H1899] and [B1903], [Eis1960]. The solvability of the SG equation was first demonstrated by [AKNS1973] in the form (1.8.-27a) (for the solvability of the SG equation (1.8.-27b) in "laboratory coordinates", see [K1975] and [KN1978a]) and by [ZTF1974] (see also [TF1974] and [TF1976]).

For the *double SG equation* see [BCG1980], and the references quoted there.

The *phi-four equation* has been mainly investigated as the simplest non trivial relativistic invariant field theoretical model (see, for instance, [DHN1974] and [GJ1975]). Numerical investigations are reported in [Au1976] and [G1976].

The general solution of the *Liouville equation* was first found by [L1853]. For its treatment in the context of the spectral transform method, see, for instance, [A1976], [CK1978], and [Ts1980].

The relativistic invariant equation (1.8.-31) is a reduced case of the system (1.8.-33), that has been introduced in [Mik1979] and [FG1980] (see also [ZhS1979], [KuW1981a] and [MOP1981]).

For the *Pohlmeyer–Lund–Regge model* (1.8.-32), see [P1976], [LR1976] and [Lu1977].

The *sigma-model* (1.8.-34) has been considered in [W1968], and then shown to be solvable in [P1976]; for other sigma-models see [O1980].

For the solvability of the *principal chiral field equation*, see [ZMi1978], and the references quoted there.

The integrability of the *Einstein equation* (1.8.-36) has been first pointed out in [Ma1979]. A general scheme for its solution has been given by [BZ1978].

The *Ernst equation* (1.8.-37) was introduced in [E1968]; for the more recent findings concerning its solvability see [Ha1978], [Ma1978], [BZ1979], [Neu1979], [Cos1980], [Ha1980], [HE1980], [MD1979a], [OW1981] and [OW1981a].

References for the solvability of the *Yang–Mills equations* (1.8.-38) are [BeZ1978] and [MZ1981]; see also [ADHM1978].

The *massive Thirring model* has been introduced in [T1958], and shown to be solvable in [Mik1976] (see also [KMi1977] and [GIK1980]).

For the *Nambu–Jona Lasinio–Vaks–Larkin model* (1.8.-40) and the *Gross–Neveu model* (1.8.-41) see the general treatment given in [ZMi1980]; for the applications to solid state physics, see [CB1981] and the literature quoted there.

CHAPTER TWO

THE SCHROEDINGER SPECTRAL PROBLEM ON THE LINE

This chapter covers in more detail the material of section 1.3, and provides in addition the basis for the results that were tersely described in sections 1.4 and 1.7 (and that are treated in more detail in chapters 3, 4 and 5), as well as for the results of chapter 6. Its organisation is clearly displayed by the titles of its sections and subsections. Certain more technical developments have been confined to appendices, which are referred to at the appropriate places.

2.1. Direct spectral problem

Consider the (singular) Sturm–Liouville problem characterized, on the whole line $-\infty < x < \infty$, by the Schroedinger equation

(1) $\quad -\psi_{xx}(x,k) + u(x)\psi(x,k) = k^2 \psi(x,k).$

We assume generally in the following $u(x)$ to be real, regular for all x (in fact, infinitely differentiable) and to vanish asymptotically, say, faster than x^{-2}, so that not only the integral of u, but also its first moment, are finite:

(2) $\quad \int_{-\infty}^{+\infty} dx \, (1+|x|)|u(x)| < \infty.$

These requirements are more stringent than those actually needed for the validity of most of the results reported below; on the other hand in some cases, that are specified below, we shall assume more stringent conditions on the asymptotic behaviour of $u(x)$, namely that it vanish exponentially or even faster.

This eigenvalue, or "scattering", problem is of course very familiar to every physicist (especially to those who happen to teach elementary quantum mechanics); but in the following we shall use it merely as a mathematical tool, ignoring its physical meaning; moreover, the evolution equations we will eventually discuss have nothing to do with the time-dependent Schroedinger equation that, in the quantum mechanical context, lies behind the "stationary" Schroedinger equation (1) and provides the appropriate background for any interpretation of the solutions of (1) in terms of (one-dimensional) scattering experiments.

The continuous part of the spectrum of the Schroedinger operator $-\partial^2/\partial x^2 + u(x)$ corresponds to k real, so that $k^2 > 0$. It can be characterized by the *reflection* and *transmission* coefficients $R(k)$ and $T(k)$, defined by the asymptotic conditions

(3a) $\quad \psi(x, k) \to T(k) \exp(-ikx), \quad x \to -\infty,$

(3b) $\quad \psi(x, k) \to \exp(-ikx) + R(k) \exp(ikx), \quad x \to +\infty$

((3a) identifies the solution $\psi(x, k)$ of (1) up to a multiplicative constant, that is fixed by the requirement that the coefficient of the first exponential in the r.h.s. of (3b) be unity).

An equivalent definition of the reflection and transmission coefficients, that is more convenient in order to analyse their analytic properties (see below), can be given through the *Jost solutions* of (1), characterized by the simple asymptotic properties

(4) $\quad f^{(\pm)}(x, k) \to \exp(\pm ikx), \quad x \to \pm \infty.$

Note incidentally that the existence of two such independent solutions indicates that the positive eigenvalue k^2 is generally two-fold degenerate, corresponding to the second-order character of (1).

Clearly the function $\psi(x, k)$ characterised by the boundary conditions (3), is related to the Jost solutions (4) by

(5) $\quad \psi(x, k) = T(k) f^{(-)}(x, k) = f^{(+)}(x, -k) + R(k) f^{(+)}(x, k).$

There immediately follows that, in terms of the Jost solutions, the transmission and reflection coefficients are given by the compact formulae

(6a) $\quad T(k) = 2ik / W[f^{(-)}(x, k), f^{(+)}(x, k)],$

(6b) $\quad R(k) = - \dfrac{W[f^{(-)}(x, k), f^{(+)}(x, -k)]}{W[f^{(-)}(x, k), f^{(+)}(x, k)]},$

(6c) $\quad R(k) = -(2ik)^{-1} T(k) W[f^{(-)}(x, k), f^{(+)}(x, -k)].$

Here, and always below, the wronskian of two functions $f^{(1)}(x)$ and $f^{(2)}(x)$

is defined by the formula

(7) $$W[f^{(1)}(x), f^{(2)}(x)] = f^{(1)}(x) f_x^{(2)}(x) - f_x^{(1)}(x) f^{(2)}(x),$$

implying of course that, if two functions are proportional to each other, their wronskian vanishes, and if they both satisfy the same ordinary second-order differential equation of type (1), their wronskian is a constant, i.e. it does not depend on x. Thus for instance

(8) $$W[f^{(+)}(x, -k), f^{(+)}(x, k)] = 2ik,$$

a formula that is instrumental in obtaining (6) from (5).

The Jost solutions $f^{(\pm)}(x, k)$ are also defined by the Volterra integral equations

(9) $$f^{(\pm)}(x, k) = \exp(\pm ikx) \\ - k^{-1} \int_x^{\pm\infty} dy \sin[k(x-y)] u(y) f^{(\pm)}(y, k),$$

that clearly correspond to (1) and (4).

The analogous formulae for $\psi(x, k)$ read

(10a) $$\psi(x, k) = T(k) \exp(-ikx) + k^{-1} \int_{-\infty}^x dy \sin[k(x-y)] u(y) \psi(y, k),$$

(10b) $$\psi(x, k) = \exp(-ikx) + R(k) \exp(ikx) \\ - k^{-1} \int_x^\infty dy \sin[k(x-y)] u(y) \psi(x, k).$$

A comparison of the asymptotic limit of (10a) as $x \to +\infty$ (or of (10b) as $x \to -\infty$) with (3b) (or (3a)) yields the formulae

(11a) $$T(k) = 1 + (2ik)^{-1} \int_{-\infty}^{+\infty} dx \exp(ikx) \psi(x, k) u(x),$$

(11b) $$R(k) = (2ik)^{-1} \int_{-\infty}^{+\infty} dx \exp(-ikx) \psi(x, k) u(x).$$

The quantities $T(k)$ and $R(k)$ have the following "physical" meaning: for positive k, they are respectively the amplitude of the wave transmitted to the left and reflected back to the right, when a wave of unit amplitude and wave number k impinges on the "potential" $u(x)$ from the right (but we repeat that this interpretation has no relevance to the developments that follow).

2.1 Direct spectral problem

The transmission and reflection coefficients are generally complex. If $u(x)$ is real, as we generally assume for simplicity, and also k is real, they satisfy the reflection properties

(12a) $\quad T(-k)=T^*(k), \quad \operatorname{Im} k=0,$
(12b) $\quad R(-k)=R^*(k), \quad \operatorname{Im} k=0,$

as well as the "unitarity" condition

(13) $\quad |T(k)|^2+|R(k)|^2=1.$

Note that (12) implies that the moduli of T and R are even functions of k:

(14a) $\quad |T(-k)|=|T(k)|,$
(14b) $\quad |R(-k)|=|R(k)|;$

their phases $\theta(k)$ and $\chi(k)$, defined by

(15a) $\quad T(k)=|T(k)|\exp[i\theta(k)], \quad \operatorname{Im} k=0,$
(15b) $\quad R(k)=|R(k)|\exp[i\chi(k)], \quad \operatorname{Im} k=0,$

are instead odd in k:

(16a) $\quad \theta(-k)=-\theta(k),$
(16b) $\quad \chi(-k)=-\chi(k).$

Of course the formulae (15) define the phases $\theta(k)$ and $\chi(k)$ only mod(2π); and the symmetry properties (16) hold only for an appropriate determination (see, for instance, (43b) below). The equations (12) are an immediate consequence of the formula

(17) $\quad \psi(x,-k)=\psi^*(x,k)$

that is implied, for real $u(x)$ and k, by (1) and (3). The unitarity equation (13) follows via (12) from the more general formula

(18) $\quad R(k)R(-k)+T(k)T(-k)=1,$

that holds even if $u(x)$ is not real, as shown in section 2.4.

All these formulae have been written here for real k; appropriate analytic continuation to complex k is generally possible (see below).

The spectral problem characterized by the Schroedinger equation (1) may also possess a finite number of discrete negative eigenvalues

(19) $\quad k_n^2 = E_n = -p_n^2, \quad n = 1, 2, \ldots, N;$

here we have of course introduced the convenient notation

(20) $\quad k_n = i p_n, \quad p_n > 0, \quad n = 1, 2, \ldots, N,$

that will be largely employed. To these eigenvalues there correspond solutions of (1), that are square integrable. We indicate with $\varphi_n(x)$ the solutions that are normalized,

(21) $\quad \int_{-\infty}^{+\infty} dx \, \varphi_n^2(x) = 1,$

and with $f_n(x)$ the solutions that are characterized by a unit coefficient in front of the exponential giving the behaviour as $x \to +\infty$:

(22) $\quad f_n(x) \to \exp(-p_n x), \quad x \to +\infty.$

(The notation $\varphi_n(x)$ rather than $\psi_n(x)$ is employed to emphasize that $\varphi_n(x)$ does not coincide with $\psi(x, i p_n)$, because of the different normalization employed; actually the quantity $\psi(x, i p_n)$ does not even exist, because the quantities $R(k)$ and $T(k)$ associated with the asymptotic normalization of $\psi(x, k)$, see (3), blow up at $k = i p_n$, see below. On the other hand clearly

(23) $\quad f_n(x) = f^{(+)}(x, i p_n),$

as implied by (4) and (20); provided the r.h.s. of the last equation is properly defined by analytic continuation, see below).

If $u(x)$ is real, the function $f_n(x)$ is also real, and $\varphi_n(x)$ can (and will) be chosen to be real as well. These two functions are clearly proportional to each other:

(24) $\quad \varphi_n(x) = c_n f_n(x).$

The (real) quantity c_n, or rather its square, the (real and positive) quantity

(25) $\quad \rho_n = c_n^2$

is called *normalization coefficient*, and it plays an important role (as already seen in chapter 1). Clearly there hold the formulae

$$(26) \qquad c_n = \lim_{x \to +\infty} \left[\exp(p_n x) \varphi_n(x) \right],$$

$$(27) \qquad p_n = \left[\int_{-\infty}^{+\infty} dx\, f_n^2(x) \right]^{-1}.$$

The reality of the eigenvalues k_n^2 is implied by the hermiticity of the Schroedinger operator $-\partial^2/\partial x^2 + u(x)$ (here the assumption that $u(x)$ is real plays an essential role); the fact that the discrete eigenvalues k_n^2 must be negative is implied by the requirement that $\varphi_n(x)$ be square-integrable (for real k, $\psi(x,k)$ oscillates asymptotically, see (3), and cannot therefore be square-integrable); the positivity of the quantities p_n is conventional; the asymptotic behaviour (22) is implied by (1) and (20), together of course with the normalizability of $f_n(x)$ and the asymptotic vanishing of $u(x)$; and the finiteness of the number of discrete eigenvalues N is implied by the assumed asymptotic vanishing of $u(x)$ faster than x^{-2} (a negative asymptotic tail of $u(x)$ extending to $+\infty$ and/or to $-\infty$ proportionally to $|x|^{-1}$ would instead yield an infinite number of discrete eigenvalues $-p_n^2$, accumulating at $p_\infty \equiv 0$). The number N of discrete eigenvalues may of course be zero; indeed $N=0$ if $u(x) \geq 0$ for $-\infty < x < \infty$; on the other hand $N > 0$ if $u(x) < 0$ for $-\infty < x < \infty$. Clearly N may be considered a global measure of the negativeness of $u(x)$; unless $v(x)$ is everywhere non-negative, the number $N(g)$ of discrete eigenvalues associated to $u(x) = g^2 v(x)$ grows, as $g \to \infty$, proportionally to g. These results on the number of discrete eigenvalues are proved in appendix A.1, or are immediately implied by the results given there. Of course the physical meaning of the discrete eigenvalues is in terms of "bound states" produced by the potential $u(x)$; but again this interpretation will have no relevance to the developments that follow.

In the following we shall occasionally refer, for shortness, to the (positive) quantities p_n as the "discrete eigenvalues" (although of course the discrete eigenvalues are actually $-p_n^2$).

The eigenfunctions of the Sturm–Liouville problem (1) satisfy of course orthogonality and completeness relations, that are easily derived by standard techniques (although some care is required to deal with the continuous spectrum). The main formulae are displayed in appendix A.2, together with a terse outline of their derivation.

The *direct spectral problem* consists in the evaluation, given a function ("potential") $u(x)$, of the various quantities defined above, namely the transmission and reflection coefficients $T(k)$ and $R(k)$, the discrete eigenvalues p_n (if any) and the corresponding normalization parameters ρ_n. The main step to implement this program is solving the Schroedinger equation (1) (a second order linear ordinary differential equation) and identifying the particular solutions characterized by the conditions (3) or (4), and by (21) (if any). Clearly, given $u(x)$, this procedure defines uniquely the quantities $T(k)$, $R(k)$, p_n and ρ_n.

It is apparent from the structure of the Volterra equations (9) that both $f^{(+)}(x,k)$ and $f^{(-)}(x,k)$ can be analytically continued in k off the real axis and are holomorphic in k in the half-plane $\operatorname{Im} k > 0$. (The analytic continuation can be prevented only by the integral ceasing to converge; a simple iteration argument shows this cannot happen for $\operatorname{Im} k > 0$, and also indicates to what extent analyticity may extend to the lower half-plane, see below.) This immediately implies (see (6a)) that $T(k)$ is also analytically continuable, has no zeros, and is meromorphic in the half-plane $\operatorname{Im} k > 0$. Its poles are moreover all located on the imaginary axis and coincide with the discrete eigenvalues, i.e. the poles of $T(k)$ occur at $k_n = i p_n$. Indeed at the discrete eigenvalue $k_n = i p_n$ both $f^{(+)}(x, i p_n)$ and $f^{(-)}(x, i p_n)$ are proportional to $\varphi_n(x)$,

(28) $\quad \varphi_n(x) = c_n^{(+)} f_n^{(+)}(x, i p_n) = c_n^{(-)} f_n^{(-)}(x, i p_n),$

as implied by the asymptotic vanishing of $\varphi_n(x)$ as $x \to \pm\infty$, see (21), and by their own asymptotic vanishing, see (4) and (20); therefore their wronskian vanishes, and T has a pole, see (6a). Note that the quantity $c_n^{(+)}$, introduced here for notational symmetry, coincides with the quantity c_n introduced above,

(29) $\quad c_n^{(+)} \equiv c_n,$

see (23) and (24). It can moreover be shown that the following formulae hold:

(30) $\quad \lim_{k \to i p_n} \left[(k - i p_n) T(k) \right] = i c_n^{(+)} c_n^{(-)},$

(31) $\quad \lim_{k \to i p_n} \left[(k - i p_n) \psi(x, k) \right] = i c_n^{(+)} \varphi_n(x).$

If the potential $u(x)$ vanishes asymptotically (at least) exponentially,

(32) $\quad \lim_{x \to \pm\infty} \left[u(x) \exp(\pm 2\mu^{(\pm)} x) \right] = 0, \quad \mu^{(\pm)} > 0,$

it is apparent from (9) that $f^{(+)}(x,k)$ resp. $f^{(-)}(x,k)$ are holomorphic also for Im $k<0$, provided Im $k>-\mu^{(+)}$ resp. Im $k>-\mu^{(-)}$ (note the factor of 2 in the exponent in (32)). Thus one may conclude (see (6a)) that $T(k)$ is meromorphic in the half-plane

(33) $\quad$ Im $k>-\mu$,

while $R(k)$ (see (6b)) is meromorphic in the *Bargmann strip*

(34) $\quad -\mu<$ Im $k<\mu^{(+)}$,

where

(35) $\quad \mu=\min(\mu^{(+)},\mu^{(-)})$

(of course in the last two equations the *largest* values of $\mu^{(+)}$ and $\mu^{(-)}$ compatible with (32) enter).

Inside the Bargmann strip, there is a one-to-one correspondence between the poles of $R(k)$ in the upper k-plane and the discrete eigenvalues: if $R(k)$ has a pole in the upper half of the complex k-plane inside the Bargmann strip, it sits on the imaginary axis and it corresponds to one of the discrete eigenvalues, $k_n=ip_n$; and if there is a discrete eigenvalue $k_n=ip_n$ with $0<p_n<\mu^{(+)}$, it also shows up as a pole of $R(k)$ (indeed in this region the analyticity of $f^{(\pm)}(x,k)$ and of $f^{(+)}(x,-k)$ implies that the only possible singularities of $R(k)$ are poles due to the vanishing of the denominator in the r.h.s. of (6b); thus the same argument given above for T, for the whole upper half k-plane, applies, within the Bargmann strip, also to R). On the other hand, above (or at) the boundary of the Bargmann strip such a correspondence need not exist: there may be an eigenvalue with $k_n=ip_n$ and $p_n\geqslant\mu^{(+)}$ without a corresponding pole of $R(k)$ (either because the numerator has no analytic continuation to that point, or because the numerator also vanishes there; that the latter cannot happen *inside* the Bargmann strip is implied by (36), see below); or there may be a pole of $R(k)$ occurring at a value $k=i\nu$ with $\nu\geqslant\mu^{(+)}$, to which there corresponds no discrete eigenvalue (in such a case the pole would appear as a singularity of the numerator rather than a zero of the denominator in the r.h.s. of (6b)). Recall in contrast that there is always a one-to-one correspondence between the zeros of $W[f^{(-)}(x,k),f^{(+)}(x,k)]$, or equivalently the poles of $T(k)$, in the upper half k-plane, and the discrete eigenvalues.

Where the one-to-one correspondence between poles of $R(k)$ and discrete eigenvalues holds, namely *inside* the Bargmann strip (34), there is also an important relation between the residue of $R(k)$ at the pole $k_n = i p_n$ and the normalization coefficient ρ_n defined by (27):

$$(36) \qquad \lim_{k \to i p_n} \left[(k - i p_n) R(k) \right] = i \rho_n.$$

This formula follows easily, for instance, from (3), (22), (24) and (31).

In writing the last equation, and some of the preceding ones, we are implying that the pole at $k_n = i p_n$ is simple. That this is indeed always the case (for real $u(x)$) is proved in appendix A.5. We have also always implied that the discrete eigenvalues are simple. This is indeed a general property of the Schroedinger spectral problem, for a second order linear ordinary differential equation such as (1) cannot possess, for $k^2 = -p^2$, two linearly independent solutions both asymptotically vanishing, since their linear combination could then not, as instead it should, represent the most general solution, that may asymptotically diverge (at either, or both, ends), rather than vanish (either behaviour must of course be exponential, as implied by (1) and the asymptotic vanishing of $u(x)$).

If the potential $u(x)$ vanishes faster than exponentially as $x \to +\infty$, namely if for any (finite) μ

$$(37) \qquad \lim_{x \to \infty} \left[u(x) \exp(\mu x) \right] = 0,$$

the Bargmann strip invades the whole upper half of the complex k-plane; for such potentials $R(k)$ shares with $T(k)$ the property of being meromorphic in the whole upper half k-plane, and having there poles (if any) only on the imaginary axis, in one-to-one correspondence with the discrete eigenvalues; and for all these poles, and all the discrete eigenvalues, the relationship (36) holds.

Note that, as is appropriate to this section in which the direct spectral problem is treated, throughout this discussion our point of view has considered the potential $u(x)$ as given; for it is the (asymptotic) properties of the potential that determine the size of the Bargmann strip. In the next section we consider the inverse problem, in which (appropriate) spectral data are given, and the corresponding potential (or potentials) must be retrieved. Then of course the structure of the Bargmann strip is not generally a given input of the problem, but rather part of the answer. We shall return to this point at the end of the following section.

As we have just seen, the basic element that determines the analyticity properties of $R(k)$ in k is the asymptotic behaviour of $u(x)$, in particular

2.1 Direct spectral problem

how fast $u(x)$ vanishes as $x \to \infty$. Another important characteristic of $T(k)$ and $R(k)$, now considered again as functions of the *real* variable k, is the behaviour as $k \to \pm\infty$. This is mainly determined by the smoothness (differentiability) of $u(x)$.

It is obvious from the definition of $T(k)$ and $R(k)$ (see, for instance, (11)) that, provided $u(x)$ is regular for $-\infty < x < \infty$,

(38a) $$\lim_{k \to \pm\infty} [T(k)] = 1,$$

(38b) $$\lim_{k \to \pm\infty} [R(k)] = 0.$$

If moreover, as we generally assume, $u(x)$ is infinitely differentiable for all (real) values of x, then $R(k)$ and the modulus of $T(k)$ tend to their asymptotic values, as $k \to \pm\infty$, faster than any (negative) power of k:

(39a) $$\lim_{k \to \pm\infty} [|k|^M (1 - |T(k)|)] = 0, \quad M < \infty, \quad \text{Im } k = 0,$$

(39b) $$\lim_{k \to \pm\infty} [|k|^M R(k)] = 0, \quad M < \infty, \quad \text{Im } k = 0.$$

The phase of $T(k)$, see (15a), has instead an asymptotic expansion in inverse (odd) powers of k:

(40) $$\theta(k) = \sum_{m=0}^{M} \theta_m (2k)^{-2m-1} + O(k^{-2M-3})$$

(note the factor of 2 that has been introduced, for notational convenience, in this formula, which of course holds mod(2π), and for real k). The coefficients θ_m are given, in terms of the potential $u(x)$, by the compact formula

(41) $$\theta_m = (-1)^{m+1}(2m+1)^{-1} \int_{-\infty}^{+\infty} dx\, L^m \cdot [2u(x) + x u_x(x)],$$

$$m = 0, 1, 2, \ldots.$$

Here the integro-differential operator L is defined by the formula

(42) $$L \cdot f(x) = f_{xx}(x) - 4u(x) f(x) + 2 u_x(x) \int_x^\infty dy\, f(y),$$

that specifies its action on a generic function $f(x)$ (vanishing as $x \to +\infty$). Note that this is the same operator L that was introduced in chapter 1, see e.g. (1.2.-2); it is an important object, and it plays a prominent role throughout this book. The formula (41) will be used, and further discussed, in chapter 5. The results we have just reported, and in particular the formula (41), are tersely proved in appendix A.3.

The analyticity and asymptotic properties of $T(k)$ described above imply the following *dispersion relations* between the modulus and the phase of $T(k)$, see (15a), and its poles on the upper imaginary axis in the complex k-plane, that as we have seen are in one-to-one correspondence with the discrete eigenvalues $k_n = i p_n$:

(43a) $$|T(k)| = \prod_{n=1}^{N} \left[1 + (p_n/k)^2\right]^{-1}$$
$$\cdot \exp\left[\pi^{-1} P \int_{-\infty}^{+\infty} dq (q^2 - k^2)^{-1} q \theta(q)\right], \quad \text{Im } k = 0,$$

(43b) $$\theta(k) = 2 \sum_{n=1}^{N} \text{Arctg}(p_n/k)$$
$$- (k/\pi) P \int_{-\infty}^{+\infty} dq (q^2 - k^2)^{-1} \ln|T(q)|, \quad \text{Im } k = 0.$$

The choice of the principal determination in the r.h.s. of (43b), indicated by the symbol Arctg, lifts the $\text{mod}(2\pi)$ ambiguity intrinsic in the definition (15a), consistently with the symmetry property (16a). If there are discrete eigenvalues ($N > 0$), the function $\theta(k)$ defined by (43b) has a discontinuity at $k = 0$, characterised by the formulae ("Levinson's theorem")

(44) $\quad \theta(0^+) = N\pi, \quad \theta(0^-) = -N\pi.$

Note incidentally that these formulae imply that knowledge of the modulus of the reflection coefficient $R(k)$ for all real values of k, and of the discrete eigenvalues p_n (if any), is sufficient to determine uniquely the transmission coefficient $T(k)$; indeed $|T(k)|$ is determined by the unitarity condition (13), and the phase $\theta(k)$ of $T(k)$ is then determined by (43b). Note that neither the phase of $R(k)$, nor the normalisation coefficients ρ_n, are required for the determination of $T(k)$; while instead they are required for the determination of the potential $u(x)$, as discussed in the following section 2.2.

From (43b) (and (40)) there follows the formula

(45) $$\theta_m = 4^m \left\{ 2(-1)^m (2m+1)^{-1} \sum_{n=1}^{N} p_n^{2m+1} + \frac{2}{\pi} \int_{-\infty}^{+\infty} dq\, q^{2m} \ln|T(q)| \right\},$$
$m = 0, 1, 2, \ldots,$

that can be considered the counterpart in spectral space of (41). Note that the integral in the r.h.s. is convergent, since we are always restricting attention to the class of smooth $u(x)$ such that $|T(k)|$ tend to unity, as $k\to\pm\infty$, faster than any inverse power of k, and this implies that $\ln|T(k)|$ vanishes, as $k\to\pm\infty$, faster than any inverse power of k.

Of course the formulae (43) hold for real k. For $\mathrm{Im}\, k>0$ there holds correspondingly the representation

$$(46) \quad T(k)=\prod_{n=1}^{N}\left[1+(p_n/k)^2\right]^{-1}\exp\left[(k/\pi)\int_{-\infty}^{+\infty}dq\,(q^2-k^2)^{-1}q\theta(q)\right].$$

The derivation of these formulae is standard; for completeness we outline it in appendix A.4.

We end this section with one caveat and one remark. *Caveat*: in our treatment we have, for simplicity, ignored the marginal case when $f^{(+)}(x,0)$ turns out to be proportional to $f^{(-)}(x,0)$, namely when for $k^2=0$ there exists a solution of (1) that does not diverge (linearly) at either end ($x\to\pm\infty$); and in the following we persevere to ignore this marginal case, that may be interpreted as corresponding to the occurrence of a discrete eigenvalue at $k=0$ (see appendix A.15). *Remark*: the spectral problem could be defined having in mind a "scattering experiment" with the incoming beam coming from the left rather than the right; this would entail some trivial changes (in (3), and in some other equations), as well as certain simple relations between the "old" and "new" reflection and transmission coefficients, whose derivation is left as an exercise for the reader.

2.1.1. Transformation properties

The Schroedinger equation (2.1.-1) is invariant under space translations and scale transformations. The corresponding transformation properties of the various quantities associated with it, which follow directly from their definitions, are reported here, to be used later.

Let $u(x)$ be the "potential" in the Schroedinger equation (2.1-1), $w(x)$ its integral, $w(x)=\int_x^\infty dy\, u(y)$, and $f^{(\pm)}(x,k)$, $\psi(x,k)$, $R(k)$, $T(k)$, $\varphi_n(x), p_n, \rho_n$ the various quantities associated with $u(x)$ as specified in section 2.1.

Translations. Let

$$(1) \quad u'(x)\equiv u(x+a), \quad w'(x)\equiv w(x+a),$$

and indicate with $f^{(\pm)\prime}(x,k)$, $\psi'(x,k)$, $R'(k)$, $T'(k)$, $\varphi'_n(x)$, p'_n, ρ'_n the corresponding quantities associated with the potential $u'(x)$. Then

(1a) $\quad f^{(\pm)\prime}(x,k) = \exp(\mp ika) f^{(\pm)}(x+a,k),$
$\quad\quad\quad \psi'(x,k) = \exp(ika) \psi(x+a,k),$
(1b) $\quad R'(k) = \exp(2ika) R(k),$
(1c) $\quad T'(k) = T(k),$
(1d) $\quad \varphi'_n(x) = \varphi_n(x+a),$
$\quad\quad\quad f^{(\pm)\prime}(x,ip_n) = \exp(\pm p_n a) f^{(\pm)}(x+a,ip_n),$
(1e) $\quad p'_n = p_n,$
(1f) $\quad \rho'_n = \exp(-2p_n a) \rho_n.$

In view of their importance, see below, we also display the variations of all these quantities corresponding to *infinitesimal translations* ($a = \varepsilon$, $\varepsilon \to 0$):

(2) $\quad \delta u(x) = \varepsilon u_x(x), \quad \delta w(x) = \varepsilon w_x(x) = -\varepsilon u(x),$
(2a) $\quad \delta f^{(\pm)}(x,k) = \varepsilon\bigl[\mp ik f^{(\pm)}(x,k) + f_x^{(\pm)}(x,k)\bigr],$
$\quad\quad\quad \delta\psi(x,k) = \varepsilon\bigl[ik\psi(x,k) + \psi_x(x,k)\bigr],$
(2b) $\quad \delta R(k) = \varepsilon 2ik R(k),$
(2c) $\quad \delta T(k) = 0,$
(2d) $\quad \delta\varphi_n(x) = \varepsilon \varphi_{nx}(x),$
$\quad\quad\quad \delta f^{(\pm)}(x,ip_n) = \varepsilon\bigl[\pm p_n f^{(\pm)}(x,ip_n) + f_x^{(\pm)}(x,ip_n)\bigr],$
(2e) $\quad \delta p_n = 0,$
(2f) $\quad \delta\rho_n = \varepsilon(-2p_n)\rho_n.$

Scale transformations. Let instead

(3) $\quad u'(x) = \mu^2 u(\mu x), \quad w'(x) = \mu w(\mu x).$

Then, with obvious notation:

(3a) $\quad f^{(\pm)\prime}(x,k) = f^{(\pm)}(\mu x, k/\mu), \quad \psi'(x,k) = \psi(\mu x, k/\mu),$
(3b) $\quad R'(k) = R(k/\mu),$
(3c) $\quad T'(k) = T(k/\mu),$
(3d) $\quad \varphi'_n(x) = \mu^{1/2} \varphi_n(\mu x),$
(3e) $\quad p'_n = \mu p_n,$
(3f) $\quad \rho'_n = \mu \rho_n.$

Again we also display the variations corresponding to *infinitesimal scale transformations* ($\mu=1+\varepsilon$, $\varepsilon\to 0$):

(4) $\quad \delta u(x)=\varepsilon[2u(x)+xu_x(x)], \qquad \delta w(x)=\varepsilon[w(x)+xw_x(x)],$

(4a) $\quad \delta f^{(\pm)}(x,k)=\varepsilon[xf_x^{(\pm)}(x,k)-kf_k^{(\pm)}(x,k)]$

$\quad\quad\ \delta\psi(x,k)=\varepsilon[x\psi_x(x,k)-k\psi_k(x,k)],$

(4b) $\quad \delta R(k)=-\varepsilon k R_k(k),$

(4c) $\quad \delta T(k)=-\varepsilon k T_k(k),$

(4d) $\quad \delta\varphi_n(x)=\varepsilon[\tfrac{1}{2}\varphi_n(x)+x\varphi_{nx}(x)],$

(4e) $\quad \delta p_n=\varepsilon p_n,$

(4f) $\quad \delta\rho_n=\varepsilon\rho_n.$

2.2. Inverse spectral problem

A more complete discussion of the inverse spectral problem is given in appendix A.5, where the proofs of the results given below (some of which have been already reported in section 1.3.2) are also to be retrieved.

The fundamental result may be formulated as follows (using the language and notation introduced in the preceding section): given, for $-\infty<k<+\infty$, a function $R(k)$, arbitrary except for the requirement that it satisfy the appropriate properties characterizing a reflection coefficient (see in particular (2.1.-12b, 13, 38b)), and given (arbitrarily) $2N$ positive numbers p_n and ρ_n, $n=1,2,\ldots,N$, there exists correspondingly (in the sense of the spectral problem of the preceding section) a unique "potential" $u(x)$. A constructive procedure to identify this function is provided by the "Gel'fand-Levitan-Marchenko" equation

(1) $\quad K(x,y)+M(x+y)+\int_x^{+\infty} dz\, K(x,z)M(z+y)=0, \quad y>x.$

In this Fredholm integral equation $M(x)$ should be considered as given, being defined in terms of the spectral data by the formula

(2) $\quad M(x)=\sum_{n=1}^{N}\rho_n\exp(-p_n x)+(2\pi)^{-1}\int_{-\infty}^{+\infty} dk\, R(k)\exp(ikx)$

(thus, apart from the contribution of the discrete spectrum parameters, M is precisely the (inverse) Fourier transform of the reflection coefficient R), while $K(x, y)$ is the (a priori unknown) function (uniquely) determined by the integral equation itself (note that the integral equation refers to the dependence of $K(x, y)$ on the second variable, y, while the dependence on the first variable x originates, as it were, parametrically, from its presence in the argument of the inhomogeneous term and as lower limit of integration). The function $K(x, y)$ determines $u(x)$ via the formulae

(3) $$w(x) = 2K(x, x+0) \equiv 2 \lim_{\varepsilon \to 0} [K(x, x+|\varepsilon|)],$$

(4a) $$w(x) = \int_x^{+\infty} dy\, u(y),$$

(4b) $$u(x) = -dw(x)/dx.$$

The function $K(x, y)$ is quite directly related also to the Jost solution $f^{(+)}(x, k)$, through the formulae

(5a) $$f^{(+)}(x, k) = \exp(ikx) + \int_x^{+\infty} dy\, K(x, y) \exp(iky),$$

(5b) $$K(x, y) = (2\pi)^{-1} \int_{-\infty}^{+\infty} dk\, [f^{(+)}(x, k) - \exp(ikx)] \exp(-iky),$$

$$y > x,$$

whose mutual consistency is guaranteed by the holomorphy of $f^{(+)}(x, k)$ in the upper half of the complex k-plane and by its asymptotic behavior there:

(6) $$\lim_{|k| \to \infty,\, \mathrm{Im}\, k \geq 0} [f^{(+)}(x, k) - \exp(ikx)] = 0.$$

(Actually the difference $f^{(+)}(x, k) - \exp(ikx)$ vanishes exponentially if $\mathrm{Im}\, k \to +\infty$ and at least proportionally to $|k|^{-1}$ if $\mathrm{Re}\, k \to \pm\infty$, $\mathrm{Im}\, k = \text{constant} \geq 0$. Note that this implies that $K(x, y)$, as defined by (5b), vanishes for $y < x$, while of course the GLM equation (1) determines $K(x, y)$ only for $y > x$; the value of $K(x, y)$ for $y = x$ relevant to the determination of $u(x)$ is obtained by continuity from the values of $K(x, y)$ for $y > x$, see (3)).

Another useful relationship between $K(x, y)$ and the Jost solutions reads

(5c) $$K(x, y) = -\sum_{n=1}^{N} \rho_n f_n(x) \exp(-p_n y)$$

$$- (2\pi)^{-1} \int_{-\infty}^{+\infty} dk\, R(k) f^{(+)}(x, k) \exp(iky).$$

Thus the collection of spectral data $\{R(k), -\infty < k < +\infty;\ p_n, \rho_n, n = 1, 2, \ldots, N\}$ determine uniquely the function $u(x)$ (and therefore these data

2.2 Inverse spectral problem

also determine $T(k)$; indeed $|T(k)|$ is immediately given by $R(k)$ through the unitarity condition (2.1.-13); and the reconstruction of the phase of $T(k)$ from its modulus can be effected by standard techniques, taking advantage of its analyticity properties described in the preceding section, see (2.1.-43b)).

As we have already mentioned (and as is indeed quite obvious), in order for the function $u(x)$ to possess all the nice properties that were prescribed in the preceding section, the corresponding reflection coefficient $R(k)$, given as (part of the) input in the inverse spectral problem, must itself possess all the nice properties that were ascribed to it in the preceding section. In particular, in order that $u(x)$ turn out to be real and regular, the modulus of $R(k)$ must not exceed unity (as required by (2.1.-13)), the reflection property (2.1.-12b) must hold (note that this implies the reality of $M(x)$; see (2)) and of course $R(k)$ must vanish asymptotically (at least fast enough to insure that its Fourier transform exists, see (2); faster than any negative power of k if the corresponding $u(x)$ is to be infinitely differentiable, see (2.1-39b)). These requirements refer only to properties of $R(k)$ for k real; indeed only these values of $R(k)$ enter as input in the inverse spectral problem (see (2)). But $R(k)$ must be analytically continuable in k off the real axis and have the appropriate analytic structure in order that the corresponding $u(x)$ have those properties that, in the context of the direct problem of the preceding section, were sufficient to guarantee the analyticity of $R(k)$. For instance, only if the reflection coefficient $R(k)$ is meromorphic in k in the whole complex plane (excluding the point at infinity), can the corresponding $u(x)$ vanish faster than exponentially as $x \to \pm \infty$; moreover this asymptotic behavior may obtain, if the input data contain also some discrete eigenvalues p_n, only if the poles of $R(k)$ occur at $k = k_n = i p_n$ and the corresponding normalization coefficients ρ_n are related to the residues at these poles by (2.1.-36). It is obvious how this analysis can be extended to cover the more general case in which there is a finite Bargmann strip; a relevant, very simple example is given below.

A more detailed analysis of the classes of functions $u(x)$ and $R(k)$ for which the main results stated at the beginning of this section take the form of rigorous existence and uniqueness theorems is beyond our scope (see, however, appendix A.16). Moreover, as will be seen below, in the context of primary interest to us, this problem generally does not arise, or, equivalently, it tends to take care of itself; we shall not be interested in solving the inverse spectral problem for arbitrarily chosen inputs but rather for spectral data that, having themselves originated rather straightforwardly from the solution of a direct problem, are, as it were automatically, guaranteed to yield reasonable outputs (see below). Eventually, however, the rigorous understanding mentioned above will have to be mastered, since it will provide an important tool of investigation. Indeed there are some cases (see for instance section 6.2) where this understanding is clearly called for.

We end this section by reporting some examples, that will play an important rôle in the following.

The simplest, and also most important example corresponds to the case in which the reflection coefficient $R(k)$ vanishes identically and there is only one discrete eigenvalue:

(7) $\quad R(k)=0, \quad -\infty<k<+\infty; \quad p_1\equiv p, \quad \rho_1\equiv\rho.$

Then the GLM equation (1) becomes separable and its solution is trivial. One thus finds

(8) $\quad K(x,y)=-p\exp[-p(y-\xi)]/\cosh[p(x-\xi)]$

where

(9) $\quad \xi=(2p)^{-1}\ln(\rho/2p).$

The corresponding function $u(x)$ reads

(10) $\quad u(x)=-2p^2/\cosh^2[p(x-\xi)].$

The Jost functions and the transmission coefficient are also easily computed:

(11) $\quad f^{(\pm)}(x,k)=(k+ip)^{-1}\{k\pm ip\,\text{tgh}[p(x-\xi)]\}\exp(\pm ikx),$
(12) $\quad T(k)=(k+ip)/(k-ip)$

((11) follows from (5a) and (8) together with the remark that $f^{(-)}(x,k)$ must be proportional to $f^{(+)}(x,-k)$ in order that (2.1.-6b) be compatible with (7); and (12) follows from the two formulae preceding it, via (2.1.-6a)). Note the consistency of (7) and (12) with all the pertinent equations of section 2.1; in particular in this case

(13) $\quad |T(k)|=1,$
(14) $\quad \theta(k)=2\arctan(p/k),$
(15) $\quad \theta_m=2(-1)^m(2p)^{2m+1}/(2m+1)$

(see (2.1.-13, 15a, 40); of course (13) and (14) are written for real k). Also note that $T(k)$ is holomorphic for $\operatorname{Im} k > -p$ (see (2.1.-33); and compare (2.1.-32) with (10)); while $R(k)$ may be considered holomorphic inside the Bargmann strip $|\operatorname{Im} k| < p$ (see (2.1.-34)), but it does not satisfy (2.1.-36) (at $k = ip$ the analytic continuation of $R(k)$, being identically zero, has certainly no pole, while there is a discrete eigenvalue; and indeed $k = ip$ does not lie *inside* the Bargmann strip, but precisely on its boundary).

Let us also report the explicit form of the normalized wave function $\varphi(x)$ of the potential (10) corresponding to the eigenvalue $E = -p^2$:

(16) $\quad \varphi(x) = (\tfrac{1}{2}p)^{1/2} / \cosh[p(x-\xi)].$

In this formula (that follows immediately from (11) and (2.1.-28)) the quantity ξ is of course always related to ρ by (9).

It is trivial (but interesting; see below) to verify that the potential (10) is related to the normalized wave function (16) by the two formulae

(17) $\quad u(x) = -4p\varphi^2(x)$

and

(18) $\quad u(x) = -dw(x)/dx$

with

(19a) $\quad w(x) = -2\rho^{1/2} \exp(-px) \varphi(x),$
(19b) $\quad w(x) = -2(2p)^{1/2} \exp[-p(x-\xi)] \varphi(x)$

(the equivalence of the last two equations is of course implied by (9)).

A more general example corresponds to the case in which, still with vanishing reflection coefficient, $R(k) = 0$, $-\infty < k < +\infty$, there are N discrete eigenvalues, with parameters p_n and ρ_n, $n = 1, 2, \ldots, N$. The corresponding functions $u(x)$ can be again calculated by solving the GLM equation (1) (that has in this case a separable kernel of rank N): and one finds

(20) $\quad u(x) = -2d^2\{\ln \det[I + C(x)]\}/dx^2,$

where I is the unit matrix (of order N) and $C(x)$ is the symmetrical matrix of order N defined by

(21) $\quad C_{mn}(x) = c_m c_n (p_m + p_n)^{-1} \exp[-(p_m + p_n)x],$

with (see 2.1.-25))

(22) $\quad c_n = \rho_n^{1/2}.$

The Jost solution $f^{(+)}(x, k)$, the solution $K(x, y)$ of the GLM equation, and the transmission coefficient $T(k)$ corresponding to the potential (19), can also be calculated. We report here the (very simple) explicit expression of the transmission coefficient,

(23) $\quad T(k) = \prod_{n=1}^{N} [(k + i p_n)/(k - i p_n)],$

and the (convenient) expressions of $f^{(+)}(x, k)$ and of $K(x, y)$ in terms of the normalized eigenfunctions $\varphi_n(x)$ corresponding to the eigenvalues $E_n = -p_n^2$ (and of course to the potential (20)):

(24) $\quad f^{(+)}(x, k) = \exp(ikx) \left[1 - \sum_{n=1}^{N} c_n (p_n - ik)^{-1} \exp(-p_n x) \varphi_n(x) \right],$

(25) $\quad K(x, y) = -\sum_{n=1}^{N} c_n \varphi_n(x) \exp(-p_n y).$

As for the eigenfunctions themselves, rather than giving their explicit expressions (easily derivable from the following), we display the system of linear equations that determines them:

(26) $\quad \varphi_n(x) + \sum_{m=1}^{N} c_n c_m (p_n + p_m)^{-1} \exp[-(p_n + p_m)x] \varphi_m(x)$
$= c_n \exp(-p_n x), \quad n = 1, 2, \ldots, N.$

Finally we report two remarkable expressions of the potential (20) itself in terms of the eigenfunctions, that are the analogs of (17) and (18). They read:

(27) $\quad u(x) = -4 \sum_{n=1}^{N} p_n \varphi_n^2(x),$

(28) $\quad u(x) = -\sum_{n=1}^{N} dw_n(x)/dx$

with

(29) $\quad w_n(x) = -2 c_n \exp(-p_n x) \varphi_n(x).$

Of course in the last formula, as well as in (24), (25) and (26), the quantities c_n are related to the ρ_n's by (22); in fact, to avoid any sign ambiguity, it may be advisable to rewrite these equations replacing everywhere the product $c_n \varphi_n(x)$ by $\rho_n f_n(x)$, see (2.1.-24). Note that these formulae are merely (1.3.3.-2) and (1.3.3.-3), for the special case considered here, characterized by the vanishing of the reflection coefficient.

2.3. The spectral transform of *bona fide* potentials

Given a function $u(x)$, defined for all (real) values of x, vanishing (sufficiently fast) as $x \to \pm\infty$, and having the other properties of regularity discussed in the preceding sections (see also below), we define its spectral transform S to be the following set of spectral data:

(1) $\qquad S = \{R(k), -\infty < k < \infty; p_n, \rho_n, n = 1, 2, \ldots, N\},$

where we are of course using the notation introduced above. The essence of the discussion given in the preceding two sections is that there exists a one-to-one correspondence between the function u and its spectral transform S. This correspondence is moreover specified by explicit and constructive prescriptions involving the solution of *linear* problems: essentially the Schroedinger equation (2.1.-1) (a second order linear ordinary differential equation) and the GLM equation (2.2.-1) (a linear integral Fredholm equation). This does not imply that the mapping $u \Leftrightarrow S$ is linear; on the contrary, this mapping is clearly nonlinear (indeed, very much so; in the loose sense in which any such qualitative assessment can be formulated).

If the function u depends on some other variable besides x, say $u \equiv u(x, y)$, then of course its spectral transform also depends on that variable,

(2) $\qquad S \equiv S(y) = \{R(k, y), -\infty < k < \infty; p_n(y), \rho_n(y);$

$\qquad\qquad\qquad n = 1, 2, \ldots, N(y)\}.$

The main focus of our considerations is the case in which the new variable is interpreted as "time" (we will accordingly generally use for it the letter t); then both u and its spectral transform S may be considered to evolve in time. The evolution of u is the main topic of our investigation; and our main tool of analysis is the spectral transform, since we identify and study precisely those time-evolutions of u which are characterized by the property that the corresponding time evolution of the spectral transform S is very simple. Thus the most appropriate way to investigate the evolution of

u consists in the study of the evolution of the corresponding spectral transform, in close analogy to the technique of solution of linear partial differential equations via the Fourier transform.

The fundamental difference between the Fourier transform and the spectral transform is that the former induces a linear mapping, while the latter is nonlinear. Moreover, the Fourier transform of the Fourier transform is trivially related to the original function, while no analogous relation obtains in the case of the spectral transform.

Because the mapping induced by the spectral transform is nonlinear, to a simple, and in particular *linear*, time evolution of the spectral transform, there corresponds a *nonlinear* time evolution of u. This explains why the spectral transform technique opens the possibility to solve and investigate certain classes of *nonlinear evolution equations*. These classes include many equations that, although nonlinear, are themselves sufficiently simple to evoke mathematical interest and to appear in many applications; this highly nontrivial circumstance constitutes the main motivation of our study.

Clearly the developments that we have now outlined suggest the need, in order to progress further, to have some method to relate change in the spectral transform to change in the corresponding function. A convenient technique to achieve this is provided in the following section. The main point is to identify those changes of $u(x)$ to which there correspond simple changes of the spectral transform. The results are then exploited in chapters 3 (infinitesimal changes) and 4 (finite changes).

It is useful to introduce the concept of *bona fide potential* for the Schroedinger spectral problem introduced and discussed in the preceding two sections. This is a function $u(x)$, real and infinitely differentiable for all (real) values of x and vanishing with all its derivatives as $x \to \pm \infty$ (faster than $|x|^{-2}$, so that (2.1.-2) hold). The corresponding spectral transform (1) is characterized by the fact that the (inverse) Fourier transform of $R(k)$,

(3) $$\hat{M}(x) = (2\pi)^{-1} \int_{-\infty}^{+\infty} dk \, R(k) \exp(ikx),$$

exists (as a function, not merely as a distribution) and is real and infinitely differentiable for all real values of x (see (2.1.-12b) and (2.1.-39b)); moreover the modulus of $R(k)$ does not exceed unity,

(4) $\quad |R(k)| < 1, \quad \text{Im } k = 0, \quad [\text{Re } k \neq 0]$

(see (2.1.-13)); and N is a finite nonnegative integer, while all the quantities p_n and ρ_n are real and positive. Our consideration is generally restricted to such functions (with some exceptions; see in particular sections 3.2.4, 6.2, 6.3, 6.4). An important problem that should be mentioned is the possibility to go beyond this functional class; for a qualitative discussion, see appendix A.16.

2.4. Formulae relating two functions to the corresponding spectral transforms

The formulae reported below relate essentially the difference and the sum of the two spectral transforms corresponding to two functions, to appropriate nonlinear combinations of these two functions. They are therefore instrumental in identifying those nonlinear, generally integrodifferential, relations between two functions, say $u^{(1)}(x)$ and $u^{(2)}(x)$, such that the relations between the corresponding spectral transforms be linear. Thus, they provide the main tool of analysis within our approach, as will be seen in detail in the following sections and chapters.

These results come in two forms, which are separately displayed below (subsections 2.4.1 and 2.4.2). The first type expresses the spectral quantities as integrals over the space coordinate x; the integrand contains the eigenfunctions $\psi^{(j)}(x, k)$ and appropriate nonlinear integrodifferential combinations of the functions $u^{(j)}(x)$. The second type expresses instead appropriate nonlinear integrodifferential combinations of the functions $u^{(j)}(x)$ as integrals over the spectrum; the integrands contain again the eigenfunctions (although a different choice than in the previous case is now more convenient; see below), and appropriate, essentially linear, combinations of the spectral quantities. We term the first type of formulae "wronskian integral relations", since their prototype is an immediate consequence of the standard wronskian theorem for second order ordinary differential equations; while for the second type we use the term "spectral integral relations".

2.4.1. Wronskian integral relations

Our starting point here is the Schroedinger spectral problem for two different "potentials":

(1) $\quad -\psi^{(j)}_{xx}(x, k) + u^{(j)}(x)\psi^{(j)}(x, k) = k^2 \psi^{(j)}(x, k), \quad j = 1, 2.$

The spectral quantities associated with the potential $u^{(j)}$ are labeled by the corresponding superscript $j = 1, 2$; otherwise the notation coincides with that used above, in particular the solutions $\psi^{(j)}(x, k)$ are characterized by the boundary conditions (2.1.-3), that are now equipped with the superscript j:

(1a) $\quad \psi^{(j)}(x,k) \to T^{(j)}(k)\exp(-ikx), \quad x \to -\infty, \quad j=1,2,$

(1b) $\quad \psi^{(j)}(x,k) \to \exp(-ikx) + R^{(j)}(k)\exp(ikx), \quad x \to +\infty, \quad j=1,2.$

An elementary consequence of the two equations (1) is the formula

(2) $\quad W[\psi^{(1)}(x,k), \psi^{(2)}(x,k)]\Big|_{x_1}^{x_2} = \int_{x_1}^{x_2} dx\, \psi^{(1)}(x,k)\psi^{(2)}(x,k)$
$\cdot [u^{(2)}(x) - u^{(1)}(x)],$

where we are using the wronskian notation (2.1.-7). In the limit $x_1 \to -\infty$, $x_2 \to +\infty$ this yields (using (1a) and (1b))

(3) $\quad 2ik[R^{(1)}(k) - R^{(2)}(k)] = \int_{-\infty}^{+\infty} dx\, \psi^{(1)}(x,k)\psi^{(2)}(x,k)$
$\cdot [u^{(1)}(x) - u^{(2)}(x)].$

We have here a first formula relating the change in the "potential" $u(x)$ to the change in the "reflection coefficient" $R(k)$.

A second formula can be obtained by appropriate manipulations, for instance by treating in an analogous fashion, say (1) for $j=1$, and the equation obtained from (1) by differentiation for $j=2$. It reads

(4a) $\quad (2ik)^2[R^{(1)}(k) + R^{(2)}(k)] = \int_{-\infty}^{+\infty} dx\, \psi^{(1)}(x,k)\psi^{(2)}(x,k)\Gamma \cdot 1,$

the operator Γ being defined by the formula

(5) $\quad \Gamma \cdot f(x) = [u_x^{(1)}(x) + u_x^{(2)}(x)]f(x)$
$+ [u^{(1)}(x) - u^{(2)}(x)] \int_x^{+\infty} dy\, [u^{(1)}(y) - u^{(2)}(y)]f(y),$

that specifies its action on the generic function $f(x)$ (which is of course assumed to possess regularity and asymptotic properties sufficient to guarantee that (5) makes good sense). Note that the operator Γ depends on $u^{(1)}$ and $u^{(2)}$, although for notational simplicity we have chosen not to indicate this explicitly. This operator plays an important rôle, and for this reason we have introduced it here; otherwise we could have written, in place of (4a), the more explicit but completely equivalent formula

(4b) $(2ik)^2[R^{(1)}(k)+R^{(2)}(k)]$
$$= \int_{-\infty}^{+\infty} dx\, \psi^{(1)}(x,k)\psi^{(2)}(x,k)\left\{u_x^{(1)}(x)+u_x^{(2)}(x)\right.$$
$$\left.+[u^{(1)}(x)-u^{(2)}(x)]\int_x^{+\infty} dy\,[u^{(1)}(y)-u^{(2)}(y)]\right\}.$$

The formulae (3) and (4) are the prototypes of the basic formulae of our approach. To generalize them we now introduce another integrodifferential operator, defined by the formula

(6) $\quad \Lambda f(x) = f_{xx}(x) - 2[u^{(1)}(x)+u^{(2)}(x)]f(x) + \Gamma \cdot \int_x^{+\infty} dy\, f(y),$

that again specifies its action on a generic function $f(x)$ (that must of course vanish as $x \to +\infty$ sufficiently quickly to insure that the integral of the last term in the r.h.s. exists, and have the other properties required to guarantee that (6) makes good sense). Note that this definition contains the operator Γ defined above; clearly Λ, as well as Γ, depend on $u^{(1)}$ and $u^{(2)}$, although we have again preferred, for notational simplicity, not to indicate this explicitly.

The operator Λ plays a central rôle in the theory, because it satisfies the formula

(7) $\quad \int_{-\infty}^{+\infty} dx\, \psi^{(1)}(x,k)\psi^{(2)}(x,k)\Lambda f(x)$
$$= -4k^2 \int_{-\infty}^{+\infty} dx\, \psi^{(1)}(x,k)\psi^{(2)}(x,k)f(x).$$

Here the function $f(x)$ is arbitrary, except for the requirement that it vanish asymptotically with its first derivative:

(8a) $\quad f(\pm\infty) = f_x(\pm\infty) = 0.$

Moreover $f(x)$ must have the regularity properties sufficient to guarantee that (7) makes good sense and that the manipulations required to derive (7) from the Schroedinger equations (1) are justified (see appendix A.6). Note that clearly (7) may be understood as if Λ were the adjoint of an operator Λ^T having the product $\psi^{(1)}(x,k)\psi^{(2)}(x,k)$ as eigenfunction, with eigenvalue

$-4k^2$ (we do not display the explicit expression of this operator, since we never use it; for definitions and discussions of the closely related operator $\tilde{\Lambda}$, see subsection 2.4.2 and appendix A.9).

The formula (7) can clearly be iterated, provided $\Lambda^m \cdot f(x)$, $m = 1, 2, \ldots$, satisfies the condition (8a). A sufficient condition for this is that the function $f(x)$ be regular for all real values of x and vanish asymptotically *with all its derivatives*:

(8b) $\quad 0 = f(\pm\infty) = f_x(\pm\infty) = f_{xx}(\pm\infty) = \cdots .$

This is assumed hereafter; it is clearly consistent (see (5) and (6)) with the notion of *bona fide* potential introduced in the preceding section 2.3. Thus (7) implies the validity of the more general formula

(9) $\quad \displaystyle\int_{-\infty}^{+\infty} dx \, \psi^{(1)}(x,k) \psi^{(2)}(x,k) g(\Lambda) \cdot f(x)$

$\qquad = g(-4k^2) \displaystyle\int_{-\infty}^{+\infty} dx \, \psi^{(1)}(x,k) \psi^{(2)}(x,k) f(x),$

where $g(z)$ is a polynomial (or even, more generally, an entire function).

This formula allows us to generalize (3) and (4), thereby obtaining the two basic formulae of our approach:

(10a) $\quad 2ikg(-4k^2)\left[R^{(1)}(k) - R^{(2)}(k)\right]$

$\qquad = \displaystyle\int_{-\infty}^{+\infty} dx \, \psi^{(1)}(x,k) \psi^{(2)}(x,k) g(\Lambda) \left[u^{(1)}(x) - u^{(2)}(x)\right],$

(10b) $\quad (2ik)^2 g(-4k^2)\left[R^{(1)}(k) + R^{(2)}(k)\right]$

$\qquad = \displaystyle\int_{-\infty}^{+\infty} dx \, \psi^{(1)}(x,k) \psi^{(2)}(x,k) g(\Lambda) \cdot \Gamma \cdot 1.$

Thus in these equations $g(z)$ is an arbitrary polynomial or even an entire function; the two functions $u^{(j)}(x)$, $j = 1, 2$, are the "potentials" in the Schroedinger equations (1), while $R^{(j)}(k)$ and $\psi^{(j)}(x,k)$ are, respectively, the corresponding reflection coefficients, and wave functions (namely, the solutions of the Schroedinger equations (1) identified by the asymptotic boundary conditions (1a) and (1b)); and the integrodifferential operators Λ and Γ are defined by (5) and (6).

An important feature of these formulae is the fact that, in the r.h.s., the only dependence on the spectral variable k appears in the wave functions

$\psi^{(j)}$. Because of this, these formulae provide a convenient tool to correlate certain changes in the potential (or equivalently certain relations between two different potentials) with the corresponding changes of the reflection coefficients. For instance they imply that, if two functions $u^{(1)}(x)$ and $u^{(2)}(x)$ are related by the (nonlinear integrodifferential) equation

$$(10c) \quad g(\Lambda)[u^{(1)}(x) - u^{(2)}(x)] + h(\Lambda) \cdot \Gamma \cdot 1 = 0,$$

where g and h are arbitrary entire functions, the corresponding reflection coefficients $R^{(1)}(k)$ and $R^{(2)}(k)$ are related by the formula

$$(10d) \quad g(-4k^2)[R^{(1)}(k) - R^{(2)}(k)] + 2ikh(-4k^2)[R^{(1)}(k) + R^{(2)}(k)] = 0,$$

or, equivalently,

$$(10e) \quad R^{(1)}(k) = R^{(2)}(k) \frac{[g(-4k^2) - 2ikh(-4k^2)]}{[g(-4k^2) + 2ikh(-4k^2)]}.$$

Note that we have in this manner identified a class of nonlinear integrodifferential relations linking $u^{(1)}(x)$ and $u^{(2)}(x)$ (namely (10c), containing the largely arbitrary functions g and h), such that the relations between the corresponding reflection coefficients take the very simple form (10e). Some implications of these results will be discussed in chapter 4 and appendix A.17.

The formulae (10) relate a *finite* change in the function u (from $u = u^{(2)}(x)$ to $u = u^{(1)}(x)$) to the corresponding change in the reflection coefficient R (from $R = R^{(2)}(k)$ to $R = R^{(1)}(k)$). A limiting case, that will play an important rôle in the following, is that of an *infinitesimal* change, that obtains setting

$$(11a) \quad u^{(2)}(x) = u(x), \quad u^{(1)}(x) = u(x) + \delta u(x)$$

and correspondingly

$$(11b) \quad R^{(2)}(k) = R(k), \quad R^{(1)}(k) = R(k) + \delta R(k),$$
$$(11c) \quad \psi^{(2)}(x, k) = \psi(x, k), \quad \psi^{(1)}(x, k) = \psi(x, k) + \delta \psi(x, k).$$

This yields

(12a) $$2ikg(-4k^2)\delta R(k) = \int_{-\infty}^{+\infty} dx\, [\psi(x,k)]^2 g(L)\delta u(x),$$

(12b) $$(2ik)^2 g(-4k^2) R(k) = \int_{-\infty}^{+\infty} dx\, [\psi(x,k)]^2 g(L) u_x(x).$$

Clearly in these formulae $\psi(x,k)$ is the solution of the Schroedinger equation (2.1.-1) characterized by the asymptotic boundary conditions (2.1.-3), while the integrodifferential operator L is the form that the operator Λ takes for $u^{(1)}(x) = u^{(2)}(x) = u(x)$. This operator plays a fundamental rôle. Its explicit definition is given by the formula (see (1.2.-2))

(13) $$Lf(x) = f_{xx}(x) - 4u(x)f(x) + 2u_x(x) \int_x^{+\infty} dy\, f(y),$$

that details its action on a generic function $f(x)$ (vanishing as $x \to \infty$ so that the integral converges). Of course it satisfies the formula

(14) $$\int_{-\infty}^{+\infty} dx\, [\psi(x,k)]^2 Lf(x) = -4k^2 \int_{-\infty}^{+\infty} dx\, [\psi(x,k)]^2 f(x),$$

that is merely the appropriate special case of (7). Here of course, as in (7), $f(x)$ is required to satisfy (8a), but is otherwise largely arbitrary. Thus L may be considered to be the adjoint of the operator L^T having the squared wave function $[\psi(x,k)]^2$ as eigenfunction, with eigenvalue $-4k^2$:

(15) $$L^T [\psi(x,k)]^2 = -4k^2 [\psi(x,k)]^2$$

(we do not display the explicit expression of this operator, since we never use it; for definitions and discussions of the closely related operator $\tilde{L}$, see subsection 2.4.2 and appendix A.9).

Note that (12) are simply the first nontrivial relations that obtain from (10) in the limit of vanishing $\delta u(x)$, $\delta R(k)$ and $\delta\psi(x,k)$; but the additional relations that could be obtained from (10) by equating differentials of higher order are less useful, since they contain $\delta\psi(x,k)$ in addition to $\delta u(x)$ and $\delta R(k)$.

Another special case of (10) obtains for $u^{(2)}(x) \equiv 0$, implying of course $R^{(2)}(k) \equiv 0$ and $\psi^{(2)}(x,k) = \exp(-ikx)$ (see 2.1.-3). They read

(16a) $$2ikg(-4k^2) R(k) = \int_{-\infty}^{+\infty} dx\, \psi(x,k) \exp(-ikx) g(\Lambda^{(0)}) u(x),$$

(16b) $$(2ik)^2 g(-4k^2) R(k) = \int_{-\infty}^{+\infty} dx\, \psi(x,k) \exp(-ikx) g(\Lambda^{(0)})$$
$$\cdot \left[u_x(x) + u(x) \int_x^{+\infty} dy\, u(y) \right],$$

the integrodifferential operator $\Lambda^{(0)}$ being now defined by the formula

(17) $$\Lambda^{(0)}f(x) = f_{xx}(x) - 2u(x)f(x) + u_x(x)\int_x^{+\infty} dy f(y)$$
$$+ u(x)\int_x^{+\infty} dy\, u(y)\int_y^{+\infty} dz\, f(z),$$

corresponding of course to (6) with $u^{(2)}(x)=0$ and $u^{(1)}(x)=u(x)$. These formulae relate the single potential $u(x)$ to its reflection coefficient $R(k)$; the function $\psi(x,k)$ in the r.h.s. of (16) is again the solution of the Schroedinger equation (2.1.-1) identified by the asymptotic boundary conditions (2.1.-3).

Of course, more general expressions of this sort obtain by inserting in (10a) and (10b) a potential $u^{(2)}(x)$ having a vanishing reflection coefficient, $R^{(2)}(k)=0$, and N discrete eigenvalues characterized by the $2N$ parameters p_n and ρ_n, $n=1,2,\ldots,N$. These formulae are not displayed here; suffice it to note that, in this case, $\psi^{(2)}(x,k) = f^{(+)}(x,-k)$, with $f^{(+)}(x,k)$ given by (2.2-4), while $u^{(2)}(x)$ has the explicit expression (2.2-20). It is remarkable that the $2N$ parameters p_n and ρ_n appear only in the r.h.s. of the formulae obtained in this manner.

The formulae we have written thus far involve only the reflection coefficient $R(k)$. Analogous formulae involving also the transmission coefficient $T(k)$ can be obtained in an analogous manner (the modification vis-à-vis the treatment discussed above should be clear from the very structure of these equations; for details, see appendix A.6). They read:

(18) $$2ikg(-4k^2)\left[R^{(1)}(k)R^{(2)}(-k) + T^{(1)}(k)T^{(2)}(-k) - 1\right]$$
$$+ 4k^2 T^{(1)}(k)T^{(2)}(-k)\int_{-\infty}^{+\infty} dx\, g^{\#}(-4k^2, \Lambda)\left[u^{(1)}(x) - u^{(2)}(x)\right]$$
$$+ 2ikT^{(1)}(k)T^{(2)}(-k)\int_{-\infty}^{+\infty} dy\left[u^{(1)}(y) - u^{(2)}(y)\right]$$
$$\cdot \int_y^{+\infty} dx\, g^{\#}(-4k^2, \Lambda)\left[u^{(1)}(x) - u^{(2)}(x)\right]$$
$$= \int_{-\infty}^{+\infty} dx\, \psi^{(1)}(x,k)\psi^{(2)}(x,-k)g(\Lambda)\left[u^{(1)}(x) - u^{(2)}(x)\right],$$

(19) $(2ik)^2 g(-4k^2)\bigl[R^{(1)}(k)R^{(2)}(-k) - T^{(1)}(k)T^{(2)}(-k) + 1\bigr]$
$+ (2ik)g(-4k^2)T^{(1)}(k)T^{(2)}(-k)\int_{-\infty}^{+\infty} dx\bigl[u^{(1)}(x) - u^{(2)}(x)\bigr]$
$+ 4k^2 T^{(1)}(k)T^{(2)}(-k)\int_{-\infty}^{+\infty} dx\, g^{\#}(-4k^2, \Lambda)\Gamma\cdot 1$
$+ 2ik T^{(1)}(k)T^{(2)}(-k)\int_{-\infty}^{+\infty} dy\bigl[u^{(1)}(y) - u^{(2)}(y)\bigr]$
$\cdot \int_{y}^{+\infty} dx\, g^{\#}(-4k^2, \Lambda)\Gamma\cdot 1$
$= \int_{-\infty}^{+\infty} dx\, \psi^{(1)}(x,k)\psi^{(2)}(x,-k) g(\Lambda)\Gamma\cdot 1.$

Most of the symbols in these equations have been defined above (including the integrodifferential operators Λ and Γ); additionally we have introduced the short-hand notation

(20) $\quad g^{\#}(z_1, z_2) \equiv [g(z_1) - g(z_2)]/(z_1 - z_2),$

that will be frequently used in the following.

Note that all the formulae given above remain valid even if $u^{(j)}(x)$ is not real. Of course, if $u^{(2)}(x)$ is real, than $T^{(2)}(-k)$, $R^{(2)}(-k)$ and $\psi^{(2)}(x,-k)$ can be replaced by $[T^{(2)}(k)]^*$, $[R^{(2)}(k)]^*$ and $[\psi^{(2)}(x,k)]^*$, as implied by (2.1.-12) and (2.1.-17). Moreover, even if $u^{(2)}(x)$ is not real, the formulae remain true after these replacement have been performed, provided $u^{(2)}(x)$ is also replaced by $[u^{(2)}(x)]^*$. Analogous remarks apply to other formulae reported below; we will not repeat them.

In these formulae, as in (10), $g(z)$ is an arbitrary entire function. In the special case $g(z)=1$, these formulae simplify, since the expression (20) vanishes:

(21) $2ik\bigl[R^{(1)}(k)R^{(2)}(-k) + T^{(1)}(k)T^{(2)}(-k) - 1\bigr]$
$= \int_{-\infty}^{+\infty} dx\, \psi^{(1)}(x,k)\psi^{(2)}(x,-k)\bigl[u^{(1)}(x) - u^{(2)}(x)\bigr],$

(22) $(2ik)^2\bigl[R^{(1)}(k)R^{(2)}(-k) - T^{(1)}(k)T^{(2)}(-k) + 1\bigr]$
$+ 2ik T^{(1)}(k)T^{(2)}(-k)\int_{-\infty}^{+\infty} dx\bigl[u^{(1)}(x) - u^{(2)}(x)\bigr]$
$= \int_{-\infty}^{+\infty} dx\, \psi^{(1)}(x,k)\psi^{(2)}(x,-k)\Gamma\cdot 1.$

These equations are clearly the analogs of (3) and (4).

2.4.1 Wronskian integral relations

It is again useful, as above, to consider also the case in which the two functions $u^{(j)}(x)$, and therefore the corresponding spectral quantities, differ only infinitesimally. In addition to the formulae (11) given above, one writes then

(23) $\qquad T^{(2)}(k)=T(k), \qquad T^{(1)}(k)=T(k)+\delta T(k).$

The formula (21) yields then, first of all

(24) $\qquad R(k)R(-k)+T(k)T(-k)=1,$

which is just the unitarity relation (2.1.-18). Then one gets from (18) and (19) the formulae

(25) $\qquad 2ikg(-4k^2)[\delta R(k)R(-k)+\delta T(k)T(-k)]$
$$+4k^2 T(k)T(-k)\int_{-\infty}^{+\infty} dx\, g^{\#}(-4k^2, L)\,\delta u(x)$$
$$=\int_{-\infty}^{+\infty} dx\, \psi(x,k)\psi(x,-k)g(L)\,\delta u(x),$$

(26) $\qquad (2ik)^2 g(-4k^2)R(k)R(-k)$
$$=\int_{-\infty}^{+\infty} dx\, \psi(x,k)\psi(x,-k)g(L)u_x(x),$$

that are clearly the analogs of (12) (indeed the remarks given after (12) apply also to these equations). Of course the symbol $g^{\#}(z_1, z_2)$ is defined by (20).

To derive (26) from (19) we have used, in addition to (24), the important property described by the formula

(27) $\qquad L^m u_x(x)=g_x^{(m)}(x), \quad m=0,1,2,\ldots,$

where $g^{(m)}$ is a polynomial of degree $m+1$ in $u(x)$ and its derivatives. For instance (see (1.2.-3))

(28a) $\qquad g^{(0)}(x)=u(x),$

(28b) $\qquad g^{(1)}(x)=u_{xx}(x)-3[u(x)]^2,$

(28c) $\qquad g^{(2)}(x)=u_{xxxx}(x)-10u_{xx}(x)u(x)-5[u_x(x)]^2-10[u(x)]^3.$

Indeed this property clearly implies

(29) $\qquad \int_{-\infty}^{+\infty} dx\, g(L)u_x(x)=0$

for any (entire) function $g(z)$, since we assume $u(x)$ to vanish asymptotically with all its derivatives, and therefore all the "currents" $g^{(m)}(x)$ also vanish asymptotically: $g^{(m)}(\pm\infty)=0$.

Let us emphasize that (27) is highly nontrivial since L is an integrodifferential operator (see (13)). Also this formula plays an important rôle (for its proof, see appendices A.8 and A.9).

We also report the analogs of (16), that follow trivially from (18) and (19) (of course, if $u^{(2)}(x)=0$, $T^{(2)}(k)=1$):

$$(30) \quad 2ikg(-4k^2)[T(k)-1]+4k^2T(k)\int_{-\infty}^{+\infty}dx\, g^{\#}(-4k^2,\Lambda^{(0)})u(x)$$

$$+2ikT(k)\int_{-\infty}^{+\infty}dy\,u(y)\int_{y}^{+\infty}dx\, g^{\#}(-4k^2,\Lambda^{(0)})u(x)$$

$$=\int_{-\infty}^{+\infty}dx\,\psi(x,k)\exp(ikx)g(\Lambda^{(0)})u(x),$$

$$(31) \quad -(2ik)^2g(-4k^2)[T(k)-1]+2ikg(-4k^2)T(k)\int_{-\infty}^{+\infty}dx\,u(x)$$

$$+4k^2T(k)\int_{-\infty}^{+\infty}dx\,g^{\#}(-4k^2,\Lambda^{(0)})\left[u_x(x)+u(x)\int_{x}^{+\infty}dy\,u(y)\right]$$

$$+2ikT(k)\int_{-\infty}^{+\infty}dy\,u(y)\int_{y}^{+\infty}dx\,g^{\#}(-4k^2,\Lambda^{(0)})$$

$$\cdot\left[u_x(x)+u(x)\int_{x}^{+\infty}dz\,u(z)\right]$$

$$=\int_{-\infty}^{+\infty}dx\,\psi(x,k)\exp(ikx)g(\Lambda^{(0)})$$

$$\cdot\left[u_x(x)+u(x)\int_{x}^{+\infty}dy\,u(y)\right],$$

where of course $\Lambda^{(0)}$ is defined by (17). We also display the simpler forms that these formulae take for $g(z)=1$:

$$(32) \quad 2ik[T(k)-1]=\int_{-\infty}^{+\infty}dx\,\psi(x,k)\exp(ikx)u(x),$$

$$(33) \quad -(2ik)^2[T(k)-1]+2ikT(k)\int_{-\infty}^{+\infty}dx\,u(x)$$

$$=\int_{-\infty}^{+\infty}dx\,\psi(x,k)\exp(ikx)\left[u_x(x)+u(x)\int_{x}^{+\infty}dy\,u(y)\right].$$

2.4.1 *Wronskian integral relations* 99

Again, more general analogous formulae are obtained by inserting in (18) and (19) $R^{(2)}(k)=0$ with $u^{(2)}(x)$ given by (2.2.-20), $T^{(2)}(k)$ given by (2.2.-23) and $\psi^{(2)}(x,-k)=f^{(+)}(x,k)$ given by (2.2.-4).

All the formulae we have written so far in this subsection refer to the continuous part of the spectrum. We report now analogous formulae for the discrete spectrum.

There are several types of such formulae. The simplest read

(34) $\quad -2p_n g(4p_n^2)c_n^{(1)} = \int_{-\infty}^{+\infty} dx \, \varphi_n^{(1)}(x)\psi^{(2)}(x, ip_n)g(\Lambda)$
$$\cdot [u^{(1)}(x) - u^{(2)}(x)],$$

(35) $\quad 4p_n^2 g(4p_n^2)c_n^{(1)} = \int_{-\infty}^{+\infty} dx \, \varphi_n^{(1)}(x)\psi^{(2)}(x, ip_n)g(\Lambda)\Gamma \cdot 1,$

and are valid provided $E_n = -p_n^2$ is a discrete eigenvalue for the potential $u^{(1)}(x)$ but *not* for the potential $u^{(2)}(x)$. The notation used here coincides with that introduced in section 2.1 (except for the superscripts, that distinguish the quantities associated with the two potentials $u^{(1)}(x)$ and $u^{(2)}(x)$; see in particular (2.1.-24), (2.1.-25) and (2.1.-3)). Note that the last of these formulae implies that $\psi^{(2)}(x, ip_n)$ vanishes proportionally to $\exp(p_n x)$ as $x \to -\infty$ and diverges precisely as $\exp(p_n x)$ as $x \to +\infty$ (it should be emphasized that, here and throughout, $p_n > 0$).

For future reference, we also rewrite here the last two equations in a more symmetrical fashion, taking advantage of the invariance of the integrodifferential operators Γ and Λ under the interchange of $u^{(1)}$ and $u^{(2)}$:

(34a) $\quad -2\left\{ p_n^{(1)} g\left[4(p_n^{(1)})^2\right]c_n^{(1)} - p_m^{(2)} g\left[4(p_m^{(2)})^2\right]c_m^{(2)} \right\}$
$$= \int_{-\infty}^{+\infty} dx \left[\varphi_n^{(1)}(x)\psi^{(2)}(x, ip_n^{(1)}) + \varphi_m^{(2)}(x)\psi^{(1)}(x, ip_m^{(2)}) \right]$$
$$\cdot g(\Lambda)[u^{(1)}(x) - u^{(2)}(x)],$$

(35a) $\quad 4\left\{ (p_n^{(1)})^2 g\left[4(p_n^{(1)})^2\right]c_n^{(1)} + (p_m^{(2)})^2 g\left[4(p_m^{(2)})^2\right]c_m^{(2)} \right\}$
$$= \int_{-\infty}^{+\infty} dx \left[\varphi_n^{(1)}(x)\psi^{(2)}(x, ip_n^{(1)}) \right.$$
$$\left. + \varphi_m^{(2)}(x)\psi^{(1)}(x, ip_m^{(2)}) \right] g(\Lambda)\Gamma \cdot 1.$$

We trust that all symbols appearing in these formulae have a self-evident significance; let us emphasize that their validity is contingent on $E_n^{(1)} = -(p_n^{(1)})^2$ being a discrete eigenvalue of the potential $u^{(1)}(x)$ but not of the potential $u^{(2)}(x)$, and $E_m^{(2)} = -(p_m^{(2)})^2$ being an eigenvalue of $u^{(2)}(x)$ but not of $u^{(1)}(x)$.

We also display the special case of (34) and (35) corresponding to $u^{(2)}(x) = 0$, $u^{(1)}(x) = u(x)$. Then, again with an obvious significance of all symbols,

$$(36) \quad -2p_n g(4p_n^2) c_n = \int_{-\infty}^{+\infty} dx\, \varphi_n(x) \exp(p_n x) g(\Lambda^{(0)}) u(x),$$

$$(37) \quad 4p_n^2 g(4p_n^2) c_n = \int_{-\infty}^{+\infty} dx\, \varphi_n(x) \exp(p_n x) g(\Lambda^{(0)})$$
$$\cdot \left[u_x(x) + u(x) \int_x^{+\infty} dy\, u(y) \right].$$

A second set of formulae reads:

$$(38) \quad \int_{-\infty}^{+\infty} dx\, \varphi^{(1)}(x) \varphi^{(2)}(x) \left[u^{(1)}(x) - u^{(2)}(x) - (E^{(1)} - E^{(2)}) \right] = 0,$$

$$(39) \quad \int_{-\infty}^{+\infty} dx\, \varphi^{(1)}(x) \varphi^{(2)}(x)$$
$$\cdot \left[u_x^{(1)}(x) + u_x^{(2)}(x) + \left[u^{(1)}(x) - u^{(2)}(x) - (E^{(1)} - E^{(2)}) \right] \right.$$
$$\left. \cdot \left\{ (E^{(1)} - E^{(2)})(x - x_1) + \int_x^{x_2} dy \left[u^{(1)}(y) - u^{(2)}(y) \right] \right\} \right] = 0,$$

$$(40) \quad \int_{-\infty}^{+\infty} dx\, \varphi^{(1)}(x) \varphi^{(2)}(x) \left\{ f_{xx}(x) + 2 \left[E^{(1)} + E^{(2)} - u^{(1)}(x) - u^{(2)}(x) \right] \right.$$
$$\cdot f(x) + \left[u_x^{(1)}(x) + u_x^{(2)}(x) \right] \int_x^{x_3} dy\, f(y)$$
$$+ \left[u^{(1)}(x) - u^{(2)}(x) - (E^{(1)} - E^{(2)}) \right]$$
$$\left. \cdot \int_x^{x_4} dy \left[u^{(1)}(y) - u^{(2)}(y) - (E^{(1)} - E^{(2)}) \right] \int_y^{x_3} dz\, f(z) \right\} = 0.$$

In these equations $E^{(1)}$ and $E^{(2)}$ are discrete eigenvalues corresponding to the potentials $u^{(1)}(x)$ and $u^{(2)}(x)$, $\varphi^{(1)}(x)$ and $\varphi^{(2)}(x)$ are the corresponding eigenfunctions, and the quantities x_1, x_2, x_3, x_4 are arbitrary (real) constants; in the last equation, $f(x)$ is an arbitrary function (not even required to vanish asymptotically). Actually these equations remain valid even if only one of the two quantities $E^{(j)}$ is a discrete eigenvalue, say if

(41) $\quad E^{(1)} = -p^2, \quad p > 0,$

is a discrete eigenvalue of the potential $u^{(1)}(x)$, provided

(42) $\quad E^{(2)} \equiv k^2$

satisfies the condition

(43) $\quad |\operatorname{Im} k| < p$

(which is of course always satisfied if k is real); then the eigenfunction $\varphi^{(2)}(x)$ should be replaced by $\psi^{(2)}(x, k)$ (note that, in the equations (38–40), the normalization of the wave function is irrelevant).

These equations are the analogs of (3), (4) and (7); and clearly, taking advantage of the arbitrariness of the function $f(x)$ in (40), we may obtain many other relations from (38) and (39), including some analogs of (10), and of (16). We will display some such equations only in the less general cases considered immediately below. But before doing this we remark that (38)–(40) remain valid even in the case of potentials that diverge asymptotically to positive infinity, and have therefore a purely discrete spectrum; indeed it was largely for this reason that we have chosen to display them here in this explicit form.

An important special case of (38)–(40) is obtained if

(44) $\quad E^{(1)} = E^{(2)} = E = -p^2,$

namely if the two potentials $u^{(1)}(x)$ and $u^{(2)}(x)$ have the *same* discrete eigenvalue. The corresponding equations read

(45) $\quad \int_{-\infty}^{+\infty} dx\, \varphi^{(1)}(x) \varphi^{(2)}(x) [u^{(1)}(x) - u^{(2)}(x)] = 0,$

(46) $\quad \int_{-\infty}^{+\infty} dx\, \varphi^{(1)}(x) \varphi^{(2)}(x) \Gamma \cdot 1 = 0,$

(47a) $\quad \int_{-\infty}^{+\infty} dx\, \varphi^{(1)}(x) \varphi^{(2)}(x) \Lambda f(x) = 4p^2 \int_{-\infty}^{+\infty} dx\, \varphi^{(1)}(x) \varphi^{(2)}(x) f(x),$

with Γ and Λ defined by (5) and (6) (to get the last two equations we have set $x_2 = x_3 = x_4 = +\infty$ in (39) and (40)). As above, these equations may be read to imply that Λ is the adjoint of the operator $\tilde{\Lambda}$ (see (9)) having the product $\varphi^{(1)}(x)\varphi^{(2)}(x)$ as eigenfunction and $4p^2 = -4E$ as eigenvalue (recall that $\varphi^{(1)}(x)$ and $\varphi^{(2)}(x)$ are by hypothesis here two eigenfunctions of the Schroedinger equations (1) corresponding to the *same* discrete eigenvalue $E=-4p^2$). It implies the more general formula

$$(47b) \quad \int_{-\infty}^{+\infty} dx\, \varphi^{(1)}(x)\varphi^{(2)}(x)g(\Lambda)f(x)$$
$$= g(4p^2)\int_{-\infty}^{+\infty} dx\, \varphi^{(1)}(x)\varphi^{(2)}(x)f(x),$$

with $g(z)$ an arbitrary entire function (this equation is, for the special case under present consideration, the analog of (9)); and this formula allows one to extend (45) and (46), yielding the formulae

$$(48) \quad \int_{-\infty}^{+\infty} dx\, \varphi^{(1)}(x)\varphi^{(2)}(x)g(\Lambda)\left[u^{(1)}(x)-u^{(2)}(x)\right]=0,$$

$$(49) \quad \int_{-\infty}^{+\infty} dx\, \varphi^{(1)}(x)\varphi^{(2)}(x)g(\Lambda)\Gamma\cdot 1=0,$$

which are the analogs of (10). Let us re-emphasize that in these equations $g(z)$ is an arbitrary entire function.

Another set of formulae obtains from (38)–(40) if one considers the case of infinitesimal change. Setting

$$(50a) \quad u^{(2)}(x)=u(x), \qquad u^{(1)}(x)=u(x)+\delta u(x)$$

and correspondingly

$$(50b) \quad E^{(2)}=E_n=-p_n^2, \qquad E^{(1)}=E_n+\delta E_n=-p_n^2-2p_n\,\delta p_n,$$
$$(50c) \quad \varphi^{(2)}(x)=\varphi_n(x), \qquad \varphi^{(1)}(x)=\varphi_n(x)+\delta\varphi_n(x),$$

where of course E_n is one of the discrete eigenvalues of the potential $u(x)$, and $\varphi_n(x)$ is the corresponding eigenfunction, one gets

$$(51) \quad \delta E_n=-2p_n\,\delta p_n=\int_{-\infty}^{+\infty} dx\,[\varphi_n(x)]^2\,\delta u(x),$$

$$(52) \quad 0=\int_{-\infty}^{+\infty} dx\,[\varphi_n(x)]^2 u_x(x),$$

$$(53) \quad \int_{-\infty}^{+\infty} dx\,[\varphi_n(x)]^2 Lf(x)=4p_n^2\int_{-\infty}^{+\infty} dx\,[\varphi_n(x)]^2 f(x).$$

To obtain the first of these formulae we have used the normalization condition (2.1.-21); to obtain the last, the choice $x_3=x_4=+\infty$ has been made in (40). As above, the last of these equations may be read to imply that L is the adjoint of the operator $\tilde{L}$ having $[\varphi_n(x)]^2$ as eigenfunction, with eigenvalue $4p_n^2 = -4E_n$. It implies the more general formula

(54) $$\int_{-\infty}^{+\infty} dx\, [\varphi_n(x)]^2 g(L) f(x) = g(4p_n^2) \int_{-\infty}^{+\infty} dx\, [\varphi_n(x)]^2 f(x),$$

with $g(z)$ an arbitrary entire function (and L defined by (14)); and this formula allows one to generalize (51) and (52) to

(55) $$-2p_n g(4p_n^2)\, \delta p_n = \int_{-\infty}^{+\infty} dx\, [\varphi_n(x)]^2 g(L)\, \delta u(x),$$

(56) $$0 = \int_{-\infty}^{+\infty} dx\, [\varphi_n(x)]^2 g(L) u_x(x).$$

These formulae are the analogs of (12); they will play an important rôle in the following.

A third set of formulae for the discrete spectrum quantities read as follows:

(57) $$-2p_n[\rho_n^{(1)} - \rho_n^{(2)}]/[\rho_n^{(1)}\rho_n^{(2)}]^{1/2} = \int_{-\infty}^{+\infty} dx\, [\varphi^{(1)}\varphi^{(2)}]_p \cdot [u^{(1)}(x) - u^{(2)}(x)],$$

(58) $$4p_n^2[\rho_n^{(1)} + \rho_n^{(2)}]/[\rho_n^{(1)}\rho_n^{(2)}]^{1/2} = \int_{-\infty}^{+\infty} dx\, [\varphi^{(1)}\varphi^{(2)}]_p \Gamma \cdot 1,$$

(59) $$\int_{-\infty}^{+\infty} dx\, [\varphi^{(1)}\varphi^{(2)}]_p \Lambda f(x) = \int_{-\infty}^{+\infty} dx\, [4p^2 \varphi^{(1)}\varphi^{(2)}]_p f(x).$$

They also refer to the special case in which the two potentials $u^{(1)}(x)$ and $u^{(2)}(x)$ have the *same* discrete eigenvalue $E_n = -p_n^2$; $\rho_n^{(j)}$ is the normalization constant of this discrete eigenvalue, for the potential $u^{(j)}(x)$ (see (2.1.-27)); Γ and Λ are the by now familiar integrodifferential operators (defined by (5) and (6)); and the symbol $[\varphi^{(1)}\varphi^{(2)}]_p$ is a shorthand notation:

(60) $$[\varphi^{(1)}\varphi^{(2)}]_p \equiv \{\partial/\partial p [\varphi^{(1)}(x,p) \varphi^{(2)}(x,p)]\}\big|_{p=p_n},$$

with

(61) $$\varphi(x,p) = c_n^{(-)} f^{(-)}(x, ip)$$

so that

(62) $\quad \varphi(x, p_n) = \varphi_n(x)$

(see (2.1.-28); and note that in the last two equations we have omitted, for notational simplicity, to indicate the superscript that labels the different potentials). In (59) the function $f(x)$ is arbitrary, except for the requirement that it vanish asymptotically with its first derivative (see (8a)) and that it have the usual regularity properties; and the symbol $[4p^2\varphi^{(1)}\varphi^{(2)}]_p$, consistently with (60), is defined by

(63) $\quad [4p^2\varphi^{(1)}\varphi^{(2)}]_p \equiv \{\partial/\partial p[4p^2\varphi^{(1)}(x,p)\varphi^{(2)}(x,p)]\}|_{p=p_n}$,

with $\varphi^{(j)}(x, p)$ defined by (61).

Clearly (57), (58) and (59) are the analogs, for the case under consideration, of (3), (4) and (7). Note moreover that (48) and (49) imply that (59) can be replaced by

(64) $\quad \int_{-\infty}^{+\infty} dx [\varphi^{(1)}\varphi^{(2)}]_p \Lambda f(x) = 4p_n^2 \int_{-\infty}^{+\infty} dx [\varphi^{(1)}\varphi^{(2)}]_p f(x)$

provided

(65) $\quad f(x) = g(\Lambda)[u^{(1)}(x) - u^{(2)}(x)] + h(\Lambda)\Gamma \cdot 1$,

$g(z)$ and $h(z)$ being arbitrary entire functions. Therefore for all functions f having the form (65) it is also permissible to generalize (64) to

(66) $\quad \int_{-\infty}^{+\infty} dx [\varphi^{(1)}\varphi^{(2)}]_p g(\Lambda) f(x) = g(4p_n^2) \int_{-\infty}^{+\infty} dx [\varphi^{(1)}\varphi^{(2)}]_p f(x)$,

where $g(z)$ indicates again an arbitrary entire function; and this implies the possibility to generalize (57) and (58) to

(67) $\quad -2p_n g(4p_n^2)[\rho_n^{(1)} - \rho_n^{(2)}]/[\rho_n^{(1)}\rho_n^{(2)}]^{1/2}$
$\quad = \int_{-\infty}^{+\infty} dx [\varphi^{(1)}\varphi^{(2)}]_p g(\Lambda)[u^{(1)}(x) - u^{(2)}(x)]$,

(68) $\quad 4p_n^2 g(4p_n^2)[\rho_n^{(1)} + \rho_n^{(2)}]/[\rho_n^{(1)}\rho_n^{(2)}]^{1/2}$
$\quad = \int_{-\infty}^{+\infty} dx [\varphi^{(1)}\varphi^{(2)}]_p g(\Lambda)\Gamma \cdot 1$.

These equations are the analogs of (10). We repeat that they hold for an arbitrary entire function $g(z)$, under the assumption that the two potentials $u^{(1)}(x)$ and $u^{(2)}(x)$ have the same discrete eigenvalue $E_n = -p_n^2$ (with the, generally different, normalization constants $\rho_n^{(1)}$ and $\rho_n^{(2)}$; and with the symbol $[\varphi^{(1)}\varphi^{(2)}]_p$ defined by (60)).

Again the case in which the two potentials differ only infinitesimally deserves special mention. The relevant formulae read

(69) $\quad -p_n g(4p_n^2) \, \delta\rho_n/\rho_n = \int_{-\infty}^{+\infty} dx \, \varphi_n(x) \varphi_p(x, p_n) g(L) \, \delta u(x),$

(70) $\quad 2p_n^2 g(4p_n^2) = \int_{-\infty}^{+\infty} dx \, \varphi_n(x) \varphi_p(x, p_n) g(L) u_x(x),$

where of course

(71) $\quad \varphi_p(x, p_n) = |\{\partial/\partial p [\varphi(x, p)]\}|_{p=p_n}$

with $\varphi(x, p)$ defined by (61), and all the other symbols have an obvious significance. These formulae are the analogs of (12a) and (12b).

We terminate at this point our analysis of generalized Wronskian relations. The main results are displayed as formulae (10), (12), (25) and (26), (29), (48) and (49), (55) and (56), (57) and (58), (69) and (70); in addition to (5), (6) and (13), which define the integrodifferential operators Γ, Λ and L. It should be emphasized that all these formulae are (rather elementary) consequences of the Schroedinger spectral problem. Due to the flexibility implied by the presence of the arbitrary function $g(z)$, they provide a powerful tool to investigate this spectral problem and to display its implications, as shown in the following.

2.4.1.1. Additional wronskian integral relations

In the preceding subsection 2.4.1 we have reported several relations involving *two* functions ("potentials") $u^{(j)}(x)$, and the corresponding spectral quantities; for instance (2.4.1-10, 18, 19, 38, 39, 40, 48, 49, 67, 68). We have also given several relations for the quantities associated with a *single* potential, including its infinitesimal variation; for instance (2.4.1-12, 16, 25, 26, 30, 31, 55, 56, 69, 70). In this subsection we report some other analogous formulae involving a *single* potential, that follow quite directly from the relations of the preceding section involving *two* potentials and from the formulae of subsection 2.1.1 that relate the changes of the potential induced by space translations and scale transformations to the associated changes of the corresponding spectral quantities.

Thus (2.4.1.-10) and (2.1.1.-1) imply

(1) $\quad -4k\sin(ka)\,g(-4k^2)R(k)$
$$= \int_{-\infty}^{+\infty} dx\,\psi(x,k)\psi(x+a,k)g(\Lambda)[u(x+a)-u(x)],$$

(2) $\quad -8k^2\cos(ka)\,g(-4k^2)R(k)$
$$= \int_{-\infty}^{+\infty} dx\,\psi(x,k)\psi(x+a,k)g(\Lambda)\Gamma\cdot 1,$$

where clearly the operators Λ and Γ are defined by (2.4.1.-6) and (2.4.1.-5) with

(3) $\quad u^{(2)}(x)=u(x), \qquad u^{(1)}(x)=u(x+a).$

Similarly, from (2.4.1.-10) and (2.1.1.-3) there follows

(4) $\quad 2ikg(-4k^2)[R(k/\mu)-R(k)]$
$$= \int_{-\infty}^{+\infty} dx\,\psi(x,k)\psi(\mu x,k/\mu)g(\Lambda)[\mu^2 u(\mu x)-u(x)],$$

(5) $\quad (2ik)^2 g(-4k^2)[R(k/\mu)+R(k)]$
$$= \int_{-\infty}^{+\infty} dx\,\psi(x,k)\psi(\mu x,k/\mu)g(\Lambda)\Gamma\cdot 1,$$

where of course the operators Λ and Γ are now defined by (2.4.1.-6) and (2.4.1.-5) with

(6) $\quad u^{(2)}(x)=u(x), \qquad u^{(1)}(x)=\mu^2 u(\mu x).$

Of course it would be easy to write a single formula that combines (1) and (4), or (2) and (5), by considering the combined effect of a translation and a scale transformation. We prefer here, and below, to write separate formulae, namely to trade off generality for the sake of simplicity.

It is of interest to consider the form that these results take for infinitesimal transformations, namely for $a=\varepsilon$ and $\mu=1+\varepsilon$ with $\varepsilon\to 0$ (see (2.1.1.-2) and (2.1.1.-4)). Then (1), (2) and (5) merely reproduce (2.4.1.-12b), while (4) yields the important novel formula

(7) $\quad -2ikg(-4k^2)kR_k(k)$
$$= \int_{-\infty}^{+\infty} dx\,[\psi(x,k)]^2 g(L)[xu_x(x)+2u(x)],$$

where of course the operator L is defined by (2.4.1.-13). Obviously this

2.4.1.1   Additional wronskian integral relations

formula could also be obtained as a special case of (2.4.1.-12a). Let us call attention to the differentiation with respect to k appearing in the l.h.s. of this equation.

Analogous equations obtain if we insert (3) or (6), and the corresponding expressions for the spectral quantities and the wave functions (see (2.1.1.-1) or (2.1.1.-3)), in (2.4.1.-18) and (2.4.1.-19). We do not report here the four formulae that obtain in this manner. As above, in the limit of infinitesimal transformations the first three of these equations merely reproduce an already known formula (namely, (2.4.1.-26)), while the last one (or, equivalently, (2.4.1.-25)) yields the important novel formula

$$(8) \quad -2ikg(-4k^2)[kR_k(k)R(-k)+kT_k(k)T(-k)]$$
$$+4k^2 T(k)T(-k)\int_{-\infty}^{+\infty} dx\, g^*(-4k^2, L)[xu_x(x)+2u(x)]$$
$$= \int_{-\infty}^{+\infty} dx\, \psi(x,k)\psi(x,-k)g(L)[xu_x(x)+2u(x)].$$

This formula, that contains the transmission coefficient in addition to the reflection coefficient, is the analog of the preceding one, in the same sense that (2.4.1.-25) is the analog of (2.4.1.-12a).

It is easy to obtain in a similar manner relations for the discrete spectrum parameters; for instance by inserting (3) or (6) in (2.4.1.-38, 39, 40) and by using (2.1.1.-1) or (2.1.1.-3). Again we do not report all the formulae that one can obtain in this manner, but only those that appear particularly remarkable. In particular the formulae

$$(9) \quad \int_{-\infty}^{+\infty} dx\, \varphi_n(x)\varphi_n(x+a)g(\Lambda)[u(x+a)-u(x)]=0,$$

$$(10) \quad \int_{-\infty}^{+\infty} dx\, \varphi_n(x)\varphi_n(x+a)g(\Lambda)\Gamma\cdot 1 = 0,$$

follow from (2.4.1.-48,49), (3) and (2.1.1.-1), and the formulae

$$(11) \quad 4p_n \sinh(p_n a)\, g(4p_n^2)$$
$$= \int_{-\infty}^{+\infty} dx\, [\varphi(x)\varphi(x+a)]_p g(\Lambda)[u(x+a)-u(x)],$$

$$(12) \quad 8p_n^2 \cosh(p_n a)\, g(4p_n^2)$$
$$= \int_{-\infty}^{+\infty} dx\, [\varphi(x)\varphi(x+a)]_p g(\Lambda)\Gamma\cdot 1$$

follow from (2.4.1.-67,68), (3) and (2.1.1.-1). Here of course the operators Γ

and Λ are defined, as in (1) and (2) above, by (2.4.1.-5,6) with (3); and the shorthand notation $[\varphi(x)\varphi(x+a)]_p$ is defined consistently with (2.4.1.-60):

(13) $\quad [\varphi(x)\varphi(x+a)]_p \equiv \{\partial/\partial p[\varphi(x,p)\varphi(x+a,p)]\}|_{p=p_n},$

with $\varphi(x, p)$ defined by (2.4.1.-61).

Note the consistency of (2.1.1.-1e) with the assumption (2.4.1.-44) required for the validity of (2.4.1.-48,49) and (2.4.1.-67,68); such consistency does not hold for (2.1.1.-3e), and therefore the scale transformations can only be combined with formulae, such as (2.4.1.-38, 39, 40), that refer to two potentials having different discrete eigenvalues. We report here only the novel formula that obtains in this manner for *infinitesimal* transformations:

(14) $\quad 2p_n^2 g(4p_n^2) = \int_{-\infty}^{+\infty} dx \, [\varphi_n(x)]^2 g(L)[xu_x(x) + 2u(x)];$

(an even simpler derivation of this equation starts from (2.4.1.-55)). We remark on the other hand that, in the limit of infinitesimal transformations, neither (9, 10) nor (11, 12) yield novel formulae, but reproduce instead (2.4.1.-56) or (2.4.1.-70).

We terminate here our analysis. Of the formulae displayed above, those that will later play a more important rôle are (7), (8) and (14).

2.4.2. Spectral integral relations

As in the preceding subsection, we consider the Schroedinger spectral problem for two different "potentials":

(1) $\quad -f_{xx}^{(j)}(x,k) + u^{(j)}(x) f^{(j)}(x,k) = k^2 f^{(j)}(x,k), \quad j=1,2.$

However, as our notation already suggests, in this subsection it is more convenient to focus on the Jost solutions of the Schroedinger equation (1), characterized by the boundary conditions

(1a) $\quad f^{(\pm)(j)}(x,k) \to \exp(\pm ikx), \quad x \to \pm\infty, \quad j=1,2,$

(1b) $\quad f^{(\pm)(j)}(x,k) \to [T^{(j)}(k)]^{-1} \exp(\pm ikx) \mp [R^{(j)}(\mp k)/T^{(j)}(\mp k)]$
$\quad \cdot \exp(\mp ikx), \quad x \to \mp\infty, \quad j=1,2.$

Note that (1a) coincides with (2.1.-4), except for the addition of the superscript j; as for (1b), it follows easily from (1a) and (2.1.-5). This

2.4.2 Spectral integral relations

(elementary) result, as well as all the others reported in this subsection, are proved in appendix A.7.

It is moreover convenient, in this subsection, to make ample use of the integrals of the functions $u^{(j)}(x)$, which are denoted as follows (consistently with the notation used throughout this book; see for instance (1.7.1.-12)):

(2a) $\quad w^{(j)}(x) = \int_x^{+\infty} dy\, u^{(j)}(y), \quad j=1,2,$

(2b) $\quad u^{(j)}(x) = -w_x^{(j)}(x), \quad j=1,2.$

Clearly this definition, together with the properties of the functions $u^{(j)}(x)$, to vanish asymptotically with their derivatives, imply

(3a) $\quad 0 = w^{(j)}(+\infty) = w_x^{(j)}(\pm\infty) = w_{xx}^{(j)}(\pm\infty) = \cdots, \quad j=1,2.$

On the other hand the quantities $w^{(j)}(-\infty)$ need not vanish:

(3b) $\quad w^{(j)}(-\infty) = \int_{-\infty}^{+\infty} dx\, u^{(j)}(x), \quad j=1,2.$

Two basic formulae read:

(4) $\quad w^{(1)}(x) - w^{(2)}(x) = -2 \sum_{n=1}^{N^{(1)}} f^{(+)(1)}(x, i p_n^{(1)}) f^{(+)(2)}(x, i p_n^{(1)}) \rho_n^{(1)}$

$\quad + 2 \sum_{n=1}^{N^{(2)}} f^{(+)(1)}(x, i p_n^{(2)}) f^{(+)(2)}(x, i p_n^{(2)}) \rho_n^{(2)}$

$\quad - \pi^{-1} \int_{-\infty}^{+\infty} dk\, f^{(+)(1)}(x,k) f^{(+)(2)}(x,k) [R^{(1)}(k) - R^{(2)}(k)],$

(5) $\quad w_x^{(1)}(x) + w_x^{(2)}(x) + \tfrac{1}{2}[w^{(1)}(x) - w^{(2)}(x)]^2$

$\quad = 4 \sum_{n=1}^{N^{(1)}} f^{(+)(1)}(x, i p_n^{(1)}) f^{(+)(2)}(x, i p_n^{(1)}) p_n^{(1)} \rho_n^{(1)}$

$\quad + 4 \sum_{n=1}^{N^{(2)}} f^{(+)(1)}(x, i p_n^{(2)}) f^{(+)(2)}(x, i p_n^{(2)}) p_n^{(2)} \rho_n^{(2)}$

$\quad - \pi^{-1} \int_{-\infty}^{+\infty} dk\, f^{(+)(1)}(x,k) f^{(+)(2)}(x,k)$

$\quad\quad \cdot 2ik[R^{(1)}(k) + R^{(2)}(k)].$

Here we are again using the notation of section 2.1, of course with the addition of the superscripts to denote the quantities corresponding to the two spectral problems. Note that $f^{(+)(j)}(x, i p_n^{(j)})$ is the eigenfunction corresponding to the discrete eigenvalue $-[p_n^{(j)}]^2$, so that

(6a) $\quad f^{(+)(j)}(x, i p_n^{(j)}) \to \exp(-p_n^{(j)} x) \to 0, \quad x \to +\infty,$

(6b) $\quad f^{(+)(j)}(x, i p_n^{(j)}) \to (c_n^{(-)(j)}/c_n^{(+)(j)}) \exp(p_n^{(j)} x) \to 0, \quad x \to -\infty$

(see (2.1.-22), (2.1.-23) and (2.1.-28)) and

(7) $\quad \rho_n^{(j)} = \left\{ \int_{-\infty}^{+\infty} dx \left[f^{(+)(j)}(x, i p_n^{(j)}) \right]^2 \right\}^{-1}$

(see (2.1.-23) and (2.1.-27)). On the other hand neither $f^{(+)(1)}(x, i p_n^{(2)})$ nor $f^{(+)(2)}(x, i p_n^{(1)})$ are generally eigenfunctions corresponding to discrete eigenvalues (unless it happens that some discrete eigenvalue corresponds both to $u^{(1)}(x)$ and to $u^{(2)}(x)$, namely that some eigenvalue with different superscripts coincide); but the analysis of section 2.1 implies that these functions are well defined by analytic continuation, from $f^{(+)(j)}(x, k)$ for real k (let us emphasize that we always assume $p_n^{(j)} > 0$). Their asymptotic behaviour can be obtained directly from (1a) and (1b):

(8a) $\quad f^{(+)(1)}(x, i p_n^{(2)}) \to \exp(-p_n^{(2)} x) \to 0, \quad x \to +\infty,$

(8b) $\quad f^{(+)(1)}(x, i p_n^{(2)}) \to \left[T^{(1)}(i p_n^{(2)}) \right]^{-1} \exp(-p_n^{(2)} x), \quad x \to -\infty$

(there is of course an analogous formula with the indices 1 and 2 exchanged). Note that the r.h.s. of (8b) diverges. Clearly the formula written above refers to the case when $-[p_n^{(2)}]^2$ is not a discrete eigenvalue of the spectral problem (1) with $j=1$, this being a necessary and sufficient condition to exclude the vanishing of the coefficient $[T^{(1)}(i p_r^{(2)})]^{-1}$ in the r.h.s. of (8b); otherwise the appropriate formulae are again (6).

The formulae (4) and (5) can easily be generalized, introducing the integrodifferential operator $\tilde{\Lambda}$ that has the product of two eigenfunctions of the two Schroedinger equations (1) as its own eigenfunction:

(9) $\quad \tilde{\Lambda}\left[f^{(+)(1)}(x, k) f^{(+)(2)}(x, k) \right] = -4k^2 \left[f^{(+)(1)}(x, k) f^{(+)(2)}(x, k) \right].$

This formula is proved in appendix A.9, where a more general one is given with the Jost solutions replaced by generic solutions of (1); in fact (9) holds only for the Jost solutions of type "plus", $f^{(+)(j)}(x, k)$, while for other solutions of (1) additional contributions appear in its r.h.s. (see (A.9.-40)).

2.4.2 Spectral integral relations

The explicit definition of the operator $\tilde{\Lambda}$ is given by the two equivalent formulae

(10a) $\quad \tilde{\Lambda} F(x) = F_{xx}(x) - 2[u^{(1)}(x) + u^{(2)}(x)] F(x)$
$$- \int_x^{+\infty} dy \, [u_y^{(1)}(y) + u_y^{(2)}(y)] F(y)$$
$$+ \int_x^{+\infty} dy \, [u^{(1)}(y) - u^{(2)}(y)]$$
$$\cdot \int_y^{+\infty} dz \, [u^{(1)}(z) - u^{(2)}(z)] F(z),$$

(10b) $\quad \tilde{\Lambda} F(x) = F_{xx}(x) + [w_x^{(1)}(x) + w_x^{(2)}(x)] F(x)$
$$- \int_x^{+\infty} dy \, [w_y^{(1)}(y) + w_y^{(2)}(y)] F_y(y)$$
$$+ \int_x^{+\infty} dy \, [w_y^{(1)}(y) - w_y^{(2)}(y)]$$
$$\cdot \int_y^{+\infty} dz \, [w_z^{(1)}(z) - w_z^{(2)}(z)] F(z),$$

that specify its action on a generic function $F(x)$. This integrodifferential operator depends on $u^{(1)}(x)$ and $u^{(2)}(x)$ (or equivalently on $w^{(1)}(x)$ and $w^{(2)}(x)$), although for notational simplicity we omit to indicate this explicitly. Note that the l.h.s. of (5) is merely $\tilde{\Lambda} \cdot 1$.

From (4), (5) and (9) there follow the more general formulae

(11) $\quad g(\tilde{\Lambda}) [w^{(1)}(x) - w^{(2)}(x)] = -2 \sum_{n=1}^{N^{(1)}} f^{(+)(1)}(x, i p_n^{(1)})$
$$\cdot f^{(+)(2)}(x, i p_n^{(1)}) g[4(p_n^{(1)})^2] \rho_n^{(1)} + 2 \sum_{n=1}^{N^{(2)}} f^{(+)(1)}(x, i p_n^{(2)})$$
$$\cdot f^{(+)(2)}(x, i p_n^{(2)}) g[4(p_n^{(2)})^2] \rho_n^{(2)} - \pi^{-1} \int_{-\infty}^{+\infty} dk \, f^{(+)(1)}(x, k)$$
$$\cdot f^{(+)(2)}(x, k) g(-4k^2) [R^{(1)}(k) - R^{(2)}(k)],$$

(12) $\quad g(\tilde{\Lambda}) \tilde{\Lambda} \cdot 1 = 4 \sum_{n=1}^{N^{(1)}} f^{(+)(1)}(x, i p_n^{(1)}) f^{(+)(2)}(x, i p_n^{(1)}) g[4(p_n^{(1)})^2] p_n^{(1)} \rho_n^{(1)}$
$$+ 4 \sum_{n=1}^{N^{(2)}} f^{(+)(1)}(x, i p_n^{(2)}) f^{(+)(2)}(x, i p_n^{(2)}) g[4(p_n^{(2)})^2] p_n^{(2)} \rho_n^{(2)}$$
$$- \pi^{-1} \int_{-\infty}^{+\infty} dk \, f^{(+)(1)}(x, k)$$
$$\cdot f^{(+)(2)}(x, k) 2ikg(-4k^2) [R^{(1)}(k) + R^{(2)}(k)].$$

In each of these two formulae g(z) is an arbitrary polynomial, or for that matter any entire function such that the integrals in the r.h.s. converge. Let us incidentally note that, at least as long as g(z) is a polynomial, convergence is indeed insured by (2.1.-39b) and by the asymptotic relation

(13) $$\lim_{k \to \pm \infty} \left[f^{(+)(j)}(x,k) \exp(-ikx) \right] = 1,$$

that is a direct consequence of the very definition of the Jost solution, see (1) and (1a) or (1b) (with (2.1.-38)); or see (2.2.-6).

Clearly these formulae, as (2.4.1.-10), provide a tool to identify those (nonlinear integrodifferential) relations between two functions $u^{(j)}(x)$, such that the relations between the corresponding spectral transforms are very simple. This is discussed in more detail in chapter 4.

There are two special cases of (11) and (12) that deserve special mention.

The first is the case when the two functions $u^{(j)}(x)$ differ only infinitesimally, say

(14a) $\quad u^{(1)}(x) = u(x) + \delta u(x), \qquad u^{(2)}(x) = u(x),$

(14b) $\quad w^{(1)}(x) = w(x) + \delta w(x), \qquad w^{(2)}(x) = w(x),$

(14c) $\quad R^{(1)}(k) = R(k) + \delta R(k), \qquad R^{(2)}(k) = R(k),$

(14d) $\quad N^{(1)} = N^{(2)} = N,$

(14e) $\quad p_n^{(1)} = p_n + \delta p_n, \quad p_n^{(2)} = p_n, \qquad n = 1, 2, \ldots, N,$

(14f) $\quad \rho_n^{(1)} = \rho_n + \delta \rho_n, \quad \rho_n^{(2)} = \rho_n, \qquad n = 1, 2, \ldots, N,$

(14g) $\quad f^{(+)(1)}(x,k) = f^{(+)}(x,k) + \delta f^{(+)}(x,k),$
$\quad f^{(+)(2)}(x,k) = f(x,k).$

Then from (11) and (12) there obtain

(15) $$g(\tilde{L}) \delta w(x) = -2 \sum_{n=1}^{N} \left\{ 2 f_p^{(+)}(x, i p_n) f^{(+)}(x, i p_n) g(4p_n^2) \rho_n \delta p_n \right.$$
$$\left. + \left[f^{(+)}(x, i p_n) \right]^2 \left[g_p(4p_n^2) \rho_n \delta p_n + g(4p_n^2) \delta \rho_n \right] \right\}$$
$$- \pi^{-1} \int_{-\infty}^{+\infty} dk \left[f^{(+)}(x,k) \right]^2 g(-4k^2) \delta R(k),$$

(16) $$g(\tilde{L}) w_x(x) = 4 \sum_{n=1}^{N} \left[f^{(+)}(x, i p_n) \right]^2 g(4p_n^2) p_n \rho_n$$
$$- \pi^{-1} \int_{-\infty}^{+\infty} dk \left[f^{(+)}(x,k) \right]^2 2ikg(-4k^2) R(k).$$

2.4.2 Spectral integral relations

Here of course

(17a) $\quad f_p^{(+)}(x, ip_n) \equiv \partial f(x, ip)/\partial p|_{p=p_n},$

(17b) $\quad g_p(4p_n^2) \equiv \partial g(4p^2)/\partial p|_{p=p_n};$

and note that the functions $f^{(+)}(x, ip_n)$ are now indeed the eigenfunctions of the Schroedinger spectral problem with potential $u(x)$, corresponding to the discrete eigenvalues $-p_n^2$ (if any).

The integrodifferential operator $\tilde{L}$ in the l.h.s. of (15) and (16) is of course the operator $\tilde{\Lambda}$ defined by (10), with $u^{(1)}(x) = u^{(2)}(x) = u(x)$. In view of its importance, we display here three, obviously equivalent, definitions of this operator:

(18a) $\quad \tilde{L}F(x) = F_{xx}(x) - 4u(x)F(x) - 2\int_x^{+\infty} dy\, u_y(y) F(y),$

(18b) $\quad \tilde{L}F(x) = F_{xx}(x) - 2u(x)F(x) + 2\int_x^{+\infty} dy\, u(y) F_y(y),$

(18c) $\quad \tilde{L}F(x) = F_{xx}(x) + 2[w(x)F(x)]_x + 2\int_x^{+\infty} dy\, w(y) F_{yy}(y).$

Clearly its characteristic property (see (9)) is to possess as eigenfunctions the squares of the Jost solutions $f^{(+)}(x, k)$ of the Schroedinger equation (2.1.-1):

(19) $\quad \tilde{L}[f^{(+)}(x, k)]^2 = -4k^2[f^{(+)}(x, k)]^2.$

Other properties of this operator are given in appendix A.9.

The special case of (16) with $g(z) = 1$ coincides with (1.3.3-2).

The other interesting special case of (11) and (12) corresponds to

(20a) $\quad u^{(1)}(x) = u(x), \qquad u^{(2)}(x) = 0,$

(20b) $\quad w^{(1)}(x) = w(x), \qquad w^{(2)}(x) = 0,$

which imply

(20c) $\quad R^{(1)}(k) = R(k), \qquad R^{(2)}(k) = 0,$

(20d) $\quad N^{(1)} = N, \qquad N^{(2)} = 0,$

(20e) $\quad p_n^{(1)} = p_n, \quad n = 1, 2, \ldots, N,$

(20f) $\quad \rho_n^{(1)} = \rho_n, \quad n = 1, 2, \ldots, N,$

(20g) $\quad f^{(+)(1)}(x, k) = f^{(+)}(x, k), \qquad f^{(+)(2)}(x, k) = \exp(ikx).$

Then (11) and (12) yield

$$(21) \quad g(\tilde{\Lambda}^{(0)})w(x) = -2 \sum_{n=1}^{N} f^{(+)}(x, ip_n) \exp(-p_n x) g(4p_n^2) \rho_n$$
$$- \pi^{-1} \int_{-\infty}^{+\infty} dk\, f^{(+)}(x,k) \exp(ikx) g(-4k^2) R(k),$$

$$(22) \quad g(\tilde{\Lambda}^{(0)})\tilde{\Lambda}^{(0)} \cdot 1 = 4 \sum_{n=1}^{N} f^{(+)}(x, ip_n) \exp(-p_n x) g(4p_n^2) p_n \rho_n$$
$$- \pi^{-1} \int_{-\infty}^{+\infty} dk\, f^{(+)}(x,k)$$
$$\cdot \exp(ikx) 2ikg(-4k^2) R(k).$$

Here of course the integrodifferential operator $\tilde{\Lambda}^{(0)}$ is just $\tilde{\Lambda}$, see (10), but with (20a, b):

$$(23a) \quad \tilde{\Lambda}^{(0)} F(x) = F_{xx}(x) - 2u(x) F(x) - \int_{x}^{+\infty} dy\, u_y(y) F(y)$$
$$+ \int_{x}^{+\infty} dy\, u(y) \int_{y}^{+\infty} dz\, u(z) F(z),$$

$$(23b) \quad \tilde{\Lambda}^{(0)} F(x) = F_{xx}(x) + w_x(x) F(x) - \int_{x}^{+\infty} dy\, w_y(y) F_y(y)$$
$$+ \int_{x}^{+\infty} dy\, w_y(y) \int_{y}^{+\infty} dz\, w_z(z) F(z).$$

In the special case $g(z)=1$, (21) yields (1.3.3.-3), while (22) reads

$$(24) \quad w_x(x) + \tfrac{1}{2}[w(x)]^2 = 4 \sum_{n=1}^{N} f^{(+)}(x, ip_n) \exp(-p_n x) p_n \rho_n$$
$$- \pi^{-1} \int_{-\infty}^{+\infty} dk\, f^{(+)}(x,k) \exp(ikx) 2ik R(k).$$

A number of important formulae can be obtained from those given here, by considering the limit as $x \to -\infty$. These formulae are displayed and derived in appendix A.9. Let us also emphasize that all the formulae written above in terms of the operators $\tilde{\Lambda}$ and $\tilde{L}$ can be rewritten in terms of the operators Λ and L using the operator identities (see (A.9.-25, 30))

$$(25) \quad g(\Lambda) D = D g(\tilde{\Lambda}), \qquad g(L) D = D g(\tilde{L}),$$

where D is the differential operator,

(26) $\quad D \equiv d/dx.$

2.4.2.1. Additional spectral integral relations

In the preceding subsection 2.4.2 we have reported two important formulae, (2.4.2.-11) and (2.4.2.-12), relating two functions ("potentials") $u^{(j)}(x)$ to the corresponding spectral quantities. We have also displayed certain formulae that follow from these, but provide relations between a single function, or its variation, and the corresponding spectral quantities (and their variation): see (2.4.2.-15, 16, 21, 22). Here we merely remark that additional relations involving a *single* function $u(x)$ and the spectral quantities related to it are also yielded directly by (2.4.2.-11) and (2.4.2.-12), together with the results of subsection 2.4.1.1 that display explicitly how the potentials, and the spectral quantities, change under translations and scale transformation. For instance setting

(1) $\quad u^{(1)}(x) = \mu^2 u(\mu x), \quad u^{(2)}(x) = u(x),$

together with the corresponding expressions, see (2.1.1.-3), in (2.4.2.-11), there obtains

(2) $\quad g(\tilde{\Lambda})[\mu w(\mu x) - w(x)]$

$$= -2 \sum_{n=1}^{N} \left[f^{(+)}(\mu x, i p_n) f^{(+)}(x, i \mu p_n) g(4\mu^2 p_n^2) \mu \rho_n \right.$$

$$\left. - f^{(+)}(\mu x, i p_n/\mu) f^{(+)}(x, i p_n) g(4 p_n^2) \rho_n \right]$$

$$- \pi^{-1} \int_{-\infty}^{+\infty} dk \, f^{(+)}(\mu x, k/\mu)$$

$$\cdot f^{(+)}(x, k) g(-4k^2) [R(k/\mu) - R(k)].$$

Here of course the operator $\tilde{\Lambda}$ is defined by (2.4.2.-10), but with (1) (and the corresponding relation for the integrals,

(1a) $\quad w^{(1)}(x) = \mu w(\mu x), \quad w^{(2)}(x) = w(x),$

see (2.1.1.-3)); and all the other quantities are related to the spectral problem with potential $u(x)$ (note incidentally that, although the discrete eigenvalues are not invariant under a scale transformation, their total number N clearly is invariant; see (2.1.1.-3e)).

The derivation of other formulae by analogous techniques is left as a simple exercise for the diligent reader. Here we display instead the special case of (2) corresponding to infinitesimal scale transformations ($\mu=1+\varepsilon$, $\varepsilon \to 0$):

(3) $\quad g(\tilde{L})[w(x)+xw_x(x)]$
$$= -2 \sum_{n=1}^{N} \left\{ 2 f_p^{(+)}(x, ip_n) f^{(+)}(x, ip_n) g(4p_n^2) p_n \rho_n \right.$$
$$+ [f^{(+)}(x, ip_n)]^2 [g_p(4p_n^2) p_n + g(4p_n^2)] \rho_n \right\}$$
$$+ \pi^{-1} \int_{-\infty}^{+\infty} dk \, [f^{(+)}(x, k)]^2 g(-4k^2) k R_k(k).$$

Here we are using the notation (2.4.2.-17), and the operator $\tilde{L}$ is defined by (2.4.2.-18). Of course this last equation follows, even more directly, from (2.4.2.-15) and (2.1.1.-4). For future reference, it is convenient to rewrite this formula in the form

(4) $\quad g(L)[xu_x(x)+2u(x)]$
$$= (d/dx) \left\{ 2 \sum_{n=1}^{N} \left\{ 2 f_p^{(+)}(x, ip_n) f^{(+)}(x, ip_n) g(4p_n^2) p_n \rho_n \right. \right.$$
$$- [f^{(+)}(x, ip_n)]^2 [g_p(4p_n^2) p_n + g(4p_n^2)] \rho_n \right\}$$
$$- \pi^{-1} \int_{-\infty}^{+\infty} dk \, [f^{(+)}(x, k)]^2 g(-4k^2) k R_k(k) \right\},$$

that follows from (3) using the operator identity (see (A.9.-30))

(5) $\quad g(L)D = Dg(\tilde{L}),$

where D is the differential operator,

(6) $\quad D \equiv d/dx$

so that in particular

(7) $\quad u(x) = -Dw(x),$
$\quad\quad\quad xu_x(x) + 2u(x) = -D[xw_x(x) + w(x)] = -D^2[xw(x)].$

2.4.3. Connection between wronskian and spectral integral relations

Clearly the results of the preceding two subsections, 2.4.1 and 2.4.2, are related to each other; they can be viewed as the analogs of the inverse and direct Fourier transform formulae, except that the rôle of the Fourier basis (the exponential, or circular, functions) is now taken over by the products of two eigenfunctions (or the square of one eigenfunction) of the Schroedinger spectral problem. The dependence of these eigenfunctions on the "potentials" $u^{(j)}(x)$ (in the framework of the direct spectral problem) or on the spectral transforms $S^{(j)} = \{R^{(j)}(k), -\infty < k < +\infty; p_n^{(j)}, \rho_n^{(j)}, n = 1,2,\ldots,N^{(j)}\}$ (in the framework of the inverse spectral problem), is of course at the origin of the nonlinear character of these formulae, and causes their eventual usefulness to solve (certain classes of) nonlinear evolution equations (see chapter 3).

Thus, the natural tool to connect the results of subsection 2.4.1 to those of subsection 2.4.2 are the closure (orthogonality and completeness) relations for the products of two eigenfunctions of the Schroedinger spectral problem, corresponding to the same eigenvalue k^2 (but possibly to two different potentials $u^{(j)}(x), j=1,2$). These products are themselves the eigenfunctions of the operator $\tilde{\Lambda}$ corresponding to the eigenvalue $-4k^2$ (see (2.4.2.-9); incidentally this integrodifferential equation can be easily transformed into an ordinary differential equation: of fourth order if $u^{(1)}(x) \neq u^{(2)}(x)$, of third order if $u^{(1)}(x) = u^{(2)}(x)$).

Several remarkable and useful properties of the integrodifferential operator $\tilde{\Lambda}$ are exhibited in appendix A.9. But we do not discuss the spectral properties of this operator and of its eigenfunctions, since they are not essential for our purposes; the interested reader is referred to the literature (see Notes). Concerning the completeness of the eigenfunctions of $\tilde{\Lambda}$, let us merely point out that the wronskian integral relations

(1)
$$2ik\{R^{(2)}(-k)/[T^{(1)}(k)T^{(2)}(-k)]$$
$$-R^{(1)}(-k)/[T^{(1)}(-k)T^{(2)}(k)]\}$$
$$= \int_{-\infty}^{+\infty} dx\, f^{(+)(1)}(x,k) f^{(+)(2)}(x,k) [u^{(1)}(x) - u^{(2)}(x)]$$

and (2.4.1.-38) imply that the function $u^{(1)}(x) - u^{(2)}(x)$ is orthogonal to every eigenfunction $f^{(+)(1)}(x,k)f^{(+)(2)}(x,k)$, $-\infty < k < +\infty$ and $\varphi_n^{(1)}(x)\varphi_n^{(2)}(x)$, $n=1,2,\ldots,N$, of the operator $\tilde{\Lambda}$ (see subsection 2.4.2, in particular (2.4.2.-9)), provided the spectral transforms of $u^{(1)}(x)$ and $u^{(2)}(x)$

differ only in the values of the normalization coefficients $\rho_n^{(1)}$ and $\rho_n^{(2)}$ ($R^{(1)}(k) = R^{(2)}(k)$, $p_n^{(1)} = p_n^{(2)}$, $n = 1, 2, \ldots, N$, and therefore, see appendix A.4, $T^{(1)}(k) = T^{(2)}(k)$); this indicates a lack of completeness of the set $\{f^{(+)(1)}(x, k) f^{(+)(2)}(x, k), -\infty < k < +\infty; \varphi_n^{(1)}(x) \varphi_n^{(2)}(x), n = 1, 2, \ldots N\}$.

2.N. Notes to chapter 2

Much of the material in this chapter is standard, and has been treated before in several books and review articles. However, the Schroedinger spectral (or "scattering") problem that has been more extensively investigated is that on the semiline ($0 \leq x < +\infty$; x is then interpreted as a radial coordinate) rather than on the whole line ($-\infty < x < +\infty$); but the two problems are sufficiently similar to allow a relatively easy transfer of results and techniques from one to the other. General references for the radial Schroedinger scattering problem are, for instance, [DAR1965], [N1966], [C1967], [GW1964], [JRT1972], [AM1963], [Mar1977] [CS1977] (the last 3 concentrate on the inverse problem). For detailed treatments of the Schroedinger problem on the whole line, see for instance, in addition to the last two references quoted above, [F1964], [DT1979] and [Ne1980]; all these papers focus on the inverse problem (but contain also a discussion of the direct problem).

2.1. All the results reported in this section are by now standard (see references quoted above), except for (2.1.-41), that has been discovered only recently [CD1978f].

2.2. For general references on the inverse spectral problem see above; some additional references are given in section A.5.N. Note that nowhere in this book we discuss *rigorously* the (direct and) inverse spectral problems; in particular we never attempt to characterize the broadest class of functions for which the spectral transform is bijective (indeed, this is largely an open problem). For recent rigorous results, see e.g. [DT1979] (note that the restriction posited in this paper,

$$\int_{-\infty}^{+\infty} dx (1 + x^2) |u(x)| < \infty,$$

is more stringent than (2.1.-2)).

2.4.1. The relevance of wronskian integral relations (or "generalized wronskian relations", as we termed them in previous publications), as a tool to investigate nonlinear evolution equations solvable by the spectral transform, has been emphasized by us in several papers; see for instance [C1975], [C1976a], [CD1976a], [CD1977a], [D1978] (the last three papers are however concentrated on more general spectral problems than that considered here; see volume II). See also [CaC1977] and [GK1980a], [GK1980b].

2.4.2. The relevance of spectral integral relations such as those presented in this subsection as a tool to study nonlinear evolution equations solvable by spectral transform techniques has been noted by many authors, for instance by the Clarkson school [AKNS1974]; they focus however on the (generalized) Zakharov–Shabat spectral problem, rather than the Schroedinger problem. Yet the collection of formulae reported in this subsection and in appendix A.9 is probably the most complete display published thus far (in particular the equations for two different potentials are not given in [AKNS1974], and they are misprinted in [DB1979]; but this case is treated in [Khr1980]).

2.4.3. The importance of the closure relations for squared eigenfunctions has been emphasized by various authors, see for instance [KN1979]; they have been given, in the context of the Zakharov–Shabat spectral problem (see volume II) by [K1976]. A rigorous treatment in the context of the Schroedinger spectral problem of interest here is given in [Khr1980]; in fact this treatment is more general than those mentioned above, since it deals with the spectral properties of the operator $\tilde{\Lambda}$, that has as eigenfunctions the products of two Schroedinger eigenfunctions (see appendix A.9), corresponding to two different potentials $u^{(1)}(x)$ and $u^{(2)}(x)$ (the squared eigenfunctions are instead the eigenfunctions of the operator $\tilde{L}$, to which $\tilde{\Lambda}$ reduces whenever $u^{(1)}(x)=u^{(2)}(x)=u(x)$). For analogous results, but in the context of the Zakharov–Shabat (or Dirac) spectral problem, see [GK1980a] and [GK1980b]; these papers treat also the evolution equations solvable via this spectral problem.

CHAPTER THREE

NONLINEAR EVOLUTION EQUATIONS SOLVABLE BY THE (SCHROEDINGER) SPECTRAL TRANSFORM

We have already described, in sections 1.4, 1.5 and 1.6, the main ideas of the technique of solution of certain classes of nonlinear evolution equations via the spectral transform. In this chapter we cover again the same ground; but we are now able to prove the basic results, that were given without proof in chapter 1. Let us recall that the crucial point is to identify those nonlinear time-evolutions of a function $u(x, t)$, to which there corresponds a *simple* evolution of the associated spectral transform $S(t)$. This we do in the following section 3.1, using as a convenient tool some of the formulae given in section 2.4, in particular those that relate infinitesimal changes of the function $u(x)$ to the corresponding changes of the spectral transform associated with it. We thus identify a class of nonlinear evolution equations solvable via the spectral transform defined by the Schroedinger spectral problem, and we discuss the solution of the Cauchy problem for this class of evolution equations. Our consideration here will be limited to equations invariant under space translations. A larger class of non-translation-invariant evolution equations, with coefficient linearly dependent on the space variable x, are also solvable via the spectral transform associated with the Schroedinger spectral problem; but their discussion is postponed until section 6.2. Other classes of nonlinear evolution equations that are solvable by other spectral transforms will be treated in volume II.

In the subsequent section 3.2, the behaviour of the solutions of the class of nonlinear evolution equations introduced in section 3.1 is analyzed (qualitatively and, whenever appropriate, quantitatively).

3.1. KdV and higher KdVs

Let u be a function of the two (real) variables x and t:

(1) $u \equiv u(x, t), \quad -\infty < x < +\infty, \quad t \geq 0.$

The general framework of our investigation is the Cauchy problem, in which u is assigned at an initial time, say $t=0$,

(2) $\quad u(x,0)=u_0(x),$

and is then determined at all later times by some evolution equation (see below). The x-dependence of $u(x,t)$ is such that, for given t, u falls in the class of potentials considered in the preceding sections; namely $u(x,t)$ is a regular function defined for all (real) values of the variable x and vanishing (sufficiently fast) asymptotically with all its derivatives,

(3) $\quad 0=u(\pm\infty,t)=u_x(\pm\infty,t)=u_{xx}(\pm\infty,t)=\cdots;$

the class of evolution equations that we discuss is moreover such that, if these properties of u hold at $t=0$, they continue to hold for all values of t; namely, it will be enough to restrict our consideration to an initial (Cauchy) datum $u_0(x)$ that qualifies as a *bona fide* potential, to be guaranteed that $u(x,t)$ for $t\geqslant 0$ (or, for that matter, for $t\leqslant 0$) is, as a function of x, also a *bona fide* potential. Thus, in one-to-one correspondence to $u(x,t)$, there exists a spectral transform,

(4) $\quad S(t)=\{R(k,t),-\infty<k<+\infty;\ p_n(t),\rho_n(t),n=1,2,\ldots,N(t)\},$

that also evolves in time, starting from the initial value

(5) $\quad S(0)=\{R_0(k)\equiv R(k,0),-\infty<k<+\infty;\ p_n(0),\rho_n(0),$
$\quad\quad\quad n=1,2,\ldots,N(0)\},$

that corresponds to the initial Cauchy datum $u_0(x)$.

Consider now the infinitesimal change in $u(x,t)$ corresponding to an infinitesimal time evolution,

(6a) $\quad \delta u(x,t)=u_t(x,t)\,dt,$

and the corresponding change in the spectral transform

(6b) $\quad \delta S(t)=\{R_t(k,t)\,dt,-\infty<k<+\infty;\ \dot p_n(t)\,dt,\dot\rho_n(t)\,dt,$
$\quad\quad\quad n=1,2,\ldots,N(t)\}.$

In writing the last equation we have used the superimposed dot to indicate time-differentiation; we will often use this notation in the following (but only for functions of the single variable t). Note moreover that we have implicity assumed that the number N of discrete eigenvalues has not changed in the infinitesimal interval dt (indeed it will be presently seen that, for the class of time evolutions under consideration here, N does not vary with time).

To correlate the (infinitesimal) change (6a) to the corresponding change (6b) we may now use the appropriate formulae of section 2.4. Indeed we have the option to use the wronskian integral relations of subsection 2.4.1 or the spectral integral relations of subsection 2.4.2. The former are actually more appropriate to ascertain which is the time evolution of the spectral transform that corresponds to an (appropriately) given time evolution of the function $u(x,t)$; the latter are instead more appropriate to determine the time evolution of the function $u(x,t)$ that corresponds to a given time evolution of the spectral transform. Although the latter approach is more compact, we begin with the former, since our main interest is indeed to determine the time evolution of the spectral transform that corresponds to a given time evolution of $u(x,t)$. But we then verify the consistency of all the results by using also the second approach.

Let us begin by considering the reflection coefficient. Then (2.4.1.-12) yield

$$(7) \qquad 2ikg(-4k^2,t)R_t(k,t) = \int_{-\infty}^{+\infty} dx\,[\psi(x,k,t)]^2 g(L,t)u_t(x,t),$$

$$(8) \qquad (2ik)^2 h(-4k^2,t)R(k,t) = \int_{-\infty}^{+\infty} dx\,[\psi(x,k,t)]^2 h(L,t)u_x(x,t).$$

Here $g(z,t)$ and $h(z,t)$ are two arbitrary functions, only required to be entire in the first variable, z; note that we have indicated explicitly that they may depend on t. The remainder of the symbols require no explanation; indeed the notation coincides with that of the preceding sections, except for the additional presence of the t variable. We do however rewrite, for the convenience of the reader, the definition of the integrodifferential operator L, by detailing its effect on the generic function $f(x)$:

$$(9) \qquad Lf(x) = f_{xx}(x) - 4u(x,t)f(x) + 2u_x(x,t)\int_x^{+\infty} dy\,f(y)$$

(of course the function f might also depend on t).

Clearly (7) and (8) imply that, if $u(x,t)$ satisfies the nonlinear evolution equation

$$(10) \qquad g(L,t)u_t(x,t) = h(L,t)u_x(x,t),$$

then the corresponding reflection coefficient $R(k,t)$ evolves according to the linear first-order ordinary differential equation

(11) $\quad g(-4k^2, t) R_t(k,t) = 2ikh(-4k^2, t) R(k,t),$

that can be immediately integrated:

(12) $\quad R(k,t) = R(k,0) \exp\left\{ 2ik \int_0^t dt' \left[h(-4k^2, t')/g(-4k^2, t') \right] \right\}.$

Note that this formula implies that, throughout the time evolution, only the phase of $R(k,t)$ changes, while its modulus is, for all values of k, time-independent; that is, provided the two functions g and h are real, namely the nonlinear evolution equation (10) is itself real (which need not necessarily imply that $u(x,t)$ itself be real; although our consideration is generally limited to this case).

We have thus identified the class (10) of nonlinear evolution equations for $u(x,t)$, characterized by the fact that the corresponding evolution of the reflection coefficient $R(k,t)$ is given by the explicit formula (12). Thus the Cauchy problem for (10) can be solved by first evaluating the transmission coefficient $R(k,0) = R_0(k)$ corresponding to the initial Cauchy datum (2), then obtaining $R(k,t)$ for $t>0$ from (12), and then reconstructing from $R(k,t)$ the function $u(x,t)$. The last step requires, however, if discrete eigenvalues are present, knowledge of the corresponding parameters $p_n(t)$ and $\rho_n(t)$. Let us therefore proceed to investigate what their time evolution is, when $u(x,t)$ evolves according to (10).

Consider first the evolution of the parameters $p_n(t)$. Then (6), together with (2.4.1.-55) and (2.4.1.-56), imply

(13) $\quad -2p_n(t) g(4p_n^2(t), t) \dot{p}_n(t) = \int_{-\infty}^{+\infty} dx \left[\varphi_n(x,t) \right]^2 g(L,t) u_t(x,t),$

(14) $\quad 0 = \int_{-\infty}^{+\infty} dx \left[\varphi_n(x,t) \right]^2 h(L,t) u_x(x,t).$

There immediately follows that, as $u(x,t)$ evolves according to (10), the discrete eigenvalues $p_n(t)$ remain constant,

(15) $\quad \dot{p}_n(t) = 0,$

or equivalently,

(16) $\quad p_n(t) = p_n(0) \equiv p_n.$

This phenomenon is quite remarkable (note that the evolution (10) need not even be autonomous, namely (10) may contain explicitly time-dependent coefficients); it implies that the Schroedinger operator

(17) $-\partial^2/\partial x^2 + u(x,t)$

undergoes an *isospectral deformation* when $u(x,t)$ evolves according to (10). Note that the time-invariance of the discrete eigenvalues implies in particular that their number does not change throughout the time evolution.

Let us now proceed to consider the time evolution of the normalization coefficient $\rho_n(t)$ associated with the eigenvalue p_n of the potential $u(x,t)$, when the latter evolves according to (10). Then (6), together with (2.4.1.-69) and (2.4.1.-70) (that we are now justified in using, since we are considering infinitesimal changes of u that do not modify the discrete eigenvalues but only the corresponding normalization coefficients), yield

(18) $-p_n g(4p_n^2, t) \dot{\rho}_n(t)/\rho_n(t)$
$$= \int_{-\infty}^{+\infty} dx\, \varphi_n(x,t) \varphi_p(x, p_n, t) g(L,t) u_t(x,t),$$

(19) $2 p_n^2 h(4p_n^2, t) = \int_{-\infty}^{+\infty} dx\, \varphi_n(x,t) \varphi_p(x, p_n, t) h(L,t) u_x(x,t),$

where of course

(20) $\varphi_p(x, p_n, t) \equiv \{\partial/\partial_p [\varphi(x, p, t)]\}\big|_{p=p_n}$

with $\varphi(x, p, t)$ defined by (2.4.1.-61) (with the t variable added as appropriate). From (18) and (19) it immediately follows that, if $u(x,t)$ evolves in time according to (10), the corresponding evolution of each of the normalization coefficients $\rho_n(t)$ is given by the ordinary first-order linear differential equation

(21) $g(4p_n^2, t) \dot{\rho}_n(t) = -2 p_n h(4p_n^2, t) \rho_n(t),$

that can be immediately integrated:

(22) $\rho_n(t) = \rho_n(0) \exp\left\{ -2 p_n \int_0^t dt'\, [h(4p_n^2, t')/g(4p_n^2, t')] \right\}.$

Before proceeding further, let us pause to note that the evolution equations (15) and (21) for the discrete spectrum parameters follow directly also from the evolution equation (11) satisfied by the reflection coefficient $R(k, t)$ by using the relation (see (2.1.-36))

(23) $\quad R(k,t) \approx i\rho_n(t)/[k-ip_n(t)],$

valid for $k \approx ip_n(t)$. Indeed, by differentiating this relation with respect to t one gets

(24) $\quad R_t(k,t) \approx i\dot{\rho}_n(t)/[k-ip_n(t)] - \dot{p}_n(t)\rho_n(t)/[k-ip_n(t)]^2,$

and by inserting this expression in (11) and then equating the coefficients of the second and first order poles at $k=ip_n(t)$ one gets indeed (15) and (21). This derivation of the evolution equations for the parameters of the discrete spectrum from the evolution of the reflection coefficient is a convenient shortcut that bypasses the use of (and therefore eliminates the need to establish) wronskian-type relations for these quantities. For this reason we will employ it occasionally in the following. This procedure is, however, questionable, because, as we know, the relation (23) is not universally valid (see sections 2.1 and 2.2); and for this reason the derivation of the evolution equations for the parameters of the discrete spectrum has been given here firstly without making use of (23). It is, however, a fact that the evolution equations for the discrete spectrum parameters generated from the evolution equation for the reflection coefficient via the shortcut procedure based on (23) are generally correct; not only in the relatively simple case discussed here, but also in the more complicated ones considered in the following.

In conclusion, the Cauchy problem for the nonlinear evolution equation (10) (or rather for the class of evolution equations (10); since a different evolution equation obtains for each specific choice of the functions g and h) is now solved by the following three steps: first evaluate the spectral transform $S(0)$ of the initial Cauchy datum $u(x,0)=u_0(x)$; then let the spectral transform evolve, namely evaluate $S(t)$ from $S(0)$ via the explicit formulae (12), (16) and (22); and finally reconstruct $u(x, t)$ from $S(t)$. The analogy of this procedure to the solution via Fourier transform of the linear evolution equations discussed in section 1.1 has already been emphasized in section 1.4, and no additional repetition of the points made there is needed here.

Let us now discuss the class of nonlinear evolution equations (10). We note first of all that our developments have been so far rather formal. Indeed, what is the significance of (10) when the functions $g(z,t)$ and $h(z,t)$ have a non-polynomial dependence on z? Moreover, even in the polynomial case, what is the significance of the formula (12) if the function $g(z,t)$ appearing in the denominator vanishes for some negative value of z and positive value of t? Or the significance of (22) if $g(4p_n^2, t)$ vanishes? Indeed, in the latter case even the derivation of (15) from (13) and (14) becomes questionable.

This is not the appropriate place to confront such questions; suffice it here to emphasize the main message implied by the technique of solution we have just described, namely that *the time evolution is much simpler in the spectral space than in configuration space*. This remark is highly relevant to the study of the possible pathologies mentioned above, as the reader may wish to experience; and it is relevant as well to investigate the physiology of the more restricted class of equations that obtain from (10) by assuming $g(z,t)$ to be unity and $h(z,t)$ to be a polynomial in z, on which our attention will be hereafter mainly focussed. But before this restriction of our scope becomes operational, let us display at least one instance of the more general class (10) with $g(z,t) \neq 1$, namely that corresponding to

(25) $\quad g(z,t) = g_0 + g_1 z, \quad h(z,t) = h_0 + h_1 z$

(g_0, g_1, h_0 and h_1 may be functions of t, although for notational simplicity we do not exhibit such dependence). It reads

(26a) $\quad g_0 u_t(x,t) + g_1 \left[u_{xxt}(x,t) - 4u_t(x,t)u(x,t) + 2u_x(x,t) \right.$

$\left. \int_x^{+\infty} dy\, u_t(y,t) \right] = \{h_0 u(x,t) + h_1 [u_{xx}(x,t) - 3u^2(x,t)]\}_x.$

This is a rather complicated integrodifferential equation; it can be reduced to the pure differential equation

(26b) $\quad g_0 w_{xt} + g_1(w_{xxxt} + 4w_{xt}w_x + 2w_{xx}w_t) = \left[h_0 w_x + h_1(w_{xxx} + 3w_x^2) \right]_x$

by the simple change of dependent variable

(27) $\quad w \equiv w(x,t) = \int_x^{+\infty} dy\, u(y,t), \quad w_x(x,t) = -u(x,t);$

but, even after this simplification, it still looks sufficiently messy to discourage further investigation. Moreover, this is not, strictly speaking, an evolution equation; so there is hardly any justification to expect the Cauchy problem to make much sense for it (yet, if one looks at the time evolution via the spectral transform, everything looks remarkably simple).

The reader interested in the class of nonlinear PDEs (10) will find some additional information in appendix A.18. Here we focus on the (more important) class of evolution equations that obtain by setting $g(z,t)=1$ in (10). We write them in the form

(28) $\quad u_t(x,t) = \alpha(\boldsymbol{L},t) u_x(x,t),$

which is obtained from (10) with the additional (notational) identification $h(z,t) = \alpha(z,t)$, introduced for the sake of consistency with the notation used in chapter 1 (see (1.2.-1) and (1.4.-2); note however that we are not excluding here the possibility that $\alpha(z,t)$ depend explicitly on the time variable). In the following we moreover generally assume $\alpha(z,t)$ to be a *polynomial* in z,

(29) $\quad \alpha(z,t) = \sum_{m=0}^{M} \alpha_m(t) z^m, \quad M < \infty;$

although much of what follows applies without change if $\alpha(z,t)$ is an entire function of z (or, for that matter, even if it is the ratio of two entire functions, thereby reintroducing again the general case (10)).

As it was noted in section 1.2, highly relevant to the structure of the class of nonlinear evolution equations (28) (with (29)) is the formula

(30) $\quad \boldsymbol{L}^m u_x(x,t) = g_x^{(m)}(x,t), \quad m = 0, 1, 2, \ldots,$

where $g^{(m)}$ is a *polynomial* of degree $m+1$ in $u(x,t)$ and the x-derivatives (up to the order $2m$) of $u(x,t)$ (see (1.2.-3) or (2.4.1.-27); for a proof, see appendix A.9). This formula implies that (28) (with (29)) is a pure *nonlinear partial differential equation* (not an integrodifferential equation), having the form

(31) $\quad u_t(x,t) = \left[\sum_{m=0}^{M} \alpha_m(t) g^{(m)}(x,t) \right]_x.$

We also report the expressions of the first few polynomials $g^{(m)}$ (see (1.2.-3) and (2.4.1.-28))

(32a) $\quad g^{(0)} = u,$

(32b) $\quad g^{(1)} = u_{xx} - 3u^2,$

(32c) $\quad g^{(2)} = u_{xxxx} - 10 u_{xx} u - 5 u_x^2 + 10 u^3.$

Thus, the simplest nontrivial evolution equation contained in the class (28) corresponds to $\alpha(z) = -z$ and reads

(33a) $\quad u_t(x,t) + [u_{xx}(x,t) - 3u^2(x,t)]_x = 0,$

or, equivalently,

(33b) $\quad u_t(x,t) + u_{xxx}(x,t) - 6u_x(x,t)u(x,t) = 0.$

As we already noted in sections 1.2 and 1.8, this is the famous Korteweg–de Vries (KdV) equation; first introduced in 1895 to investigate "the change of form of long waves advancing in a rectangular canal" [KdV1895], it has subsequently appeared in many contexts of application; and a good case can even be made to justify a priori the relevance of its study from a purely mathematical point of view (see for instance [K1974a]). As we noted in section 1.8, arbitrary constants can be introduced in front of each of the terms in (33b) by trivial rescaling of the dependent and independent variables; we see now, from the developments reported above, that the presence of a time dependent coefficient in front of the first term in (33b), and/or, equivalently, in front of the last two terms, can be handled without difficulty (indeed it corresponds merely to a redefinition of the time variable); but, if such time-dependent factors are present, those multiplying the second and third terms in (33b) must coincide (otherwise the equation cannot be brought back into the class (10)). Note moreover that an additional term proportional to $u_x(x,t)$ in (33b) (corresponding to the $m=0$ term in the r.h.s. of (31)) can be eliminated by an appropriate transformation to a moving frame of reference, namely by the change of variable $x' = x + vt$.

The nonlinear evolution equations (31) with $M > 1$ are generally termed "higher KdV equations"; sometimes, more specifically, this name is given only to the nonlinear evolution equations that obtain from the class (31) when all but one of the coefficients α_m vanish (and the remaining one is, without loss of generality, set equal to unity), namely to the equations

(34) $\quad u_t(x,t) = g_x^{(m)}(x,t), \quad m = 2,3,4,\ldots;$

in other cases, less specifically, this name is employed to refer to the whole class (10).

The very simplest equation of the class (28) or (31) obtains for $\alpha(z) = 1$, and is merely the first-order wave equation

(35) $\quad u_t(x,t) = u_x(x,t),$

whose general solution is of course

(36) $\quad u(x,t)=f(x+t).$

Thus, one certainly does not need the spectral transform technique to solve this equation (but advantage will be taken later of the presence of the simple wave equation (35) in the class (28)).

The class (31) (or, for that matter, (28)) of nonlinear evolution equations can be written in compact form for the dependent variable $w(x,t)$, see (27), taking advantage of the *operator identity* (see A9.-30, 5, 4, 3)

(37) $\quad \int_x^{+\infty} dy\, g(L) f_y(y) = -g(\tilde{L}) f(x).$

Here $g(z)$ is an arbitrary polynomial (or, for that matter, any entire function), L is the operator defined by (9) (with x replaced by y, to be sure), $f(x)$ is an arbitrary function (vanishing as $x \to +\infty$) and $\tilde{L}$ is the operator defined by the formula (see (2.4.2.-18) and (27))

(38) $\quad \tilde{L} F(x) = F_{xx}(x) + 4 w_x(x,t) F(x) + 2 \int_x^{+\infty} dy\, w_{yy}(y,t) F(y),$

that specifies its action on a generic function $F(x)$.

A special case of (37) reads (see (27))

(39) $\quad \int_x^{+\infty} dy\, a(L,t) u_y(y,t) = a(\tilde{L},t) w_x(x,t),$

and this formula, together with (27) (that of course implies $w(+\infty, t) = 0$), allows recasting (28) into the form

(40) $\quad w_t(x,t) = a(\tilde{L},t) w_x(x,t).$

Note that, for any polynomial choice of $a(z)$, this is a (nonlinear) partial differential equation (not an integrodifferential equation; see (31)). The main difference between (28) and (40) is the replacement of L by $\tilde{L}$; moreover, while we have always assumed the function $u(x,t)$ to vanish asymptotically with all its derivatives at both ends (see (3)), the function $w(x,t)$ has derivatives that vanish asymptotically at both ends, but itself need vanish only as $x \to +\infty$ (see (27)):

(41) $\quad 0 = w(+\infty, t) = w_x(\pm\infty, t) = w_{xx}(\pm\infty, t) = \cdots.$

So far, we have assumed that the function $u(x,t)$, or equivalently $w(x,t)$, satisfies a nonlinear evolution equation, say (28) or (40), and we have ascertained what is the time evolution of the corresponding spectral transform,

(42a) $\quad R_t(k,t) = 2ik\alpha(-4k^2, t) R(k,t),$

(42b) $\quad \dot{p}_n(t) = 0,$

(42c) $\quad \dot{\rho}_n(t) = -2 p_n \alpha(4p_n^2, t) \rho_n(t).$

Let us now verify that, if the spectral transform evolves according to (42), $w(x,t)$ does indeed evolve according to (40), and correspondingly $u(x,t)$ according to (28) (this is merely a check of consistency, since this result is implied by the one-to-one correspondence between any function, within the class of *bona fide* potentials, and its spectral transform). This is most easily done by using the spectral integrals relations of section 2.4.2, in particular those accounting for the case of infinitesimal change, (2.4.2.-15) and (2.4.2.-16), which may in the present context be rewritten in the form

(43) $\quad w_t(x,t) = -2 \sum_{n=1}^{N} \{ 2 f_p^{(+)}(x, ip_n, t) f^{(+)}(x, ip_n, t) \rho_n(t) \dot{p}_n(t)$

$\quad + [f^{(+)}(x, ip_n, t)]^2 \dot{\rho}_n \} - \pi^{-1} \int_{-\infty}^{+\infty} dk \, [f^{(+)}(x, k, t)]^2 R_t(k,t),$

(44) $\quad \alpha(\tilde{L}, t) w_x(x,t) = 4 \sum_{n=1}^{N} [f^{(+)}(x, ip_n, t)]^2 \alpha(4p_n^2, t) p_n(t) \rho_n(t)$

$\quad - \pi^{-1} \int_{-\infty}^{+\infty} dk \, [f^{(+)}(x, k, t)]^2 2ik\alpha(-4k^2, t) R(k,t).$

It is then clear by inspection that (42) yields (40).

Let us finally reemphasize that, at least for real equations, namely if $\alpha(z,t)$ in (28) or (40) is real, then (42a) implies that only the phase of $R(k,t)$ evolves in time, but not its modulus:

(45a) $\quad R(k,t) = R(k,0) \exp\left[2ik \int_0^t dt' \alpha(-4k^2, t') \right],$

(45b) $\quad |R(k,t)| = |R(k,0)|, \quad \text{Im } k = 0.$

This has a number of (desirable) consequences. First of all it guarantees that the two important properties of $R(k,t)$, that reflect its suitability as a reflection coefficient, namely to vanish asymptotically ($k \to \pm \infty$) and to have a modulus that does not exceed unity, are preserved by the time evolution; and also preserved is the symmetry property (2.1-12b) (namely, all these properties hold for all time if they hold at the initial time). It is thus seen that, if $u(x,0)$ is a *bona fide* potential, so is $u(x,t)$ for $t \neq 0$; a neat proof of the existence and uniqueness of the solutions of the class of nonlinear evolution equations (28) or (40) (within the functional class of *bona fide* potentials).

Moreover, since the transmission coefficient is determined once the modulus of the reflection coefficient and the discrete eigenvalues are assigned (see the end of section 2.1), and since all these quantities are time-independent (see (42b) and (45b)), there follows that the transmission coefficient $T(k)$ corresponding to $u(x,t)$ is itself time-independent when $u(x,t)$ evolves in time according to (28). Indeed a direct proof of this important result can be easily given using (2.4.1-25) and (2.4.1.-26). But we postpone it to chapter 5, where we shall use this result to obtain an infinite set of conservation laws for the class of evolution equations (28).

3.2. Analysis of the solutions

In this section we discuss the qualitative and quantitative information on the behaviour of the solutions of the class of evolution equations discussed in the preceding section 3.1. We assume however that no explicit time dependence occurs in the equation, since we are interested to understand the type of time evolution induced by the structure of the evolution equations rather than that due to an explicit time dependence of the coefficients. Thus we restrict our attention throughout this section to the class of evolution equations

(1a) $\quad u_t(x,t) = \alpha(L) u_x(x,t)$

or, equivalently,

(1b) $\quad w_t(x,t) = \alpha(\tilde{L}) w_x(x,t)$,

(with L and $\tilde{L}$ defined by (3.1.-9) and (3.1.-38), and with $u(x,t)$ and $w(x,t)$ related by (3.1.-27) and characterized by the boundary conditions (3.1.-3) and (3.1.-41)). Let us recall that, for any polynomial choice of $\alpha(z)$, both (1a) and (1b) are nonlinear partial differential equations (not integrodifferential equations). The corresponding evolution of the spectral transform reads (see (3.1.-42)):

(2a) $\quad R(k,t) = R(k,0) \exp[2\mathrm{i} k \alpha(-4k^2)t],$

(2b) $\quad p_n(t) = p_n(0) = p_n,$

(2c) $\quad \rho_n(t) = \rho_n(0) \exp[-2p_n \alpha(4p_n^2)t].$

We do not repeat here the points already made in section 1.6, that the meticulous reader may wish to reread before proceeding further. In subsection 3.2.1 we discuss the *solitons* in somewhat more detail than it was done in subsection 1.6.1; we also present a number of graphs, and discuss the distinction between *solitons* and *solitary waves* (or *solitrons*, as we propose to call them). In subsection 3.2.2 we take up again the "weak field" limit of these nonlinear evolution equations, mainly to supplement the treatment of section 1.5. In subsection 3.2.3 we review tersely the qualitative behaviour of the generic solution of (1) as time evolves, with particular reference to the asymptotic outcome as $t \to \infty$. Finally in subsection 3.2.4 we discuss certain special solutions of the class of evolution equations (1a), or rather of the simplest nontrivial member of this class, the KdV equation (3.1.-33).

3.2.1. Solitons and solitrons

At the end of subsection 1.6.1 the possibility was mentioned to identify at any time one part of the solution of (3.2.-1) as being associated with one particular soliton. Let us recall that there is no difficulty to do this in the spectral space; indeed there each soliton is clearly associated with one discrete eigenvalue, p_n (note however that the actual position of the nth soliton in x-space, say in the asymptotic $t \to \pm \infty$ limit when it generally separates out, does not depend only on the nth normalization coefficient $\rho_n(t)$, as evidenced for instance by the shift in the asymptotic position of each soliton produced by its interaction with other solitons, see section 1.6.1; or see the explicit analysis of the 2-soliton solution in appendix A.10). But the nonlinear character of the relation between a function and its spectral transform implies that there is no clear univocous definition of any part of a function $u(x)$ as corresponding to one component of the spectral transform.

A pseudo-linear relation between a function and its spectral transform can however be written, by introducing the eigenfunctions of the spectral problem. Particularly appealing from this point of view are the two formulae (1.3.3.-2) and (1.3.3.-3), that we now write in the form

(1) $$u(x,t) = -4 \sum_{n=1}^{N} p_n [\varphi_n(x,t)]^2$$
$$+ \pi^{-1} \int_{-\infty}^{+\infty} dk \, [f^{(+)}(x,k,t)]^2 2ikR(k,t),$$

(2) $$w(x,t) = -2 \sum_{n=1}^{N} f_n(x,t) \exp(-p_n x) \rho_n(t)$$
$$- \pi^{-1} \int_{-\infty}^{+\infty} dk \, f^{(+)}(x,k,t) \exp(ikx) R(k,t).$$

These formulae are the special cases of (2.4.2.-16) and (2.4.2.-21) corresponding to $g(z) = 1$; of course $\{R(k,t), -\infty < k < +\infty; p_n, \rho_n(t), n = 1, 2, \ldots, N\}$ is the spectral transform of $u(x,t)$, $f^{(+)}(x,k,t)$ is the Jost solution corresponding to the "potential" $u(x,t)$, see for instance (2.1.-1) and (2.1.-4), while $f_n(x,t)$ and $\varphi_n(x,t)$ are the eigenfunctions corresponding to the discrete eigenvalue p_n, see for instance (2.1.-21), (2.1.-22), (2.1.-23) and (2.1.-24).

The appealing aspect of (1) and (2) is that the r.h.s. has an additive character; in particular the contributions of the different discrete eigenvalues are neatly summed up. It is therefore suggestive to interpret each term in the sum in the r.h.s. as representing one soliton (in x-space, for all time). A first difficulty is the availability of two different options, based respectively on (1) or (2). Note that they are not equivalent, at least from the point of view of interest here; for, while it is undoubtedly true that, consistently with the identity

$$w(x,t) = \int_x^{+\infty} dy \, u(y,t), \qquad u(x,t) = -w_x(x,t),$$

the derivative of the r.h.s. of (2) equals (up to a change of sign) the r.h.s. of (1), the equality holds only globally, but it does not hold term by term (except in the very special case of the single-soliton solution, when the integrals in the r.h.s. of (1) and (2) disappear and the sums contain one term only; see (2.2.-17), (2.2.-18) and (2.2.-19)).

A second difficulty is that the additive character of the r.h.s. of (1) and (2) is somewhat deceptive, for the intrinsic nonlinearity of the relationship between a function and its spectral transform shows up through the dependence of the eigenfunctions on the global potential (for instance, $\varphi_n(x,t)$ is the normalized eigenfunction corresponding to the eigenvalue $-p_n^2$ of the Schroedinger equation with the overall potential $u(x,t)$).

Whether or not each term in the r.h.s. of (1) or (2) can be *defined* to provide the representation of a soliton in x-space throughout the time evolution is of course a matter of choice. To ascertain which, if any, of the two possible definitions, based on (1) or (2), appears more appropriate, we display in figures 1–4 the two-soliton solution of the KdV equation

$$(3) \quad u_t + u_{xxx} - 6uu_x = 0, \quad u \equiv u(x,t),$$

in the two forms

$$(4) \quad u(x,t) = -4 \sum_{n=1}^{2} p_n \rho_n(t) [f_n(x,t)]^2,$$

$$(5) \quad u(x,t) = 2 \sum_{n=1}^{2} \rho_n(t) [f_n(x,t) \exp(-p_n x)]_x,$$

corresponding respectively to (1) (recall (2.1.-24) and (2.1.-25)) and (2). Here of course (see section 1.6.1. or (2.2.-26, 27, 28, 29) and (2.1.-24, 25))

$$(6) \quad \rho_n(t) = 2p_n \exp[2p_n \xi_n(t)], \quad n = 1,2,$$

$$(7) \quad \xi_n(t) = \xi_{0n} + 4p_n^2 t, \quad n = 1,2,$$

$$(8) \quad f_n(x,t) + \sum_{m=1}^{2} \rho_m(t)(p_n + p_m)^{-1} \exp[-(p_n + p_m)x] f_m(x,t)$$
$$= \exp(-p_n x), \quad n = 1,2.$$

It is then easily seen that neither one of the two definitions is entirely satisfactory; certainly not the second, that appears particularly inappropriate since the actual solution emerges after important cancellations of the two terms in the r.h.s. of (5); but the first is also unsatisfactory, as it is for instance seen in the sixth snapshot of the series of figs. 1 or 3, where the graph that should represent *one* soliton has clearly two peaks. Of course, when the solitons are well separated (and therefore, in particular, in the remote future and past), each term in the two sums coincides with a single-soliton expression (appropriately localized); it is instead when the solitons are close to each other that the relevant terms get deformed, since the wave function of each "bound state" in the Schroedinger spectral

3.2.1

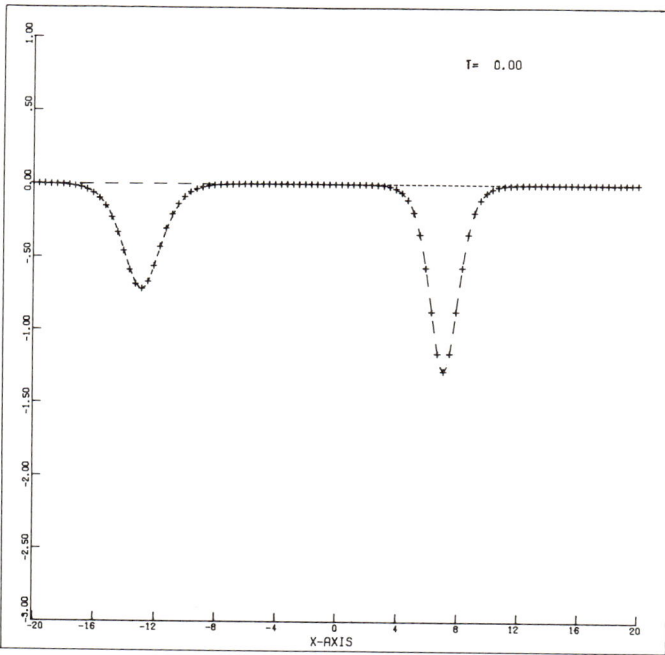

Figure 1. 11 snapshots equally spaced in time of the two-soliton solution of the KdV equation (3), as seen in the CM frame of reference, moving with speed $v=4(p_1^3+p_2^3)/(p_1+p_2)=4(p_1^2-p_1p_2+p_2^2)$. The soliton parameters are $p_1=\frac{3}{5}$, $p_2=\frac{4}{5}$, $\xi_{01}=-9.6$, $\xi_{02}=7.2$. The crosses indicate the complete solution; each of the other two graphs (broken lines) represent one term in the r.h.s. of (4).

problem is influenced, in a nonlinear fashion, by the overall structure of the "potential" $u(x, t)$ (this does not happen when $u(x, t)$ factors into the sum of N separated wells, with one "bound state" localized in each well; note incidentally that this requires that the distance between such wells be large not only in comparison to their size, which is clearly of order p_n^{-1}, but also relative to the differences $|p_n-p_m|^{-1}$; see appendix A.10).

The sequence of snapshots of figs. 1 and 3 (or, equivalently, figs. 2 and 4) are also worthy of scrutiny inasmuch as they display the phenomenon of (elastic) soliton collision (to see this one should now ignore the broken-line graphs and look only at the crosses). The two examples have actually been chosen so as to display the two typical modes according to which the crossover of two solitons occurs; either by an "exchange of identity" (as the solitons get close, the taller and thinner one decreases in height and grows fatter, while the smaller and fatter one grows in height and becomes thinner;

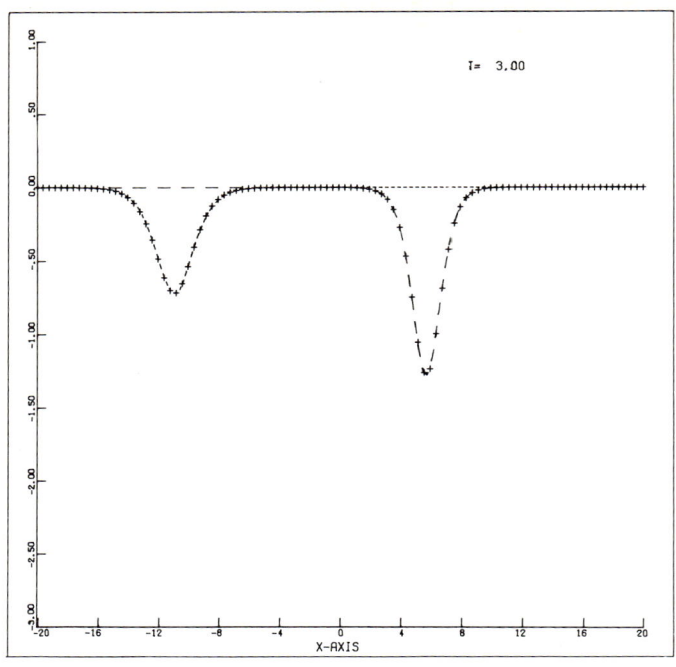

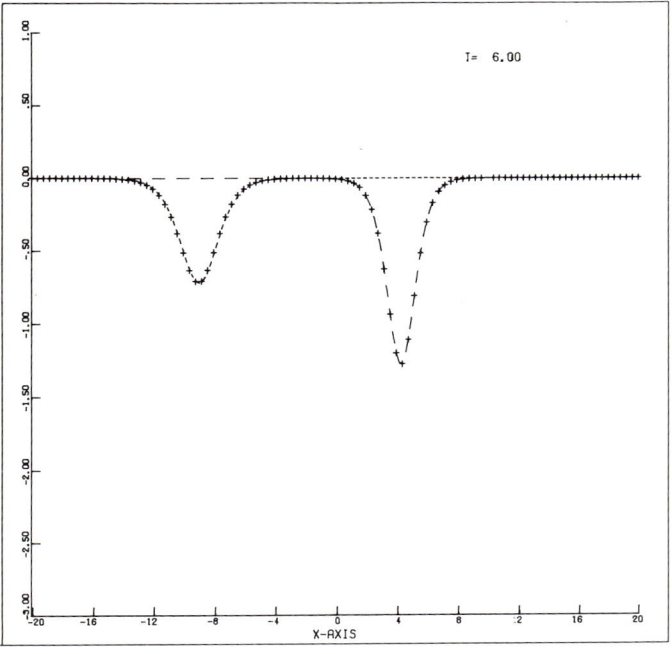

Figure 1 (contd.).

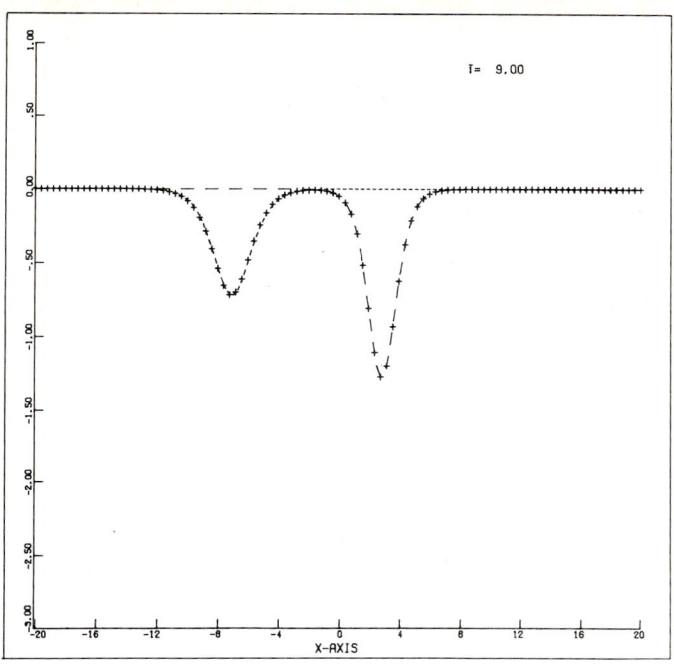

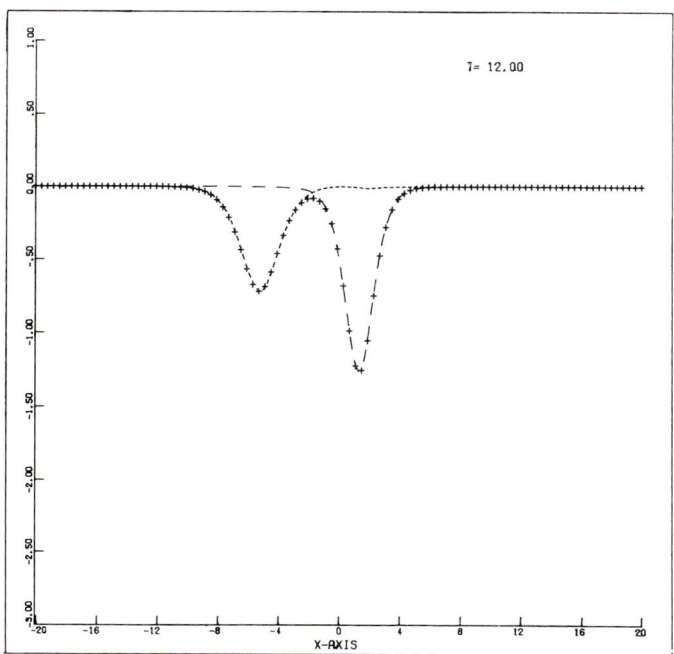

Figure 1 (contd.).

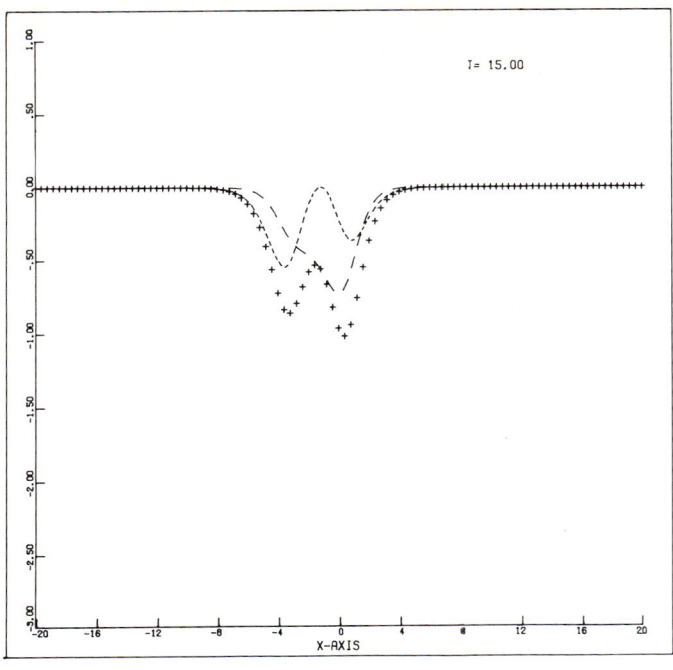

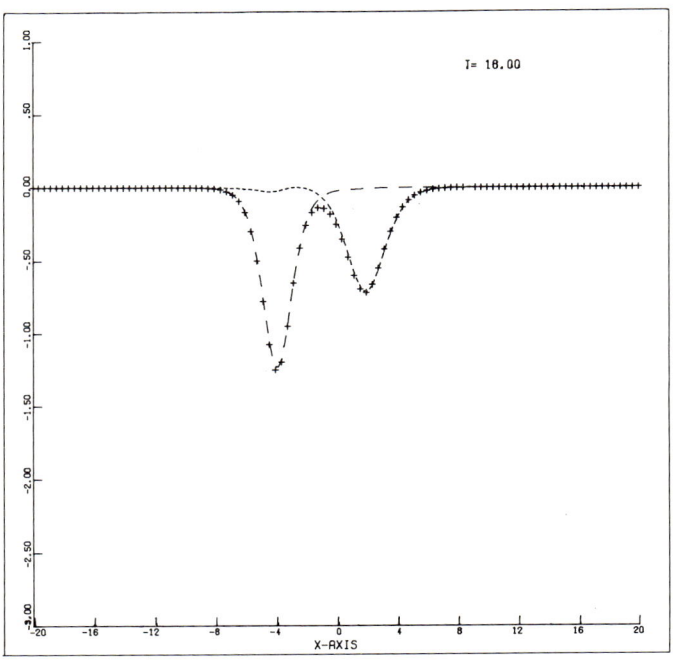

Figure 1 (contd.).

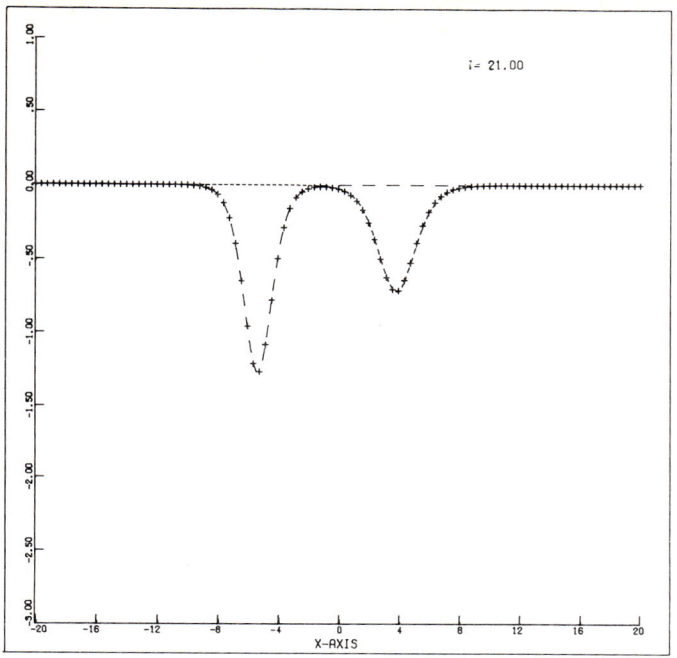

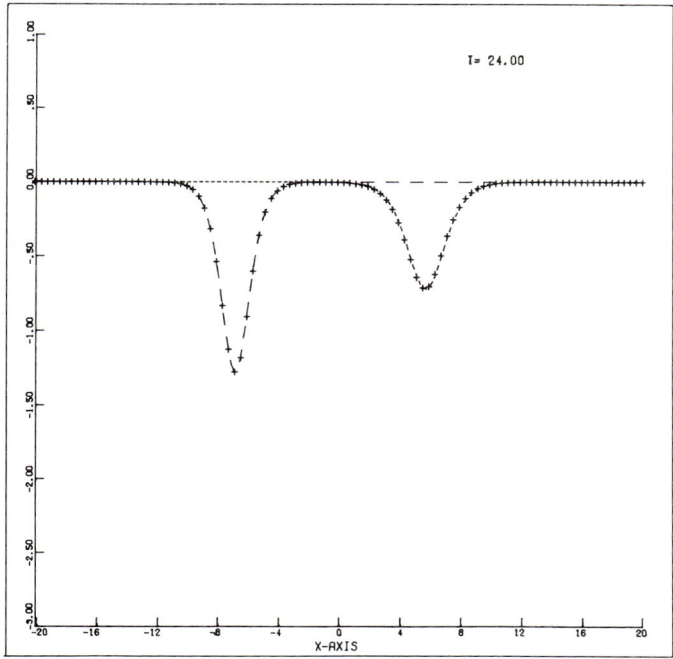

Figure 1 (contd.).

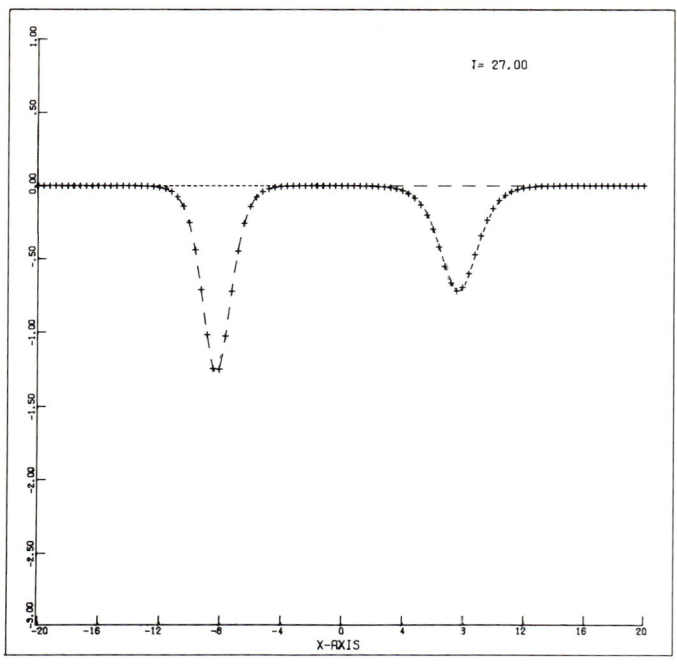
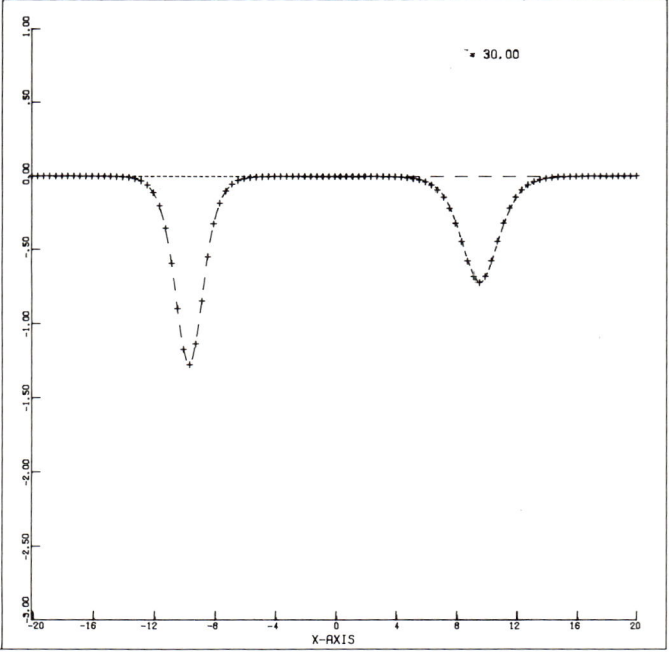

Figure 1 (contd.).

3.2.1

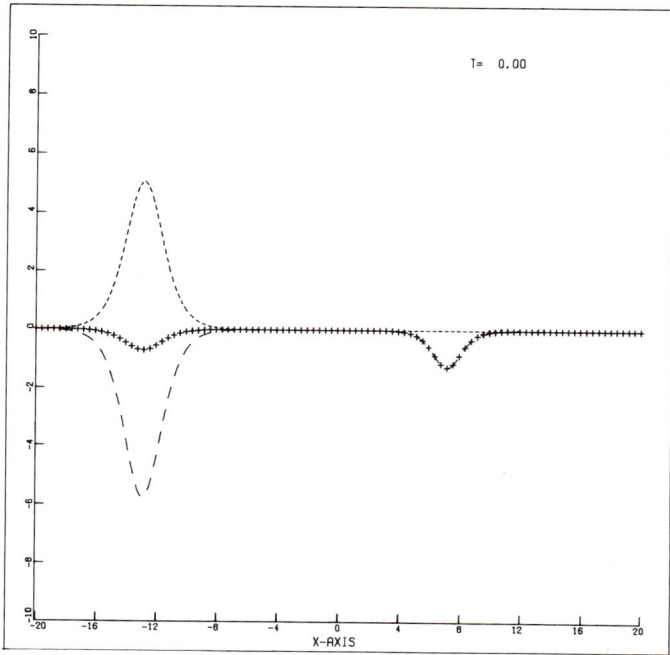

Figure 2. This represents the same solution as in figure 1, but decomposed according to (5) rather than (4). Note also the change of the vertical scale.

then they separate, without having merged; see figs. 1 or 2); or through a "loss of identity" (the two solitons coalesce temporarily into a single lump, then emerge out of it again, apparently going through each other; see figs. 3 or 4).

Clearly the occurence of solutions, such as those we are discussing here, of nonlinear field theories, that behave asymptotically as a collection of separated localized entities moving as free particles, may appear quite appealing in the context of elementary particle physics; it should be emphasized that one witnesses here a mechanism yielding a particle-like behaviour out of a classical (i.e., non quantized) field theory. Most likely Einstein, who to the end of his life was skeptical about the need to quantize, would have fancied this development. Moreover, even if one insists that the basis of elementary particle theory is some sort of (second) quantized nonlinear field theory, the question arises, if the classical counterpart of that field theory possesses solutions displaying properties of localization and permanence analogous to those described above, whether such solutions are not given to play a special rôle also in the quantized context; that this may

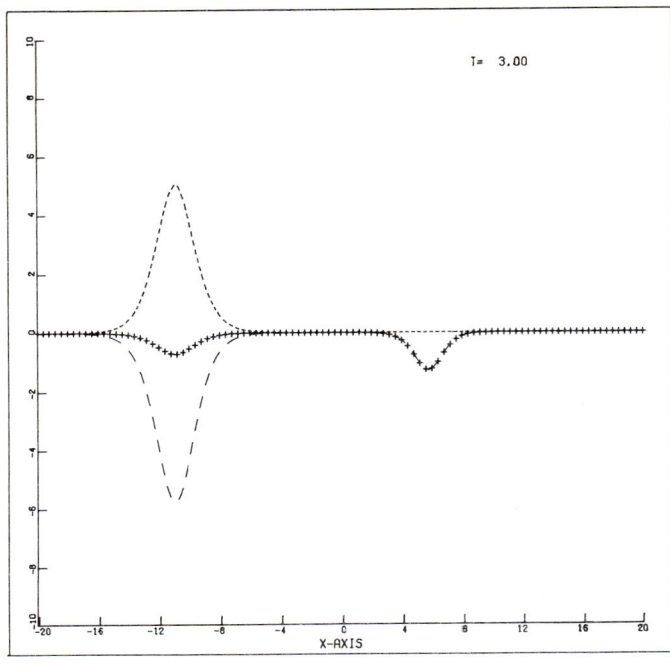

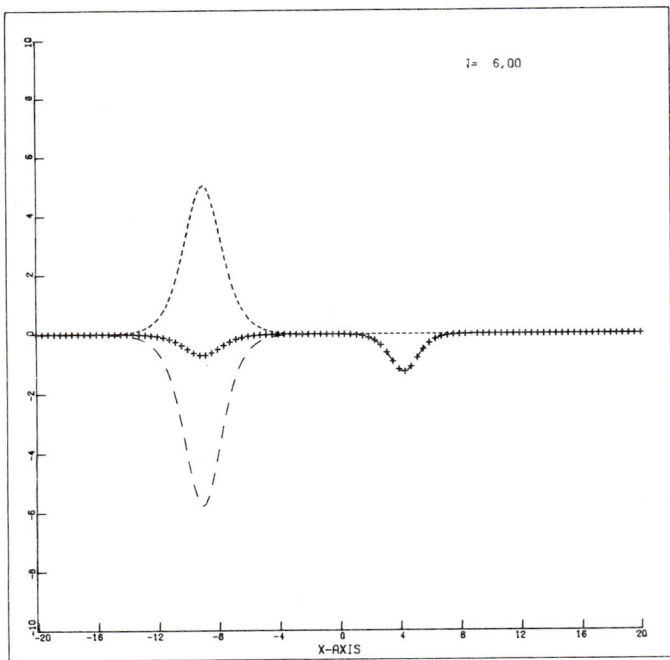

Figure 2 (contd.).

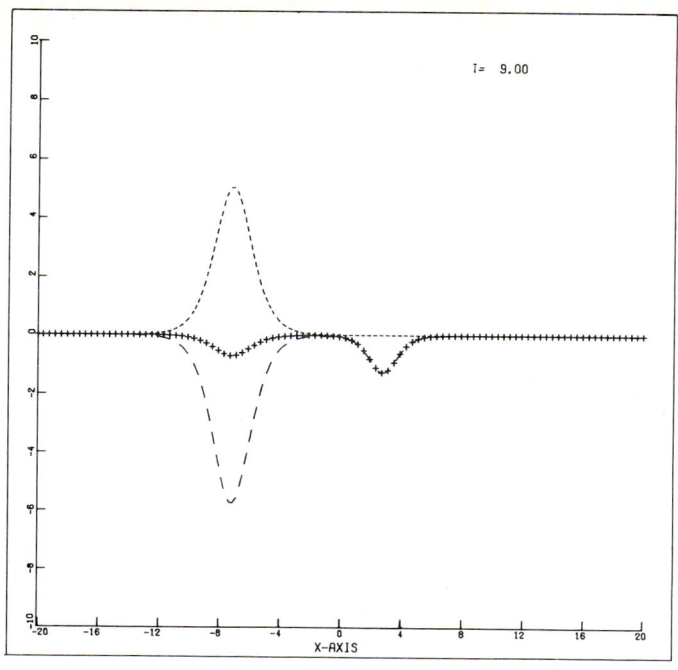

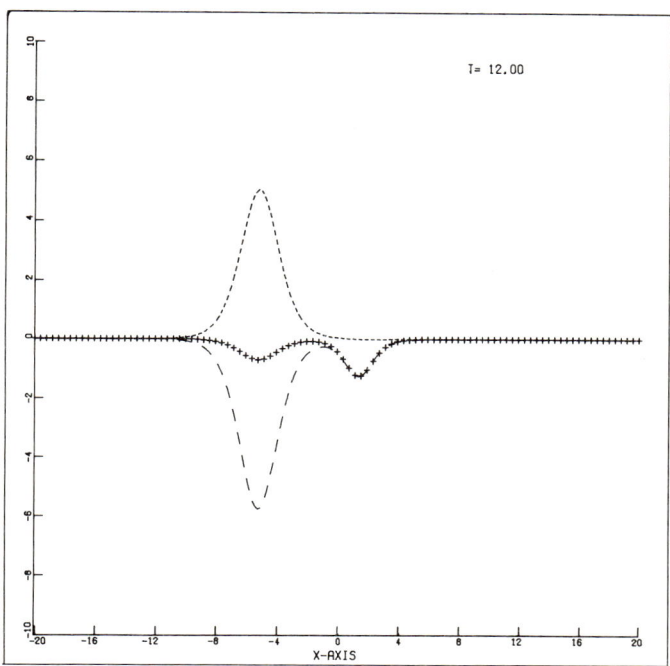

Figure 2 (contd.).

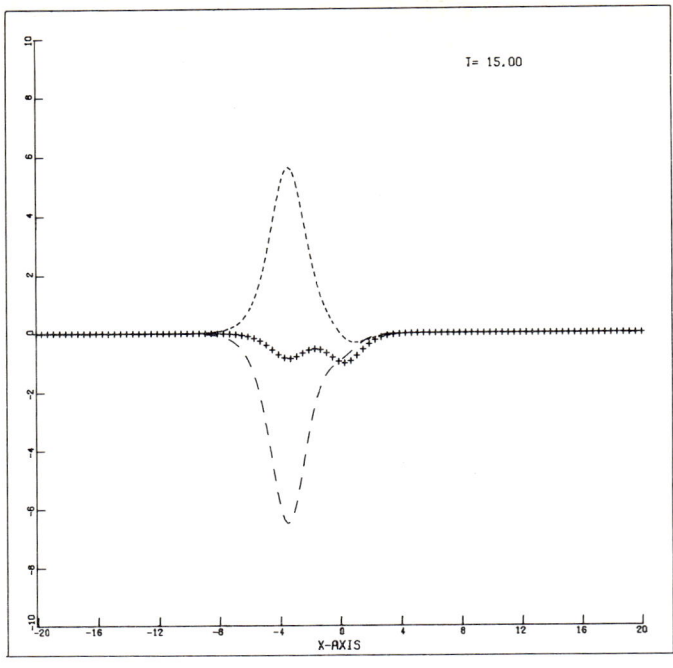

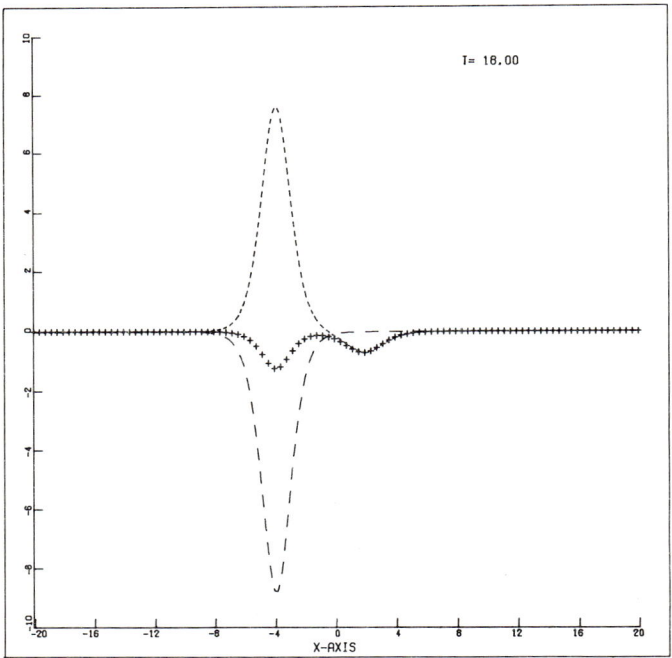

Figure 2 (contd.).

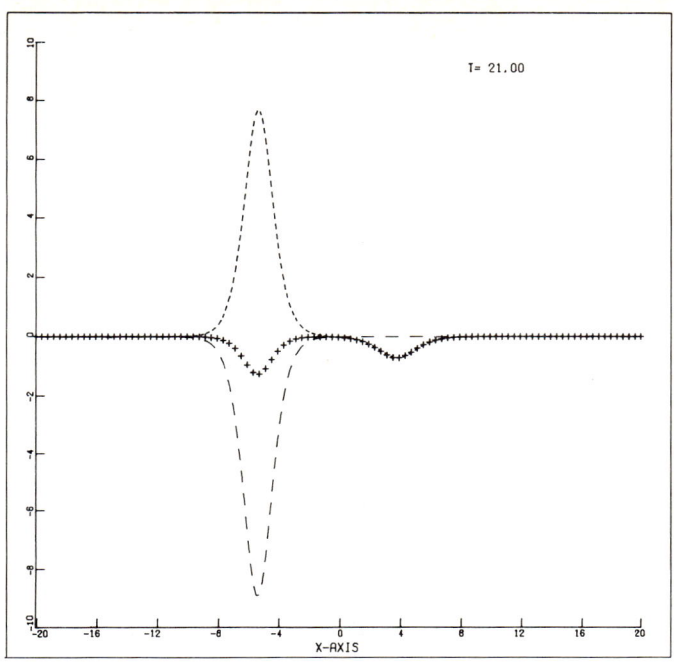

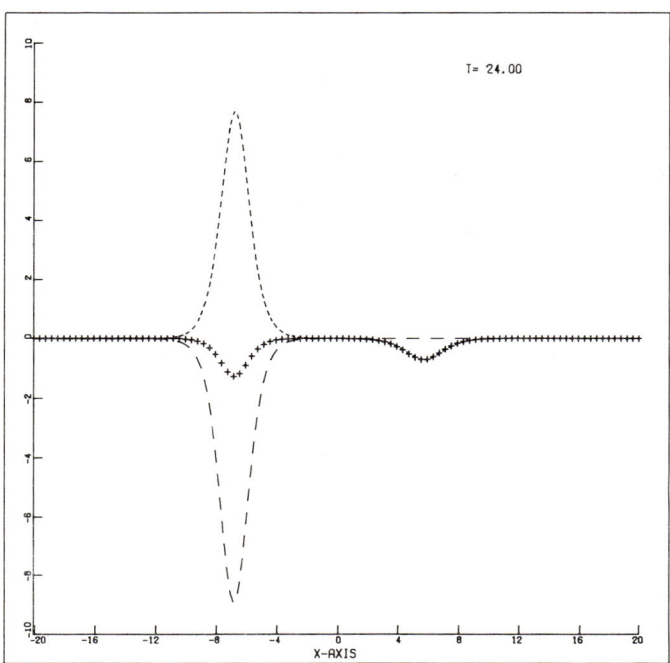

Figure 2 (contd.).

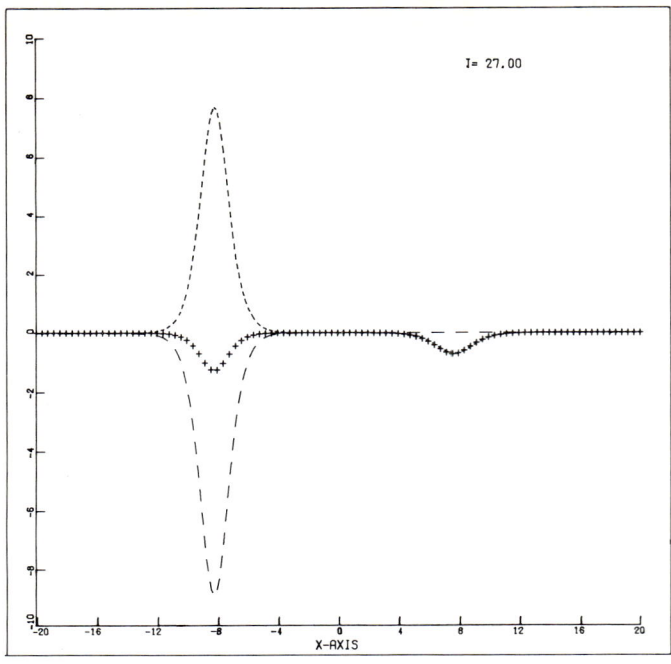
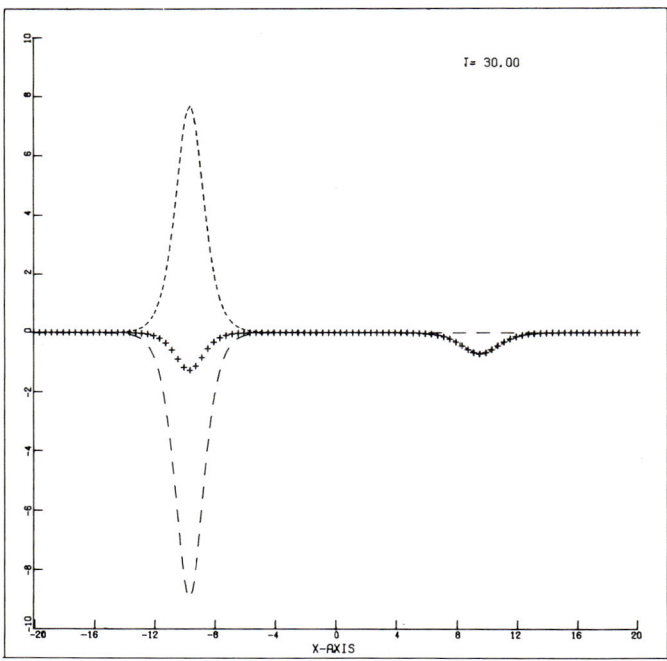

Figure 2 (contd.).

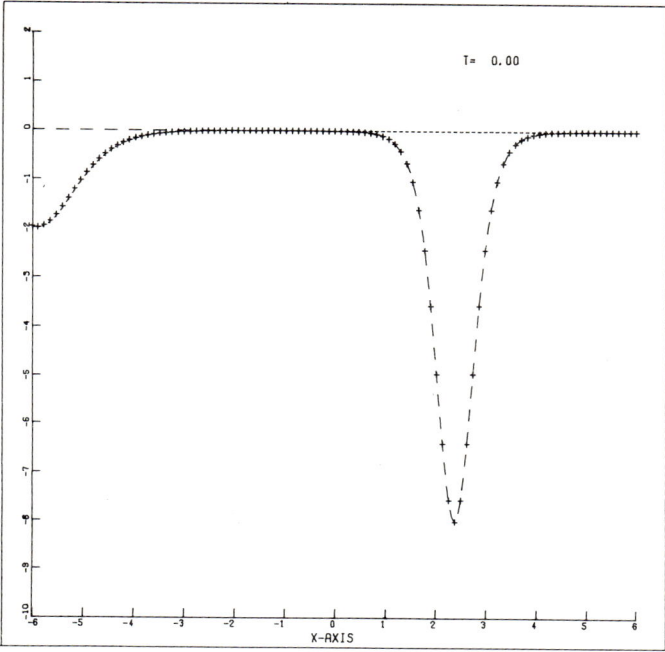

Figure 3. Same as figure 1, but for different soliton parameters:
$p_1=1, p_2=2, \xi_{01}=-4.8, \xi_{02}=2.4$.

indeed be the case is confirmed by the investigation of certain examples, notably the so-called Sine–Gordon equation.

In the following, we will have little to say about elementary particles, since this topic, as well as all other applications, are outside the scope of this treatment. Yet some additional words are perhaps needed here, to clear up a widespread misunderstanding.

As should be clear from the terse discussion outlined above, the essential element for (prospective) elementary particle applications is the existence of solutions that remain localized throughout the time-evolution (moreover, of course, one should actually work in 3-dimensional, rather than 1-dimensional, space; but we ignore this detail here). Recalling the treatment of *linear* evolution equations given in section 1, we see that this phenomenon can occur, for the class of evolution equations of type

(9) $\quad u_t = F[u, u_x, u_{xx}, \ldots; t], \quad u \equiv u(x, t),$

only if F is nonlinear in u and/or its derivatives (except for the single trivial

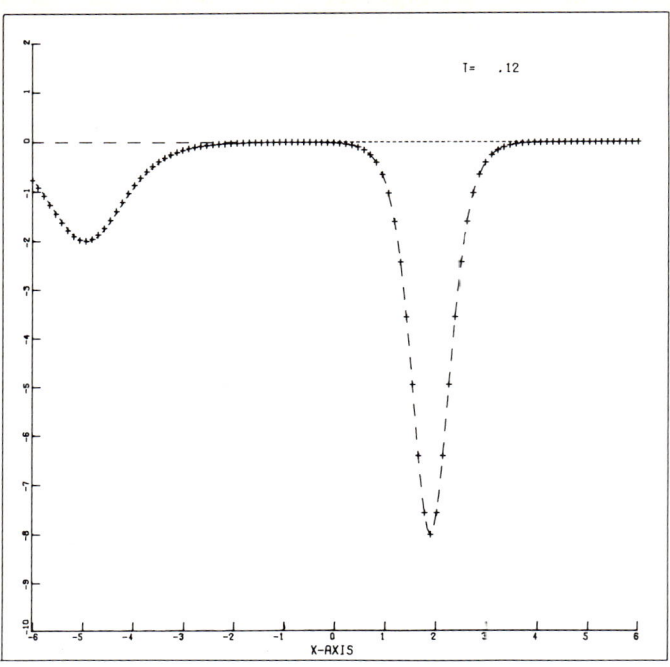

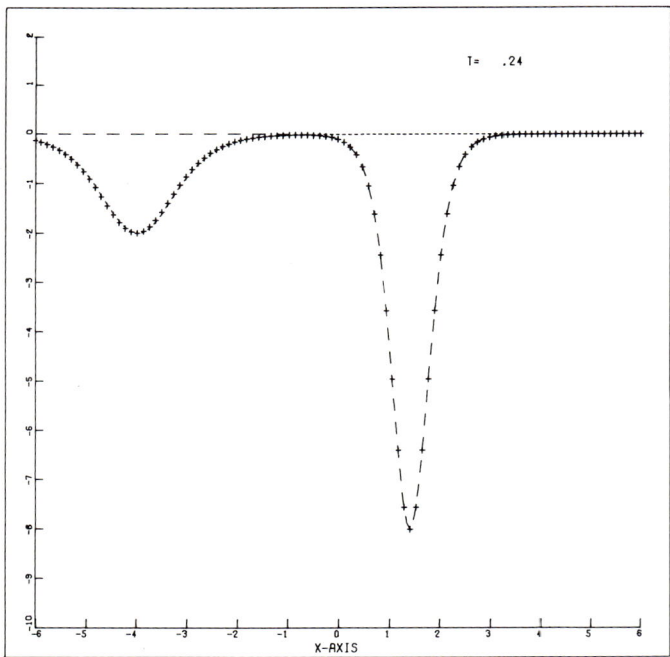

Figure 3 (contd.).

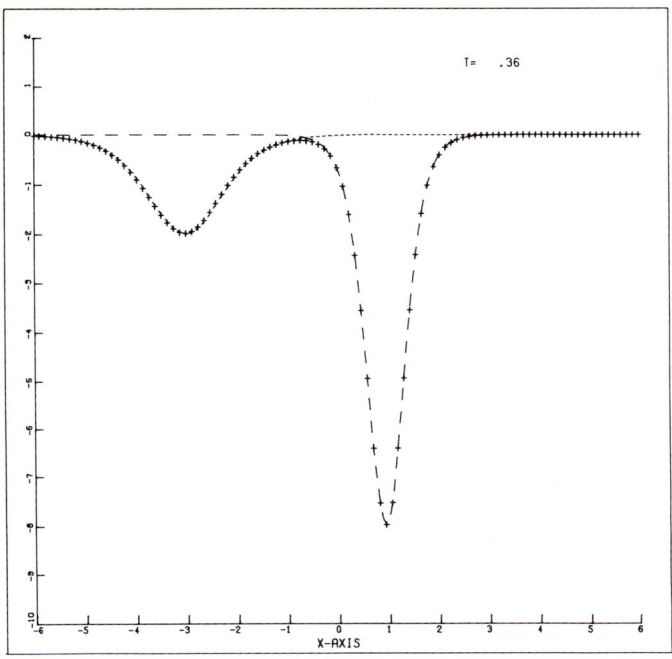

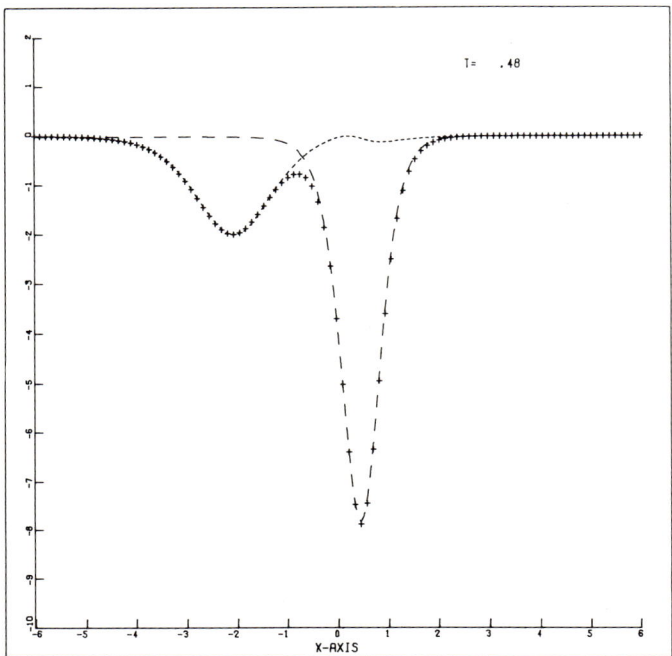

Figure 3 (contd.).

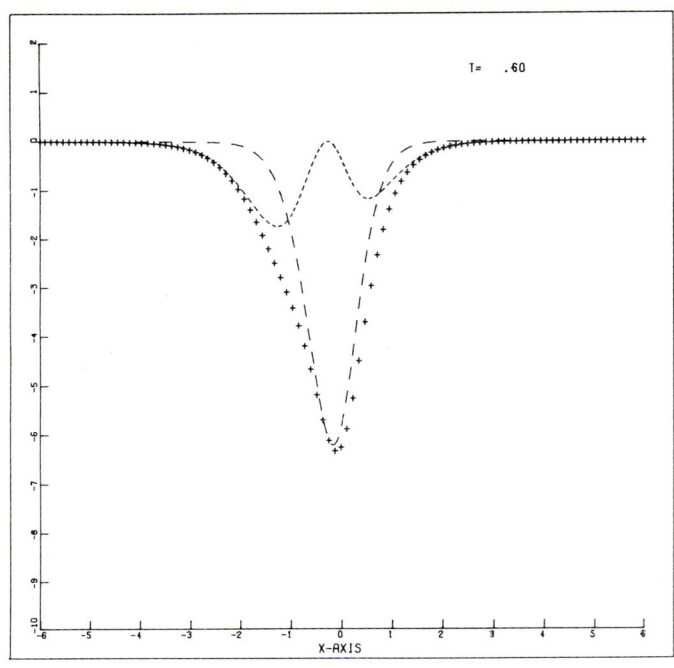
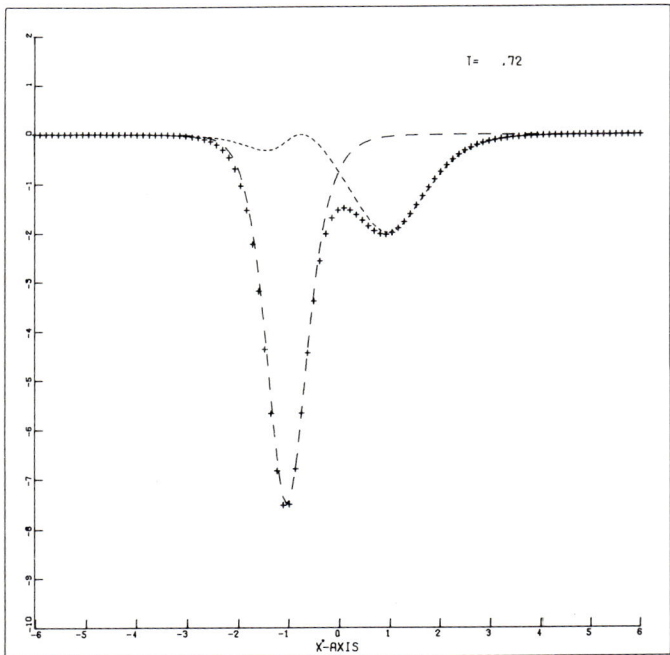

Figure 3 (contd.).

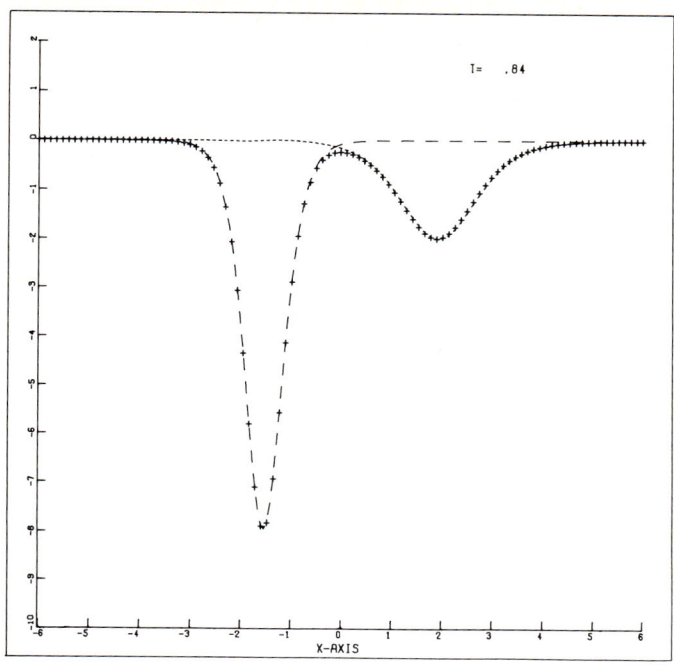
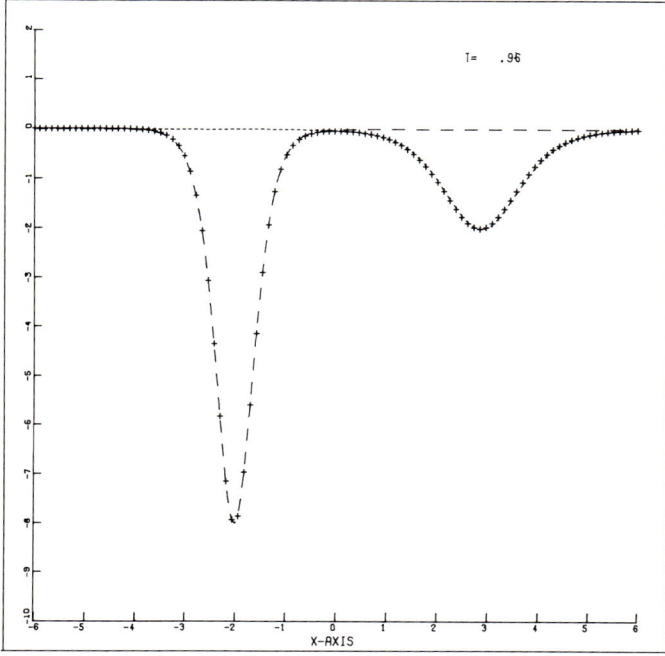

Figure 3 (contd.).

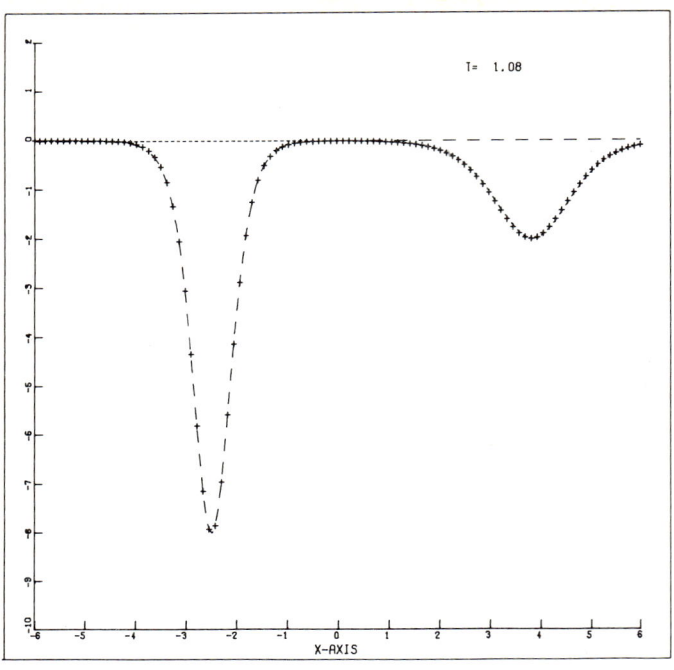

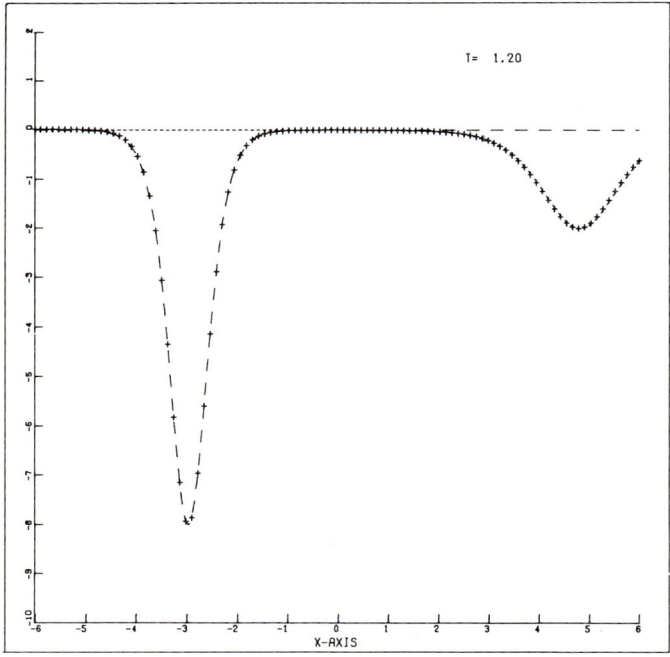

Figure 3 (contd.).

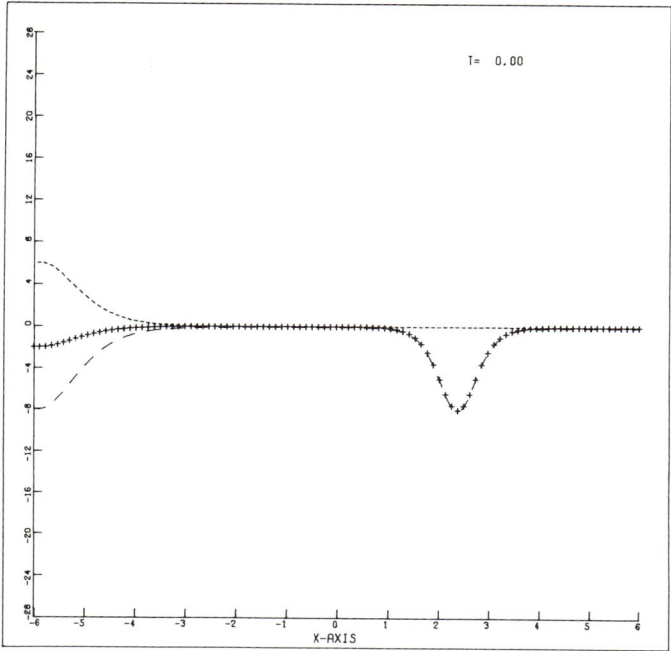

Figure 4. Same as figure 3, but decomposed according to (5) rather than (4). Note again the change of the vertical scale.

case of the first order wave equation). Indeed the mechanism whereby solutions of (9) remain localized throughout their time-evolution may be viewed as due to a compensation between dispersive and nonlinear effects, the former tending generally to delocalize the solution, the latter having (in appropriate cases) precisely the opposite influence. For instance, in the case of the KdV equation (3), elimination of the nonlinear part yields

(10) $\quad u_t = -u_{xxx}, \quad u \equiv u(x,t),$

that has the form (1.1.-1) with $\omega(z) = -z^3$; thus all solutions of this equation have the typical dispersive behaviour, namely they spread out as $t \to \infty$ (so that their amplitude vanishes proportionally to $t^{-1/2}$). On the other hand, elimination of the (linear) dispersive part in the r.h.s. of (3) yields the "shock equation"

(11) $\quad u_t = 6u_x u, \quad u \equiv u(x,t),$

whose solution, given in terms of the initial Cauchy datum $u(x,0) = u_0(x)$

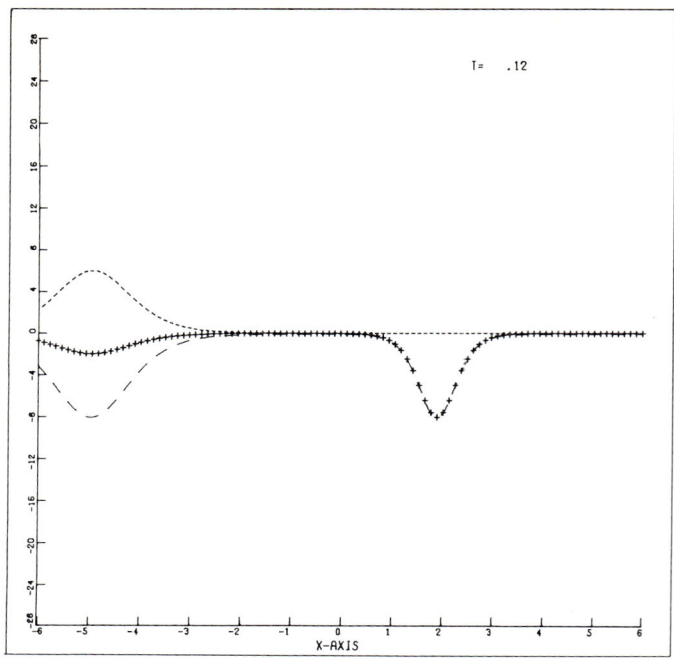

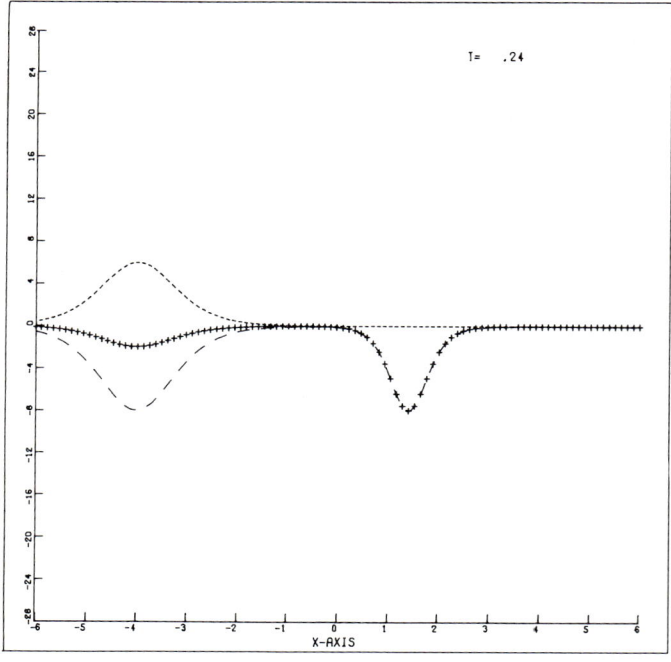

Figure 4 (contd.).

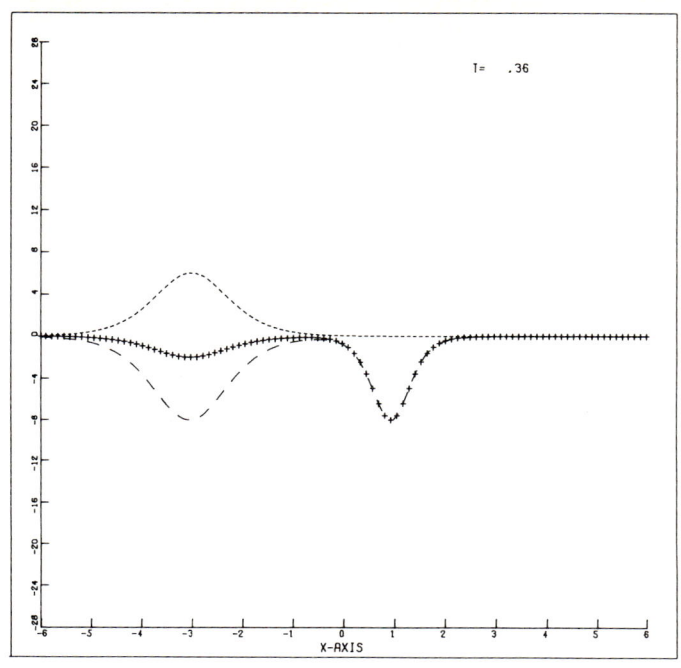

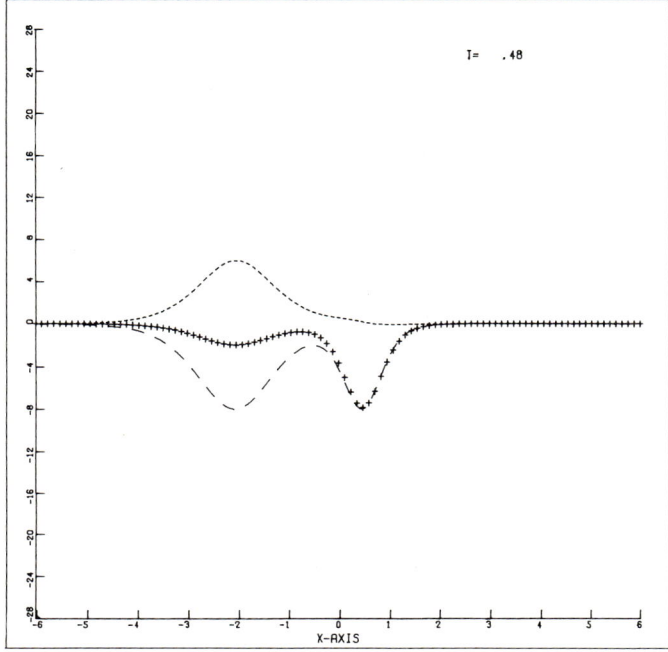

Figure 4 (contd.).

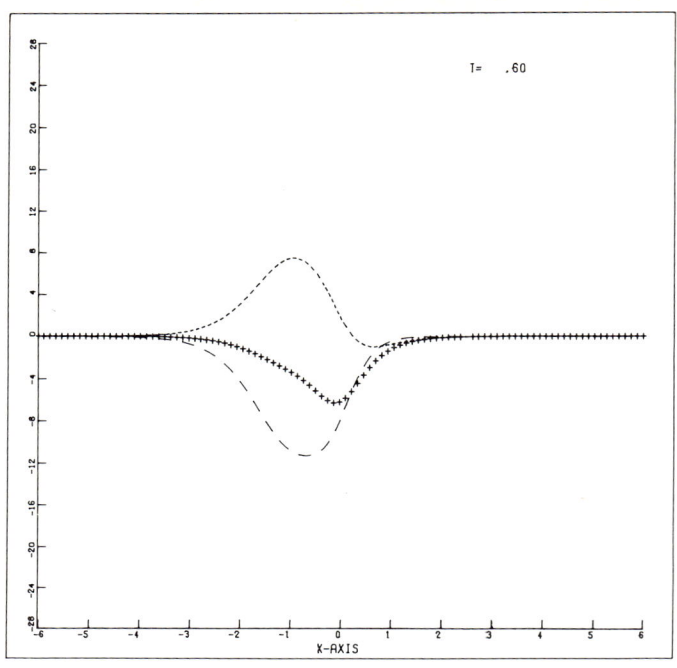

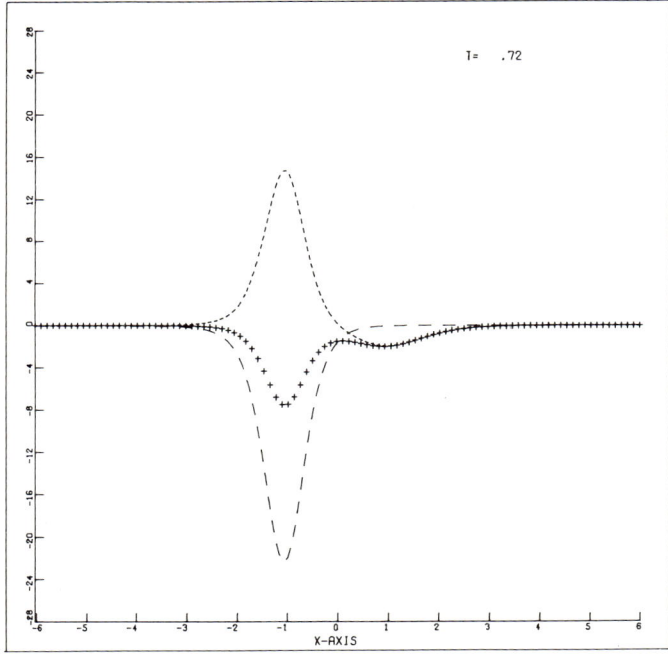

Figure 4 (contd.).

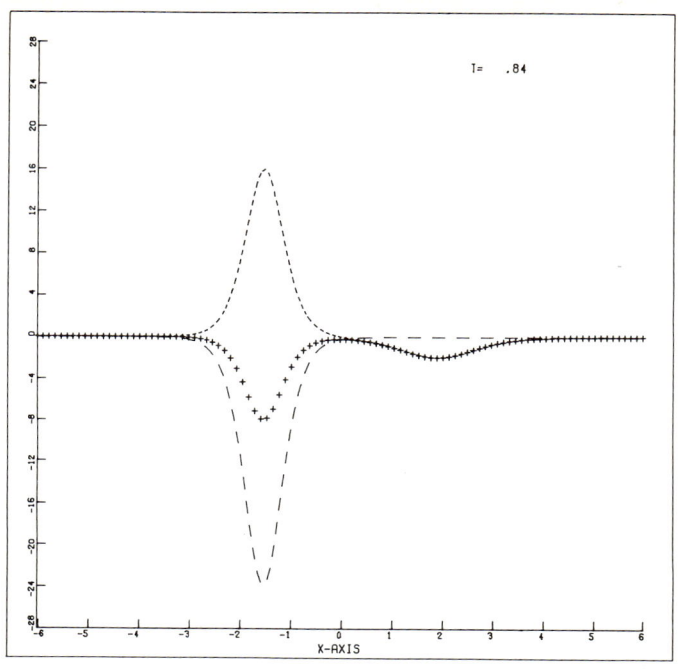

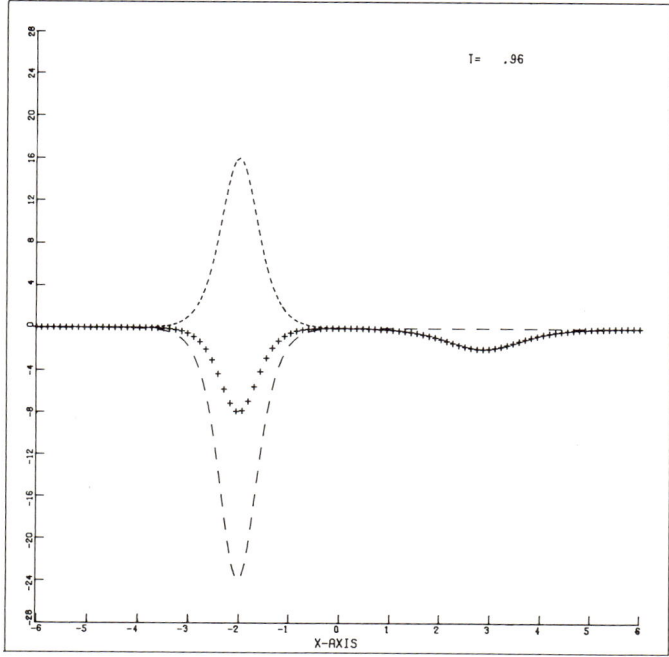

Figure 4 (contd.).

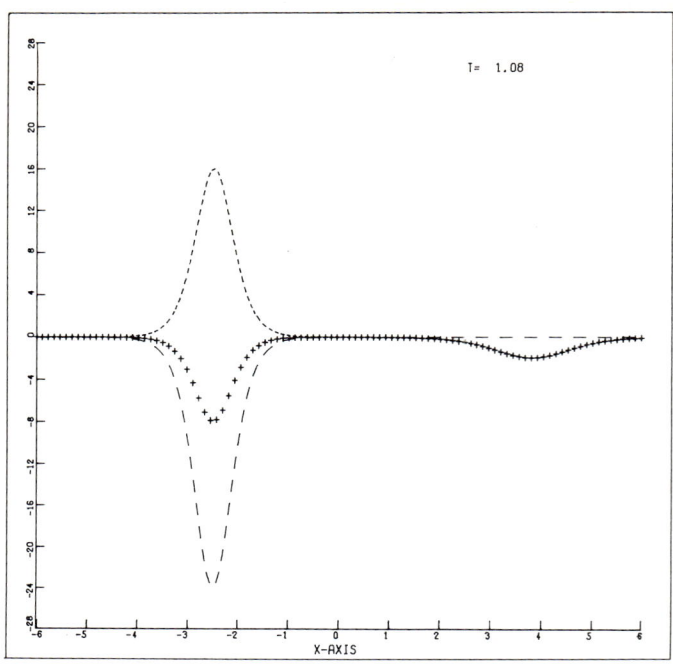
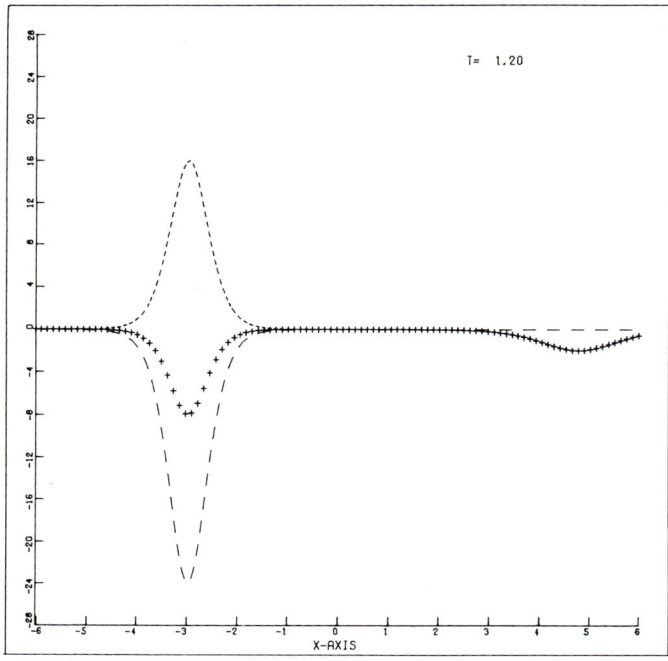

Figure 4 (contd.).

by the "explicit–implicit" formula

(12) $\quad u(x,t) = u_0[x + 6tu(x,t)],$

if initially localized, develops a steeper and steeper profile so that its x-derivative eventually blows up after some (finite) time.

The possibility that this mechanism of compensation operate is clearly a rather general feature; thus the class of nonlinear evolution equations of type (9) that do possess solutions that evolve in time as *solitary waves*, traveling without change of form or speed, is rather large. Moreover the problem of identifying such solutions is generally reducible to that of solving an *ordinary* nonlinear differential equation. We illustrate this below by considering two specific examples; we trust this suffices to convey the general applicability of the approach.

Let us first consider again the KdV equation, writing it however now in the form

(13) $\quad u_t = -u_x - u_{xxx} + 6u_x u, \quad u \equiv u(x,t).$

The additional term $-u_x$ in the r.h.s. has been inserted to facilitate comparison with the case treated below; we have already mentioned that this modification of the KdV equation is a trivial one, since the additional term is eliminated by the transformation to the novel space variable $x' = x - t$.

We search now for a solution of this equation having the form

(14) $\quad u(x,t) = f(x - vt),$

$f(z)$ being a function to be determined. Insertion of this *ansatz* in (13) yields for $f(z)$ the ordinary nonlinear differential equation

(15) $\quad f''' + (1-v)f' - 6f'f = 0, \quad f \equiv f(z).$

Here primes indicate differentiation. This ODE is easily integrated: a first time, by inspection; a second time, again by inspection, after multiplication by f'; and finally by separation (do as exercise!). It is thereby seen that the only solution that is regular for all real values of z and that vanishes as $z \to \pm \infty$ has the form

(16) $\quad f(z) = A / \cosh^2[p(z - z_0)],$

z_0 being an arbitrary real constant, and the two constants A and p being related to each other and to v by

(17a) $\quad A=-2p^2$,

(17b) $\quad v=4p^2+1$.

We have thus recovered the single-soliton solution (1.6.1.-5) (with (1.6.1.-6) and (1.6.1.-8)) of the KdV equation (the additional term $+1$ in the r.h.s. of (17b), as compared to (1.6.1.-8), takes account of the additional term in the r.h.s. of (13), as compared to (3)).

Note incidentally that (15) possesses in fact a more general class of solutions that are both real and regular for all (real) values of z, namely

(18) $\quad f(z)=A\operatorname{cn}^2[p(z-z_0),k], \quad A=-2p^2k^2$,
$\quad\quad\quad v=4p^2(2k^2-1)+1$,

which contain the arbitrary positive constant $k^2\leqslant 1$ (and reduce to (16, 17) for $k^2=1$). Here of course $\operatorname{cn}(y,k)$ is the (second) Jacobian elliptic function. Thus these solutions yield, via (14), other solutions of the KdV equation, namely the "cnoidal waves". These solutions, however, do not vanish asymptotically ($x\rightarrow\pm\infty$), but are instead periodic; we therefore limit our consideration of them to this terse mention here (the study of the solutions of the class of nonlinear evolution equations (3.2.-1) characterized by periodic, rather than asymptotically vanishing, boundary conditions, is actually a field of research in its own right, where much work has been done, with important ramifications in algebraic geometry, topology, differential geometry).

Let us now proceed and consider another nonlinear partial differential equation of evolution type, namely the so-called "regularized long wave" (RLW) equation (see section 1.8), reading

(19) $\quad u_t=-u_x+u_{xxt}+6u_xu, \quad u\equiv u(x,t)$.

The search for a solitary-wave solution of type (14) of the RLW equation (19) can now proceed as above, and it is immediately seen that it leads again to (16), but now with

(20a) $\quad A=-2p^2/(1-4p^2)$,

(20b) $\quad v=1/(1-4p^2)$.

Note incidentally that, if the parameter p is small, these formulae go over (up to corrections of order p^4) into (17); indeed, for small p, the solution (16) is a slowly varying function of x and t, and therefore the difference between (19) and (13), that refers only to the thrice-differentiated terms, loses importance.

We have thus seen that also the RLW equation (19) possesses a solitary wave solution, indeed one that resembles closely the single-soliton solution of the KdV equation (13). However, in contrast to the KdV case, no analytic solution of the RLW equation is known that might be considered an analog of the KdV two-soliton solution; nor is there available, for the solution of the Cauchy problem for the RLW equation, any scheme comparable to the spectral transform method of solution that we now possess for the KdV equation. Of course the sum of two localized solutions of type (14) (with (16) and (20); the parameter p may take two different values for these two solutions) does provide, as long as they are localized far apart so that their overlap is negligible, an (approximate) solution to (19). But it is no more true that a solution, given initially as the sum of two solitary waves far apart from each other but moving with different speeds so that they approach each other, reduces in the remote future to the sum of the same two solitary waves; the collision of solitary waves is now, as it were, *inelastic*, and it may change their properties, give rise to other solitary waves, and give out some additional stuff that disperses away. This is actually shown in fig. 5, that displays the time evolution of the solution of the RLW equation (19) characterized by the initial condition

$$(21) \quad u(x,0) = \sum_{j=1}^{2} A_j / \cosh^2\left[p_j(x - x_j) \right],$$

with A_j related to p_j by (20a) (clearly this initial condition represents the sum of two solitary waves located at x_j and characterized by the parameters p_j). Note that the final picture clearly suggests the birth of a third soliton, and perhaps of a fourth one as well.

The purists reserve the term *soliton* for those cases in which only *elastic* collisions occur, as discussed above. It is believed that all nonlinear evolution equations giving rise to the soliton phenomenology (in this strict sense of the word) are also solvable by some appropriate variant of the spectral transform method. The term *solitary wave* is instead used by the purists to denote solutions of type (14) (with $f(z)$ vanishing asymptotically, $f(\pm \infty) = 0$); the class of nonlinear evolution equations admitting such solutions includes (and is indeed *much larger* than—whatever this means!) the class possessing solitons. The purists are, however, in the minority; the term

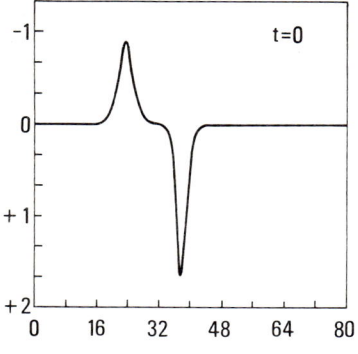
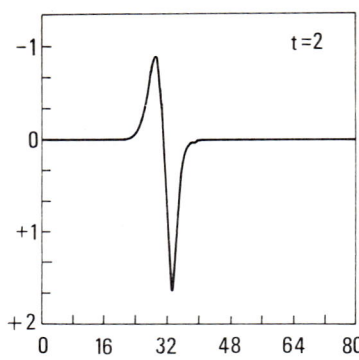
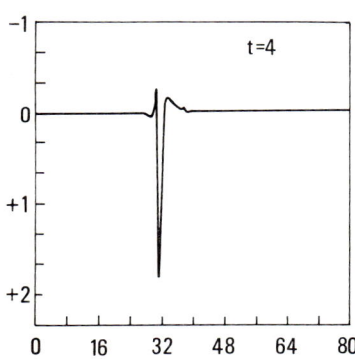
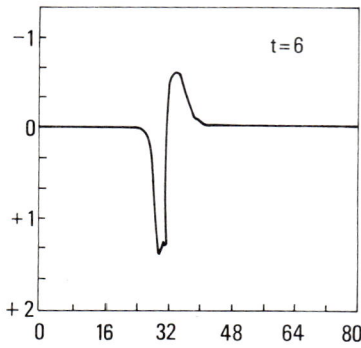
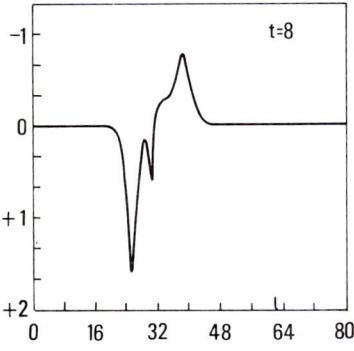

Figure 5. 9 snapshots equally spaced in time (from $t=0$ to $t=16$) of the solution of the RLW equation (19) with the initial condition (21) with $x_1 = 23$, $x_2 = 35$, $p_1 = \frac{2}{5}$, $p_2 = \frac{3}{5}$ (implying $v_1 = \frac{25}{9}$, $v_2 = -\frac{25}{11}$; see (20b)). Note that the ordinates are oriented downwards.

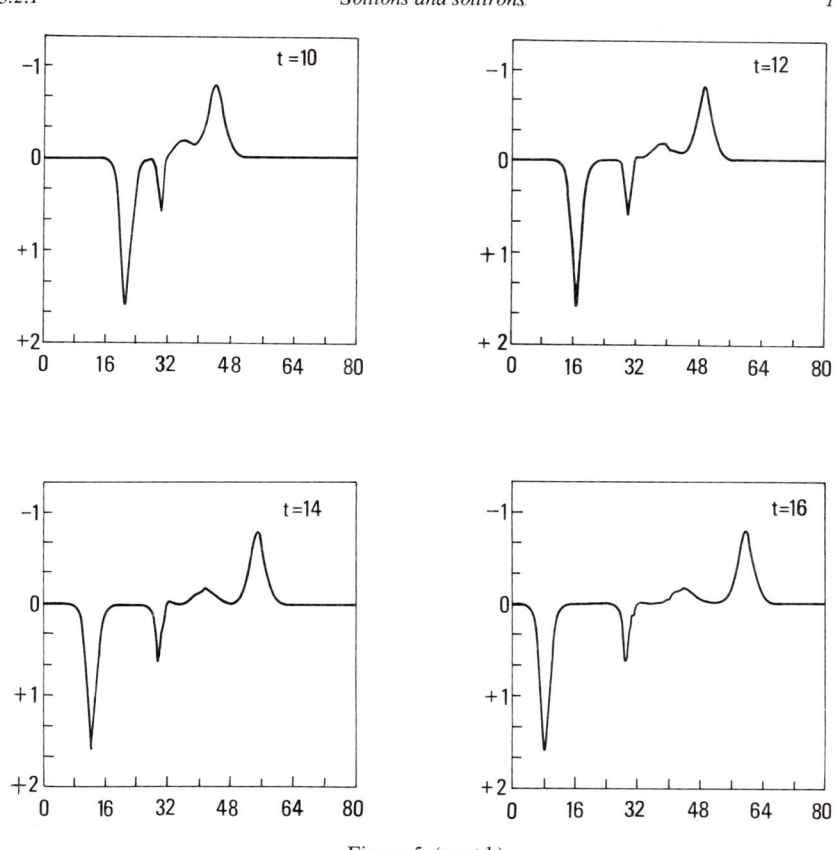

Figure 5 (contd.).

soliton is now rather generally used (especially, but not exclusively, by elementary particle physicists) to denote solitary waves. This may be deplorable; but it is a fact of life, about which we felt duty bound to warn the reader, lest he gets confused by the titles, if not the content, of a large number of the papers that are now circulating.

Actually the term *solitary wave* is itself rather awkward; not only because it is made up of two words rather than only one, but more seriously because the characteristic distinction between *solitons* and *solitary waves* is their different behaviours when they collide; but the discussion of the collision of the solitary waves has an implicit semantic stridency (if they encounter, how can they be solitary?). Thus, aware as we are of the difficulty to dictate semantic usage on the basis of reasonable conventions rather than historic habits, we propose here the use of the term "*solitron*" to replace "*solitary wave*". This name should identify any solution of a nonlinear evolution equation or field theory that remains localized throughout the time evolution, thus possessing the characteristic behaviour of a particle-like object (the term *solitron* is suggestive of this through its analogy to the name of elementary particles: electron, positron, neutron, etc). On the other hand the term *soliton* should be reserved for those special cases in which the localized solutions have the remarkable property to undergo only elastic collisions; thus it would be associated essentially only with solutions of the class of equations solvable by spectral transform techniques (note however that the results reported in volume II imply that *solitons* are not required to move with constant speed).

3.2.2. The weak field limit

In section 1.5 the weak field limit of the class of nonlinear evolution equations (3.2.-1a) was discussed, and it was shown that, just as in this limit the equations can be approximated by retaining only the linear part, so the method of solution via the spectral transform goes over into the method of solution, via Fourier transform, appropriate to linear evolution equations. Two qualifications were however mentioned at the end of that subsection. Here we take up these points once again, to clarify the situation. For the sake of simplicity, we focus on the simplest nontrivial equation of the class (3.2.-1a), namely the KdV equation

(1) $\quad u_t + u_{xxx} - 6u_x u = 0.$

Both qualifications raised at the end of section 1.5 refer to components of the solution $u(x, t)$ that are slowly varying in x; this is clearly the case for the Fourier components at or near $k=0$; it is obviously also the case for the component of $u(x, t)$ corresponding to the single discrete eigenvalue whose presence may be compatible with the assumed weakness of $u(x, t)$, since such a discrete eigenvalue must clearly be very small, being therefore associated with a slowly varying eigensolution of the Schroedinger equation.

Thus the point is, that the neglect of the third term in (1) relative to the second is justified when u is small, provided however it is not too slowly varying. Indeed the origin of this qualification is clear; if u is a slowly varying function of x, the third derivative in the second term introduces an element of smallness that may be comparable, or even more significant, than the smallness characteristic of the third (nonlinear) term due to the smallness of u itself. Note that the third term in (1) contains also a derivative in x, but only one, while the first term has a third derivative; this suggests, as a rough quantitative estimate of which of the two terms may be dominant, to compare the magnitude of the Fourier component $\hat{u}(k,t)$ to the quantity k^2; clearly the former is generally smaller if $u(x,t)$ itself is very small, except for the Fourier components at or near $k=0$.

In conclusion, it is now clear why the replacement of a nonlinear evolution equation of the class

(3) $\qquad u_t = \alpha(L)u_x, \quad u \equiv u(x,t),$

with $\alpha(z)$ a polynomial and L the differential operator (3.1.-9), by its linearized version

(4a) $\qquad u_t = \alpha(\partial^2/\partial x^2)u_x, \quad u \equiv u(x,t),$

or equivalently by

(4b) $\qquad u_t = -i\omega(\partial/\partial x)u, \quad u \equiv u(x,t)$

with $\omega(z)$ related to $\alpha(z)$ by (1.5.-3), is a weak field approximation, that breaks down for the more slowly varying components of $u(x,t)$ (considered as a function of x); and the discussion above provides some quantitative indication of the applicability of this criterium.

3.2.3. Generic solution

Not much can be added here to the content of section 1.6.3. It should however be emphasized that the qualitative analysis of the long-time behavior of the solutions of the class of nonlinear evolution equations (3.2.-1) given there can be backed by a quantitative treatment. This however requires the separate study of each equation of the class (3.2.-1); and even for the simplest nontrivial member of this class, the KdV equation (3.1.-33), a full quantitative analysis is a task that requires considerable stamina.

A convenient starting point is the representation (3.2.1.-1) of the generic solution. There remains then to investigate the form that the eigenfunctions $\varphi_n(x,t)$ and $f^{(+)}(x,k,t)$ of the Schroedinger problem take as $t \to \infty$. This can be done, by taking advantage of the simple explicit time evolution of the spectral transform, see (3.2.-2), and by using the formalism of the inverse spectral problem, see section 2.2 and appendix A.5. The guiding idea is of course that asymptotically ($t \to \infty$), in the region to the right (large positive values of x), the potential $u(x,t)$ becomes, up to locally small corrections, just the multisoliton solution, namely it reduces only to the *sum* in the r.h.s.

of (3.2.1.-1), where moreover each eigenfunction $\varphi_n(x, t)$ corresponds self-consistently (again up to small corrections) only to the potential

(1) $\quad -4p_n[\varphi_n(x,t)]^2 \approx -2p_n^2/\cosh^2\{p_n[x-\xi_n(t)]\}.$

An important rôle in the analysis is played by the fact that the functions (1) vanish exponentially as $|x-\xi_n(t)| \to \infty$, and that the distances between the solitons, $|\xi_n(t)-\xi_m(t)|$, diverge (linearly in t) as $t \to \infty$. Much also depends on the nature of the initial datum,

(2) $\quad u(x,0)=u_0(x).$

For instance, to analyze the emergence of the solitons, the asymptotic behavior of $u_0(x)$ as $x \to +\infty$ is highly relevant. Indeed, if $u_0(x)$ vanishes in that limit faster than exponentially, then eventually the solitons emerge, as $t \to \infty$, clean up to exponentially small corrections (namely, at any fixed distance from the position of each soliton, that moves out with constant speed to the right, the solution is represented by the single-soliton form, up to corrections that vanish exponentially); if instead $u_0(x)$ vanishes as $x \to +\infty$ slower than exponentially, then there are corrections that account, even in the soliton region, for the (disappearing) background associated with the tail of the initial datum.

The discussion outlined above refers to the behaviour of the generic solution $u(x,t)$ of the KdV equation as $t \to +\infty$ and as $x \to +\infty$ proportionally to t,

(3) $\quad x=vt+a, \quad v>0$

(the discussion implies of course that it is crucially important whether v does or does not coincide with the velocity $v_n = 4p_n^2$ of a soliton contained in the solution under consideration). As for the behaviour, always for $t \to +\infty$, but for fixed x or for $x \to -\infty$, it is rather more complicated, being dominated by the background part of the solution; it is largely determined by the linearized version of the equation, as already mentioned in subsections 1.6.2 and 1.6.3, but there are some qualifications (for instance, clearly this cannot be the case if no background at all is present, namely for the pure multi-soliton solutions). An important rôle in this connection is also played by the so-called "similarity solutions" (see subsection 3.2.4.3).

3.2.4. Special solutions of the KdV equation

In this subsection, which is conveniently subdivided into three parts, we discuss certain special solutions of the KdV equation

(1) $\quad u_t + u_{xxx} - 6u_x u = 0, \quad u \equiv u(x, t),$

that do not belong to the class of functions ("*bona fide* potentials") discussed so far.

Some of the developments described below (especially those of subsection 3.2.4.1) might naturally be extended to include in the discussion the whole class of higher KdV equations, say (3.2.-1); but for simplicity we restrict our consideration to the KdV equation (1).

In subsection 3.2.4.1 we discuss rational solutions of (1); in subsection 3.2.4.2, solutions that diverge linearly in x as $x \to \pm \infty$; and in subsection 3.2.4.3, the so-called similarity solutions of (1), and the relationship instituted by such solutions between the KdV partial differential equation (1) and an ordinary (nonlinear) differential equation of Painlevé type.

3.2.4.1. Rational solutions

The rational solutions of the KdV equation

(1) $\quad u_t + u_{xxx} - 6u_x u = 0, \quad u \equiv u(x, t)$

that we discuss here are generally complex and/or singular. The simpler of these solutions reads

(2) $\quad u(x, t) = 2(x - \xi)^{-2},$

where ξ indicates an arbitrary constant, possibly complex (note that, unless ξ is complex, (2) is singular for real x). The fact that (2) satisfies the KdV equation (1) can be easily verified by explicit computation.

One looks then for solutions of the KdV equation (1) of the form

(3) $\quad u(x, t) = 2 \sum_{j=1}^{N} \left[x - \xi_j(t) \right]^{-2}$

(it is indeed easily seen, by analyzing the behavior of $u(x, t)$ in the neighborhood of its poles, that any more general *ansatz* that represents $u(x, t)$ as a rational function of x must reduce to (3) in order to satisfy (1)).

It is then also easy to see, by matching the coefficients of the poles of different orders, that the following 2 conditions are necessary and sufficient in order that (3) satisfy (1):

(4) $$\sum_{k=1}^{N}{}' [\xi_j(t)-\xi_k(t)]^{-3}=0, \quad j=1,2,\ldots,N,$$

(5) $$\dot{\xi}_j(t)=-12\sum_{k=1}^{N}{}' [\xi_j(t)-\xi_k(t)]^{-2}, \quad j=1,2,\ldots,N$$

(here, and always below, the prime attached to the symbol of summation indicates that the singular term is omitted). Moreover, differentiating (4) and using (5) and some algebra, it is easily seen that the two conditions (4) and (5) are compatible; thus if (4) is satisfied at any one time (say, initially, at $t=0$), then it is always satisfied, that is, as long as the quantities $\xi_j(t)$ evolve in time according to (5).

Note that one has thus established a connection between the nonlinear partial differential equation (1) and the dynamical system with the n degrees of freedom ξ_j whose time evolution is described by (5) (with the constraint (4)). This is, of course, interesting only if the condition (4) admits some solution, namely if the algebraic manifold defined by the condition

(4a) $$\sum_{k=1}^{N}{}' (\xi_j-\xi_k)^{-3}=0, \quad j=1,2,\ldots,N,$$

is not empty.

It is easy to convince oneself that (4a) admits no solution (for finite N) with all the quantities ξ_j being real. Consider indeed the one-dimensional N-body problem defined by the hamiltonian

(6) $$H=\tfrac{1}{2}\sum_{j=1}^{N} p_j^2 + \sum_{j>k=1}^{N}(\xi_j-\xi_k)^{-2},$$

that yields the equations of motion

(7) $$\dot{\xi}_j=p_j, \quad \dot{p}_j=2\sum_{k=1}^{N}{}'(\xi_j-\xi_k)^{-3}, \quad j=1,2,\ldots,N,$$

or, equivalently

(8) $$\ddot{\xi}_j=2\sum_{k=1}^{N}{}'(\xi_j-\xi_k)^{-3}, \quad j=1,2,\ldots,N.$$

Thus this N-body problem describes the time evolution on the line of N unit-mass classical particles interacting pairwise via a repulsive force proportional to the inverse cube of their mutual distance. Clearly such a system cannot possess (for finite $N>1$) any equilibrium configuration. Yet the equations that characterize the equilibrium configuration are precisely (4a).

The above "physical" argument, however, applies only to real values of the particle coordinates; it does not exclude the possibility that there be complex solutions of (4a). And indeed it has been shown that such solutions exist iff

(9) $\quad N=\tfrac{1}{2}m(m+1), \quad m=1,2,3,\dots$.

For instance, the solution (2) obtains for $m=1$ ($N=1$); the next one, for $m=2$, has three poles ($N=3$) and reads

(10) $\quad u(x,t)=6x(x^3-24t)/(x^3+12t)^2,$

the pole positions being

(11) $\quad \xi_j(t)=-\exp(2\pi i j/3)(12t)^{1/3}, \quad j=1,2,3.$

Moreover the motion of the poles determined by (5) turns out to be closely related to the N-body system that we have just introduced (being now formally extended to complex values of the variables), that constitutes in itself a mathematically remarkable problem. The interested reader may pursue these developments in the literature (see Notes).

The structure of (3) clearly suggests that $u(x,t)$ can also be written in the form

(12) $\quad u(x,t)=-2\partial^2[\ln P_N(x,t)]/\partial x^2,$

where $P_N(x,t)$ is a polynomial of degree N in x having the $\xi_j(t)$'s as zeros:

(13) $\quad P_N(x,t)=\prod_{j=1}^{N}[x-\xi_j(t)].$

Also these polynomials (that are defined only for the values of N given by (9)) have been studied in much detail; one important result obtained in this manner is the discovery that the solution $u(x,t)$ given by (3) (or equivalently by (12) and (13)) with (4) and (5) is rational not only in x, but also in t.

The formula (12) is clearly analogous to the multisoliton formula (1.6.1.-9), and indeed it can be obtained from it in the (nontrivial) limit in which all the parameters p_n and ρ_n tend to zero in an appropriate way (for instance the 1-pole solution (2) obtains in the $p=0$ limit from the solution $u(x,t) = 2p^2/\sinh^2[p(x-\xi)]$, that is just the single-soliton solution (1.6.1.-5) with the soliton position ξ shifted off the real axis by $i\pi/(2p)$).

On the other hand (4) has indeed a simple solution if one allows the number N of poles to be infinitely large; for instance

$$(14) \quad \xi_j = \xi_0' + aj, \quad j=0, \pm 1, \pm 2, \ldots,$$

satisfies (4a), for obvious symmetry reasons (ξ_0' and a being two arbitrary, possibly even complex, numbers). Indeed, through (5) and (3), this reproduces merely the single-soliton solution (1.6.1.-5) (with (1.6.1.-6b) and (1.6.1.-8)), with $p = i\pi/a$; of course a must be chosen imaginary to get a real solution), since

$$\sum_{j=-\infty}^{+\infty} \left[z - i\pi(j+\tfrac{1}{2})\right]^{-2} = -(\cosh z)^{-2}.$$

The fact that the KdV equation (1) possesses rational solutions is one of the many highly nontrivial properties of this equation. To demonstrate this consider the nonlinear evolution equation

$$(15) \quad u_t + u_{xxx} - 6u_x u - 24cu = 0,$$

where c is a nonvanishing constant. It is then easy to prove that this equation has no solutions that are rational in x (for all t; of course one can always force the solution to be rational at one particular time, for instance initially). Indeed the same analysis outlined above implies that such solutions should again have the form (3) with (5), but with (4) replaced now by

$$(16) \quad \sum_{k=1}^{N}{}' (\xi_j - \xi_k)^{-3} = c, \quad j = 1, 2, \ldots, N.$$

This system of equations has, however, no solution, as is immediately seen by summing them all (i.e. summing over the index j from 1 to N), whereby the double sum in the l.h.s. vanishes (due to the antisymmetry of the summand), while the r.h.s. equals cN.

This should suffice to convince the reader that the property to possess rational solutions is, for a nonlinear partial differential equation, definitely unusual. On the other hand, in every instance in which such solutions do exist, they provide, as indicated above, a (generally interesting) mapping of the time evolution described by the nonlinear partial differential equation into the time evolution of a dynamical system with a finite number of degrees of freedom.

Finally let us point out that, if instead of the infinity of poles equally spaced along a straight line given by (14), one were to consider the (double) infinity of poles located on a rectangular infinite lattice, the solution (3) would become the (doubly periodic) Weierstrass function; one has recovered in this manner a (special) *periodic* solution of the KdV equation.

3.2.4.2. Asymptotically diverging solutions

Thus far, our consideration has always been limited to functions u that vanish asymptotically, namely as $x \to \pm \infty$ (except for a terse mention of periodic solutions in section 3.2.1, see (3.2.1.-18)). Actually the nonlinear evolution equations that we are studying also possess solutions that diverge asymptotically; it is for instance easy to verify that the simple function

(1) $\qquad \bar{u}(x,t) = -\tfrac{1}{6}(x-x_0)/(t-t_0)$

satisfies the KdV equation (3.2.4.-1). This solution is, however, real (for real x and t) only if the two (a priori arbitrary) constants x_0 and t_0 are both real; but it is then singular at a real value of the time variable, namely at $t = t_0$.

There exists in fact a whole class of solutions of the KdV equation having the form

(2) $\qquad u(x,t) = \bar{u}(x,t) + v(x,t),$

with $\bar{u}(x,t)$ given by (1) and $v(x,t)$ vanishing asymptotically ($x \to \pm \infty$). It is possible to solve the Cauchy problem for this class of solutions by a (nontrivial) variant of the techniques introduced above; and this turns out to be relevant for the solution of the so-called cylindrical KdV equation (already mentioned in section 1.8, see (1.8.-4)). We therefore treat this item separately below (section 6.3).

3.2.4.3. Similarity solutions and ODEs of Painlevé type

The term "similarity solution" refers generally to a solution that depends (in a nontrivial way) on less independent variables than the complete solution—thus, for the partial differential equations in two variables, x and

t, that we are considering here, a similarity solution depends essentially only on one variable, say y, being an appropriate combination of x and t (see below). The name presumably originates from the fact that, if x and t vary so that their combination y remain constant, then the similarity solution does not change at all, or only in a trivial fashion (another way to denote these solutions, that reflects more clearly this feature, is by the qualifying adjective "self-similar").

Consider for instance the KdV equation

(1) $\quad u_t + u_{xxx} - 6u_x u = 0, \quad u \equiv u(x,t),$

and set formally

(2) $\quad u(x,t) = [3(t-t_0)]^{-2/3} f(y), \quad y = [3(t-t_0)]^{-1/3}(x-x_0),$

with t_0 and x_0 arbitrary constants. It is then easily seen that (2) satisfies (1) provided $f(y)$ satisfies the nonlinear ordinary differential equation

(3) $\quad f_{yyy} - y f_y - 2f - 6 f_y f = 0, \quad f \equiv f(y).$

This equation can actually be integrated once, by noticing that the l.h.s. of (3) multiplied by $(y+2f)$ is a perfect differential. There obtains

(4) $\quad (y+2f)[f_{yy} - (y+2f)f] - (1+f_y)f_y = c,$

where c is an integration constant. This can be written in slightly neater form via the position

(5) $\quad h(y) = y + 2f(y),$

that yields

(6) $\quad 2h[h_{yy} + (y-h)h] + (1 - h_y^2) = 4c.$

Actually a more transparent form obtains via the Miura transformation, see appendix A.11. Indeed it is easily seen that the modified KdV equation

(7) $\quad v_t + v_{xxx} - 6 v_x v^2 = 0, \quad v \equiv v(x,t),$

is satisfied by the similarity solution

(8) $\quad v(x,t) = [3(t-t_0)]^{-1/3} g(y), \quad y = [3(t-t_0)]^{-1/3}(x-x_0),$

3.2.4.3 Similarity solutions

provided $g(y)$ is a solution of the nonlinear ordinary differential equation

(9) $\quad g_{yyy} - 6g_y g^2 - y g_y - g = 0.$

This is easily integrated to

(10) $\quad g_{yy} - 2g^3 - yg = a$

where a is a constant. Thus any solution g of this equation yields a solution of (4) via the formula

(11) $\quad f(y) = g_y(y) + g^2(y),$

that is clearly just the Miura transformation (A.11.-3) for the similarity solutions (2) and (8). This remarkable property of the Miura transformation (11) is explicitly exhibited by the formula

(12) $\quad \{(y+2f)[f_{yy} - (y+2f)f] - (1+f_y)f_y\}_y$
$= (y + 2g_y + 2g^2)\{(g_{yy} - 2g^3 - yg)_{yy} + 2g(g_{yy} - 2g^3 - yg)_y\},$

that is implied by (11).

These similarity solutions generally do not belong to the class of *bona fide* potentials on which our considerations has been until now mainly concentrated; for instance (2) and (8), that generally are real only if x_0 and t_0 are real, are generally singular at $t = t_0$; moreover there is no guarantee that the solutions of (4) or (10) vanish asymptotically ($y \to \pm \infty$) (note incidentally that (3.2.4.2.-1) is indeed a similarity solution of type (2); it corresponds to $f(y) = -\frac{1}{2}y$, that is clearly a solution of (4), with $c = \frac{1}{4}$). These similarity solutions are however important for several reasons.

Firstly, they may yield explicit solutions (always a nice thing to have). For instance (10) (with $a = 2^{-2/3}$) has the solution

(13) $\quad g(y) = d[\ln F(-2^{-1/3}y)]/dy,$

where $F(z)$ is any solution of the Airy equation

(14) $\quad F_{zz}(z) = zF(z),$

as can be easily verified. Through (8), this yields a nontrivial, but generally singular, solution of the mKdV equation (7) (it also yields, through (11) and

(2), a solution of the KdV equation (1); that however coincides simply with (3.2.4.2-1)).

Secondly, these similarity solutions generally play a rôle in the study of the long-time behavior of solutions of the nonlinear evolution equations under consideration (in the region where the background contribution is nonnegligible).

Thirdly, they provide a relationship between nonlinear PDEs and ODEs, that has displayed some remarkable features. It appears in fact, not altogether surprisingly, that to the special nonlinear PDEs that are solvable by the spectral technique, such as the KdV equation (1) (or the mKdV equation (7); see appendix A.11 and section 1.8), there correspond ODEs that are also in some sense special. And the class that appears to be singled out as particularly significant is that introduced at the turn of the century by Painlevé, Gambier and others, namely the class of nonlinear ODEs whose solutions have no moving singularities other than poles. In particular the class of ODEs of the form

(15) $\quad w'' = A(z,w)(w')^2 + B(z,w)w' + C(z,w), \quad w \equiv w(z),$

was analyzed (here the primes indicate derivatives; and A, B and C are rational functions of their two arguments); and all equations of this form having no movable critical points (i.e., such that their solutions have no branch points or essential singularities in z whose location depends on the constants of integration) were identified and classified (a list with 50 entries). Only 6 of these equations originate, through their solutions, novel special functions, the so-called *Painlevé transcendents*. The first two of these equations read:

(16) $\quad w'' = 6w^2 + z,$

(17) $\quad w'' = 2w^3 + zw + a,$

and their solutions are generally denoted as the first and second Painlevé transcendents; each comprises of course a two-parameter family of functions, since the general solution of these second-order ODEs contains generally two integration constants (the second Painlevé transcendent depends moreover on the constant a in (17)).

Clearly, except for notational changes, the equations for the second Painlevé transcendent, (17), coincides with (10). This remarkable discovery (that has been shown not to be just a fluke) has opened a field of investigation that has yielded a number of interesting results, including novel properties of the Painlevé transcendents; indeed perhaps more has been learned about ODEs, from what is now known about PDEs in this particular field, then the other way round (especially by using the type of results discussed in the following chapter 4).

We end this subsection reporting another property of this kind possessed by the KdV equation; it relates another type of similarity solution directly to the first Painlevé transcendent. It is displayed by the formula

$$(18) \quad u(x,t) = 2[h(z) - t], \quad z = x - 6t^2;$$

for it is immediately verified that, if u satisfies the KdV equation (1), $h(z)$ satisfies (16) (up to a translation in the independent variable z); and vice versa. Note however that, once again, this similarity solution cannot qualify as a *bona fide* potential, since the expression (18) is clearly inconsistent with the asymptotic property $u(\pm\infty, t) = 0$.

3.N. Notes to chapter 3

The original idea of the spectral transform method to solve nonlinear evolution equations was introduced in [GGKM1967]; see also [Mi1968], [MGK1968], [SG1969], [G1971], [KMGZ1970], [GGKM1974]. The first idea to employ wronskian-type relations to generate solvable nonlinear evolution equation was introduced in [C1975].

3.1. The isospectrality of the time evolution of the Schroedinger operator (3.1.-17) has been pointed out for the first time in [L1968]. This property plays a central rôle in other approaches to the general class of problems treated in this book (see, for instance, [ZS1974], [AKNS1974]).

Some of the problems connected with the more general class of evolution equations (3.1.-10) (in particular, those associated with the real zeros of g) are discussed in [KN1979] (in the context of the class of nonlinear evolution equations solvable via the Zakharov–Shabat spectral problem; see volume II).

The nonlinear equations of the class (3.1.-40) are sometimes distinguished from their analogs in the class (3.1.-28) by the term "potential" (used generally as a qualifying adjective). This confusing terminology originates presumably from the analogy of the relation between u and w, see (3.1.-27), to that between a force and its potential (in one-dimensional Newtonian dynamics).

3.2.1. The original suggestion to evince from (3.2.1.-1) a definition of the soliton in x-space for all time can be found in [K1974], [K1974a]. The graphs that display the detail of soliton collisions, figs. 1−4, have been computer-produced by J.K. Drohm and L.P. Kok at the Institute of Theoretical Physics in Groningen. An analytic analysis of the behavior of the KdV solitons as they collide has been given in [L1968]. For film displays see [ZDK1968] and [T1971].

The literature on elementary particles and "solitons" has had an exaggerated growth in the last few years, reflecting the contemporary tendency of theoretical research in elementary particle physics to progress through stampedes determined by fashion; the interested reader is referred for instance to the review paper [R1975] and to the popularized account [Re1979]. Curiously enough, the main idea concerning the possible connections of "solitons" to elementary particles had in fact been put forward several years ago, before the actual discovery of the soliton, see [Sky1958] and [Sky1961]; but they had then gone practically unnoticed.

The literature on the KdV equation, and all other "soliton" equations, with *periodic boundary conditions*, is vast; see for instance [DN1974], [Mar1974], [N1974], [D1975], [DMN1976], [Mat1976] and [Man1978].

The RLW equation (3.2.1.-19) is sometimes referred to as the BBM or PBBM equation, see [P1966] and [BBM1972]. It is a matter of dispute (see, for instance, [K1974]) whether this equation or KdV is preferable in applications (in particular, to describe long water waves). A conjecture that the *solitrons* of the RLW equation be actually *solitons*, namely interact elastically, had been put forward in [EMG1975], [EMG1977], but was shown to be invalid in [ABM1976] (that displays a small but significant inelastic effect). The macroscopic effect displayed in the graphs of fig. 5 is due to a more appropriate choice of the solitron parameters. These graphs are taken from [Sa1978]; they incidentally display one of the first explicitly computed examples of solitron creation (see also [OB1978]). See also [GF1980].

The suggestion to use the term "solitron" in place of "solitary wave" was originally presented by one of us (FC) at a banquet in Jadwisin near Warsaw, held on August 31st, 1979, at the end of an international meeting on solitons and related topics. Both Kruskal and Zabusky were present (the term "soliton" was first introduced by them [ZK1965]; but originally Zabusky had used the term "solitron", and only at the last moment he changed it to "soliton" to avoid copyright difficulties that might be caused by the word "solitron" being used as the trade mark of an electronics firm).

3.2.2. For a discussion of the opposite limit, in which (for the KdV equation (3.2.2.-1)) the dispersive term u_{xxx} is small relative to the nonlinear term $u_x u$, see [LL1979].

3.2.3. The long-time behaviour of solutions of various equations solvable by spectral transform techniques has been investigated by several authors, see for instance [AN1973], [Seg1976a], [ZM1976], [AS1977a], [San1978], [San1979], [MST1980]. A rigorous recent treatment on the emergence of solitons as $t \to +\infty$ in the KdV equation with arbitrary initial conditions (vanishing as $x \to \pm \infty$) has been given by [ES1982]; see also [ZMNP1980] and [EVH1981].

3.2.4.1. The first idea to investigate the time evolution of the position of the poles (in the complex x-plane) of solutions of the KdV equation goes back to [K1974a]. This investigation was pursued in [Th1976], and was greatly advanced by [AMM1977] and [CC1977] (see also [DMN1976]). The most complete discussion of the manifold of solutions of (3.2.4.1.-4a) is given in [AMM1977]. The polynomials of (3.2.4.1.-13) have been investigated in [AM1978]. The hamiltonian system (3.2.4.1.-6) was originally solved in the quantal case in [C1971], and in the classical case in [Mo1975a] and [OP1976] (see also [A1977], [A1977a], [KKS1978]). There has been recently considerable progress in the study of these types of integrable dynamical systems; see for instance the review papers [C1977a], [C1979b], [C1980b], [OP1981], and the papers [Kri1978] and [Kri1980]. The idea to relate integrable dynamical systems and partial differential equations has been pursued in [C1978a]; a spin-off of these investigations are results on special functions [ABCOP1979] and on linear algebra [C1980a], [C1980b], [C1980c], [BC1981], [C1981a], [C1981b], [C1982].

The derivation of (3.2.4.1.-12) from the multisoliton formula (1.6.1.-9) can be for instance achieved by techniques analogous to those discussed in [Sym1978]; see also [AM1978].

Techniques analogous to those discussed in this subsection have been recently used to discover the multisoliton solution of the Benjamin–Ono equation (see section 1.8) [CLP1978], as well as special solutions of the KdV equation with damping [COP1979].

3.2.4.2. The remark that solutions of the ordinary and cylindrical KdV equations may be related by (3.2.4.2.-2), together with an appropriate change of variable, is due to [LL1964]. For additional references on the cylindrical KdV equation see the Notes to section 6.3.

3.2.4.3. The connection between nonlinear partial differential equations solvable by spectral transform techniques and nonlinear ordinary differential equations of Painlevé type was first noted in [N1975] (see also [Ab1977a]). There have been several additional contributions, see for instance [AS1977b], [ARS1978], [ARS1980a], [ARS1980b], [SA1979], [BP1979], [Ab1979], [FN1979], [BP1980].

For the connection between the asymptotic behaviour of solutions of nonlinear PDEs and Painlevé transcendents see e.g. [SA1979].

The discovery of the similarity solution (3.2.4.3.-18) is due to A. Fokas (private communication, summer 1980; see [FA1982]).

CHAPTER FOUR

BÄCKLUND TRANSFORMATIONS AND RELATED RESULTS

In the preceding chapter we have discussed a class of nonlinear evolution equations solvable via the spectral transform associated with the Schroedinger spectral problem. The main tool used to identify this class of nonlinear evolution equations are the wronskian integral relations of subsection 2.4.1 and the spectral integral relations of subsection 2.4.2; in particular, in both cases, the formulae that relate *infinitesimal* changes of a function $u(x)$ to the corresponding (infinitesimal) changes of its spectral transform S. In this chapter we rather exploit the analogous, but more general, formulae, that relate *finite* changes of $u(x)$ to the corresponding (finite) changes of its spectral transform; or, more specifically, we use the appropriate formulae of subsections 2.4.1 and 2.4.2, to identify certain relations between two different functions, $u^{(1)}(x)$ and $u^{(2)}(x)$, such that the relations between the corresponding spectral transforms are very simple (essentially linear). Thus this approach identifies a large class of "Bäcklund transformations", namely formulae that relate two functions, say $u^{(1)}(x, t)$ and $u^{(2)}(x, t)$, in such a way that, if one of them satisfies a nonlinear evolution equation (say, an equation of the class (3.1.-10)), then the other does so too.

Such transformations were introduced in differential geometry, and their name goes back to this origin; while closely related formulae ("Darboux transformations"; see below) have originated in the (closely related) context of ordinary linear differential equations. Their analogs in the linear case are the (relatively trivial) transformations discussed in section 1.1 (see in particular (1.1.-10)).

The structure and properties of Bäcklund transformations are discussed in the following section 4.1. The important property of commutativity they possess, and a remarkable consequence of it—often referred to as *nonlinear superposition principle*—are displayed and discussed in section 4.2 (thus the material in these two sections treats somewhat more in depth the ground already covered in subsections 1.7.1 and 1.7.2). A formal extension, applicable to the class of nonlinear evolution equations (3.1.-10), of the *resolvent formula* (1.1.-13) is then given in section 4.3, and it is used in section 4.4 to obtain certain intriguing nonlinear operator identities. Finally, in section 4.5, relations that connect solutions of *different* nonlinear evolution equations (belonging to the same class (3.1.-10)) are exhibited; the results of this last section encompass those of the two sections that precede it, but the order of presentation we have selected is (presumably) didactically preferable (it also reflects the historical development).

All the results of this chapter can be easily established in the framework of the class of nonlinear evolution equations (3.1.-10), as we have stated above. In the following, however, for the mere sake of notational simplicity, we restrict our attention to the class of nonlinear evolution equations (3.2.-1a), namely

(1) $\quad u_t(x,t) = \alpha(L) u_x(x,t),$

with

(2) $\quad Lf(x) = f_{xx}(x) - 4u(x,t)f(x) + 2u_x(x,t) \int_x^{+\infty} dy f(y).$

Let us recall that the corresponding time evolution of the spectral transform of $u(x,t)$ is then given by (3.2.-2), namely

(3a) $\quad R(k,t) = R(k,0) \exp[2ik\alpha(-4k^2)t],$
(3b) $\quad p_n(t) = p_n(0) \equiv p_n,$
(3c) $\quad \rho_n(t) = \rho_n(0) \exp[-2p_n\alpha(4p_n^2)t].$

In fact, in this chapter it is often more convenient to work with

(4) $\quad w(x,t) = \int_x^{+\infty} dy\, u(y,t),$

rather than with $u(x,t)$. Let us recall that the evolution equation corresponding to (1) reads then (see (3.2.-1b))

(5) $\quad w_t(x,t) = \alpha(\tilde{L}) w_x(x,t),$

with (see (3.1.-38))

(6) $\quad \tilde{L}f(x) = f_{xx}(x) + 4w_x(x,t)f(x) + 2\int_x^{+\infty} dy\, w_{yy}(y,t)f(y).$

4.1. Bäcklund transformations

As in chapter 3 (see section 3.1), we have here the option to use as our main tool of analysis the (appropriate) wronskian integral relations of subsection 2.4.1 or the spectral integral relations of subsection 2.4.2. In fact, these two sets of formulae are again better used in a complementary fashion. Thus we first utilize the wronskian integral equations of subsection 2.4.1 to establish the main results, and then use the spectral integral relations of subsection 2.4.2 to verify their consistency. We thus take as our starting point the formulae (2.4.1.-10c) and (2.4.1.-10e), that we rewrite here:

(1) $\quad g(\Lambda)\left[u^{(1)}(x,t) - u^{(2)}(x,t)\right] + h(\Lambda)\Gamma \cdot 1 = 0,$

(2) $\quad R^{(1)}(k,t) = R^{(2)}(k,t) \dfrac{\left[g(-4k^2) - 2ikh(-4k^2)\right]}{\left[g(-4k^2) + 2ikh(-4k^2)\right]}.$

The integrodifferential operators Γ and Λ are defined by (2.4.1.-5) and (2.4.1.-6), namely

(3) $\quad \Gamma f(x) = \left[u_x^{(1)}(x,t) + u_x^{(2)}(x,t)\right] f(x) + \left[u^{(1)}(x,t) - u^{(2)}(x,t)\right]$
$\quad \cdot \int_x^{+\infty} dy \left[u^{(1)}(y,t) - u^{(2)}(y,t)\right] f(y),$

(4) $\quad \Lambda f(x) = f_{xx}(x) - 2\left[u^{(1)}(x,t) + u^{(2)}(x,t)\right] f(x) + \Gamma \int_x^{+\infty} dy\, f(y).$

In all these equations $u^{(1)}(x,t)$ and $u^{(2)}(x,t)$ are two functions (belonging to the class of *bona fide* potentials as far as their x-dependence is concerned, but otherwise arbitrary); $R^{(1)}(k,t)$ and $R^{(2)}(k,t)$ are the corresponding reflection coefficients; the (entire) functions g and h in (1) and (2) are arbitrary (but we assume here that they are time-independent), and $f(x)$ in (3) and (4) indicates of course an arbitrary function (that may also depend on time). Note that the only (notational) change that we have introduced here with respect to the formulae of section 2.4.1 is the explicit indication that the $u^{(j)}$'s (and therefore also the corresponding $R^{(j)}$'s) may depend on the time variable, t.

Assume now that one of the two functions $u^{(j)}$, say $u^{(2)}(x,t)$, evolves in time according to the nonlinear evolution equation (4.-1). Hence $R^{(2)}(k,t)$ evolves according to (4.-3a). But then $R^{(1)}(k,t)$, being related to $R^{(2)}(k,t)$ by (2), evolves in time according precisely to the same law (here the assumed time independence of g and h is essential). On the other hand, if we assume for a moment that there are no discrete eigenvalues associated with the potential $u^{(1)}(x, t)$, then we know that there is a one-to-one correspondence between $u^{(1)}(x, t)$ and $R^{(1)}(k, t)$; thus the fact that $R^{(1)}(k, t)$ evolves according to (4.-3a) implies that $u^{(1)}(x, t)$ evolves according to (4.-1). We conclude therefore that, if two functions $u^{(1)}(x, t)$ and $u^{(2)}(x, t)$ are related by the integrodifferential relation (1), then if one of them satisfies (4.-1), the other also does.

To complete the argument we also consider the time evolution of the parameters associated with the discrete part of the spectral transform. The first thing to prove is that the discrete eigenvalues associated with the function $u^{(1)}(x, t)$ are time-independent, if $u^{(1)}$ is related to $u^{(2)}$ by (1) and if the discrete eigenvalues $p_n^{(2)}$ associated with $u^{(2)}$ are themselves time-independent. In fact we now prove much more, namely that the eigenvalues of $u^{(1)}(x, t)$ either coincide with those of $u^{(2)}(x, t)$, or with the *positive* roots (if any) of the (time-independent) equation

(5) $\quad g(4p^2) - 2ph(4p^2) = 0.$

Let indeed $p > 0$ be an eigenvalue of $u^{(1)}$ but not of $u^{(2)}$; and indicate, in the notation of section 2.1, by $c = \rho^{1/2}$ the corresponding normalization coefficient (note that, by convention, both p and c are positive). Then rewrite (2.4.1.-34) and (2.4.1.-35) to apply to this case, replace the (arbitrary) function $g(z)$ in the second of these equations by the (arbitrary) function $h(z)$, sum the two equations, use (1) and the positivity of c; and thereby obtain precisely (5). (Note that here, and below, we use for brevity the locution "eigenvalue of $u(x)$", meaning of course "discrete eigenvalue of the Schroedinger operator $-d^2/dx^2 + u(x)$"; moreover, the actual eigenvalue is of course $-p^2$, but for brevity we refer to p as the eigenvalue).

There remains the comparison of the time evolution of the normalization coefficients. We limit our consideration here to those eigenvalues $p_n^{(1)} = p_n^{(2)} = p_n$ that are common to $u^{(1)}$ and $u^{(2)}$ (since we have not established in section 2.4 wronskian relations that are adequate to deal with the case of different eigenvalues). Then, using as above (2.4.1.-67) and (2.4.1.-68), we obtain the formula

(6) $\quad \rho_n^{(1)}(t) = \rho_n^{(2)}(t) \dfrac{\left[g(4p_n^2) + 2p_n h(4p_h^2)\right]}{\left[g(4p_n^2) - 2p_n h(4p_n^2)\right]},$

and the rest of the argument is completely analogous to that given above for the reflection coefficients.

Note that this formula could have been obtained directly from (2) and (3.1.-23); as well as the identification of the discrete eigenvalues of $u^{(1)}(x,t)$ either with those of $u^{(2)}(x,t)$ or with the positive roots of (5) (that clearly coincide with the positive imaginary zeros, $k_n = i p_n$, $p_n > 0$, of the denominator in the r.h.s. of (2)). It is easily seen, either by the (more rigorous) argument used above or by the (shortcut) argument based on (3.1.-23), that the only possible eigenvalues of $u^{(2)}(x,t)$ that are not also eigenvalues of $u^{(1)}(x,t)$ are the positive roots of

(7) $\quad g(4p^2) + 2ph(4p^2) = 0$

(note the change of sign relative to (5)). In fact it is easily seen that these arguments imply a bit more, namely the following: if the positive number $p_n^{(2)}$ is a discrete eigenvalue of $u^{(2)}(x,t)$ and it is not a root of (7), it is then necessarily an eigenvalue also of $u^{(1)}(x,t)$; and similarly, if $p_n^{(1)}$ is a discrete eigenvalue of $u^{(1)}(x,t)$ and it is not a root of (5), it is then necessarily an eigenvalue also of $u^{(2)}(x,t)$; and the corresponding normalization coefficients are, in both cases, related by (6). On the other hand, if a positive number p is not an eigenvalue of $u^{(2)}$ but it is a root of (5), then it may (but it need not) be an eigenvalue of $u^{(1)}$; and conversely, if a positive number p is not a discrete eigenvalue of $u^{(1)}$ but it satisfies (7), then it may (but it need not) be an eigenvalue of $u^{(2)}$.

Note, incidentally, that the above implies that, in the present context, the only case in which (2.4.1-34a) and (2.4.1.-35a) may be applicable (by replacing g by h in the second equation, and then summing the two equations) yields merely the identity $0=0$.

Of course, all these conclusions apply provided both functions $u^{(1)}$ and $u^{(2)}$ qualify as *bona fide* potentials. That this will not be compatible with an arbitrary choice of g, h and, say, $u^{(2)}$, is evident, for instance, by the requirement that (6) be consistent with the positivity of both $\rho^{(1)}$ and $\rho^{(2)}$ (see below for examples).

On the other hand, the condition that two functions, related by (1), satisfy the same equation of type (4.-1), has clearly a more general validity; indeed the property to satisfy (4.-1) is a local condition (since (4.-1) is generally a pure partial differential equation), hence it is largely independent of the asymptotic or regularity properties of $u^{(1)}(x,t)$ and $u^{(2)}(x,t)$, that are required to qualify them as *bona fide* potentials (see examples below).

Summarizing we conclude that (1) is a *Bäcklund transformation* (or rather, given the arbitrariness of g and h, it provides *a class of Bäcklund transformations*) for the class of nonlinear evolution equations (4.-1): namely, it is a relation between $u^{(1)}(x,t)$ and $u^{(2)}(x,t)$ such that, if one of these two functions satisfies (4.-1), the other also does. If moreover both $u^{(1)}(x,t)$ and $u^{(2)}(x,t)$ belong to the class of *bona fide* potentials, so that they both possess a spectral transform, then (2) provides the relation between their reflection coefficients, and (6) the relation between the normalization coefficients of the discrete eigenvalues that belong both to $u^{(1)}$ and to $u^{(2)}$ (if any); furthermore, if the positive numbers $p_n^{(2)}$ are the eigenvalues of $u^{(2)}$, then only those (if any) that satisfy the condition (7) can fail to appear also as eigenvalues of $u^{(1)}$, and only positive numbers p that satisfy (5) (if any) can provide *additional* eigenvalues of $u^{(1)}$, i.e. eigenvalues different from the $p_n^{(2)}$'s (clearly this same statement applies if one exchanges $u^{(1)}$ and $u^{(2)}$, and (5) and (7)).

Before proceeding to consider some simple examples, let us take a second look at these results, in the framework of the spectral integral relations of subsection 2.4.2. We thus assume now (appropriate) relations between two spectral transforms, and obtain relations for the corresponding functions. Assume then the two reflection coefficients, $R^{(1)}$ and $R^{(2)}$, to be related by (2), and moreover the two sets of discrete eigenvalues (if any) to coincide, $p_n^{(1)} = p_n^{(2)}$, and the corresponding normalization coefficients to be related by (6). It is then sufficient to sum (2.4.2.-11) and (2.4.2.-12) (the latter with $g(z)$ replaced by $h(z)$) to obtain the relation

$$(8) \quad g(\tilde{\Lambda})[w^{(1)}(x,t) - w^{(2)}(x,t)] + h(\tilde{\Lambda})\tilde{\Lambda} \cdot 1 = 0.$$

We next use the formulae

$$(9) \quad g(\tilde{\Lambda}) = -Ig(\Lambda)D, \quad h(\tilde{\Lambda}) = -Ih(\Lambda)D,$$

implied by (A.9.-25), to recast (8) in the form

$$(10) \quad I\{g(\Lambda)D[w^{(1)}(x,t) - w^{(2)}(x,t)] + h(\Lambda)D\tilde{\Lambda} \cdot 1\} = 0.$$

(Here of course I and D are the operators of integration and differentiation; see (A.9.-4) and (A.9.-3). It is then sufficient to apply the differentiation operator D to this formula and to use (A.9.-5) and (A.9.-25) and the relation

$$(11) \quad u^{(j)}(x,t) = -Dw^{(j)}(x,t), \quad j=1,2,$$

implied by (4.-4) and (A.9.-3), to recover (1). It is moreover easily seen that the same result (1) obtains if the two sets of discrete eigenvalues do not match, provided any $p_n^{(1)}$ that is not a discrete eigenvalue for $u^{(2)}$ is instead a root of (5), and any $p_n^{(2)}$ that is not a discrete eigenvalue of $u^{(1)}$ is instead a root of (7). The previous results are therefore fully confirmed.

4.1 Bäcklund transformations

Let us now proceed to consider some examples.

The simplest instance of Bäcklund transformation corresponds to constant g and h. It is then notationally convenient to set

(12) $\quad g = 2ph,$

so that (2) reads

(13) $\quad R^{(1)}(k,t) = -R^{(2)}(k,t)(k+ip)/(k-ip),$

and (1) becomes

(14) $\quad 2p[u^{(1)}(x,t) - u^{(2)}(x,t)] + u_x^{(1)}(x,t) + u_x^{(2)}(x,t)$
$\qquad + [u^{(1)}(x,t) - u^{(2)}(x,t)] \int_x^{+\infty} dy [u^{(1)}(y,t) - u^{(2)}(y,t)] = 0.$

It is convenient to rewrite the last equation in the form

(15) $\quad w_x^{(1)} + w_x^{(2)} = -\tfrac{1}{2}(w^{(1)} - w^{(2)})(4p + w^{(1)} - w^{(2)}),$

that obtains by introducing the usual notation (see (4.-4))

(16a) $\quad w^{(j)} \equiv w^{(j)}(x,t) = \int_x^{+\infty} dy\, u^{(j)}(y,t), \quad j=1,2,$

(16b) $\quad w_x^{(j)}(x,t) = -u^{(j)}(x,t), \quad j=1,2.$

To get (15) from (14), one integration has been performed; the vanishing of $w^{(j)}$ at $+\infty$,

(17) $\quad w^{(j)}(+\infty,t) = 0, \quad j=1,2,$

implied by the definition (16a), as well as the asymptotic vanishing of $w_x^{(j)}$, implied by (16b), have moreover been invoked to set to zero the integration constant.

The simplest Bäcklund transformation (15) is characterized by the single parameter p, that appears in a simple fashion in the formula (13) relating $R^{(1)}$ to $R^{(2)}$, as well as in the formula (see (6) and (12)),

(18) $\quad \rho_n^{(1)}(t) = \rho_n^{(2)}(t)(p+p_n)/(p-p_n),$

that relates the normalization coefficients for the discrete eigenvalues p_n that belong both to $u^{(1)}$ and to $u^{(2)}$ (if any). Indeed this formula (as well as the discussion given above of the relationship between the discrete eigenvalues of $u^{(1)}$ and $u^{(2)}$) implies for the simplest Bäcklund transformation (15) the following simple conclusions. Assume $u^{(2)}$ is a *bona fide* potential, and indicate by the N positive numbers p_n, labeled in increasing order,

(19) $\quad 0 < p_1 < p_2 < \cdots < p_N$,

its discrete eigenvalues (if any). Then validity of one or other of the following inequalities is a necessary condition in order that $u^{(1)}$ be also a *bona fide* potential:

(20a) $\quad p > p_N$,

(20b) $\quad p \leq -p_N$.

Moreover, if the first (strict) inequality holds, $u^{(1)}$ has all the eigenvalues of $u^{(2)}$ and generally, in addition, the extra eigenvalue p; if the second inequality prevails in strict form (namely, excluding the equality sign), $u^{(1)}$ has precisely the same eigenvalues as $u^{(2)}$; finally, if $p = -p_N$, then $u^{(1)}$ has only the $N-1$ eigenvalues $p_1, p_2, \ldots, p_{N-1}$. (The diligent reader may verify these results with the treatment of appendix A.12).

Let us emphasize however that these conditions are necessary, not sufficient, for $u^{(1)}$ to be a *bona fide* potential; while the conclusions about the discrete eigenvalues of $u^{(1)}$ hold of course under the assumption that $u^{(1)}$ is indeed a *bona fide* potential. Finally note that, if $N = 0$ (namely, if $u^{(2)}$ has no discrete eigenvalues), much of the above analysis applies, with p_N replaced by 0.

In this analysis we have always referred to the "discrete eigenvalues p_n", in the conventional sense indicated above (the actual discrete eigenvalues are of course $E_n = -p_n^2$; and by convention always $p_n > 0$). In fact, in the present context, we might as well have talked, instead of the discrete eigenvalue p_n, of the *soliton* characterized by the parameter p_n (recall that there indeed is a one-to-one correspondence between solitons and discrete eigenvalues, for any solution $u(x, t)$ of the class of nonlinear evolution equations (4.-1) that qualifies as a *bona fide* potential; see section 3.2.1). Thus we may conclude that the Bäcklund transformation (14) (or (15)), from $u^{(2)}(x, t)$ to $u^{(1)}(x, t)$, with the positive parameter p satisfying (20a), adds the soliton characterized by the parameter p. Note that in this case the integration of the Bäcklund transformation (15) introduces generally one x-independent quantity ("integration constant"; see appendix A.12). The value of this quantity is directly related to the value of the normalization constant ρ associated with the additional discrete eigenvalue p; thus, in the sense that has been clarified in section 3.2.1 and in appendix A.10, it is related with the localization of the novel soliton. Of course the "integration constant" must be equipped with an appropriate time-dependence, as implied by its relation with $\rho(t)$, in order that $w^{(1)}(x,t)$ satisfy the same equation of the class (4.-5) as $w^{(2)}(x,t)$ does (see appendix A.12).

Essentially the same method used to relate (by (2)) the reflection coefficients of two potentials $u^{(1)}$ and $u^{(2)}$ linked by (1) can be used to relate the corresponding transmission coefficients. The basic formulae to do this, (2.4.1.-18) and (2.4.1.-19), are however more complicated than their counterparts for the reflection coefficients, (2.4.1.-10). We report here only the neat formula corresponding to the simplest Bäcklund transformation (15), that is easily obtained using the (simpler) formulae (2.4.1.-21) and (2.4.1.-22). It reads:

(21) $\quad T^{(1)}(k) = T^{(2)}(k)(k+\mathrm{i}p)/(k\pm\mathrm{i}p),$

the upper or lower sign prevailing, depending on whether the upper or lower sign prevails in the formula

(22) $\quad \int_{-\infty}^{+\infty} \mathrm{d}x \left[u^{(1)}(x,t) - u^{(2)}(x,t) \right] = 2p(\pm 1 - 1).$

As for this equation, that clearly implies a relationship between $u^{(1)}$ and $u^{(2)}$, it is easily seen that it is a consequence of (15) (the l.h.s. of (15) vanishes as $x \to -\infty$, and therefore also its r.h.s. must vanish, yielding (22)); thus its validity is automatically guaranteed if both $u^{(1)}$ and $u^{(2)}$ are *bona fide* potentials related by (15).

Note that, if the upper sign prevails in (21), then $T^{(1)}$ is simply identical to $T^{(2)}$; while if the lower sign prevails, the relation between $T^{(1)}$ and $T^{(2)}$ is, except for the overall sign, just the same as (13), that relates $R^{(1)}$ to $R^{(2)}$. Which sign is actually appropriate in (21) and (22) depends on the parameter p and on the constant introduced by integrating (15). In fact, it is easily found (see appendix A.12) that if (20a) holds and the integration constant yields a nonvanishing normalization constant $\rho(t)$, then (21) and (22) read with the lower sign; this is also the case when $p = -p_N$, i.e. when (20b) holds with the equality sign. The upper sign prevails instead in (21) and (22) when the strict inequality (20b) holds (i.e. $p < -p_N$) and also when $p > p_N$ but the associated normalization constant ρ vanishes.

In this discussion we have considered the formula (15) as a (Bäcklund) transformation from $w^{(2)}$ to $w^{(1)}$. Of course an analogous analysis could have been done by considering (15) as a (Bäcklund) transformation from

$w^{(1)}$ to $w^{(2)}$, with the modifications obviously implied by the invariance of (13), (14), (15), and (18) under the simultaneous exchange of the superscripts 1 and 2 and of p with $-p$. (Incidentally, to compare with the treatment of subsection 1.7.1, note that one considers there a transformation from $w^{(0)}$ to $w^{(1)}$).

The simplest application of the simplest Bäcklund transformation (15) consists in the evaluation of the potential $u^{(1)} \equiv u$ corresponding, via (15) and (16), to $u^{(2)} = 0$ (implying $w^{(2)} = 0$), that clearly is a (trivial) solution of (4.-1). An elementary computation yields

$$(23a) \quad w(x,t) = -2p\left(1 - \tgh\{p[x - \xi(t)]\}\right),$$

$$(23b) \quad u(x,t) = -2p^2/\cosh^2\{p[x - \xi(t)]\}.$$

It should be noted that this result obtains for positive p (although the expression (23b) does not depend on the sign of p); for negative p only the vanishing solution is consistent with (17). The quantity $\xi(t)$ originates as an integration constant, which means, in the present context, that it does not depend on x ((15) is a differential equation in the x variable); this however does not prevent it from depending on t, as we have explicitly indicated.

Clearly (23) is just the single-soliton solution; the time dependence of $\xi(t)$ can be determined by inserting it in (4.-1), whereupon, using the formula

$$(24a) \quad Lu_x(x,t) = 4p^2 u_x(x,t),$$

that is easily verified from (23b) and (4.-2), and that clearly implies

$$(24b) \quad \alpha(L)u_x(x,t) = \alpha(4p^2)u_x(x,t),$$

there obtains

$$(25) \quad \dot{\xi}(t) = -\alpha(4p^2),$$

consistently with (1.6.1.-6b,7). Note that the fact that (23) satisfies, for an appropriate $\xi(t)$, the nonlinear evolution equation (4.-1), is nontrivial; indeed, it reflects its being obtained via the Bäcklund transformation (15).

Let us also reiterate that a Bäcklund transformation may yield a singular or a regular solution; this also depends on the choice of the integration constant. Indeed (25) yields

$$(26) \quad \xi(t) = \xi_0 - \alpha(4p^2)t,$$

with ξ_0 arbitrary. For real ξ_0, (23) is real and regular (for all real x); but if

(27) $\quad \xi_0 = \xi_0' + i\pi/2p$

with ξ_0' real, then (23), although still real, becomes singular at the real value $x = \xi_0' - \alpha(4p^2)t$. Nevertheless it does always provide a solution to (4.-1), although not one acceptable as a *bona fide* potential (see also appendix A.12).

It is instructive (and therefore left as an exercise for the reader) to calculate $u^{(1)} \equiv u$ from the same Bäcklund transformation (15) (with p replaced, for notational neatness, by p_1; see below), but using as input, instead of a vanishing $u^{(2)}$, the single-soliton solution

(28) $\quad u^{(2)}(x,t) = -2p_2^2/\cosh^2\{p_2[x - \xi_2(t)]\},$

where of course

(29) $\quad \xi_2(t) = \xi_2(0) - \alpha(4p_2^2)t.$

Again the computation proves feasible, and an integration constant is introduced, whose time dependence can eventually be ascertained by requiring u to satisfy (4.-1). Of course in this manner one gets precisely the two-soliton solution with parameters p_1 and p_2; that is, provided $p_1 > p_2$ (see (20a)), and if the arbitrary constants that this solution u of (4.-1) turns out to contain are chosen appropriately, namely so that u be real and regular for all real values of x.

A shortcut procedure to perform the calculation we have just described is provided in the following section by means of the superposition principle already introduced in subsection 1.7.2.

Associated with the Bäcklund transformation (14) or (15) is the so-called "Darboux transformation" (see appendix A.12)

(30) $\quad f^{(\pm)(1)}(x,k) = \pm(p - ik)^{-1} f^{(\pm)(2)}(x,k)(d/dx)\ln\{[\rho f^{(+)(2)}(x, ip)$
$\qquad + 2pT^{(2)}(ip) f^{(-)(2)}(x, ip)]/f^{(\pm)(2)}(x,k)\}.$

This formula provides explicitly the Jost solutions $f^{(\pm)(1)}(x,k)$ associated with the potential $u^{(1)}(x)$ in terms of the Jost solutions $f^{(\pm)(2)}(x,k)$ and the transmission coefficient $T^{(2)}(k)$ for the potential $u^{(2)}(x)$. (Here we are of course using the notation of section 2.1). The corresponding spectral transforms are related by (13) and (18). There is moreover one additional discrete eigenvalue, p, associated with $u^{(1)}$, and the corresponding normalization coefficient ρ is the quantity entering explicitly in the r.h.s. of (30); with the condition (20a) to guarantee consistency within the class of *bona fide* potentials.

Let us emphasize that the relationship between $u^{(1)}$ and $u^{(2)}$ that corresponds to (30) is precisely the Bäcklund transformation (14) (or equivalently (15)). Note however that these formulae, (14) and (15), do not contain any information on the normalization coefficient ρ associated with the additional discrete eigenvalue p; indeed, as we already noted above, ρ is related to the "constant of integration" that enters when (15) is solved to obtain $w^{(1)}$. This is in contrast to the Darboux transformation (30), that provides explicitly, i.e. without any additional integration, the Jost solutions $f^{(\pm)(1)}$ (if $f^{(\pm)(2)}$ are given), and therefore also $u^{(1)}$, that is easily obtained from $f^{(\pm)(1)}$, by using the Schroedinger equation (for any value of k; for instance, for $k=0$); and it can be easily verified that, under the conditions given above, there obtains a real and regular $u^{(1)}$.

For the special case $u^{(2)}(x)=0$, implying $T^{(2)}=1$ and $f^{(\pm)(2)}(x,k)=\exp(\pm ikx)$, the Darboux transformation (30) yields

(31) $\quad f^{(\pm)(1)}(x,k) = \pm(p-ik)^{-1}\{p\,\mathrm{tgh}[p(x-\xi)] \mp ik\}\exp(\pm ikx)$

with

(32) $\quad \rho = 2p\exp(2p\xi);$

and (see (2.2.-11)) the corresponding $u^{(1)}(x)$ is of course the single-soliton potential (see (2.2.-10))

(33) $\quad u^{(1)}(x) = -2p^2/\cosh^2[p(x-\xi)].$

Some additional considerations on the simplest Bäcklund transformation (15) are confined to appendix A.12. They deal with the uniqueness of the function $u^{(1)}(x)$ yielded by (15) for given p and $u^{(2)}(x)$ (within the class of *bona fide* potentials; or possibly outside), and on the (related) questions of the asymptotic behaviours of these functions as $x \to \infty$ and of the analyticity properties of the corresponding reflection coefficients.

More complicated Bäcklund transformations than (15) obtain if g and h are not constant. For instance the choice

(34) $\quad g(z) = 4p_1 p_2 + z, \qquad h(z) = 2(p_1 + p_2),$

when inserted into (2), yields

(35) $R^{(1)}(k,t) = R^{(2)}(k,t)(k+\mathrm{i}p_1)(k+\mathrm{i}p_2)/[(k-\mathrm{i}p_1)(k-\mathrm{i}p_2)],$

while its insertion into (1) yields a fairly complicated equation that we do not write, but whose solution with the input $u^{(2)}=0$ is feasible, yielding again the two-soliton solution for $u^{(1)}(x)$, with parameters p_1 and p_2; note that the expression for $u^{(1)}$ resulting from the integration of (1) with (34) will contain now two time-dependent integration constants, whose time evolution is determined up to two arbitrary constants (independent of x and t), by the requirement that $u^{(1)}$ satisfy (4.-1). However, as suggested by a comparison of (35) with (13), and as further discussed in the following section, a Bäcklund transformation such as (1), together with (34), may be interpreted as the result of two sequential Bäcklund transformations of the simplest kind, namely those given explicitly by (15) and (16). There are, nevertheless, transformations of type (34), that carry real *bona fide* potentials into real *bona fide* potentials, but that cannot be factored into two sequential transformations each of which has this property. A typical such case obtains if p_1 and p_2 in (34) are not real, but they are complex conjugate, $p_1^* = p_2$ (so that (34) remains real, and the factor multiplying $R^{(2)}$ in the r.h.s. of (35) maintains a unit modulus). Such a transformation adds no discrete eigenvalues (no solitons) but rather, in the language of scattering theory, a "resonance". Thus, within the class of *bona fide* potentials, it has no room for any arbitrary (integration) constant; given $u^{(2)}$, and therefore its spectral transform $S^{(2)}$, the spectral transform $S^{(1)}$ is completely determined (by (2) and (6), with (34)), and therefore $u^{(1)}$ is also uniquely determined.

As we have already mentioned above, the Bäcklund transformations (1) for the class of nonlinear evolution equations (4.-1) can be viewed as the analog of the transformations (1.1.-10) relating solutions of the linear evolution equations (1.1.-1). Indeed it is immediately seen that, in the weak field limit in which all nonlinear terms are neglected, in the same way as (4.-1) goes into (1.1.-1), so (1) goes into (1.1.-10) (up to notational changes).

4.2. Commutativity of Bäcklund transformations and nonlinear superposition principle

In this section we cover the same ground as in the last part of subsection 1.7.1 and in subsection 1.7.2. The main, important, addition relative to the (incomplete) treatment given there is consideration of the rôle played by the part of the spectral transform associated with discrete eigenvalues. For simplicity of presentation, we provide here a self-consistent, if terse, treatment, with no reference to the previous discussion; nevertheless the reader may wish to have a second look at that material before proceeding further, to recover quickly the main idea.

Although the discussion that follows could be easily made in the context of the general Bäcklund transformation (4.1.-1), we restrict for simplicity our consideration to the simpler transformation (4.1.-15) (with $p>0$), that provides an adequate vehicle for the notions we wish to convey.

Consider the Bäcklund transformation (4.1.-15), with parameter p_1, from $w^{(0)}(x)$ to $w^{(1)}(x)$; and the Bäcklund transformation (4.1.-15), with parameter p_2, from $w^{(1)}(x)$ to $w^{(12)}(x)$:

(1) $\quad w_x^{(1)} + w_x^{(0)} = -\tfrac{1}{2}(w^{(1)} - w^{(0)})(4p_1 + w^{(1)} - w^{(0)}),$

(2) $\quad w_x^{(12)} + w_x^{(1)} = -\tfrac{1}{2}(w^{(12)} - w^{(1)})(4p_2 + w^{(12)} - w^{(1)}).$

Let us investigate the corresponding transformation from the point of view of the spectral transform. The effect on the reflection coefficients presents no problem:

(3) $\quad R^{(1)}(k) = -R^{(0)}(k)(k+ip_1)/(k-ip_1),$

(4) $\quad R^{(12)}(k) = R^{(0)}(k)(k+ip_1)(k+ip_2)/[(k-ip_1)(k-ip_2)].$

The effect on the component of the spectral transform associated with the discrete spectrum is a bit more delicate. Let us for definiteness assume that there are N discrete eigenvalues associated with $w^{(0)}(x)$, say

(5) $\quad p_1^{(0)} < p_2^{(0)} < \cdots < p_N^{(0)},$

and moreover that

(6) $\quad p_N^{(0)} < p_1 < p_2.$

Then the spectral transform associated with $w^{(1)}(x)$ has $N+1$ discrete eigenvalues,

(7a) $\quad p_n^{(1)} = p_n^{(0)}, \quad n = 1, 2, \ldots, N,$

(7b) $\quad p_{N+1}^{(1)} = p_1,$

and the corresponding normalization coefficients $\rho_n^{(1)}$ are related to those of $w^{(0)}(x)$ by the formulae

(8a) $\quad \rho_n^{(1)} = \rho_n^{(0)}(p_1 + p_n^{(0)})/(p_1 - p_n^{(0)}), \quad n=1,2,\ldots,N,$

(8b) $\quad \rho_{N+1}^{(1)} = \rho'.$

Here ρ' indicates a quantity whose value is associated with that of the integration constant that characterizes the solution $w^{(1)}(x)$ obtained by solving (1) (see appendix A.12).

As for the spectral transform of $w^{(12)}(x)$, it possesses $N+2$ discrete eigenvalues, with the following values (we trust the notation is self-explanatory):

(9a) $\quad p_n^{(12)} = p_n^{(0)}, \quad n=1,2,\ldots,N,$

(9b) $\quad p_{N+1}^{(12)} = p_1, \quad p_{N+2}^{(12)} = p_2,$

(10a) $\quad \rho_n^{(12)} = \rho_n^{(0)}(p_1 + p_n^{(0)})(p_2 + p_n^{(0)})/[(p_1 - p_n^{(0)})(p_2 - p_n^{(0)})],$

$\quad\quad\quad n=1,2,\ldots,N,$

(10b) $\quad \rho_{N+1}^{(12)} = \rho'(p_2 + p_1)/(p_2 - p_1),$

(10c) $\quad \rho_{N+2}^{(12)} = \rho''.$

Here again ρ'' indicates a quantity whose value is now associated with that of the integration constant that characterizes the solution $w^{(12)}(x)$ obtained by solving (2) (for given $w^{(1)}(x)$).

In conclusion, given $w^{(0)}(x)$ and the two parameters p_1 and p_2, there obtains via the two Bäcklund transformations (1) and (2) a function $w^{(12)}(x)$ that contains two integration constants, whose spectral transform is given, in terms of that of $w^{(0)}(x)$ and of p_1 and p_2, by the formulae (4), (9) and (10). The latter contains the two constants ρ' and ρ'', whose values are related to those of the integration constants contained in $w^{(12)}(x)$.

Suppose now to start again from $w^{(0)}(x)$ and apply again two sequential Bäcklund transformations, of the simpler type (4.1.-15), but now first with parameter p_2 and next with parameter p_1:

(11) $\quad w_x^{(2)} + w_x^{(0)} = -\tfrac{1}{2}(w^{(2)} - w^{(0)})(4p_2 + w^{(2)} - w^{(0)}),$

(12) $\quad w_x^{(21)} + w_x^{(2)} = -\tfrac{1}{2}(w^{(21)} - w^{(2)})(4p_1 + w^{(21)} - w^{(2)}).$

The corresponding transformations for the spectral transforms read (in self-evident notation):

(13) $\quad R^{(2)}(k) = -R^{(0)}(k)(k+ip_2)/(k-ip_2),$

(14) $\quad R^{(21)}(k) = R^{(0)}(k)(k+ip_2)(k+ip_1)/[(k-ip_2)(k-ip_1)],$

(15a) $\quad p_n^{(2)} = p_n^{(0)}, \quad n=1,2,\ldots,N,$

(15b) $\quad p_{N+1}^{(2)} = p_2,$

(16a) $\quad \rho_n^{(2)} = \rho_n^{(0)}(p_2+p_n^{(0)})/(p_2-p_n^{(0)}), \quad n=1,2,\ldots,N,$

(16b) $\quad \rho_{N+1}^{(2)} = \bar{\rho},$

(17a) $\quad p_n^{(21)} = p_n^{(0)}, \quad n=1,2,\ldots,N,$

(17b) $\quad p_{N+1}^{(21)} = p_2, \quad p_{N+2}^{(21)} = p_1,$

(18a) $\quad \rho_n^{(21)} = \rho_n^{(0)}(p_2+p_n^{(0)})(p_1+p_n^{(0)})/[(p_2-p_n^{(0)})(p_1-p_n^{(0)})],$

$$n=1,2,\ldots,N,$$

(18b) $\quad \rho_{N+1}^{(21)} = \bar{\rho}(p_1+p_2)/(p_1-p_2),$

(18c) $\quad \rho_{N+2}^{(21)} = \bar{\bar{\rho}}.$

Here $\bar{\rho}$ and $\bar{\bar{\rho}}$ are two quantities, whose values are related to the values of the two integration constants possessed by the function $w^{(21)}(x)$ obtained by integrating (11) and (12).

Comparison of (4), (9) and (10) with (14), (17) and (18), implies that, provided

(19a) $\quad \bar{\bar{\rho}} = \rho'(p_2+p_1)/(p_2-p_1),$

(19b) $\quad \bar{\rho} = -\rho''(p_2-p_1)/(p_2+p_1),$

the spectral transforms of $w^{(12)}(x)$ and $w^{(21)}(x)$ coincide; and this in its turn implies that $w^{(12)}(x)$ and $w^{(21)}(x)$ themselves coincide,

(20) $\quad w^{(12)}(x) = w^{(21)}(x).$

This conclusion is nontrivial, in view of the nonlinear character of Bäcklund transformations; it can be synthetized in the statement that *Bäcklund transformations commute* (see the graphical representation of this assertion displayed in subsection 1.7.1). A more precise formulation of this assertion, based on the analysis we have just given, can be summarized as follows. Given a function $w^{(0)}(x)$, and two Bäcklund transformations of type (4.1.-15), say BT1 and BT2 characterized respectively by the parameters p_1 and p_2, there obtain two functions $w^{(12)}(x)$ and $w^{(21)}(x)$, the former by applying to $w^{(0)}(x)$ first BT1 and then BT2, the latter by applying to $w^{(0)}(x)$ first BT2 and then BT1. Each of these two functions, $w^{(12)}(x)$ and $w^{(21)}(x)$, contains generally two integration constants, since it is obtained by integrating sequentially two first order ordinary differential equations; and the two functions $w^{(12)}(x)$ and $w^{(21)}(x)$ may only differ due to different choices of these integration constants. Indeed, for any choice of the integration constants in one of these functions, there is an appropriate choice of integration constants in the other, that bring about the coincidence of these two functions, see (20).

This conclusion has been reached above by looking at the (much simpler) effect of Bäcklund transformations in the spectral space rather than in x-space, under the (more or less implicit) assumption that the functions under consideration belong to the class of *bona fide* potentials and are therefore endowed with well-defined spectral transforms. But the Bäcklund transform equation (4.1.-15) has clearly a more general validity, since it can be applied also to functions $w^{(j)}(x)$ that do not have all the properties required of *bona fide* potentials (or rather, of integrals of *bona fide* potentials; see (4.-4)), namely also to functions which may not have an associated spectral transform. It is on the other hand fairly obvious that the property of commutativity that we have just discussed holds also in this more general context.

The same applies of course to the characteristic property of Bäcklund transformations, to transform solutions of the evolution equation (4.-1) (or, equivalently, (4.-5)), into solutions of the same equation (see the preceding section 4.1). In this context let us draw attention to the fact that all the considerations given above, in this section, apply just as well if the functions $w^{(j)}$ depend on other variables besides x (and in particular on "time" t); merely for notational simplicity such possible dependence has been ignored here.

An important consequence of the commutativity property (20) of Bäcklund transformations is now easily obtained. Consider indeed a function

(21) $\qquad w'(x,t) = w^{(12)}(x,t) = w^{(21)}(x,t)$

that has been obtained from a solution $w^{(0)}(x,t)$ of (4.-5) by the procedure described above, with a choice of integration constants such that (20) hold. It is then a matter of trivial algebra to derive from (1), (2), (11) and (12) the formula

$$(22) \quad w'(x,t) = w^{(0)}(x,t) - \frac{(p_1+p_2)[w^{(1)}(x,t)-w^{(2)}(x,t)]}{[p_1-p_2+\frac{1}{2}(w^{(1)}(x,t)-w^{(2)}(x,t))]}.$$

Let us emphasize that this is a completely explicit expression of w', in terms of $w^{(0)}$ and of the two functions $w^{(1)}$ and $w^{(2)}$ related to $w^{(0)}$ by the Bäcklund transformations with parameters p_1 and p_2, (1) and (11). Both $w^{(1)}$ and $w^{(2)}$ contain an integration constant (related respectively to ρ' and $\bar{\rho}$, in the notation used above: see (8b) and (16b)); thus also w' contains these two constants.

Let us incidentally note a remarkable property of (22), namely that from $w^{(j)}(-\infty, t) = w^{(0)}(-\infty, t) - 4p_j\theta_j$, $j=1,2$, where $\theta_j=0$ or $\theta_j=1$ (see (4.1.-22)), there follows $w'(-\infty, t) = w^{(0)}(-\infty, t) - 4(p_1\theta_1 + p_2\theta_2)$.

The treatment of the preceding section 4.1 implies that, if $w^{(0)}(x,t)$ is a solution of (4.-5) (and therefore the corresponding $u^{(0)}(x,t) = -w_x^{(0)}(x,t)$ a solution of (4.-1)), then $w^{(1)}(x,t)$ and $w^{(2)}(x,t)$ are also solutions of (4.-5), provided the constants of integration that these functions contain are equipped with an appropriate time-dependence (note that, in the present context, a "constant of integration" is an "x-independent" quantity, but does instead generally depend on time); indeed the time-dependence required in order that $w^{(1)}(x,t)$ and $w^{(2)}(x,t)$ satisfy (4.-5) is easily given in explicit form in terms of the quantities ρ' and $\bar{\rho}$,

$$(23a) \quad \rho'(t) = \rho'(0) \exp[-2p_1\alpha(4p_1^2)t],$$

$$(23b) \quad \bar{\rho}(t) = \bar{\rho}(0) \exp[-2p_2\alpha(4p_2^2)t];$$

(see (4.-3c), together with (7b), (8b), (15b) and (16b)). And moreover $w'(x,t)$ itself is also, under these conditions, a solution of (4.-5).

For this reason the formula (22) is often (perhaps improperly) referred to as a *nonlinear superposition principle* for the nonlinear evolution equation (4.-5). Let us emphasize that this formula provides a completely explicit novel solution $w'(x,t)$ of (4.-5) in terms of 3 other solutions $w^{(0)}(x,t)$, $w^{(1)}(x,t)$ and $w^{(2)}(x,t)$, the two latter solutions, $w^{(1)}$ and $w^{(2)}$, being related to the former solution, $w^{(0)}$, by Bäcklund transformations of type (4.1.-15) with parameters p_1 and p_2.

Let us reiterate that, although in the derivation of these results we have relied on the spectral transforms of the functions under consideration, assuming therefore implicitly to work within the class of *bona fide* potentials, the results have clearly a more general validity; for instance the nonlinear superposition principle (22) yields a novel solution of (4.-5) out of 3 given solutions (related via Bäcklund transformations of type (4.1.-15) as detailed above) even if some of the 4 functions $w^{(0)}$, $w^{(1)}$, $w^{(2)}$ or w' are singular (for real x), or if the conditions (5) and (6) are violated. Indeed, it is easily seen that, if $w^{(0)}$ is a *bona fide* potential and (5) and (6) hold, at least one of the 3 functions $w^{(1)}$, $w^{(2)}$ and w' must fall outside the class of *bona fide* potentials, since the requirement that ρ', $\bar{\rho}$, ρ'' and $\bar{\bar{\rho}}$ be all positive (see (8b), (10c), (16b) and (18c)) is clearly incompatible with the constraints (19). This analysis suggests that, under these circumstances, in order for w' to fall within the class of *bona fide* potentials (i.e., be regular for all real values of x), $w^{(1)}$ should be regular (namely $\rho'>0$) but $w^{(2)}$ singular ($\bar{\rho}<0$). Here we are implying a correspondence between the occurrence of singularities in configuration space for real x and the appearance of negative normalization coefficients, that is suggested by the findings discussed in the preceding section 4.1 and in appendix A.12 (see also below), but that would require an extension of the concept of the spectral transform beyond the class of *bona fide* potentials to be put on a firm basis (a task that, as we have already explained in section 2.3, falls outside of the scope of this book; see, however, appendix A.16).

It is moreover clear (see also the examples below) that, if $w^{(0)}$ is a *bona fide* potential and (5) and (6) hold, and if w' given by (22) is also a *bona fide* potential, then it possesses the same solitons as $w^{(0)}$ and two more in addition, associated with the two extra discrete eigenvalues p_1 and p_2.

Let us indeed consider, as a simple example, the application of the nonlinear superposition principle (22) taking as starting point

(24) $\quad w^{(0)}(x,t)=0.$

The results of the preceding section 4.1 then yield (see (4.1.-23a))

(25) $\quad w^{(j)}(x,t)=-2p_j\left(1-\operatorname{tgh}\left\{p_j\left[x-\xi^{(j)}(t)\right]\right\}\right), \quad j=1,2,$

with

(26) $\quad \xi^{(j)}(t)=\bar{\xi}_0^{(j)}-\alpha\left(4p_j^2\right)t.$

And it is easily verified that insertion of these expressions in the r.h.s. of (22) yields precisely the regular two-soliton expression (see appendix A.10), provided

(27a) $\quad \bar{\xi}_0^{(1)}=\xi_0^{(1)},$

(27b) $\quad \bar{\xi}_0^{(2)}=\xi_0^{(2)}+i\pi/2p_2,$

with $\xi_0^{(1)}$ and $\xi_0^{(2)}$ real constants (independent of x and of t). Note the consistency of (27) with the positivity of ρ' and negativity of $\bar{\rho}$ (see preceding discussion), as implied by the formulae (see (7b), (8b), (15b), (16b) and, say, (3.2.1.-6))

(28a) $\quad \rho' = 2p_1 \exp[2p_1 \xi^{(1)}(t)],$

(28b) $\quad \bar{\rho} = 2p_2 \exp[2p_2 \xi^{(2)}(t)].$

It is possible to continue this procedure, by taking next the single-soliton solution as starting point,

(29) $\quad w^{(0)}(x,t) = -2p_1\left(1 - \mathrm{tgh}\{p_1[x - \xi^{(1)}(t)]\}\right),$

(30) $\quad \xi^{(1)}(t) = \xi_0^{(1)} - \alpha(4p_1^2)t,$

and the two-soliton solutions, with parameters respectively p_1, p_2 and p_1, p_3, for $w^{(1)}$ and $w^{(2)}$. Insertion of these expressions in the r.h.s. of (22) (with, moreover, p_1 and p_2 replaced by p_2 and p_3), yields the 3-soliton solution; and it is easy, on the basis of the relationship that each Bäcklund transformation induces for the normalization coefficients, to figure out the conditions that are required in order that the solution thus obtained be regular (see below).

This procedure can be continued; it provides a purely algebraic technique to construct multisoliton solutions (it is sometimes referred to as *soliton ladder*). It is easily seen, on the basis of the example we have discussed above, that if the parameters p_j are chosen in increasing order,

(31) $\quad p_j < p_{j+1}, \quad j = 1, 2, \ldots, N-1,$

then the appropriate prescription, in terms of the corresponding "normalization coefficients" ρ_j appearing in the intermediate solutions, in order that the final expression be the *regular* N-soliton solution, is simply given by the alternating sign rule

(32) $\quad \mathrm{sign}\, \rho_j = (-)^{j+1}, \quad j = 1, 2, \ldots, N,$

or equivalently, for the quantities $\bar{\xi}_j$ related to ρ_j by the formula

(33) $\quad \rho_j = 2p_j \exp(2p_j \bar{\xi}_j),$

by either one of the following equivalent prescriptions,

(34a) $\quad \bar{\xi}_j = \xi_j + i(j+1)\pi/2 p_j,$

(34b) $\quad \bar{\xi}_j = \xi_j + \left[1 + (-)^j\right] i\pi/4 p_j,$

with ξ_j real.

The nonlinear superposition formula (22) could also be formally used with two complex conjugate parameters,

(35) $\quad p_1 = p_2^*.$

Assuming $w^{(0)}$ to be real, one could then also get

(36) $\quad w^{(1)} = w^{(2)*},$

yielding finally, through (22), a real w'. It is however easy to verify (and quite obvious from the point of view of the spectral transform) that, if $w^{(0)} = 0$, the corresponding w' cannot be a *bona fide* potential.

As for the case $p_1 = p_2 = p$, yielding (see (22))

(37) $\quad w'(x,t) = w^{(0)}(x,t) - 4p,$

it is clearly inconsistent with the vanishing as $x \to +\infty$ of both w' and $w^{(0)}$, that is implied by the definition (4.-4) of w. But on the other hand it is trivially obvious that, if $w^{(0)}(x,t)$ satisfies (locally) the nonlinear partial differential equation (4.-5), so does any other function differing from $w^{(0)}(x,t)$ only by an additive constant (independent of x and t).

4.3. Resolvent formula

Let $u(x,t)$ be a solution of (4.-1), so that its reflection coefficient evolves according to (4.-3a). Introduce now a function $u'(x,t)$ related to $u(x,0)$ by the transformation

(1) $\quad g(\Lambda, t)[u'(x,t) - u(x,0)] + h(\Lambda, t)\Gamma \cdot 1 = 0,$

where

(2a) $\quad g(z,t) = \cos\left[(-z)^{1/2}\alpha(z)t/2\right],$

(2b) $\quad h(z,t) = -(-z)^{-1/2}\sin\left[(-z)^{1/2}\alpha(z)t/2\right],$

and of course Γ and Λ are defined by (2.4.1.-5) and (2.4.1.-6) with the identification

(3a) $\quad u^{(1)}(x)=u'(x,t), \qquad u^{(2)}(x)=u(x,0)$.

Note that the transformation (1) is now explicitly time-dependent; moreover the functions g and h are not polynomials (but they are clearly entire in z). The function $\alpha(z)$ in (2) is of course the same one characterizing the nonlinear evolution (4.-1) satisfied by $u(x,t)$. In view of the similarity of (1) and (4.1.-1), one may call (1) a Bäcklund transformation; although, for the reasons that we have just stated, it does not belong to the class considered in section 4.1.

Now use (2.4.1.-10c) and (2.4.1.-10e), with the identification (3a) and, correspondingly,

(3b) $\quad R^{(1)}(k)=R'(k,t), \qquad R^{(2)}(k)=R(k,0);$

and of course with g and h given by (2) (note that the derivation of (2.4.1-10e) is perfectly consistent with g and h being time-dependent). This yields:

(4) $\quad R'(k,t)=R(k,0)\exp\left[2ik\alpha(-4k^2)t\right].$

However, a comparison of this formula with (4.-3a) implies

(5) $\quad R'(k,t)=R(k,t),$

and this implies

(6) $\quad u'(x,t)=u(x,t).$

(Here we are, for simplicity, assuming that it is enough to show that two functions are associated with the same reflection coefficient, to conclude that they coincide. This is of course incorrect, since one should verify that the discrete parts of the spectral transforms also coincide. It is easy to prove this in this case, by first showing that the transformation (1) with (2) cannot change the number of discrete eigenvalues, since (4.1.-5) and (4.1.-7) take now the form

(7) $\quad \exp\left[\pm p\alpha(4p^2)t\right]=0$

and cannot therefore be satisfied; and then by using (4.1.-6). We leave as a task for the diligent reader to fill in properly the details of this proof).

In conclusion, we have thus seen that (1) (with (2) and (6)) provides a (highly implicit) relationship between $u(x, t)$ and $u(x, 0)$. This is the analog, for the nonlinear evolution equation (4.-1), of the resolvent formula (1.1.-13) valid for the linear evolution equation (1.1.-1); it is indeed easily seen that, in the weak field limit in which all nonlinear contributions are neglected, just as (4.-1) reduces to (1.1.-1) (with the identification (1.2.-6b)), so (1) (with (2), (6) and again (1.2.-6b)) reduces to (1.1.-13).

Actually a slightly more general, and more elegant, presentation of this same result obtains by setting

(8) $\qquad g(\Lambda, t-t')[u(x, t) - u(x, t')] + h(\Lambda, t-t')\Gamma \cdot 1 = 0,$

with g and h still defined by (2) and with Γ and Λ always defined by (2.4.1.-5) and (2.4.1.-6), but now with the identification

(9) $\qquad u^{(1)}(x) = u(x, t), \qquad u^{(2)}(x) = u(x, t').$

The result consists then in the statement that $u(x, t)$ and $u(x, t')$, as the notation suggests, are the *same solution* of the nonlinear evolution equation (4.-1), evaluated at the different times t and t'. This shows that the time evolution itself can be viewed as a Bäcklund transformation; indeed it relates functions that are solutions of the same nonlinear evolution (4.-1) (both $u(x, t)$ and $u(x, t')$ clearly satisfy (4.-1)).

The relation (8) (with (2)) can be rewritten in the form

(10) $\qquad u(x, t) = u(x, t') + \tau(\Lambda, t-t')\Gamma \cdot 1$

with

(11) $\qquad \tau(z, t) = (-z)^{-1/2} \text{tg}[(-z)^{1/2}\alpha(z)t/2].$

Note that, by taking the limit in which t' differs only infinitesimally from t, (10) reproduces (4.-1), as indeed it should.

4.4. Nonlinear operator identities

Consider the special case of (4.-1) with $\alpha(z) = 1$. Then the evolution equation becomes simply

(1) $\qquad u_t(x, t) = u_x(x, t)$

and has the general solution

(2) $\quad u(x,t)=f(x+t)$

with f arbitrary.

Insert now these results in the formulae of the preceding section, for instance in (4.3.-10). This yields the formula (for notational convenience we have set $t=a$)

(3) $\quad f(x+a)=f(x)+(-\Lambda)^{-1/2}\operatorname{tg}\left[(-\Lambda)^{1/2}a\right]\Gamma\cdot 1,$

with Γ and Λ defined by (2.4.1.-5) and (2.4.1.-6), with the identification

(4) $\quad u^{(1)}(x)=f(x+a), \quad u^{(2)}(x)=f(x).$

Note that in these equations $f(x)$ is now any arbitrary function (incidentally, not to be confused with the, also arbitrary, function appearing in (2.4.1.-5) and (2.4.1.-6)); thus (3) is an (intriguing) nonlinear operator identity.

To make contact with known results it is convenient to replace f in (3) and (4) by λf, expand in powers of λ, and equate the terms of equal power. The terms linear in λ yield then just the well-known operator formula (1.1.-16); the terms of higher order in λ yield an endless sequence of highly nonlinear and implicit operator identities. Whether they will ever be of any use is an open question.

Note incidentally that the technique whereby these identities have been derived here is quite similar to the technique used in section 1.1 to prove (1.1.-16).

4.5. Generalized Bäcklund transformations and resolvent formula

Consider two functions $u'(x,t')$ and $u(x,t)$ related by the *generalized Bäcklund transformation*

(1) $\quad u'(x,t')=u(x,t)+\tau(\Lambda, t'-t'_0, t-t_0)\Gamma\cdot 1,$

where

(2) $\quad \tau(z,t',t)=(-z)^{-1/2}\operatorname{tg}\{(-z)^{1/2}[t'\alpha'(z)-t\alpha(z)]/2+\varphi(z)\},$

and Γ and Λ are defined as usual by (2.4.1.-5) and (2.4.1.-6), but now with the identification

(3) $\quad u^{(1)}(x) = u'(x, t'), \quad u^{(2)}(x) = u(x, t).$

In (2) the functions α', α and φ are arbitrary, except for the requirement that they make sense out of (1).

The generalized Bäcklund transformation (1) is easily seen (by using (2.4.1.-10c) and (2.4.1.-10e), as above) to imply the following relation between the reflection coefficients $R'(k, t')$, associated with $u'(x, t')$, and $R(k, t)$, associated with $u(x, t)$:

(4) $\quad R'(k, t') = R(k, t) \exp\{2ik[(t' - t'_0)\alpha'(-4k^2)$
$\qquad\qquad - (t - t_0)\alpha(-4k^2)] + 2\varphi(-4k^2)\};$

and from this formula it is concluded, as in section 4.3, that, if $u(x, t)$ obeys (4.-1), then $u'(x, t)$ obeys the analog of (4.-1) with α replaced by α':

(5a) $\quad u_t(x, t) = \alpha(L)u_x(x, t),$
(5b) $\quad u'_t(x, t) = \alpha'(L')u'_x(x, t)$

(in the last equation, L' is the operator L defined by (4.-2), but with $u(x, t)$ replaced by $u'(x, t)$).

Thus the generalized Bäcklund formula (1) transforms solutions of the nonlinear evolution equation (4.-1) into solutions of *another* equation of the same class (namely an equation having again the form (4.-1), except for the replacement of the function $\alpha(z)$ by another function $\alpha'(z)$).

Note that this generalized Bäcklund transformation contains moreover the arbitrary function $\varphi(z)$, whose presence accounts for the possibility to transform, via the ordinary Bäcklund transformations of section 4.1, solutions of one equation of the class (4.-1) into other solutions of the *same* equation. Indeed the simplest Bäcklund transformation (4.1.-14) obtains from (1) for $\alpha'(z) = \alpha(z)$, $t' - t'_0 = t - t_0$ and $\varphi(z) = \text{arctg}[(-z)^{1/2}/(2p)]$. On the other hand if $\alpha'(z)$ differs from $\alpha(z)$, namely whenever (1) relates solutions of *different* evolution equations, the transformation is always of infinite order, i.e. it cannot be put into a form analogous to (4.1.-1) with g and h polynomials in z.

For

(6) $\quad \varphi(z) = 0$

one clearly has that

(7) $\quad u'(x, t'_0) = u(x, t_0)$

(see (4)). Assume moreover $a'(z)$ to be constant, say

(8) $\quad a'(z) = v$,

so that (5b) becomes the linear wave equation

(9) $\quad u'_t(x, t) = v u'_x(x, t)$

and therefore

(10) $\quad u'(x, t') = u'(x + \Delta x, t'_0)$,

where we have set

(11) $\quad \Delta x = v(t' - t'_0)$.

But (10) and (7) combine into

(12) $\quad u'(x, t') = u(x + \Delta x, t_0)$.

Setting $t_0 - t = \Delta t$ and inserting (12) into (1) (together with (2), (6), (8) and (11)), one obtains finally the following formula that relates the solution $u(x, t)$ of (4.-1) to the *same* solution, but at a different time and with shifted space-coordinate:

(13) $\quad u(x + \Delta x, t + \Delta t)$

$= u(x, t) + (-\Lambda)^{-1/2} \text{tg}\{(-\Lambda)^{1/2}[\Delta x + \alpha(\Lambda)\Delta t]/2\}\Gamma \cdot 1$.

Here Γ and Λ are defined by (2.4.1.-5) and (2.4.1.-6), with the identification

(14) $\quad u^{(1)}(x) = u(x + \Delta x, t + \Delta t), \quad u^{(2)}(x) = u(x, t)$.

Note that the two quantities Δx and Δt are arbitrary. For $\Delta x = 0$ (13) reproduces the result (4.3.-10) of section 4.3; for this reason (see also below) (13) is referred to as the *generalized resolvent formula*. For $\Delta t = 0$, (13) reproduces the result (4.4.-3) of section 4.4 (indeed for $\Delta t = 0$ the t-dependence in (13) can be ignored, since u is everywhere evaluated at the same time t; therefore $u(x, t)$ can be chosen as an arbitrary function of x, as implied by the possibility to choose arbitrarily the initial Cauchy datum).

It is easily seen, by using (1.1.-16), that, in the weak field limit, (13) reproduces again the resolvent formula (1.1.-13). As for the more general formula (1) with (2) (but, for simplicity, with $\varphi(z)=0$), it is also easily seen that, in the weak field limit, it corresponds to the formula

$$(15) \quad u'(x,t') = \exp\{-i[(t'-t'_0)\omega'(-i\partial/\partial x) - (t-t_0)\omega(-i\partial/\partial x)]\}u(x,t),$$

that relates the solution $u(x,t)$ of

$$(16) \quad u_t(x,t) = -i\omega(-i\partial/\partial x)u(x,t)$$

to the solution $u'(x,t')$ of

$$(17) \quad u'_{t'}(x,t') = -i\omega'(-i\partial/\partial x)u'(x,t'),$$

with moreover

$$(18) \quad u(x,t_0) = u'(x,t'_0)$$

(and of course with the relationship (1.2.-6b) between the functions $\alpha(z)$ and $\omega(z)$, and an analogous relationship for the same quantities with primes).

4.N. Notes to chapter 4

A fairly complete coverage of the research on Bäcklund transformations and related topics up to 1974 is contained in the papers collected in [Mi1976]. These include a terse historical review [Lam1974a], with an extensive bibliography of early contributions including the original works by Bäcklund and by Darboux.

In addition to the approach adopted in this book, Bäcklund transformations (BTs) have been largely investigated by differential geometric techniques. In fact the BT for the KdV equation has been first introduced via the so-called prolongation structure technique in [WE1973]. For a formulation and account (including bibliography) of BTs in the context of differential geometry see [PRS1979].

4.1. It should be noted that what we term "Bäcklund transformation" is often referred to (especially in the context of differential geometry) as only the "first half" of a Bäcklund transformation. To explain this point, consider the Bäcklund transformation (4.1.-15),

(1) $\quad w_x^{(1)} + w_x^{(2)} = -\frac{1}{2}(w^{(1)} - w^{(2)})(4p + w^{(1)} - w^{(2)}),$

and assume that $w^{(1)}$ and $w^{(2)}$ satisfy the KdV equation ((3.1.-40) with $\alpha(z) = -z$),

(2) $\quad w_t + w_{xxx} - 3w_x^2 = 0;$

differentiating then both sides of (1) with respect to t and using (2) and again (1) it is easily seen that the r.h.s. takes the form of a perfect x-derivative. It is then possible to integrate both sides, obtaining thereby the "second half" of the Bäcklund transformation, namely

(3) $\quad w_t^{(1)} + w_t^{(2)} = (w_{xx}^{(1)} - w_{xx}^{(2)})(2p + w^{(1)} - w^{(2)})$
$\qquad\qquad + (w_x^{(1)})^2 + (w_x^{(2)})^2 + 4w_x^{(1)} w_x^{(2)}.$

The system of two coupled equations (1) and (3) has the following properties: (i) if, say, $w^{(2)}$ satisfies the KdV equation (2), so does also $w^{(1)}$; (ii) equation (1) is of first order and equation (3) is of second order, while the KdV equation (2) is of third order. Note also the local character of these results (no conditions on the asymptotic behaviours of the $w^{(j)}$'s, or for that matter on their regularity, are required); moreover, for given $w^{(2)}$, the solution $w^{(1)}$ of (1) and (3) is now completely determined, up (at most) to two constants (independent both of x and t).

Our main motivation for focussing on (1) only is the clear spectral significance of this equation. Moreover only (1) applies to the whole class of nonlinear evolution equations (3.1.-40) (or equivalently (3.1.-28), see (3.1.-27)); there is instead a different form of the "second half" of the Bäcklund transformation, for each of the different equations of the class (3.1.-40) (or (3.1.-28); see (3.1.-27)), corresponding to the possible different choices of $\alpha(z)$.

The fact that one and the same Bäcklund transformation applies to a whole class of nonlinear evolution equations was first emphasized in [Ch1974]. The first analysis of Bäcklund transformations from the point of view of the spectral transform is given in [FML1974]. The wronskian technique to understand Bäcklund transformations, and the related rôle of the operator Λ, were introduced in [C1975a]. The extent to which these matters have been clarified here goes somewhat beyond any previous treatment. The rôle of the "second half" of Bäcklund transformations in the wronskian technique context has been discussed by [CaC1977]. The (general) Bäcklund transformation (4.1.-1) can be derived also within the approach based on the algebra of differential operators reported in appendix A.20. This treatment, that extends the Lax approach introduced in [L1968] to BTs for the nonlinear evolution equation (3.1.-28), is due to [BR1980c]. The relevance of Darboux transformations and their relation to Bäcklund transformations has been emphasized in several recent papers by Matveev (see, for instance, [Mat1979a], [Mat1979b], [MS1979] and [Mat1979c]).

4.2. The relevance of the commutativity of Bäcklund transformations has been pointed out in [Lam1971] and [Lam1974b] (see also [Lam1974a]); the picture that visualizes this property (see subsection 1.7.1) is accordingly often termed "Lamb diagram". The first display of the "nonlinear superposition principle" for solutions of the KdV equation, as well as its usefulness to construct algebraically multisoliton solutions (via the "soliton ladder") was given in [WE1973]. For a discussion of the superposition formula within the so-called direct method, see [HS1978].

4.3. This idea was first introduced in [C1975a].

4.4. These operator identities were first introduced in [C1975a].

4.5. The results of this section are reported from [CD1976a]. Let us emphasize that there also exist other, much simpler, transformations that relate solutions of different nonlinear evolution equations solvable by spectral transform techniques. The prototype of such formulae is the Miura transformation (see appendix A.11). The appropriate context to understand the origin of such formulae are the matrix spectral problems treated in volume II, and in particular the reduction technique (see volume II, where additional examples of such transformations are reported).

CHAPTER FIVE

CONSERVATION LAWS

An important property of the class of nonlinear evolution equations solvable via the spectral transform is the existence of an infinite sequence of local conservation laws. Each of these conservation laws yields, for the class of asymptotically ($x \to \pm \infty$) vanishing solutions on which our consideration has been almost exclusively focussed up to now (and to which our consideration is also restricted in this chapter), a conserved quantity expressed as the integral over all space of an appropriate nonlinear (polynomial) combination of the field $u(x,t)$ and its derivatives.

As we now show, the derivation of such an infinite sequence of conserved quantities is a very simple task, in the framework of the treatment given above; and this property, remarkable as it is, appears here as a secondary feature of this class of nonlinear evolution equations. In other approaches to this subject, the existence of an infinite number of conservation laws plays instead a fundamental rôle; indeed the discovery of several independent conservation laws for the KdV equation was the first known indication of its peculiarity, and it was actually in the process of trying to understand this property that the spectral transform technique was invented.

Throughout this chapter we focus on the class of evolution equations (3.1.-28),

(1) $\quad u_t(x,t) = \alpha(L,t) u_x(x,t),$

with $\alpha(z,t)$ polynomial in z,

(2) $\quad \alpha(z,t) = \sum_{m=0}^{M} \alpha_m(t) z^m.$

(The more general class (3.1.-10) is discussed in appendix A.18). Let us recall the corresponding time-evolution of the spectral transform:

(1a) $\quad R_t(k,t) = 2ik\alpha(-4k^2, t)R(k,t),$

(1b) $\quad \dot{p}_n(t) = 0, \quad n = 1, 2, \ldots, N,$

(1c) $\quad \dot{\rho}_n(t) = -2p_n\alpha(4p_n^2, t)\rho_n(t), \quad n = 1, 2, \ldots, N,$

implying of course

(3a) $\quad R(k,t) = R(k,0) \exp\left[2ik \int_0^t dt' \, \alpha(-4k^2, t')\right],$

(3b) $\quad p_n(t) = p_n(0) = p_n, \quad n = 1, 2, \ldots, N,$

(3c) $\quad \rho_n(t) = \rho_n(0) \exp\left[-2p_n \int_0^t dt' \, \alpha(4p_n^2, t')\right], \quad n = 1, 2, \ldots, N.$

Actually the evolution equation (1) has itself (see section 3.1) the structure of a local conservation law,

(4) $\quad u_t(x,t) = \gamma_x^{(0)}(x,t).$

Moreover $\gamma^{(0)}(x,t)$, being a polynomial in u and its x-derivatives, vanishes asymptotically,

(5) $\quad \gamma^{(0)}(\pm\infty, t) = 0.$

Thus the very structure of the nonlinear evolution equation (1) implies that the quantity

(6) $\quad C_0 = c_0 = \int_{-\infty}^{+\infty} dx \, u(x,t)$

is constant for the flow (1).

It is easy to show directly from (1), using the operator identities established in appendices A.8 and A.9, that this quantity C_0 is merely the first member of an infinite sequence of conserved quantities. We postpone however such a direct derivation to the very end of this chapter, since a treatment based on the spectral transform is more appropriate, and perhaps more enlightening, within the approach developed in this book. The reader who thinks otherwise may go directly to the end of this chapter.

Let us then shift our attention from configuration space to spectral space, namely from $u(x,t)$ to its spectral transform, $S(t)$. Indeed we already know (see (1b)) that the discrete eigenvalues (if any) associated with $u(x,t)$ are time-independent when $u(x,t)$ evolves according to (1); thus they do provide other conserved quantities, that however cannot be written in explicit form in terms of the function $u(x,t)$ and its derivatives.

As it is clearly implied by (3a), an analogous situation obtains for the quantity $R(k,t)R(-k,t)$ (that coincides with the modulus square of $R(k,t)$ if $u(x,t)$ is real; see (2.1.-12)); but again there is no way to express, *in explicit form*, this quantity in terms of $u(x,t)$ and its x-derivatives (the formula (2.4.1.-26), that provides an explicit expression of $R(k,t)R(-k,t)$, involves, in addition to the function $u(x,t)$, the Schroedinger eigenfunctions $\psi(x,\pm k,t)$). Note however that the time-independence of $R(k,t)R(-k,t)$ holds for all values of k, thereby providing a nondenumerable infinity of conserved quantities (one for each value of k).

The situation is more favorable if one looks at the time evolution of the transmission coefficient $T(k,t)$. It is first of all easy to show that this time-evolution is of utmost simplicity: $T(k,t)$ is time-independent! This is easily proved using the same technique that has yielded (see section 3.1) the evolution equation for the reflection coefficient $R(k,t)$. Indeed the appropriate analogs of (3.1.-7) and (3.1.-8) read now

(7) $$2ik\left[R_t(k,t)R(-k,t)+T_t(k,t)T(-k,t)\right]$$
$$=\int_{-\infty}^{+\infty}dx\,\psi(x,k,t)\psi(x,-k,t)u_t(x,t),$$

(8) $$(2ik)^2\alpha(-4k^2,t)R(k,t)R(-k,t)$$
$$=\int_{-\infty}^{+\infty}dx\,\psi(x,k,t)\psi(x,-k,t)\alpha(L,t)u_x(x,t),$$

as implied by (2.4.1.-25) (with $g(z)=1$) and (2.4.1.-26) (with $g(z)=\alpha(z,t)$; we trust the notation here to be self-evident, in view of the preceding developments, see for instance section 3.1). Subtraction of (8) from (7), and use of (1) and (1a), then yields

(9) $$T_t(k,t)=0.$$

The time-independence of the transmission coefficient that has just been proved,

(10) $$T(k,t)=T(k,0)=T(k),$$

could also have been inferred (at least if $u(x,t)$ is real) directly from the fact that the modulus of $R(k, t)$ is time-independent, since $T(k, t)$ is determined by $|R(k, t)|$, and by the discrete eigenvalues p_n that are also time-independent (see appendix A.4). It corresponds again to a nondenumerable infinity of conserved quantities (one for each value of k); and again, since $T(k)$ cannot be expressed in explicit form in terms of u and its derivatives, this appears to lack practicality. But there is now a convenient way to by-pass this difficulty. First of all one notices that the time-independence of T implies of course that both the modulus and the phase of T are time-independent:

(11a) $\quad T(k) = |T(k)| \exp[i\theta(k)], \quad \text{Im } k = 0,$

or, more generally

(11b) $\quad T(k) = T(-k) \exp[2i\theta(k)], \quad \text{Im } k = 0$

(the second definition, (11b), of the phase $\theta(k)$ applies even if $u(x)$ is not real; it reduces to the first, (11a), if $u(x)$ is real, see (2.1.-12a)). Then one recalls (see section 2.1 and appendix A.3) that the phase $\theta(k)$ of $T(k)$ admits generally an asymptotic expansion in inverse powers of k, and that the coefficients of this asymptotic expansion can be written as integrals over all space of u and its x-derivatives (see (2.1.-41)). Now of course if T, and therefore its phase θ, are time-independent, certainly the coefficients of the asymptotic expansion of θ are time-independent as well. This immediately implies, through (2.1.-41), that the endless sequence of quantities

(12) $\quad C_m = (-1)^m (2m+1)^{-1} \int_{-\infty}^{+\infty} dx \, L^m [x u_x(x, t) + 2u(x, t)]$

$\quad (= -\theta_m), \quad m = 0, 1, 2, \ldots,$

are constants, when $u(x, t)$ evolves in time according to (1). The operator L appearing in this formula is precisely the same as that appearing in (1), defined by a formula that we have already written many times but for the convenience of the reader we report once more here:

(13) $\quad Lf(x) = f_{xx}(x) - 4u(x, t)f(x) + 2u_x(x, t) \int_x^{+\infty} dy f(y).$

The explicit expressions of the first 3 C_m's have already been reported (see (1.7.3.-12a, b, c, and (A.9.-58)) and are therefore not displayed here. Let us also recall that an alternative expression of the conserved quantities C_m is provided by (A.9.-57b).

An equivalent, but even more direct, proof of the time-independence of the C_m's is provided by (A.9.-57c), that has moreover the merit to display the significance of these quantities in spectral terms:

$$(14a) \quad C_m = (2m+1)^{-1}\left\{(-)^{m+1}2\sum_{n=1}^{N}(2p_n)^{2m+1}\right.$$

$$\left. + (2\pi)^{-1}\int_{-\infty}^{+\infty}dk\,(2k)^{2m+1}(d/dk)\ln[T(k)T(-k)]\right\},$$

$$(14b) \quad C_m = (-1)^{m+1}2(2m+1)^{-1}\sum_{n=1}^{N}(2p_n)^{2m+1}$$

$$-\pi^{-1}\int_{-\infty}^{+\infty}dk\,(2k)^{2m}\ln[T(k)T(-k)].$$

This definition in terms of spectral quantities is alternative to that given above, see (12), in terms of the coefficients θ_m of the asymptotic expansion at large k of the phase $\theta(k)$ of the transmission coefficient $T(k)$. The equivalence of these two definitions is guaranteed by the results of appendix A.9, see in particular (A.9.-78b, 56).

Note that, for pure multisoliton solutions, $T(k)T(-k)=1$ (see (2.2.-23)), and therefore

$$(15) \quad C_m = (-1)^{m+1}2(2m+1)^{-1}\sum_{n=1}^{N}(2p_n)^{2m+1}.$$

Thus in this case C_m appears as a sum of contributions, each of which is identified with a different soliton; this could have been anticipated by observing that C_m, being time-independent, can be computed for $t\to\pm\infty$, when $u(x,t)$ does indeed separate into N single-soliton contributions, see e.g. (1.6.1.-14).

An equivalent series of conserved quantities is provided by the formula

$$(16) \quad c_m = (-1)^m \int_{-\infty}^{+\infty} dx\, u(x,t) M^{2m}\cdot 1, \quad m=0,1,2,\ldots,$$

where the integrodifferential operator M is defined by

$$(17) \quad Mf(x) = f_x(x) - \int_{-\infty}^{x} dy\, u(y,t) f(y),$$

that specifies its action on the generic function $f(x)$. The derivation of this set of conserved quantities is analogous to that given above, for it can be shown (see appendix A.13) that c_m is the coefficient of the term $(2k)^{-2m}$ in the asymptotic expansion, as $k \to \infty$, of the quantity $2ik[1-1/T(k)]$.

The two sets (12) and (16) of conserved quantities are completely equivalent. Indeed, using the property (2.1.-39a), the relationship between the constants C_m and c_m implied by their respective origins can be condensed in the compact formula (see appendices A.9 and A.13)

(18) $$\sum_{m=0} c_m z^{2m+1} = \sin\left[\sum_{m=0} C_m z^{2m+1}\right],$$

which is of course to be understood by reexpanding the r.h.s. and equating the coefficients of equal powers of z.

Once the time-independence of the constants of the motion (12) and (16) has been shown to follow directly from the time-independence of the transmission coefficient $T(k)$, it is interesting to display alternative expressions of these quantities. They read (see appendix A.9):

(19) $$C_m = (-1)^m \int_0^1 dy \int_{-\infty}^{+\infty} dx [L(yu)]^m u(x,t), \quad m=0,1,2,\ldots,$$

(20) $$C_m = (-1)^m (2m+1)^{-1} \int_{-\infty}^{+\infty} dx \tilde{L}^m u(x,t), \quad m=0,1,2,\ldots,$$

(21) $$c_m = (-1)^m \int_{-\infty}^{+\infty} dx \Lambda^{(0)m} u(x,t), \quad m=0,1,2,\ldots .$$

Here the integrodifferential operator $L(yu)$ is defined by (13) with $u(x,t)$ replaced by $yu(x,t)$, while $\tilde{L}$ and $\Lambda^{(0)}$ are defined by the formulae (see (A.9.-12) and (A.9.-51), or (2.4.2.-18) and (2.4.1.-17))

(22) $$\tilde{L}f(x) = f_{xx}(x) - 2u(x,t)f(x) + 2\int_x^{+\infty} dy\, u(y,t)f_y(y),$$

(23) $$\Lambda^{(0)}f(x) = f_{xx}(x) - 2u(x,t)f(x) + u_x(x,t)\int_x^{+\infty} dy\, f(y)$$
$$+ u(x,t)\int_x^{+\infty} dy\, u(y,t)\int_y^{+\infty} dz\, f(z).$$

It should be emphasized that these conserved quantities are constant as $u(x,t)$ evolves according to any equation of the class (1); and their definitions contain no reference to the polynomial (in z) $\alpha(z,t)$ in (1) (see (2)), whose choice remains arbitrary (including its time dependence!). On the other hand, a dependence on α (namely on the particular evolution equation of the class (1)) appears if these conservation laws are written in *local* (i.e. differential rather than integral) form, namely as a "continuity equation". Let us illustrate this point by focussing, for simplicity, on the definitions (12) and (20). We write the corresponding local conservation laws in the form

(24) $\quad f_t^{(m)}(x,t) = \varphi_x^{(m)}(x,t), \quad m=0,1,2,\ldots,$

(25) $\quad g_t^{(m)}(x,t) = \gamma_x^{(m)}(x,t), \quad m=0,1,2,\ldots,$

where of course the "densities" $f^{(m)}$ and $g^{(m)}$ are precisely the integrands in (12) and (20),

(26) $\quad f^{(m)}(x,t) \equiv L^m[xu_x(x,t) + 2u(x,t)], \quad m=0,1,2,\ldots,$

(27) $\quad g^{(m)}(x,t) \equiv \tilde{L}^m u(x,t), \quad m=0,1,2,\ldots$

(the constancy of the integrals (12) and (20) is then obviously implied by the asymptotic vanishing, as $x \to \pm\infty$, of the "currents" $\varphi^{(m)}$ and $\gamma^{(m)}$; see below). The first three densities of the sequence (27) have been already given by (3.1.-32) (see (3.1.-30) and (A.9.-30)) and are not reported here. The explicit expressions of the first three densities of the other set, (26), read:

(28a) $\quad f^{(0)} = (xu)_x + u,$

(28b) $\quad f^{(1)} = \left[x(u_{xx} - 3u^2) + 3u_x + 2u \int_x^{+\infty} dy\, u(y,t) \right]_x - 3u^2,$

(28c) $\quad f^{(2)} = \left[(f^{(1)} - 6u^2)_x + x(-4u_{xx}u + u_x^2 + 10u^3) \right.$

$\left. \qquad - 5u^2 \int_x^{+\infty} dy\, u(y,t) - 2u \int_x^{+\infty} dy\, u^2(y,t) \right]_x + 5(u_x^2 + 3u^3)$

(here the dependence on x and t has not been indicated explicitly whenever this is possible without generating confusion). Note that these expressions have been written so as to display separately the part that can be integrated away, being a perfect differential; this of course always includes the contribution containing the explicit (linear) x-dependence.

As it has been repeatedly emphasized, the densities $g^{(m)}$ are polynomials in u and its x-derivatives (this is displayed by the explicit expressions (3.1.-32), and it has been proved quite generally in appendix A.9); indeed $g^{(m)}$ is a (linear) combination of monomials of "rank" (or "dimension")

5. Conservation laws

$m+1$, the rank (or dimension) being an additive index, associated with each monomial, to whose value each power of u (or its derivatives) contributes one unit while each operator of differentiation contributes half a unit (e.g., the rank of $u_{xx}^3 u^5$ is 11). The densities $f^{(m)}$ contain instead, for $m>0$, also integral expressions.

As implied by their definitions, see (26) and (27), the conserved densities $f^{(m)}$ and $g^{(m)}$ are obviously independent of which particular evolution equation of the class (1) is being considered. As indicated above, this is not the case of the currents $\varphi^{(m)}$ and $\gamma^{(m)}$, whose definitions depend on the polynomial (in z) $\alpha(z,t)$ that identifies each evolution equation of the class (1), as shown by the explicit expressions of these quantities, that we now report (for the corresponding derivations, see appendix A.21).

The current $\gamma^{(m)}$ reads

$$(29) \quad \gamma^{(m)} = \alpha(\tilde{L}, t) g^{(m)} + 2 \sum_{k=1}^{M} \alpha_k(t)$$

$$\cdot \sum_{j=0}^{k-1} [G^{(k-j-1, m+j)} - G^{(m+j, k-j-1)}], \quad m=0,1,2,\ldots .$$

Here the quantities $\alpha_k(t)$ and $g^{(m)}$ are defined by (2) and (27), while the quantities $G^{(j,k)}$ are related to the quantities $g^{(m)}$ by the formula

$$(30a) \quad G_x^{(j,k)}(x,t) = -g^{(j)}(x,t) g_x^{(k)}(x,t), \quad j,k=0,1,2,\ldots,$$

implying

$$(30b) \quad G^{(j,k)}(x,t) = \int_x^{+\infty} dy \, g^{(j)}(y,t) g_y^{(k)}(y,t), \quad j,k=0,1,2,\ldots .$$

But in spite of the integration in the r.h.s. of this formula, $G^{(j,k)}$ is in fact merely a polynomial, of rank $j+k+2$, in u and its x-derivatives; this is proved in appendix A.9 (see the discussion following (A.9.-98)), where the explicit expressions of the first few $G^{(j,k)}$ are also recorded. Thus, for instance, in the (trivial) case $\alpha(z,t)=1$, corresponding to the wave equation $u_t = u_x$, the currents $\gamma^{(m)}$ take the simple form

$$(31) \quad \gamma^{(m)} = g^{(m)}, \quad m=0,1,2,\ldots, \quad [u_t = u_x];$$

while in the case $\alpha(z,t) = -z$, corresponding to the KdV equation $u_t + u_{xxx}$

$-6u_x u = 0$, the currents $\gamma^{(m)}$ read

(32) $\quad \gamma^{(m)} = -g^{(m+1)} + 2[G^{(m,0)} - G^{(0,m)}], \quad m = 0, 1, 2, \ldots,$
$\quad\quad\quad [u_t + u_{xxx} - 6u_x u = 0],$

with

(33a) $\quad G^{(m,0)} = \sum_{j=0}^{m-1} \{g^{(j)}[g_{xx}^{(m-j-1)} - 2ug^{(m-j-1)}] - \tfrac{1}{2}g_x^{(j)}g_x^{(m-j-1)}\}$
$\quad\quad\quad\quad - \tfrac{1}{2} \sum_{j=0}^{m} g^{(j)}g^{(m-j)}, \quad m = 1, 2, 3, \ldots,$

(33b) $\quad G^{(0,m)} = -G^{(m,0)} - ug^{(m)}, \quad m = 0, 1, 2, \ldots,$

(33c) $\quad G^{(0,0)} = -\tfrac{1}{2}u^2.$

The first two currents of the KdV set (32) are explicitly displayed by (1.7.3.-21).

The currents $\varphi^{(m)}$, in contrast to the currents $\gamma^{(m)}$, do not reduce merely to polynomials in u and its x-derivatives; some integrals linger. Their explicit expression reads

(34) $\quad \varphi^{(m)} = Pf^{(m)} + \tilde{L}^m \{[x\alpha(\tilde{L}, t)u]_x - P(xu_x \pm 2u)\}, \quad m = 0, 1, 2, \ldots,$

with $f^{(m)}$, $\tilde{L}$ and $\alpha(z, t)$ defined by (26) (see also (28)), (22) and (1), and the differential operator P defined by the formula (see (A.21.-39, 30))

(35) $\quad P = \alpha(L, t) - 2 \sum_{m=1}^{M} \alpha_m(t) \sum_{j=0}^{m-1} (g^{(m-1-j)} + g_x^{(m-1-j)}I)L^j,$

where the $\alpha_m(t)$ are defined by (2) and the multiplicative operators $g^{(m)}$ and $g_x^{(m)}$ are defined by the formulae (see (27))

(36) $\quad g^{(m)}f(x) = g^{(m)}(x, t)f(x), \quad\quad g_x^{(m)}f(x) = g_x^{(m)}(x, t)f(x),$

which specify (see (27)) their action on a generic function $f(x)$. Thus, in the trivial case $\alpha(z, t) = 1$, the currents (34) read $\varphi^{(m)} = f^{(m)} - g^{(m)}$, while for the KdV case ($\alpha(z, t) = -z$) they read

(37) $\quad \varphi^{(m)} = -f_{xx}^{(m)} + 6uf^{(m)} + 3g^{(m+1)}, \quad m = 0, 1, 2, \ldots$
$\quad\quad\quad [u_t + u_{xxx} - 6u_x u = 0].$

5. Conservation laws

The first of these KdV currents thus reads

(38) $\quad \varphi^{(0)} = -(xu)_{xxx} - 2u_{xx} + 6xu_x u + 3u^2, \qquad [u_t + u_{xxx} - 6u_x u = 0]$.

In the preceding chapter 4 we have seen (mainly in section 4.1) that the class of nonlinear evolution equations (1) is invariant under a large class of (Bäcklund) transformations; here we have seen that this class of nonlinear evolution equations possesses an infinite sequence of conservation laws. Clearly these two properties are related; and there is a considerable body of recent literature (see Notes below) that analyzes this connection in the framework of the classical Noether's theorem (that associates, in the context of Lagrangian field theory, invariance properties with conservation laws). The conserved quantities (20) introduced above have moreover a clear significance ("hamiltonian functionals") if the evolution equations under consideration are viewed in the framework of Hamiltonian dynamics; indeed these evolution equations can be identified with Hamiltonian flows (see appendix A.23).

While we do not delve in this book into the items mentioned above, we do provide here an example of derivation of an infinite sequence of conserved quantities based on Bäcklund transformations. This we do for the simplest nontrivial equation of the class (1), namely the KdV equation

(39a) $\quad u_t + u_{xxx} - 6u_x u = 0, \quad u \equiv u(x, t)$,

corresponding to $\alpha(z, t) = -z$ in (1).

It is convenient to work with the integral of u,

(40) $\quad w(x, t) = \int_x^{+\infty} dy \, u(y, t)$,

rather than with u itself. In terms of this dependent variable the KdV equation (39a) reads

(39b) $\quad w_t = -w_{xxx} - 3w_x^2, \quad w \equiv w(x, t)$.

Consider now two solutions of (39b), say $w(x, t)$ and $w'(x, t)$, related to each other by the (simplest) Bäcklund transformation (4.1.-15),

(41) $\quad w'_x + w_x = -\tfrac{1}{2}(w' - w)^2 - \varepsilon^{-1}(w' - w)$

(here we have used the convenient notational identifications $w' = w^{(1)}, w = w^{(2)}, \varepsilon = (2p)^{-1}$). It is then easily seen (subtract (39b) from the same equation for w' and use (41)) that one can write

(42) $\quad y_t = \eta_x, \quad y \equiv y(x,t), \quad \eta \equiv \eta(x,t),$

with the definitions

(43) $\quad y \equiv (w'-w)/2\varepsilon,$
(44) $\quad \eta \equiv -y_{xx} + 3y^2 + 2\varepsilon^2 y^3.$

Introduce now the two asymptotic expansions

(45) $\quad y = \sum_{m=0}^{M} \varepsilon^m y^{(m)} + o(\varepsilon^M),$

(46) $\quad \eta = \sum_{m=0}^{M} \varepsilon^m \eta^{(m)} + o(\varepsilon^M).$

It is then easily seen that (41), that can be rewritten in the form

(47) $\quad y = u - \varepsilon y_x - \varepsilon^2 y^2,$

yields the relations

(48a) $\quad y^{(0)} = u,$
(48b) $\quad y^{(1)} = -u_x,$
(49) $\quad y^{(m+1)} = -y_x^{(m)} - \sum_{j=0}^{m-1} y^{(j)} y^{(m-j-1)}, \quad m = 1, 2, 3, \ldots.$

These equations are easily solved by recursion:

(50a) $\quad y^{(2)} = u_{xx} - u^2,$
(50b) $\quad y^{(3)} = (-u_{xx} + 2u^2)_x,$
(50c) $\quad y^{(4)} = (u_{xx} - 3u^2)_{xx} + u_x^2 + 2u^3,$

and so on. Note in particular that they imply

(51) $\quad \lim_{x \to \pm\infty} [y^{(m)}] = 0, \quad m = 0, 1, 2, 3, \ldots.$

5. Conservation laws

On the other hand from (44) and (45) there follows that

(52a) $\quad \eta^{(0)} = -u_{xx} + 3u^2,$

(52b) $\quad \eta^{(1)} = (u_{xx} - 3u^2)_x,$

(52c) $\quad \eta^{(m)} = -y_{xx}^{(m)} + 3\sum_{j=0}^{m} y^{(j)} y^{(m-j)}$

$$+ 2\sum_{k=0}^{m-2} \sum_{j=0}^{m-k-2} y^{(k)} y^{(j)} y^{(m-k-j-2)}, \quad m=2,3,4,\ldots,$$

and this, together with (51), clearly also implies

(53) $\quad \lim_{x \to \pm\infty} [\eta^{(m)}] = 0, \quad m = 0, 1, 2, \ldots .$

But (42), (45) and (46) imply

(54) $\quad y_t^{(m)} = \eta_x^{(m)}, \quad m = 0, 1, 2, \ldots,$

and this formula, together with (53), implies that the quantities

(55) $\quad a_m = \int_{-\infty}^{+\infty} dx \, y^{(m)}(x, t), \quad m = 0, 1, 2, \ldots,$

are time-independent.

It is thus seen, in conclusion, that (48), (49) and (55) generate an endless sequence of conserved quantities. Actually, however, the odd-numbered a_m's vanish, because the odd-numbered $y^{(m)}$'s are pure divergences (see (48b) and (50b); for a general proof see appendix A.11); as for the even-numbered a_m's, they reproduce the conserved quantities introduced above, indeed (see (48a), (50a), (50c) and (55))

(56) $\quad a_0 = C_0, \quad a_2 = -C_1, \quad a_4 = C_2.$

Note that this approach yields also a constructive procedure to compute the currents $\eta^{(m)}$'s (see (54)); indeed, once the $y^{(j)}$'s are known (for $0 \leq j \leq m$), $\eta^{(m)}$ is given by the explicit formulae (52).

Clearly, the technique used here has taken, as its starting point, the "infinitesimal" Bäcklund transformation (41), with $\varepsilon \to 0$. It is indeed clear, from (47) and (43), that, in this limit, w' and w coincide; this is the case, however, only as long as the solution w' of the Bäcklund transformation (41) does not diverge as $\varepsilon \to 0$. Indeed, for any value of $\varepsilon > 0$, w' has generally one more soliton than w, this soliton becoming infinitely tall and narrow as $\varepsilon \to 0$; in order to avoid the diverging contribution from this soliton, as $\varepsilon \to 0$ one has to push the soliton farther and farther away to the left (its position depends on the integration constant). This procedure guarantees that, even though the soliton is always there as long as ε is not exactly zero, the difference between $w'(x, t)$ and $w(x, t)$ (see (43)) vanishes, at fixed x, as $\varepsilon \to 0$. Thus a certain care must be exercised to avoid paradoxical results; for instance it should be kept in mind that (45) and (51) do not imply that y vanishes as $x \to -\infty$; indeed we know from the results of section 4.1 (see for instance (4.1.-22)) that the outcome

(57) $\quad y(-\infty, t) = -4p^2 = -\varepsilon^{-2}$

is (in addition to $y(-\infty, t) = 0$) consistent with the Bäcklund transformation (41); and this quantity not only differs from zero, but indeed diverges as $\varepsilon \to 0$.

We terminate here our terse discussion of the relationship between Bäcklund transformations and conservation laws. An almost identical treatment, but from a somewhat different point of view, is given in appendix A.11.

As noted at the end of subsection 1.7.3, the KdV equation (39a) has an additional conserved quantity

(58) $\quad C = \int_{-\infty}^{+\infty} dx \left[xu(x, t) + 3tu^2(x, t) \right].$

The corresponding local conservation law reads

(59) $\quad (xu + 3tu^2)_t = \left[u_x - x(u_{xx} - 3u^2) + 3t(u_x^2 - 2uu_{xx} + 4u^3) \right]_x,$

and is associated with the property of invariance of the KdV equation (39a) under the transformation

(60) $\quad u(x, t) \to u'(x, t) = u(x - vt, t) - \tfrac{1}{6}v,$

v being an arbitrary real number. This transformation is generally called a "Galileo" transformation; the term is particularly appropriate if u itself can be interpreted as a velocity in the x direction. Note however that this transformation carries u outside the class of asymptotically vanishing functions.

5.   *Conservation laws*

We noted already in subsection 1.7.3 that the time-independence of C, C_0 and C_1 implies that, for the generic solution of the KdV equation (39a), the center of mass

(61) $\quad X(t) = \int_{-\infty}^{+\infty} dx\, x u(x,t) \Big/ \int_{-\infty}^{+\infty} dx\, u(x,t)$

moves with constant speed,

(62) $\quad X(t) = X_0 + Vt, \qquad X_0 = C/C_0, \qquad V = -3C_1/C_0.$

A larger class of "KdV-like" equations (not necessarily solvable by spectral transform techniques), for which somewhat analogous results hold, is discussed in appendix A.14.

It is instructive to verify by a direct explicit computation the time-independence of C and the validity of (62) for the single-soliton solution

(63) $\quad u(x,t) = -2p^2/\cosh^2\{p[x-\xi(t)]\},$
(64) $\quad \xi(t) = \xi_0 + vt,$
(65) $\quad v = 4p^2.$

One easily finds

(66) $\quad C = p[vt - \xi(t)] = -p\xi_0,$
(67) $\quad X(t) = \xi(t) = \xi_0 + 4p^2 t.$

The first of these two formulae displays the time-independence of C; the consistency of the second with (62), (64) and (65), is implied by (see (15))

(68) $\quad C_0 = -p, \qquad C_1 = \tfrac{4}{3}p^3.$

For the multisoliton solution (1.6.1.-9),

(69) $\quad C = \sum_{n=1}^{N} p_n \xi_n^{(\pm)},$

with the quantities $\xi_n^{(\pm)}$ defined by (1.6.1.-14). Note that (1.6.1.-15), (1.6.1.-16) and (1.6.1.-17) imply that the same result obtains whether one inserts the $\xi_n^{(+)}$'s or the $\xi_n^{(-)}$'s in the above expression (this verifies the fact that the constant of the motion C has the same value whether it is evaluated at $t = +\infty$ or $t = -\infty$). Note moreover that for this multisoliton solution,

(1.6.1.-9), the speed of the center of mass $X(t)$ (see (61), (62) and (15)) is given by the formula

(70) $$V = 4 \sum_{n=1}^{N} p_n^3 \Big/ \sum_{n=1}^{N} p_n.$$

In the spirit of the analogy of the behaviour of solitons to that of classical particles, this formula suggests the attribution, to the soliton of parameter p_n, of a mass

(71) $$m_n = \mu p_n$$

with μ a universal (scaling) constant, so that (70) read

(72) $$V = \sum_{n=1}^{N} m_n v_n \Big/ \sum_{n=1}^{N} m_n,$$

where of course the velocity $v_n = 4p_n^2$ is precisely that pertaining to the nth soliton (see (65)). Thus (72) takes indeed the appropriate form to express the velocity of the center of mass of N points, of mass m_n and velocities v_n. It is moreover gratifying to note that, with the definition (71), the mass of the soliton is proportional to the modulus of its integral extended over all space (see (6) and (15)), which is quite a reasonable measure of its overall "size".

The motion of the center of mass (61) associated with the generic solution of (1) for an arbitrary polynomial $\alpha(z, t)$ (see (2)) can be similarly analyzed. In fact, from (25) and (29), together with (27) and (2), it follows that the local conservation law

(73) $$\left[x g^{(0)} + \sum_{m=0}^{M} \left(\int_{t_0}^{t} dt' \, \alpha_m(t') \right) g^{(m)} \right]_t$$
$$= \left[x \gamma^{(0)} + \sum_{m=0}^{M} \left(\int_{t_0}^{t} dt' \, \alpha_m(t') \right) \gamma^{(m)} \right]_x$$

holds; this implies (see (27) and (2)) that the quantity

(74) $$C = \int_{-\infty}^{+\infty} dx \left[x u(x, t) + \int_{t_0}^{t} dt' \, \alpha(\tilde{L}, t') u(x, t) \right]$$

is time-independent if $u(x, t)$ satisfies (1). There immediately follows that the position $X(t)$ of the center of mass, see (61), is explicitly given by the

5. Conservation laws

formula

$$(75) \quad X(t) = X_0 + \sum_{m=0}^{M} (-)^{m+1}(2m+1)(C_m/C_0) \int_{t_0}^{t} dt'\, \alpha_m(t'),$$

$$X_0 = C/C_0,$$

where the constants of the motion C_m are of course those introduced above (see (12), (14), (19) and (20)) and the quantities $\alpha_m(t)$ are defined by (2). Note that this implies that the center of mass $X(t)$ moves uniformly (i.e., with constant speed) for the generic solution of every "autonomous" evolution equation of the class (1) (i.e., any evolution equation of the class (1) that does not contain an explicit time-dependence, namely such that $\alpha(z,t) \equiv \alpha(z)$, and therefore also $\alpha_m(t) \equiv \alpha_m$, see (2)); and the velocity of the center of mass is then given by the explicit formula

$$(76) \quad V = \sum_{m=0}^{M} (-1)^{m+1}(2m+1)\alpha_m C_m/C_0.$$

The formula (75) implies moreover that the center of mass $X(t)$ moves uniformly even for a subclass of solutions of the non-autonomous evolution equation (1), namely the subclass identified by the requirement that C_m vanish if $\alpha_m(t)$ is not time-independent (note however that such a requirement might be inconsistent with the condition that $u(x,t)$ be real; see for instance the explicit expression (1.7.3.-12b) (or (A.9-58b)) of C_1).

The easy, but instructive, extension to the general evolution equation (1) of the terse discussion of multisoliton solutions given above (after (68)) in the context of the KdV equation, is left as an exercise.

Let us finally end this chapter by deriving the infinite sequence (12) of conservation laws directly from the evolution equation (1), as it was promised at the beginning of this chapter.

The idea is to use the identity

$$(77) \quad dC_m/dt = (-1)^m \int_{-\infty}^{+\infty} dx\, L^m u_t(x,t), \quad m = 0, 1, 2, \ldots,$$

that, with the definition (12) of C_m, is merely the special case of the operator identity (A.8.-8) or (A.9.-63a) corresponding to an (infinitesimal) variation in time. Let us emphasize that this formula is merely an operator *identity*, its validity depending only on the asymptotic vanishing of $u(x,t)$ and its x-derivatives as $x \to \pm\infty$ (and of course on the definition (13) of L).

Using now the evolution equation (1) with (2), and the operator identity (see (A.8.-6) or (A.9.-54a))

(78) $$\int_{-\infty}^{+\infty} dx\, L^m u_x(x,t) = 0, \quad m = 0, 1, 2, \ldots,$$

there immediately follows that all the quantities C_m are time-independent, q.e.d.

5.N. Notes to chapter 5

The infinite sequence of conserved quantities for the KdV equation (5.-39a) was first given in [MGK1968]; see also [KMGZ1970] and the review [Mi1976a]. This early derivation of the sequence of conserved quantities was mainly based on the Gardner one-parameter transformation, as discussed in appendix A.11; this approach has been recently investigated in a broader mathematical context in [Kup1980a]. A largely equivalent method is based on the invariance under Bäcklund transformations, as described also in this chapter; see, e.g., [S1974], [W1977] and [W1978].

The idea to exploit the time-independence of the transmission coefficient $T(k)$ associated with a solution $u(x,t)$ of the KdV equation has been introduced in [ZF1971]. An advantage of this approach is to prove directly that the same sequence of conserved quantities is time-independent for the whole class (5.-1) of evolution equations. The compact expression (5.-12) was first given in [CD1978f], the expression (5.-16) in [C1977] and the formulae (5.-29) and (5.-34) in [D1982].

For an early treatment in a field-theoretical context of the relation between conservation laws and invariance properties of the KdV equation see [W1977]. Let us recall in this connection that an (evolution) equation is said to possess a "symmetry" if there exists a transformation that maps solutions of this equation into other solutions of the *same* equation. If the transformation depends on a parameter so that the corresponding infinitesimal transformation can be introduced, and if moreover the evolution equation can be derived, in a field-theoretical context, from a Lagrangian (or Hamiltonian), then the classical Noether's theorem implies that to each symmetry property there corresponds a conservation law (see, f.i., [Fok1979]). This is indeed the case for the KdV equation, that in the version (see (5.-39, 40))

(1) $\quad w_t + w_{xxx} + 3w_x^2 = 0$

follows from the variational principle

(2) $$\delta \int_{t_1}^{t_2} dt \int_{-\infty}^{+\infty} dx \, \mathcal{L}(w_t, w_x, w_{xx}) = 0,$$

with the Lagrangian density

(3) $$\mathcal{L}(w_t, w_x, w_{xx}) = w_t w_x - w_{xx}^2 + 2w_x^3.$$

An infinite sequence of conserved quantities for the KdV equation can therefore be associated, in this context, with the infinite sequence of symmetries possessed by this evolution equation (see, f.i., [Ol1977]), that correspond to the Bäcklund transformations of chapter 4; for such a derivation and analysis see [St1975] and [McG1978]. A thorough investigation of the KdV equation in this context can also be found in [W1978]; and for a general formulation of the symmetry approach as applied to a whole class of evolution equations (including the KdV equation) see [Fok1980].

Another procedure to construct the conserved quantities associated with the flow (5.-1), based on the asymptotic expansion of the resolvent of the Schroedinger operator $-d^2/dx^2 + u(x,t) - k^2$ in inverse powers of the spectral parameter k (much in the same spirit as the treatment discussed above), is given in [GD1975].

The observation that the KdV equation, as well as all the evolution equations of the class (5.-1), can be interpreted as hamiltonian flows, has shed new light on the occurrence of infinitely many constants of the motion, that can be interpreted in this context as action variables. This topic is tersely reviewed in appendix A.23.

CHAPTER SIX

EXTENSIONS

In this chapter we consider a number of extensions of the approach developed so far. The main purpose is to enlarge the class of nonlinear evolution equations that are solvable via the spectral transform associated with the Schroedinger spectral problem. Other classes of nonlinear evolution equations, that are also solvable by the spectral transform technique but in the framework of different spectral problems, will be treated in the second volume.

In section 6.1 the possibility to treat equations with additional independent variables besides x and t is considered, and a class of such equations solvable via the spectral transform based on the Schroedinger spectral problem is identified.

In section 6.2 a class of nonlinear evolution equations with coefficients linearly dependent on x is introduced, that can also be solved via the spectral transform. It may be considered a nonlinear generalization of the class of linear equations (1.1.-20). The generic solution of some of the equations of this class is, however, plagued by singularities, whose nature is not yet fully understood.

In section 6.3 a more general definition of the spectral transform of a function $u(x)$ is introduced by associating with $u(x)$ the spectral problem based on the ODE

(1) $\quad -\psi_{xx}(x)+r(x)\psi(x)+u(x)\psi(x)=k^2\psi(x),$

$r(x)$ being a given reference potential; of course the spectral transform we have dealt with in the preceding chapters corresponds to the special choice $r(x)=0$. The subsequent discussion focuses on two choices of the reference potential $r(x)$. In one case $r(x)$ depends linearly on x, and the corresponding spectral problem constitutes the appropriate tool to investigate the class of asymptotically divergent solutions of the KdV equation; their interest is underscored by relating them to the asymptotically vanishing, and therefore "physically significant", solutions of the so-called "cylindrical KdV equation". This connection makes it possible to obtain some solutions of the cylindrical KdV equation, and, what is more important, to solve the Cauchy problem for this equation. The relevant results are treated in the 4 subsections of section 6.3, that include a general treatment of the cylindrical KdV equation, including the analysis of the conserved quantities associated with it.

The second reference potential that is considered below is proportional to $(x-\bar{x})^{-2}$; it is therefore singular (as a double pole) at $x=\bar{x}$. This spectral problem is convenient to discuss solutions of the KdV equation with one real double pole (with a "natural" prescription to continue such solutions across the singularity). This topic is tersely treated in section 6.4.

Finally, in section 6.5, the class of evolution equations is discussed, that is related to the spectral problem characterized by the generalized Schroedinger equation

$$(2) \quad -\psi_{xx}(x)+u(x)\psi(x)=k^2\rho^2(x)\psi(x).$$

Since this can be reduced to the standard Schroedinger equation (corresponding to $\rho(x)=1$) by a change of variables, the results of this section might also be obtained from previous results, by a change of variables. They are, nevertheless, in our opinion sufficiently interesting (and the change of variables in question is sufficiently nontrivial) to justify an explicit, if terse, treatment.

6.1. More variables

Let us consider a function u that depends on one additional variable, say y, besides x and t:

$$(1) \quad u \equiv u(x, y, t).$$

The restriction to only one extra variable is merely for simplicity; all the developments of this section are easily extended to the case in which there are n additional variables y_m, $m=1,2,\ldots,n$, or equivalently a vector variable $\vec{y} \equiv (y_1, y_2, \ldots, y_n)$.

Of course the spectral transform S associated with u depends now on y and t:

$$(2) \quad S = \{R(k, y, t); p_n(y, t), \rho_n(y, t), n=1,2,\ldots,N(y,t)\}.$$

We proceed now in close analogy to the treatment of chapter 3; and therefore our presentation is terse, at least for all developments that are closely patterned after the previous treatment.

The following integral wronskian relations (see (2.4.1.-12)) may be taken as starting point for the analysis:

(3) $\quad 2ikg(-4k^2, y, t)R_t(k, y, t)$
$$= \int_{-\infty}^{+\infty} dx [\psi(x, k, y, t)]^2 g(L, y, t) u_t(x, y, t),$$

(4) $\quad 2ikg(-4k^2, y, t)R_y(k, y, t)$
$$= \int_{-\infty}^{+\infty} dx [\psi(x, k, y, t)]^2 g(L, y, t) u_y(x, y, t),$$

(5) $\quad (2ik)^2 g(-4k^2, y, t) R(k, y, t)$
$$= \int_{-\infty}^{+\infty} dx [\psi(x, k, y, t)]^2 g(L, y, t) u_x(x, y, t).$$

Here the function $g(z, y, t)$ is essentially arbitrary in its dependence on y and t, and is entire (say, a polynomial) in z. All the other symbols should require no definition (see for instance chapter 2); of course in the definition of the integrodifferential operator L (say, (2.4.1.-13)), $u \equiv u(x, y, t)$.

These formulae clearly imply that, if u evolves according to the nonlinear equation

(6) $\quad g_1(L, y, t) u_t(x, y, t) + g_2(L, y, t) u_y(x, y, t)$
$\quad\quad + g_3(L, y, t) u_x(x, y, t) = 0,$

then the corresponding reflection coefficient evolves according to the linear partial differential equation

(7) $\quad g_1(-4k^2, y, t) R_t(k, y, t) + g_2(-4k^2, y, t) R_y(k, y, t)$
$\quad\quad + 2ikg_3(-4k^2, y, t) R(k, y, t) = 0.$

Hence the possibility to solve (6) by the spectral transform technique.

We limit hereafter our consideration to the simpler case with

(8) $\quad g_1(z, y, t) = 1, \quad\quad g_2(z, y, t) = -\gamma(z), \quad\quad g_3(z, y, t) = -\alpha(z),$

so that the nonlinear evolution equation reads

(9) $\quad u_t(x, y, t) = \gamma(L) u_y(x, y, t) + \alpha(L) u_x(x, y, t),$

6.1 More variables

and the corresponding equation for R takes the form

(10) $\quad R_t(k, y, t) = \gamma(-4k^2) R_y(k, y, t) + 2ik\alpha(-4k^2) R(k, y, t).$

This equation can be immediately integrated

(11) $\quad R(k, y, t) = R_0\bigl(k, y + \gamma(-4k^2)t\bigr) \exp\bigl[2ik\alpha(-4k^2)t\bigr],$

where of course

(12) $\quad R_0(k, y) \equiv R(k, y, 0).$

Before proceeding to consider the discrete part of the spectral transform, whose time evolution must also be ascertained in order to complete the solution via the spectral transform of the nonlinear evolution equation (9), let us display a single instance of nonlinear evolution equation belonging to the class (9), namely that corresponding to the choice

(13) $\quad \alpha(z) = \alpha_0 + \alpha_1 z, \qquad \gamma(z) = \gamma_0 + \gamma_1 z$

(note incidentally that the choice $\gamma(z) = \gamma_0$ with arbitrary $\alpha(z)$ yields an equation that can be reduced to the class of section 3.1, see (3.1.-28), by the change of variable $y' = y + \gamma_0 t$). It reads:

(14) $\quad u_t(x, y, t) = \alpha_0 u_x(x, y, t)$
$\qquad\qquad + \alpha_1 [u_{xxx}(x, y, t) - 6u_x(x, y, t) u(x, y, t)]$
$\qquad\qquad + \gamma_0 u_y(x, y, t)$
$\qquad\qquad + \gamma_1 \Bigl[u_{xxy}(x, y, t) - 4u_y(x, y, t) u(x, y, t)$
$\qquad\qquad\qquad + 2u_x(x, y, t) \int_x^\infty dz\, u_y(z, y, t) \Bigr].$

This is a nonlinear integrodifferential equation, that can however be reduced to a pure differential equation by the usual change of dependent variable,

(15a) $\quad w(x, y, t) = \int_x^\infty dz\, u(z, y, t),$

(15b) $\quad u(x, y, t) = -w_x(x, y, t).$

Then (14) reads

(16a) $\quad w_{xt} = \alpha_0 w_{xx} + \alpha_1(w_{xxxx} + 6w_{xx}w_x) + \gamma_0 w_{xy}$
$\qquad + \gamma_1(w_{xxxy} + 4w_{xy}w_x + 2w_{xx}w_y),$

(16b) $\quad w_{xt} = [\alpha_0 w_x + \alpha_1(w_{xxx} + 3w_x^2) + \gamma_0 w_y$
$\qquad + \gamma_1(w_{xxy} + 2w_x w_y)]_x + \gamma_1(w_x^2)_y.$

In this equation of course $w \equiv w(x, y, t)$ satisfies the boundary conditions

(17) $\quad w(+\infty, y, t) = w_x(\pm\infty, y, t) = w_{xx}(\pm\infty, y, t) = \cdots = 0.$

We are not aware of any natural phenomenon to whose description the nonlinear evolution equation (16) (with $\gamma_1 \neq 0$) is relevant; it would be quite interesting to find one.

The most straightforward way to find the time evolution of the discrete part of the spectral transform exploits the relationship

(18) $\quad R(k, y, t) \approx i\rho(y, t)/[k - ip(y, t)],$

valid for $k \approx ip(y, t)$ (here p is positive; $-p^2$ is the discrete eigenvalue, ρ the corresponding normalization coefficient; all these quantities, see below, depend now both on y and on t; moreover there should be an index n to label the different discrete eigenvalues, that has however been dropped for notational simplicity). As we have explained in section 3.1 (see (3.1.-23) and the discussion following this equation) this procedure is hardly correct, but it does yield the right answer, as can be verified by using the appropriate formulae of section 2.4.

Differentiating (18) with respect to t and to y and then using (10), one gets

(19) $\quad p_t(y, t) = \gamma[4p^2(y, t)] p_y(y, t),$

(20) $\quad \rho_t(y, t) = \gamma[4p^2(y, t)] \rho_y(y, t) - 2p\rho(y, t)$
$\qquad \cdot \{\alpha[4p^2(y, t)] + 8p_y(y, t) p(y, t) \gamma_z[4p^2(y, t)]\}.$

Note that, in the last equation, the term containing $\gamma_z(z) \equiv d\gamma(z)/dz$ results from the need, in order to equate the coefficients of the first-order pole, to re-expand the factor $\gamma(-4k^2)$ in the r.h.s. of (10) when it multiplies the double pole term that results from differentiating (18) with respect to y.

Both these equations are easily integrated (in fact the more convenient procedure to do this is by inserting (18) directly into (11) rather than into (10)):

(21) $\quad p(y,t) = p_0\bigl(y + \gamma[4p^2(y,t)]t\bigr),$

(22) $\quad \rho(y,t) = \rho_0\bigl(y + \gamma[4p^2(y,t)]t\bigr)$
$$\cdot \bigl\{1 - 8tp(y,t)\gamma_z[4p^2(y,t)] p_{0,y}\bigl(y + \gamma[4p^2(y,t)]t\bigr)\bigr\}^{-1}$$
$$\cdot \exp\bigl\{-2p(y,t)\alpha[4p^2(y,t)]t\bigr\}.$$

Here of course

(23) $\quad p_0(y) \equiv p(y,0)$

and

(24) $\quad \rho_0(y) \equiv \rho(y,0)$

are the initial data, while

(25) $\quad \gamma_z[4p^2(y,t)] \equiv d\gamma(z)/dz\big|_{z=4p^2(y,t)},$

(26) $\quad p_{0,y}\bigl(y + \gamma[4p^2(y,t)]t\bigr) \equiv dp_0(z)/dz\big|_{z=y+\gamma[4p^2(y,t)]t}.$

Note that (22) provides an *explicit* expression of $\rho(y,t)$ in terms of $\rho_0(y)$, $p_0(y)$ and $p(y,t)$, while (21) is an *implicit* definition of $p(y,t)$ in terms of $p_0(y)$. The fact that the discrete eigenvalues now move as time progresses should be emphasized.

The additional "space" variable y has been introduced, precisely like the "time" variable t, by treating it as an external parameter in the spectral problem. But the variables y and t play a very different rôle in the context of the Cauchy problem, where

(27) $\quad u_0(x,y) \equiv u(x,y,0)$

is the given initial datum, and $u(x,y,t)$ has to be evaluated (say, for $t>0$). Let us emphasize that the procedure based on the spectral transform that we have just described is indeed adequate to solve this problem, provided $u_0(x,y)$ is, as a function of x for every fixed (real) value of y, a *bona fide* potential. The Cauchy input (27) defines then uniquely the spectral transform at $t=0$, namely the quantities $R(k, y, 0) \equiv R_0(k, y)$, $p_n(y, 0) \equiv p_{n,0}(y)$ and $\rho_n(y, 0) \equiv \rho_{n,0}(y)$ (here y has a purely parametric rôle, the spectral problem referring only to the variable x); from these quantities, $R(k, y, t)$, $p_n(y, t)$ and $\rho_n(y, t)$ are calculated by (11), (21), and (22); and the inverse spectral problem with these inputs (y and t being again fixed parameters) yields precisely the solution $u(x, y, t)$ of (9) at time t that corresponds to the Cauchy datum (27). Of course this technique works provided, for all (fixed) values of y and of $t \geq 0$, $u(x, y, t)$ is a *bona fide* potential (considered as a function of x). Conditions sufficient to guarantee this are easily evinced from (11), (21) and (22), and by recalling what are the properties characterizing the spectral transform of a *bona fide* potential.

It is also easy to ascertain the time evolution of the transmission coefficient $T(k, y, t)$. Using techniques analogous to those of chapter 5 and appendix A.18 one finds, with obvious notation

(28) $\quad |T(k, y, t)| = |T_0(k, y + \gamma(-4k^2)t)|,$

consistently with unitarity and (11), that of course implies

(29) $\quad |R(k, y, t)| = |R_0(k, y + \gamma(-4k^2)t)|,$

and

(30) $\quad \theta_t(k, y, t) = \gamma(-4k^2) \theta_y(k, y, t)$
$$+ 2k \int_{-\infty}^{+\infty} dx\, \gamma^\#(-4k^2, L) u_y(x, y, t)$$

where, as usual, $\theta(k, y, t)$ is the phase of $T(k, y, t)$,

(31) $\quad T(k, y, t) = |T(k, y, t)| \exp[i\theta(k, y, t)], \quad \text{Im}\, k = 0,$

and

(32) $\quad \gamma^\#(z_1, z_2) \equiv [\gamma(z_1) - \gamma(z_2)]/(z_1 - z_2).$

From these equations, following the treatment of chapter 5, one could obtain a set of evolution equations for the quantities

(33a) $\quad C_m(y, t) = (-1)^m (2m+1)^{-1}$
$$\cdot \int_{-\infty}^{+\infty} dx\, L^m [x u_x(x, y, t) + 2u(x, y, t)], \quad m = 0, 1, 2, \ldots,$$

(33b) $\quad \tilde{C}_m(y, t) = (-1)^m (2m+1)^{-1} \int_{-\infty}^{+\infty} dx\, \tilde{L}^m u(x, y, t), \quad m = 0, 1, 2, \ldots$

(for the equivalence of these expressions, and the notation, see chapter 5 and appendix A.9). More directly, these results are obtained by the technique described at the end of chapter 5, namely noting that the operator identity (A.8.-8) or (A.9-.63b) implies

$$\partial C_m(y,t)/\partial t = (-1)^m \int_{-\infty}^{+\infty} dx\, L^m u_t(x,y,t), \quad m=0,1,2,\ldots, \tag{34}$$

$$\partial C_m(y,t)/\partial y = (-1)^m \int_{-\infty}^{+\infty} dx\, L^m u_y(x,y,t), \quad m=0,1,2,\ldots. \tag{35}$$

Then using in the r.h.s. of (34) the evolution equation (9), where we now assume $\gamma(z)$ and $\alpha(z)$ to be polynomials,

$$\gamma(z) = \sum_{m=0}^{M} \gamma_m z^m, \tag{36a}$$

$$\alpha(z) = \sum_{m=0}^{M} \alpha_m z^m, \tag{36b}$$

and using (35) and the identity (see (A.8.-6) or (A.9.-54a))

$$\int_{-\infty}^{+\infty} dx\, L^m u_x(x,y,t) = 0, \quad m=0,1,2,\ldots, \tag{37}$$

there obtains the infinite set of linear equations

$$\partial C_n(y,t)/\partial t = (\partial/\partial y) \sum_{m=0}^{M} (-1)^m \gamma_m C_{m+n}(y,t), \quad n=0,1,2,\ldots. \tag{38}$$

Note incidentally that the derivation of these relations does not require that the coefficients α_m in (36b), and γ_m in (36a) and (38), be time-independent.

If the function $u(x,y,t)$ has the property to vanish, with all its x-derivatives, as $y \to \pm\infty$, then clearly (38) implies the existence of the infinite set of conserved quantities,

$$\Gamma_m = \int_{-\infty}^{+\infty} dy\, C_m(y,t), \quad m=0,1,2,\ldots, \tag{39a}$$

$$\Gamma_m = (-1)^m (2m+1)^{-1} \tag{39b}$$
$$\cdot \int_{-\infty}^{+\infty} dx\,dy\, L^m [xu_x(x,y,t) + 2u(x,y,t)], \quad m=0,1,2,\ldots,$$

$$\Gamma_m = (-1)^m (2m+1)^{-1} \int_{-\infty}^{+\infty} dx\,dy\, \tilde{L}^m u(x,y,t), \quad m=0,1,2,\ldots, \tag{39c}$$

for the class of evolution equations (9) (with (36)).

6.2. Coefficients depending linearly on x

In this section we treat a class of nonlinear evolution equations with coefficients linearly dependent on x, that are solvable via the spectral transform based on the Schroedinger spectral problem. This class may be considered a nonlinear extension of the class of linear partial differential equations (1.1.-20), in the same sense as the class of evolution equations (3.1.-28) can be considered a nonlinear extension of the class of linear partial differential equations (1.1.-1). In the first part of this section our treatment will be formal, namely we implicitly assume that the solutions of the equations under consideration remain for all time in the class of *bona fide* potentials. In the second part of this section this assumption is questioned (starting with the discussion of some examples), and we analyse the difficulties that plague some of the nonlinear evolution equations of this class (as well as the corresponding linear evolution equations of type (1.1.-20)).

Let us thus consider, as usual, a function $u(x, t)$, that is assumed to behave, as a function of x for fixed t, as a *bona fide* potential, namely to be regular (infinitely differentiable) for all x and to vanish (sufficiently fast) as $x \to \pm\infty$. (Let us repeat: the consistency of this assumption with the time evolution of $u(x, t)$ that we are now going to investigate will be discussed only later). Let $R(k, t)$ indicate as usual the reflection coefficient corresponding to $u(x, t)$. The starting point of our analysis are the usual wronskian relations (3.1.-7) and (3.1.-8), but in addition we now use also (2.4.1.1.-7). Thus we write

(1) $\quad 2ikg(-4k^2, t)R_t(k, t)$
$$= \int_{-\infty}^{+\infty} dx \, [\psi(x, k, t)]^2 g(L, t) u_t(x, t),$$

(2) $\quad (2ik)^2 h(-4k^2, t) R(k, t)$
$$= \int_{-\infty}^{+\infty} dx \, [\psi(x, k, t)]^2 h(L, t) u_x(x, t),$$

(3) $\quad -2ikf(-4k^2, t) k R_k(k, t)$
$$= \int_{-\infty}^{+\infty} dx \, [\psi(x, k, t)]^2 f(L, t) [x u_x(x, t) + 2u(x, t)].$$

Here as usual g, h and f are largely arbitrary functions (entire in their first argument). Note the presence in the l.h.s. of (3) of the (partial) differentiation with respect to k. Of course here, and always below, L is the usual integrodifferential operator, see for instance (1.2.-2).

Clearly these equations imply that, if $u(x,t)$ evolves in time according to the equation

(4) $$g(L,t)u_t(x,t) + h(L,t)u_x(x,t)$$
$$+ f(L,t)[xu_x(x,t) + 2u(x,t)] = 0,$$

the corresponding reflection coefficient $R(k,t)$ evolves according to the linear partial differential equation

(5) $$g(-4k^2,t)R_t(k,t) + 2ikh(-4k^2,t)R(k,t)$$
$$- f(-4k^2,t)kR_k(k,t) = 0.$$

Hereafter, for notational simplicity, we set

(6) $$\alpha(z,t) = -h(z,t)/g(z,t), \qquad \beta(z,t) = -f(z,t)/g(z,t),$$

so that (4) reads

(7) $$u_t(x,t) = \alpha(L,t)u_x(x,t) + \beta(L,t)[xu_x(x,t) + 2u(x,t)],$$

and correspondingly in place of (5) we have

(8) $$R_t(k,t) + k\beta(-4k^2,t)R_k(k,t) = 2ik\alpha(-4k^2,t)R(k,t).$$

Note, however, that in writing (7) we implicitly assume that the expressions $\alpha(L,t)u_x(x,t)$ and $\beta(L,t)[xu_x(x,t) + 2u(x,t)]$ make good sense. For the first of these two expressions, this question is treated in appendix A.18; let us recall in particular that a condition sufficient to guarantee this, is that the function $\alpha(z,t)$ be regular for all *real* values of z, namely that $g(z,t) \neq 0$ if Im $z = 0$ (see (6)). It is easy to convince oneself, by a treatment analogous to that of appendix A.18 (but exploiting (2.4.2.1.-3) rather than (2.4.2.-16)), that the same condition is sufficient to guarantee that $\beta(L,t)[xu_x(x,t) + 2u(x,t)]$ makes good sense as well. (Indeed such a treatment also indicates for which subclass of initial data would the Cauchy problem for (7) be still OK, even if the above conditions were violated, for instance be allowing $g(z,t)$ to vanish for some positive value of z; but we forsake to discuss any more this point here, since it would be easy, and instructive, for the reader to fill this gap).

The linear first-order partial differential equation (8) can now be integrated by the method of characteristics. Indeed, in terms of the initial datum

(9) $R_0(k) \equiv R(k,0)$,

the explicit solution reads

(10) $R(k,t) = R_0[k_0(t,k)] \exp\left[2i \int_0^t dt' \chi \alpha(-4\chi^2, t')\right]$.

In this formula

(11) $\chi \equiv \chi[t', k_0(t,k)]$,

the function $\chi(t, k_0)$ being defined by the (ordinary) nonlinear differential equation

(12) $\chi_t(t, k_0) = \chi(t, k_0) \beta[-4\chi^2(t, k_0), t]$

and by the boundary condition

(13) $\chi(0, k_0) = k_0$,

while the function $k_0(t, k)$ is (implicitly) defined from χ through the formula

(14) $\chi(t, k_0) = k$.

To solve (7) by the spectral transform technique it is also necessary to ascertain the corresponding evolution of the parameters of the discrete spectrum. The easiest way to obtain this is by relying on (3.1.-23). In this manner one finds

(15) $\dot{p}(t) = p(t) \beta[4p^2(t), t]$,

(16) $\rho(t) = \rho_0 \chi_{k_0}(t, ip_0) \exp\left\{-2\int_0^t dt' p(t') \alpha[4p^2(t'), t']\right\}$.

In these equations we have omitted for notational simplicity to indicate the index n labelling different discrete eigenvalues. The first of these two equations is a nonlinear ordinary differential equation, that defines $p(t)$ in conjunction with the initial datum

(17) $p_0 \equiv p(0)$;

(incidentally, it is obvious that, once the function $\chi(t, k_0)$ has been computed from (12) and (13), the solution of (15) with (17) is $p(t) = -i\chi(t, ip_0)$). The second formula provides an explicit expression of $\rho(t)$ in terms of the initial datum

(18) $\quad \rho_0 \equiv \rho(0),$

of the function $p(t)$ (defined by (15) and (17)), and of the function $\chi(t, k_0)$ defined by (12) and (13) (the symbol $\chi_{k_0}(t, ip_0)$ in the r.h.s. of (16) indicates of course the partial derivative of $\chi(t, k_0)$ with respect to k_0, evaluated for $k_0 = ip_0$).

These equations provide all the information needed to solve, by the technique described in section 3.1, the Cauchy problem for the class of nonlinear evolution equations (7): note that now the discrete eigenvalues are time-dependent (the flow (7) is no longer isospectral, for the Schroedinger operator (3.1.-17)). To obtain their explicit time-evolution it is necessary to solve an ordinary nonlinear differential equation (this is generally feasible in explicit form, at least for the simple choices of the function β in (7) that yield reasonably simple equations; see below).

The equations of the class (7) possess of course single-soliton solutions, that now read

(19) $\quad u(x, t) = -2p^2(t)/\cosh^2\{p(t)[x - \xi(t)]\},$

with $p(t)$ determined by (15) and (17), and $\xi(t)$ determined by its relation to $\rho(t)$ and $p(t)$ (see (2.2.-9)),

(20) $\quad \xi(t) = [2p(t)]^{-1} \ln\{\rho(t)/[2p(t)]\}.$

The fact that $p(t)$ is now time-dependent, and the complexity of the time dependence of $\rho(t)$, see (16), imply a more complicated behaviour of these solitons (examples are discussed below).

In the standard manner it is also possible to construct multisoliton solutions; indeed they are still given by (1.6.1.-9, 10), except that now the p_m's are time-dependent (according to (15) and (17), obviously with the index m attached to p), while the time dependence of the ρ_m's is now given by (16) (again with the label m suitably attached to every ρ and p).

In the weak-field limit, namely if all powers of u are neglected, the class (7) reproduces the class (1.1.-20), with the identification

(21a) $\quad \omega(z, t) = -z\alpha(-z^2, t) + 2i[\beta(-z^2, t) + \beta'(-z^2, t)],$
(21b) $\quad \omega'(z, t) = -z\beta(-z^2, t);$

note that, in the r.h.s. of (21a), the prime indicates differentiation with respect to the first argument.

For completeness, let us also report the formulae that detail the time evolution of the transmission coefficient:

(22) $\quad |T(k,t)| = |T_0[k_0(t,k)]|,$

(23) $\quad \theta_t(k,t) + k\beta(-4k^2, t)\theta_k(k,t)$
$$= -2k \int_{-\infty}^{+\infty} dx\, \beta^\#(-4k^2, L, t)[xu_x(x,t) + 2u(x,t)].$$

Here of course

(24) $\quad \beta^\#(z_1, z_2, t) \equiv [\beta(z_1, t) - \beta(z_2, t)]/(z_1 - z_2)$

and $T_0(k)$ is the transmission coefficient at $t=0$:

(25) $\quad T_0(k) = T(k, 0).$

The modulus and phase of the transmission coefficient are defined as usual,

(26) $\quad T(k,t) = |T(k,t)| \exp[i\theta(k,t)], \quad \operatorname{Im} k = 0;$

(let us recall that these formulae refer to the case of real $u(x,t)$, but they are easily extended to the more general case of complex $u(x,t)$ by replacing $|T(k,t)|^2$ by $T(k,t)T(-k,t)$ and by using the definition (A.3.-4) in place of (26)).

The proof of (22) and (23) is closely analogous to the proof of (5.-9), and is therefore left as an exercise for the diligent reader. Note that one must use the formula

(27) $\quad \int_{-\infty}^{+\infty} dx\, \alpha^\#(-4k^2, L, t) u_x(x,t) = 0,$

whose validity is trivial (see (A.9.-54a)) if $\alpha(z,t)$ is a polynomial in z, but is also implied, in the more general case (6), by the discussion of appendix A.18.

Incidentally, the validity of (22) also follows, quite straightforwardly, from (10), that clearly implies

(28) $\quad |R(k,t)| = |R_0[k_0(t,k)]|,$

and from the unitarity relation (2.1.-13).

It is now easy to investigate the time-evolution of the quantities

(29) $$C_m(t) = (-1)^m (2m+1)^{-1} \int_{-\infty}^{+\infty} dx\, L^m [xu_x(x,t) + 2u(x,t)],$$

$$m = 0, 1, 2, \ldots,$$

by starting from (23) and by proceeding in close analogy to the treatment of chapter 5 (in this manner one would also recover some of the operator equations of appendix A.9; incidentally, note the coincidence of (29) with (A.9.-57a), and the possibility to provide alternative, equivalent, expressions of the quantities $C_m(t)$, as shown by the various formulae (A.9.-57)). This is also left as an exercise for the more diligent reader, since a more straightforward derivation of the relevant formulae can be effected, in close analogy to the treatment given at the very end of chapter 5, as we now indicate.

The starting point is the formula

(30) $$\dot{C}_n(t) = (-1)^n \int_{-\infty}^{+\infty} dx\, L^n u_t(x,t), \quad n = 0, 1, 2, \ldots.$$

Here, and always below, the superimposed dot indicates time differentiation. Let us recall that, with the definition (29), this formula is an identity, being merely a special case of (A.9.-63a).

We now use the evolution equation (7), together with the formula

(31) $$\int_{-\infty}^{+\infty} dx\, L^m \alpha(L, t) u_x(x, t) = 0, \quad m = 0, 1, 2, \ldots,$$

for whose justification, in the general case (6), we refer again to appendix A.18. It is also convenient, at this stage, to introduce the expansion in powers of z of $\beta(z, t)$,

(32) $$\beta(z, t) = \sum_m \beta_m(t) z^m.$$

Note that we are not requiring, at this stage, that the sum in the r.h.s. of this formula contain only nonnegative powers, or only a finite number of terms; although of course only nonnegative powers should be present if $\beta(z, t)$ is to be holomorphic at $z = 0$ (but, as the discussion outlined above and in appendix A.18 implies, even if $\beta(z, t)$ is singular for some real value of z the evolution equation (7), or rather (4), may have solutions, although in such a case there should be some restriction on the initial datum in order for the Cauchy problem to be OK).

In this manner (and also using once more the definition (29)) we get

(33) $$\dot{C}_n(t) = \sum_m (-1)^m (2n+2m+1)\beta_m(t) C_{n+m}(t), \quad n=0,1,2,\ldots$$

This is, in general, an infinite system of linear first-order ordinary differential equations; it is however, as we presently show, possible to solve it, at least formally. Before doing this, let us however first consider the simpler case when the function $\beta(z,t)$ is a (time-independent) polynomial, of degree, say, M, so that

(34a) $\quad \beta_m(t) = 0, \quad m<0, m>M,$

(34b) $\quad \beta_m(t) = \beta_m, \quad 0 \leqslant m \leqslant M.$

Then the system (33) becomes triangular,

(35) $$\dot{C}_n(t) = \sum_{m=0}^{M} (-1)^m (2n+2m+1)\beta_m C_{n+m}(t), \quad n=0,1,2,\ldots,$$

and this immediately implies that, provided $\beta_0 \neq 0$, its solution has the structure

(36) $$C_n(t) = \sum_{m=n}^{\infty} c_{nm} \exp[(2m+1)\beta_0 t], \quad n=0,1,2,\ldots,$$

the (time-independent!) quantities c_{nm} being determined by the β_m's and by the initial values $C_n(0)$. This formula implies that in this case, which is also that yielding the simpler nonlinear evolution equations of the class (7), all the quantities $C_n(t)$ are periodic in t with the same period T,

(37) $\quad C_n(t+T) = C_n(t), \quad n=0,1,2,\ldots,$

iff β_0 is pure imaginary,

(38) $\quad \beta_0 = 2\pi i/T;$

thus for this class of nonlinear evolution equations the phenomenon of periodicity of the quantities $C_m(t)$ never occurs if the consideration is restricted to real equations (but the fact that it does occur if (38) holds is quite remarkable). The results given below (or, for that matter, the very structure of the system (35)) imply that the phenomenon of periodicity of the integrals $C_m(t)$ might occur, for real and autonomous equations (i.e., for a real time-independent β in (7)), only if in the r.h.s. of (32) there are non-vanishing coefficients β_m both for positive and negative values of m. Note that these conclusions need not require that M (see (34)) be finite, but only that $\beta(z,t)$ be holomorphic in z at $z=0$.

Perhaps it should be emphasized at this point that periodicity of all the integrals $C_m(t)$, see (29), does not imply that $u(x,t)$ be itself periodic in t (just as the constancy of all the integrals C_m if $\beta=0$ in (7) does not imply that $u(x,t)$ is time-independent!).

Another case worth mentioning, and indeed simpler (see below), is when the function β in (7) is a time-independent polynomial in z^{-1},

(39a) $\beta_m(t)=0, \quad m>0, m<-M,$

(39b) $\beta_m(t)=\beta_m, \quad -M\leqslant m\leqslant 0.$

Then the system (33) is again triangular, and its solution has, provided $\beta_0 \neq 0$, the simple structure

(40) $$C_n(t) = \sum_{m=0}^{n} c_{nm} \exp[(2m+1)\beta_0 t], \quad n=0,1,2,\ldots,$$

the constants c_{nm} being again determined by the coefficients β_m and by the initial values $C_m(0)$; for instance

(41a) $C_0(t)=C_0(0)\exp(\beta_0 t),$

(41b) $C_1(t)=C_1(0)\exp(3\beta_0 t)$
$\qquad -\frac{1}{2}[\beta_{-1}C_0(0)/\beta_0]\exp(\beta_0 t)[\exp(2\beta_0 t)-1],$

and so on. Thus in this case the time evolution of the integrals $C_n(t)$ is very simple indeed (and again periodic iff β_0 is pure imaginary). Moreover, from these formulae, explicit conservation laws are easily derived, by solving for $C_n(0)$ in terms of the $C_m(t)$'s. For instance from (41a) and (29) there follows that the quantity

(41c) $$C_0(0) = \exp(-\beta_0 t) \int_{-\infty}^{+\infty} dx\, u(x,t)$$

is a constant of the motion for the nonlinear evolution equation (7), provided $\beta(L,t)$ is given by (32) and (39); and another conserved quantity can be easily extracted from (41b); and so on (if need be). Note that these conclusions need not require that M, see (39), be finite, but only that $\beta(z,t)$

be holomorphic at $z=\infty$. Indeed, if M is finite, $\beta(z,t)$ violates for sure the rules stipulated above (see the discussion following (8)).

Let us now return to a discussion of the system (33) in the general case. A formal solution is then given by the formula

$$(42) \quad C_n(t) = (-1)^n (2n+1)^{-1} \int_{-\infty}^{+\infty} dx \left[\theta(t,t_0; L_0)\right]^{2n+1}$$
$$\cdot L_0^n \left[xu_x(x,t_0) + 2u(x,t_0)\right], \quad n=0,1,2,\ldots.$$

Here the function $\theta(t,t_0; \lambda_0)$ is defined by the ordinary differential equation

$$(43) \quad \theta_t(t,t_0; \lambda_0) = \theta(t,t_0; \lambda_0) \beta\left[\lambda_0 \theta^2(t,t_0; \lambda_0), t\right]$$

and by the boundary condition

$$(44) \quad \theta(t_0, t_0; \lambda_0) = 1;$$

thus (see (12) and (13))

$$(45) \quad \theta(t, 0; -4k_0^2) = \chi(t, k_0)/k_0.$$

As for the integrodifferential operator L_0 in (42), it is defined by the usual formula, say (2.4.1.-13), but with $u(x) = u(x,t_0)$. Note that we are assuming here that the initial condition is assigned at the time t_0 rather than 0; this notational change is convenient in the following discussion, see below.

The proof that (42) solves (33) is straightforward. For $t=t_0$, (44) implies that (42) is consistent with the definition (29). And by inserting (42) in the l.h.s. of (33) one can easily see, by using (43) and (32), that precisely the r.h.s. of (33) is reproduced (of course this assumes the freedom to exchange the order of the x-integration and the t-differentiation; moreover the operator L_0 must be treated as if it were an ordinary variable, which can be justified by performing a power expansion, as indicated below).

It should be emphasized that the function $\theta(t,t_0; \lambda_0)$ can, in the interesting cases, be calculated explicitly by solving (43) and (44) (see examples below). It may then be also possible to evaluate explicitly the coefficients $\theta_{n,m}(t,t_0)$ of the expansion

$$(46) \quad \left[\theta(t,t_0; \lambda_0)\right]^{2n+1} = \sum_{m=-\infty}^{+\infty} \theta_{n,m}(t,t_0) \lambda_0^m, \quad n=0,1,2,\ldots,$$

and by using this formula and the definition (29) it is possible to rewrite

6.2 Coefficients depending linearly on x

(42) in the form

$$(47) \quad C_n(t) = \sum_{m=-n}^{\infty} (-1)^m \frac{(2n+2m+1)}{(2n+1)} \theta_{n,m}(t, t_0) C_{n+m}(t_0),$$

$$n = 0, 1, 2, \ldots.$$

The formula (42), as well as this last formula (if applicable), can be interpreted in two ways. Firstly, they display explicitly the time evolution of each of the integrals $C_n(t)$, once the (initial) values of the quantities $C_m(t_0)$ are given. Note that, should the function $\theta(t, t_0; \lambda_0)$ (and therefore also the coefficients $\theta_{n,m}(t, t_0)$; see (46)) turn out to be periodic in t, then the same property would hold for the $C_n(t)$'s, independently of the initial values $C_m(t_0)$; on the other hand, the properties of $\theta(t, t_0; \lambda_0)$ depend only on the structure of the ordinary differential equation, (43) with (44), that defines it (cases in which $\theta(t, t_0; \lambda_0)$ is indeed periodic have been mentioned above).

A second interpretation of (42) or (47) results from the observation that clearly these formulae remain valid if the rôles of t and t_0 are exchanged, so that in place of (42) one has

$$(48) \quad C_n(t_0) = (-1)^n (2n+1)^{-1} \int_{-\infty}^{+\infty} dx \left[\theta(t_0, t; L) \right]^{2n+1}$$
$$\cdot L^n \left[xu_x(x,t) + 2u(x,t) \right], \quad n = 0, 1, 2, \ldots,$$

and in place of (47),

$$(49) \quad C_n(t_0) = \sum_{m=-n}^{\infty} (-1)^m \frac{(2n+2m+1)}{(2n+1)} \theta_{n,m}(t_0, t) C_{n+m}(t),$$

$$n = 0, 1, 2, \ldots.$$

Note that the operator L in (48) is now defined by (2.4.1.-13) with $u(x) = u(x, t)$; while of course the function θ in (48), and the coefficients $\theta_{n,m}$ in (49), are defined as above, namely by (43), (44) and (46).

The last two formulae define now quantities that are time-independent, as the field $u(x, t)$ evolves according to (7); thus, an infinite sequence of constants of the motion has been obtained also for this class of evolution equations. These formulae are particularly informative if the sums in the r.h.s. of (49) and (47) contain only a finite number of terms; the discussion given above, as well as the structure of (43) (see examples below), implies that this is indeed the case if $\beta(z, t)$ is holomorphic in z at $z = \infty$. When instead an infinite number of terms enters in the r.h.s. of (49) and (47) (as is for instance the case if $\beta(z,t)$ is a polynomial in z), the theoretical significance of the constants of the motion (48) or (49) is less transparent, and their practical usefulness moot.

On the other hand we re-emphasize that the formulae (42) and (47) are highly informative in all cases in which $\theta(t,t_0;\lambda_0)$ is periodic in t, independently of the nature of the dependence on λ_0.

A particularly simple case occurs if $\beta(z,t)$ is independent of z,

(50) $\quad \beta(z,t)=\beta_0(t).$

Then the system (33) is immediately solved:

(51) $\quad C_n(t)=C_n(0)\exp\left[(2n+1)\int_0^t dt'\,\beta_0(t')\right].$

This case, however, does not correspond to a real extension of the class of solvable nonlinear evolution equation, since it is then possible to reduce (7) back to the old class, say (3.1.-28), merely by a change of variables, as shown by the following formulae:

(52) $\quad u_t=\alpha(L,t)u_x+\beta_0(t)(xu_x+2u), \quad u\equiv u(x,t),$

(53) $\quad \bar{u}_t=\bar{\alpha}(\bar{L},t)\bar{u}_y, \quad \bar{u}\equiv\bar{u}(y,t),$

(54) $\quad \bar{u}(y,t)=[f(t)]^{-2}u(x,t),$

(55) $\quad y=xf(t),$

(56) $\quad f_t(t)/f(t)=\beta_0(t),$

(57) $\quad \bar{\alpha}(z,t)=f(t)\alpha[f^2(t)z,t];$

of course the operator $\bar{L}$ in (53) is the usual operator L, but with u replaced by $\bar{u}$ (and x by y).

Let us now consider some examples. We first focus on one belonging to the class (39):

(58) $\quad \alpha(z,t)=(\alpha_{-1}+\alpha_0 z)/z, \quad \beta(z,t)=(\beta_{-1}+\beta_0 z)/z.$

Note that we omit for simplicity any explicit t-dependence in the coefficients. Note moreover that (58) violates the conditions mentioned in the discussion following (8), since both $\alpha(z,t)$ and $\beta(z,t)$ are singular at $z=0$; the troubles that may result from this cavalier choice are discussed below.

6.2 Coefficients depending linearly on x

The nonlinear evolution equation that results from (58) is more conveniently written in terms of the dependent variable w, defined by

(59a) $\quad w(x,t) = \int_{x}^{+\infty} dy\, u(y,t),$

(59b) $\quad u(x,t) = -w_x(x,t).$

It reads:

(60) $\quad (w_{xxx} + w_x^2)_t + 2(w_x w_t)_x = [\alpha_{-1} w_x + \alpha_0(w_{xxx} + 3w_x^2)$
$+ \beta_{-1}(xw_x + w) + \beta_0(xw_{xxx} + 3w_{xx} + 2w_x w + 3xw_x^2)]_x + 3\beta_0 w_x^2.$

The various functions that characterize the solutions of this equation, as defined above, are in this case straightforwardly evaluated:

(61) $\quad \chi(t, k_0) = k_0 \exp(\beta_0 t) \{1 - (q/k_0)^2 [1 - \exp(-2\beta_0 t)]\}^{1/2},$

(62) $\quad k_0(t, k) = k \exp(-\beta_0 t) \{1 - (q/k)^2 [1 - \exp(2\beta_0 t)]\}^{1/2},$

(63) $\quad \theta(t, t_0; \lambda_0) = \exp[\beta_0(t - t_0)]$
$\cdot \{1 + (4q^2/\lambda_0)[1 - \exp(-2\beta_0(t - t_0))]\}^{1/2},$

(64) $\quad \theta_{n,m}(t, t_0) = \theta_{n,m}(t - t_0),$

(65a) $\quad \theta_{n,m}(t) = 0, \quad m > 0,$

(65b) $\quad \theta_{n,-m}(t) = \exp[(2n+1)\beta_0 t][1 - \exp(-2\beta_0 t)]^m$
$\cdot (2q)^{2m} \binom{n + \frac{1}{2}}{m}, \quad m \geq 0,$

(66) $\quad p(t) = p_0 \exp(\beta_0 t) \{1 + (q/p_0)^2 [1 - \exp(-2\beta_0 t)]\}^{1/2},$

(67) $\quad \xi(t) = [p_0/p(t)][\xi_0 - (2p_0)^{-1}$
$\cdot \ln\{1 + (q/p_0)^2 [1 - \exp(-2\beta_0 t)]\}] + \tilde{\xi}(t),$

(68) $\quad \tilde{\xi}(t) = [2\beta_0 p(t)]^{-1}[(2q)^{-1}\alpha_{-1}$
$\cdot \{\arcsin[M] - \arcsin[M\exp(-\beta_0 t)]\}$
$+ 2q\alpha_0\{[M^2 \exp(2\beta_0 t) - 1]^{1/2} - [M^2 - 1]^{1/2}$
$- \operatorname{arctg}[M^2 \exp(2\beta_0 t) - 1]^{1/2} + \operatorname{arctg}[M^2 - 1]^{1/2}\}].$

In these formulae

(69) $\quad q^2 = \beta_{-1}/4\beta_0$

and

(70) $\quad M^2 = q^2/(q^2 + p_0^2);$

the formulae are written in a fashion appropriate for positive q^2, but they are also valid if q^2 is negative (it is easily seen that they yield real quantities also in this case; and also, by taking appropriate limits, for $\beta_0 = 0$).

Thus in this case the infinite sequence of conserved quantities is given by the explicit formula

(71) $\quad C_n(0) = \sum_{m=0}^{n} (-1)^m \frac{(2n-2m+1)}{(2n+1)} \theta_{n,-m}(-t) C_{n-m}(t),$

$$n = 0, 1, 2, \ldots,$$

where of course $\theta_{n,-m}$ is given by (65b) (but note the minus sign in front of t) and the $C_m(t)$'s are the integrals (29). For instance

(72a) $\quad C_0(0) = C_0(t) \exp(-\beta_0 t),$

(72b) $\quad C_1(0) = C_1(t) \exp(-3\beta_0 t) + 2q^2 C_0(t) \exp(-\beta_0 t)$
$\quad \cdot [1 - \exp(-2\beta_0 t)],$

and so on.

Let us also analyze the behaviour of the solitons in this case (recall that the single-soliton solution is always given by (19), but now with p and ξ given by (66) and (67), (68)). Since all the relevant information is contained in the completely explicit formulae (66), (67) and (68), we limit our discussion to the consideration of some cases only, leaving a more thorough treatment as a task for the diligent reader. In particular we set $\alpha_{-1} = \alpha_0 = 0$ (implying $\tilde{\xi} = 0$; see (68)).

We assume first of all that

(73) $\quad \beta_{-1} > 0;$

it is easily seen that this excludes the occurrence of singularities for $t \geq 0$. Then we find, as $t \to +\infty$,

(74a) $\quad p(t) \approx (p_0^2 + q^2)^{1/2} \exp(\beta_0 t) \to +\infty, \quad \beta_0 > 0,$

(74b) $\quad p(t) \approx (\beta_{-1} t/2)^{1/2} \to +\infty, \quad \beta_0 = 0,$

(74c) $\quad p(t) \to |q| = \frac{1}{2}(\beta_{-1}/|\beta_0|)^{1/2}, \quad \beta_0 < 0,$

and correspondingly

(75a) $\quad \xi(t) \to 0, \quad \beta_0 \geq 0,$
(75b) $\quad \xi(t) \approx (p_0/|q|)\xi_0 + |q|^{-1}\ln(p_0/|q|) - vt \to -\infty, \quad \beta_0 < 0$

with

(76) $\quad v = -\beta_0/|q| = 2(|\beta_0|^3/\beta_{-1})^{1/2}.$

On the other hand, for negative t, a singularity arises at $t = -t_s$, with

(77) $\quad t_s = (2\beta_0)^{-1}\ln[1 + (p_0/q)^2]$

if $\beta_0 \geq 0$ (so that q^2 is positive), and also if $\beta_0 < 0$, provided $p_0 \leq |q|$; if instead $\beta_0 < 0$ but $p_0 > |q|$, then the single-soliton solution is OK for all negative times, and as $t \to -\infty$

(78) $\quad p(t) \approx [p_0^2 + q^2]^{1/2}\exp(\beta_0 t) \to \infty, \quad (\beta_0 < 0, p_0 > |q|),$
(79) $\quad \xi(t) \to 0, \quad\quad\quad\quad\quad\quad\quad\quad\quad\quad\quad (\beta_0 < 0, p_0 > |q|).$

Particularly interesting is the case with $\beta_0 < 0$; then every single-soliton solution of (60), as $t \to +\infty$, acquires precisely the same shape, including the identification of amplitude and width (see (74c) and recall (19); note that this conclusion is independent of our assumption that α_{-1} and α_0 vanish, indeed it is independent of the choice of $\alpha(z,t)$ in (7), since the evolution of $p(t)$ depends only on $\beta(z,t)$). Also the behaviour of $\xi(t)$ as $t \to \infty$, see (75b), is in this case notable; note that the velocity v, see (76), is independent of p_0, so that, for all single-soliton solutions, the asymptotic motion is in this case uniform, and with a common speed that depends only on the parameters in the nonlinear evolution equation.

These results also imply that, provided the inequality (73) holds, for $t \geq 0$ all multisoliton solutions are OK, although their behaviour as $t \to +\infty$ is less easily ascertained than that of the single-soliton solution, especially in the $\beta_0 < 0$ case.

If instead β_{-1} is negative, the single-soliton solution becomes singular at the *positive* time $t_s = -(2\beta_0)^{-1}\ln[1 + (p_0/q)^2]$, if the argument of this logarithm is positive (this is certainly the case if β_0 is also negative, so that

q^2 is positive, see (69); otherwise the single-soliton solution runs into a singularity at a positive time only if $p_0 \leq |q|$); and analogous conclusions hold for multisoliton solutions.

Incidentally, the importance of the sign of β_{-1} in characterizing the behaviour of the soliton solutions of (60) is remarkable also in view of the apparently unimportant rôle that the term proportional to β_{-1} plays in (60); but of course we also know, see (52)–(57), that the presence of β_{-1} is indeed essential to cause the novel phenomenology.

Let us turn now to a discussion of the generic solution (characterized by a nonvanishing reflection coefficient). Then, immediately as the time evolution begins, something goes wrong. Indeed, if $\beta_{-1} < 0$, the quantity $k_0(t, k)$, see (62) and (69), loses the property to be real for all values of k, as soon as $t > 0$. If instead $\beta_{-1} > 0$, $k_0(t, k)$ does remain real for all real values of k, but as k spans the intervals from 0 to $\pm \infty$, $k_0(t, k)$ only spans the intervals from $k_0(t, 0\pm) = \pm \frac{1}{2} \{\beta_{-1}[1 - \exp(-2\beta_0 t)]/\beta_0\}^{1/2}$ to $\pm \infty$. Thus $R(k, t)$ develops a (finite) discontinuity at $k = 0$, see (10) (except for the special case with $\alpha(z) = 0$, and with $R_0(k)$ real so that $R_0(k) = R_0(-k)$); but even in this case, the generic property $R(0, t) = -1$ is violated for $t > 0$, thereby implying $u(x, t)$ to belong, immediately as t exceeds zero, to the class of potentials possessing a "zero-energy bound-state"; this, however, need not entail any pathology, since there is such a potential in the immediate neighborhood of any *bona fide* potential; see appendix A.15).

In conclusion, for the nonlinear evolution equation (60) on the whole line, $-\infty < x < +\infty$, it appears that the only solutions that are OK for (at least) a finite positive time interval, are the pure multisoliton solutions; all other solutions (possibly up to marginal exceptions) get into trouble immediately as the time evolution begins. As for the precise nature of the "trouble", suffice to note here that, at least in the $\beta_{-1} > 0$ case, it is reasonable to expect that it affects the asymptotic (large x) behaviour of the solutions, rather than it manifests itself as a local singularity. Moreover, as it was already hinted above, the source of the trouble for this equation has to do with the fact that the choice (58) violates the rules intimated in the discussion following (8), and not only with the specific novelty that we are discussing in this section (namely, the enlarged class of nonlinear evolution equations due to the presence of the term β in the r.h.s. of (7)). Thus the discussion of the pathologies of this particular equation (60) is not sufficiently enlightening to justify devoting any more space to it. We therefore terminate here the treatment of this example, with a final reiteration of the fact that, even though the generic solution of (60) may develop pathologies, there exists a class of multisoliton solutions that do satisfy (60), as well as the conservation laws (48).

6.2 Coefficients depending linearly on x

Let us turn then to another example, that does not violate the restrictions intimated after (8), and indeed belongs to the class (34):

(80) $\quad \alpha(z,t) = \alpha_0 + \alpha_1 z, \quad \beta(z,t) = \beta_0 + \beta_1 z.$

Note that we have again omitted any explicit time dependence.

The corresponding evolution equation is again preferably written in terms of w rather than u (see (59)). It reads:

(81) $\quad w_{xt} = \left[\alpha_0 w_x + \alpha_1 (w_{xxx} + 3w_x^2) + \beta_0 (xw_x + w) \right.$
$\left. + \beta_1 (xw_{xxx} + 3w_{xx} + 3xw_x^2 + 2w_x w) \right]_x + 3\beta_1 w_x^2.$

We now report the expressions of the various quantities relevant to the solution of the Cauchy problem for this equation via the spectral transform technique, as explained above:

(82) $\quad \chi(t, k_0) = k_0 \exp(\beta_0 t) \left\{ 1 - (k_0/q)^2 [1 - \exp(2\beta_0 t)] \right\}^{-1/2},$

(83) $\quad k_0(t, k) = k \exp(-\beta_0 t) \left\{ 1 - (k/q)^2 [1 - \exp(-2\beta_0 t)] \right\}^{-1/2},$

(84) $\quad \theta(t, t_0; \lambda_0) = \exp[\beta_0 (t - t_0)]$
$\cdot \left\{ 1 + \lambda_0 (2q)^{-2} [1 - \exp(2\beta_0 (t - t_0))] \right\}^{-1/2},$

(85) $\quad \theta_{n,m}(t, t_0) = \theta_{n,m}(t - t_0),$

(86a) $\quad \theta_{n,m}(t) = 0, \quad m < 0,$

(86b) $\quad \theta_{n,m}(t) = \exp[(2n+1)\beta_0 t] [1 - \exp(2\beta_0 t)]^m (2q)^{-2m} \binom{-n-\frac{1}{2}}{m},$
$m \geq 0,$

(87) $\quad p(t) = p_0 \exp(\beta_0 t) \left\{ 1 + (p_0/q)^2 [1 - \exp(2\beta_0 t)] \right\}^{-1/2},$

(88) $\quad \xi(t) = [p_0/p(t)] [\xi_0 - (2p_0)^{-1} \ln\{1 + (p_0/q)^2 [1 - \exp(2\beta_0 t)]\}]$
$+ \tilde{\xi}(t),$

(89) $\quad \tilde{\xi}(t) = (\alpha_1/\beta_1) \{[p_0/p(t)] - 1\} - [q/p(t)][(\alpha_0/\beta_0) - (\alpha_1/\beta_1)]$
$\cdot \text{arctg}\{q[p(t) - p_0]/[q^2 + p_0 p(t)]\}.$

In these equations

(90) $\quad q^2 = \beta_0/4\beta_1;$

note that these formulae remain valid (and real) also if β_0 and β_1 have opposite signs, so that q^2 is negative.

In this case the conserved quantities (48) and (49) have a much more complicated structure, since the sum in the r.h.s. of (49) extends now all the way to infinity (see (86)). We therefore do not discuss them any further, and proceed immediately to analyze the single-soliton solution (19), which is now complemented by (87), (88) and (89).

Let us assume first that β_1 is negative,

(91) $\quad \beta_1 < 0,$

this inequality being sufficient to guarantee that, for every $t > 0$, the single-soliton solution be singularity-free; and let us investigate what happens in the asymptotic limit $t \to +\infty$. Then

(92a) $\quad p(t) \to |q| = [\beta_0/4|\beta_1|]^{1/2}, \quad \beta_0 > 0,$

(92b) $\quad p(t) = p_0[1 - 8\beta_1 p_0^2 t]^{-1/2} \to 0, \quad \beta_0 = 0,$

(92c) $\quad p(t) \approx p_0[1 + (p_0/q)^2]^{-1/2} \exp(\beta_0 t) \to 0, \quad \beta_0 < 0,$

and

(93a) $\quad \xi(t) \approx \xi_\infty + vt \to \text{sign}(v) \infty, \quad \beta_0 > 0,$

(93b) $\quad \xi(t) = -2\alpha_0 t + (1 - 8\beta_1 p_0^2 t)^{1/2}[b - (2p_0)^{-1} \ln(1 - 8\beta_1 p_0^2 t)]$
$\quad\quad + \xi_0 - b \to -\text{sign}(\alpha_0) \infty, \quad \beta_0 = 0,$

(93c) $\quad \xi(t) \approx c \exp(-\beta_0 t) \to \text{sign}(c) \infty, \quad \beta_0 < 0.$

The constants ξ_∞, v, b and c in the last 3 formulae are defined as follows

(94) $\quad \xi_\infty = \xi_0 p_0/|q| + |q|^{-1} \ln(|q|/p_0) - (\alpha_1/\beta_1)(1 - p_0/|q|)$
$\quad\quad + [(\alpha_0/\beta_0) - (\alpha_1/\beta_1)] \ln[(p_0 + |q|)/(2p_0)],$

(95) $\quad v = -2|\beta_0 \beta_1|^{1/2}\{1 + |q|[(\alpha_0/\beta_0) - (\alpha_1/\beta_1)]\},$

(96) $\quad b = \xi_0 + [\alpha_1 - \alpha_0/(2p_0)^2]/\beta_1,$

(97) $\quad c = [1 + (p_0/q)^2]^{1/2}\{\xi_0 - (2p_0)^{-1} \ln[1 + (p_0/q)^2]$
$\quad\quad + \alpha_1/\beta_1 + [(\alpha_0/\beta_0) - (\alpha_1/\beta_1)](q/p_0) \text{arctg}(p_0/q)\}.$

There may instead occur a singularity for $t < 0$. Indeed, if $\beta_0 \leq 0$, or if $\beta_0 > 0$

but $p_0 \geq |q|$, a singularity occurs at the *negative* time $t = t_s$,

(98) $\quad t_s = (2\beta_0)^{-1} \ln[1 + (q/p_0)^2]$;

if instead $\beta_0 > 0$ and $p_0 < |q|$, no singularity affects the single-soliton solution.

Let us turn now to the case $\beta_1 > 0$. It is then easily seen that the single-soliton solution becomes singular at the *positive* time t_s if β_0 is positive, and also if β_0 is negative provided $p_0 \geq |q|$, while, if β_0 is negative and $p_0 < |q|$, there is no singularity at any positive time.

These results are sufficiently simple not to require any comment; nor shall we dwell on the similarity to, and the differences from, the preceding example. We merely emphasize that, if the inequality (91) holds, no single-soliton solution runs into difficulty for $t > 0$; and this property is clearly shared by all pure multisoliton solutions.

On the other hand the generic solution of (81) runs into trouble immediately the time evolution begins. Indeed, if β_1 is positive, for $t > 0$ the functions $k_0(t, k)$ becomes imaginary for some real values of k. If instead β_1 is negative (namely if (91) holds), $k_0(t, k)$ does remain real for $t > 0$ and k real, but as k spans the intervals from 0 to $\pm\infty$, it varies only between 0 and

(99) $\quad k_0(t, \pm\infty) = \pm |4\beta_1|^{-1/2} \{\beta_0 / [\exp(2\beta_0 t) - 1]\}^{1/2}$.

As a consequence, the reflection coefficient, $R(k, t)$, see (10), loses the property to vanish as $k \to \infty$, as soon as t exceeds zero; and this indicates that $u(x, t)$ develops a singularity.

That such a phenomenon could happen to solutions of the nonlinear partial differential equation (81) is not surprising, since the term with the differentiation of highest order in (81) has the coefficient $\alpha_1 + \beta_1 x$, and this vanishes at the real value $x_s = -\alpha_1 / \beta_1$. And in fact even the linearized equation that obtains from (81) by neglecting all nonlinear terms has the same problem. This equation is more simply written for the dependent variable u rather than w (see (59)), and it reads

(100) $\quad u_t = \alpha_0 u_x + \alpha_1 u_{xxx} + \beta_0(xu_x + 2u) + \beta_1(xu_{xxx} + 4u_{xx})$.

It can therefore be identified with (1.1.-20) with the definitions (see (21))

(101a) $\quad \omega(z) = -z(\alpha_0 - \alpha_1 z^2) + 2i(\beta_0 - 2\beta_1 z^2)$,

(101b) $\quad \omega'(z) = -z(\beta_0 - \beta_1 z^2)$.

The solution (1.1.-22) is then (formally) applicable, with

$$(102) \quad \chi(t, k_0) = k_0 \exp(\beta_0 t) \{1 + \beta_1 k_0^2 [\exp(2\beta_0 t) - 1]/\beta_0\}^{-1/2}$$

and

$$(103) \quad k_0(t, k) = k \exp(-\beta_0 t) \{1 - \beta_1 k^2 [1 - \exp(-2\beta_0 t)]/\beta_0\}^{-1/2}.$$

Thus the same difficulty arises; while $k_0(t, k)$ is well-defined and real for $t \geq 0$ and for all real k (we are assuming $\beta_1 < 0$), it loses as soon as t exceeds zero the property to diverge when k diverges. As a consequence the Fourier transform $\hat{u}(k, t)$ given by (1.1.-22), whose modulus now reads

$$(104) \quad |\hat{u}(k, t)| = [k/k_0(t, k)] |\hat{u}_0[k_0(t, k)]|,$$

does not vanish any more as $k \to \infty$, but in fact diverges (to obtain this equation, we have assumed that α_0 and α_1 are real, so that they only contribute to the phase of $\hat{u}(k, t)$; see (1.1.-22) and (101a)). This of course implies that the derivation of (1.1.-21) from (1.1.-20) is itself questionable; therefore any conclusion based on (1.1.-22) is also open to doubt. Nevertheless this treatment, that could be made rigorous by working with distributions rather than functions, indicates that also the solutions of the linearized equation (100) develop a singularity (obviously at $x = x_s$) as soon as the time evolution begins.

Thus, it is not surprising at all that the solutions of the nonlinear evolution equation (81) (that linearizes to (100) in the weak field limit) also develop a singularity as soon as the time evolution begins (while the nature of this singularity can presumably be investigated using the spectral transform technique, this is a nontrivial task that has not yet been accomplished; see appendix A.16). What is remarkable is the existence of a whole class of nonsingular solutions, namely the pure multisoliton solutions; note however that their existence is not in contradiction to the singularity presumably characterizing *all* solutions of the linearized equation (100), in view of their essentially nonlinear nature.

It is clear that the behaviours that we have displayed by the analysis of the two examples that have been treated in this section are not peculiar to them, but are rather typical for the whole class of nonlinear evolution equations (7), with $\beta(z, t)$ belonging respectively to the classes (39) (first example) or (34) (second example) (in both cases with M finite). A natural question that comes then to mind is, whether there exist evolution equations of the class (7) which are good, namely such that the Cauchy problem yields solutions that remain for all time in the class of *bona fide* potentials, for any initial datum in that class. On the basis of the evidence discussed so far, it is clear that the answer to this question is positive; in fact, it is easy to characterize explicitly the properties of the functions $\alpha(z,t)$ and $\beta(z,t)$ in (7) that are sufficient to guarantee that (7) be OK, in the sense specified above.

First of all $\alpha(z,t)$ and $\beta(z,t)$ should be (for all values of t) holomorphic in z in the neighborhood of all real values of z; see the discussion following (8) above, and appendix A.18.

Secondly, the function $\beta(z,t)$ should be such to induce (via (12), (13) and (14)) a bijective mapping between k and $k_0(t,k)$ that preserves the interval from $-\infty$ to $+\infty$, namely such that, as k varies from $-\infty$ to $+\infty$, $k_0(t,k)$ varies uniformly in the same interval (for all t; and, to be on the safe side, let us add the specific requirement

$$(105) \quad \lim_{k\to\pm\infty} [k_0(t,k)/k] = a_\pm(t), \quad 0 < a_\pm(t) < \infty.)$$

It is easily seen that it is sufficient, in order that this be the case, that the function $\beta(z,t)$, for all values of t, be holomorphic for all *real* values of z (already required above), and moreover also at $z=\infty$.

Let us conclude this section by a terse display of the results for such an equation, that is perhaps not quite the simplest that one could write, but that has nevertheless the advantage to yield formulae more readily expressed in completely explicit form. It corresponds to the choice (see (4))

$$(106) \quad g(z,t) = z^2 + 16, \quad h(z,t) = 0, \quad f(z,t) = 2z,$$

namely (see (6))

$$(107) \quad \alpha(z,t) = 0, \quad \beta(z,t) = -2z/(z^2+16).$$

Note that, as usual, we have omitted any explicit time dependence; and we have moreover introduced some numerical constants whose values have been chosen to simplify the formulae written below (arbitrary constants may be easily reintroduced by rescaling the dependent and independent variables). The corresponding (integrodifferential) equation (7) can be written, via (59) and the position

$$(108) \quad v(x,t) = \int_x^{+\infty} dy\, u^2(y,t),$$

as the following system of coupled nonlinear PDEs:

(109a) $v_x = -w_x^2$,

(109b) $(w_{xxxxx} + 16w_x - 4v_{xxx} - 2w_{xx}^2 + 16w_x^3/3)_t + 2(w_{xxxx} - 3v_{xx})w_t$
$- 2w_{xx}v_t + 2[xw_{xxx} + 3w_{xx} + 2w_xw - 3(xv_x + v)]_x = 0.$

The results read:

(110) $\chi(t, k_0) = 2^{-1/2}k_0^{-1}\{[(k_0^4 - tk_0^2 - 1)^2 + 4k_0^4]^{1/2} + k_0^4 - tk_0^2 - 1\}^{1/2}$,

(111) $k_0(t, k) = \chi(-t, k)$,

(112) $p(t) = -i\chi(t, ip_0)$,

(113) $\rho(t) = \rho_0[p(t)/p_0]\{1 + tp_0^2[2(p_0^4 - 1) + tp_0^2]/(p_0^4 + 1)\}^{-1/2}$,

(114) $\theta(t, t_0; \lambda_0) = \chi(t - t_0, k_0)/k_0$, $\lambda_0 \equiv -4k_0^2$.

It is readily seen that $p(t)$ and $\rho(t)$ are nonsingular for all (real) values of t, and moreover that, again for all (real) values of t, the mapping between the real variables k and k_0 is bijective and has the properties

(115) $\lim_{k \to 0} [k_0(t, k)/k] = 1$,

(116) $\lim_{k \to \pm\infty} [k_0(t, k)/k] = 1$.

Thus, via (10), the reflection coefficient $R(k, t)$ is OK for all time if $R_0(k) = R(k, 0)$ was OK to begin with. It is therefore confirmed that, for any initial datum in the class of *bona fide* potentials, the nonlinear evolution equation (109) can be solved via the spectral transform technique and the solution $u(x, t)$, see (59) and (108), always remains in the class of *bona fide* potentials.

As for the behaviour of the solitons, let us merely note that (112) implies

(117) $\lim_{t \to \pm\infty} [p(t)/|t|^{\pm 1/2}] = 1$,

while from (112), (113) and (20) there obtains

(118) $\xi(t) \approx \frac{1}{2}|t|^{\mp 1/2}\{-\ln|t| + 2p_0\xi(0) + \frac{1}{2}\ln[(1 + p_0^4)/p_0^4]\}$ as $t \to \pm\infty$.

A more detailed analysis of the solutions is left as an exercise for the diligent reader.

6.3. Solutions of the KdV equation that are asymptotically linear in x

We noted already in section 3.2.4.2 that the KdV equation,

(1) $\quad u_t + u_{xxx} - 6uu_x = 0, \quad u \equiv u(x,t),$

possesses the very simple solution (3.2.4.2.-1), that we report here in the simpler form

(2) $\quad \bar{u}(x,t) = -x/6t.$

This form obtains setting to zero the two arbitrary constants x_0 and t_0 in (3.2.4.2-1); since these constants can always be trivially reintroduced by shifting the x and t variables, we refer hereafter, for notational simplicity, to the form (2).

We also noted in section 3.2.4.2 that there exists a whole class of solutions of the KdV equation, having the form

(3) $\quad u(x,t) = \bar{u}(x,t) + v(x,t),$

with $\bar{u}(x,t)$ given by (2) and $v(x,t)$ characterized by the property to vanish asymptotically ($x \to \pm \infty$),

(4) $\quad v(\pm \infty, t) = 0.$

Note that the asymptotic condition (4) is consistent with the time evolution, namely it is automatically satisfied at all future times, $t > t_0$, if it holds at the initial time t_0 (here we are assuming $t_0 > 0$; the validity of (4) for $0 < t < t_0$ is also implied by its validity at $t = t_0$; while the singularity of $\bar{u}(x,t)$ at $t = 0$ does not allow one to make statements about the behaviour of the solution for $t < 0$, if it is assigned at $t = t_0 > 0$). A direct way to prove this assertion is by inserting the ansatz (3) (with (2)) in (1), obtaining thereby the following nonlinear evolution equation for $v(x,t)$:

(5) $\quad v_t + [v_{xx} - 3v^2 + (x/t)v]_x = 0, \quad v \equiv v(x,t)$

(we are, of course, always assuming that v vanishes asymptotically with its derivatives sufficiently fast to overcompensate the growth of x).

The nonlinear equation (5) is itself an interesting evolution equation, due to its relation to the so-called cylindrical KdV (cKdV) equation, namely

(6) $\quad q_\tau + q_{\xi\xi\xi} - 6q_\xi q + (2\tau)^{-1} q = 0, \quad q \equiv q(\xi, \tau).$

The connection is given by the formulae

(7) $\quad v(x,t) = -2\tau q(\xi, \tau),$

(8) $\quad \xi = x/t,$

(9) $\quad \tau = -(2t^2)^{-1}.$

Thus, the solution of the Cauchy problem for the cylindrical KdV equation (6) is, up to changes of dependent and independent variables, equivalent to the solution of the Cauchy problem for the KdV equation (1), but for the class of functions u defined by (3) with (2) and (4). This class is clearly outside the family of *bona fide* potentials, and therefore the spectral transform technique as developed thus far is not applicable to treat such solutions. A modification of this technique, amounting to the introduction of a different spectral transform, is required; this is developed in the following subsection 6.3.1.

The Bäcklund transformations of section 4.1 can be rewritten in local terms (see below), and can be thereby extended outside of the class of functions that qualify as *bona fide* potentials. Indeed a Bäcklund transformation relating two functions $u^{(1)}(x,t)$ and $u^{(2)}(x,t)$ in the neighborhood of an arbitrary point, say x_0, still takes the expression (4.1.-1), provided the integrals from x to $+\infty$ (wherever they appear, see the definitions (4.1.-3) and (4.1.-4)) are replaced by integrals running from x to x_0. The validity of this simple rule is easily proved, for instance using the operator technique of appendix A.20.

By this trick the Bäcklund transformations of chapter 4 become applicable even in the present context. The novelty, however, is that a Bäcklund transformation maps a solution $u^{(2)}(x,t)$ of (1) belonging to the class defined by (3) with (2) and (4), into a solution $u^{(1)}(x,t)$ that is not in this class. In order to exhibit this feature, consider for instance the Bäcklund transformation that obtains from (4.1.-1) with (4.1.-12) (and with the change of the integral operator described above); this reads (see (4.1.-14))

(10) $\quad 2p\left[u^{(1)}(x,t) - u^{(2)}(x,t)\right] + u_x^{(1)}(x,t) + u_x^{(2)}(x,t)$
$\quad\quad + \left[u^{(1)}(x,t) - u^{(2)}(x,t)\right] \int_x^{x_0} dy \left[u^{(1)}(y,t) - u^{(2)}(y,t)\right] = 0,$

where x_0 is an arbitrary (real) constant (independent of x and t), while p is x-independent (but now it depends on t; see (13) below). An equivalent formula reads

(11) $$u^{(1)}(x,t) = u^{(2)}(x,t) - 2\{\ln[f^{(2)}(x,z,t)]\}_{xx},$$

where $f^{(2)}(x,z,t)$ is any solution of the Schroedinger equation

(12) $$f^{(2)}_{xx}(x,z,t) = [u^{(2)}(x,t) + z]f^{(2)}(x,z,t),$$

the parameter z being an arbitrary (real and t-independent) constant; the equivalence of (10) and (11) requires moreover that p and $f^{(2)}$ are related by the formula

(13) $$f^{(2)}_x(x_0,z,t) = pf^{(2)}(x_0,z,t).$$

Inserting now in (12) and (11), in place of $u^{(2)}(x,t)$, the diverging solution $\bar{u}(x,t)$, see (2), of the KdV equation (1), one obtains

(14) $$u^{(1)}(x,t) = -\bar{u}(x,t) - 2z + 2(-6t)^{-2/3} \cdot \frac{[\cos\gamma \text{Ai}'(y) + \sin\gamma \text{Bi}'(y)]^2}{[\cos\gamma \text{Ai}(y) + \sin\gamma \text{Bi}(y)]^2},$$

where γ is an arbitrary constant (i.e. x and t independent), $\text{Ai}(y)$ and $\text{Bi}(y)$ are the Airy functions (see the following subsection 6.3.1), and the prime means differentiation with respect to the variable

(15) $$y = (-6t)^{-1/3}(x - 6zt).$$

This new solution of the KdV equation (1) has infinitely many (double) poles for real values of x, and its asymptotic behaviour is

(16) $$u^{(1)}(x,t) \to \bar{u}(x,t)[1 - (-6tx^{-3})^{1/2}], \quad x \to -\text{sign}(t)\infty,$$

(17) $$u^{(1)}(x,t) \to \bar{u}(x,t)\{1 - 2\sin^{-2}[\tfrac{2}{3}(-y)^{3/2} + \gamma + \tfrac{1}{4}\pi]\}, \quad x \to \text{sign}(t)\infty,$$

this last formula shows that the solution $u^{(1)}(x,t)$ does not belong to the class of functions characterized by the properties (3) and (4). Analogous solutions, affected by such pathologies, can be generated by repeated use of the Bäcklund transformation (11) and (12), with different values of the parameter z. A general formula for such solutions, that bears some analogy to the pure multisoliton solution (1.6.1.-9,10), is given in [BM1979]. Real solutions, that are instead not marred by singularities, and that belong to the class defined by (3) with (2) and (4), can be obtained by the spectral transform approach described in the following subsection, that moreover provides the appropriate technique to solve the Cauchy problem associated with a class of nonlinear evolution equations that includes (5) (and therefore also the cKdV equation (6)) as a special case. This is reported in subsection 6.3.2, while in subsection 6.3.3 the associated (infinite) set of conserved quantities is derived. The last subsection, 6.3.4, is then devoted to a terse discussion of the cKdV equation (6), as well as a somewhat more general evolution equation of which the cKdV equation (6) is a special case.

6.3.1. The Schroedinger spectral problem with an additional linear potential

Consider the Schroedinger equation

(1) $\quad -\psi_{xx}(x,z) + [x+u(x)]\psi(x,z) = z\psi(x,z);$

note that the term linear in x cannot be absorbed in the "potential" $u(x)$, since we assume $u(x)$ to vanish asymptotically,

(2) $\quad u(\pm\infty) = 0.$

More detailed information on the rate of vanishing of u as x diverges is given below.

When $u(x)$ vanishes,

(3) $\quad u(x) = 0,$

(1) becomes the Airy equation (in the independent variable $y = x - z$); thus it has as independent (real) solutions the Airy functions $\text{Ai}(x-z)$ and $\text{Bi}(x-z)$. Let us recall that these functions are characterized by the following formulae:

(4) $\quad \text{Ai}''(y) = y\,\text{Ai}(y),$

(5) $\quad \text{Bi}''(y) = y\,\text{Bi}(y),$

(6) $\quad \text{Ai}'(y)\text{Bi}(y) - \text{Ai}(y)\text{Bi}'(y) = -1/\pi,$

(7a) $\quad \text{Ai}(y) = \pi^{-1}(\tfrac{1}{3}y)^{1/2} K_{1/3}(\eta),$

(7b) $\quad \text{Ai}(y) = \tfrac{1}{3}y^{1/2}[I_{-1/3}(\eta) - I_{1/3}(\eta)],$

(8) $\quad \text{Bi}(y) = (\tfrac{1}{3}y)^{1/2}[I_{-1/3}(\eta) + I_{1/3}(\eta)].$

In the first 3 formulae the primes indicate of course differentiation; in the

6.3.1 Addition of a linear potential

last 3 formulae, and also below,

(9) $\quad \eta \equiv \tfrac{2}{3} y^{3/2},$

while $K_\nu(\eta)$ and $I_\nu(\eta)$ are modified Bessel functions. The asymptotic behaviour of the Airy functions is displayed by the expansions

(10) $\quad \mathrm{Ai}(y) = \tfrac{1}{2} \pi^{-1/2} y^{-1/4} \exp(-\eta) \left[\sum_{m=0}^{M} (-1)^m c_m \eta^{-m} + O(\eta^{-M-1}) \right],$

$$|\arg y| < \pi,$$

(11) $\quad \mathrm{Ai}(-y) = \pi^{-1/2} y^{-1/4}$

$$\cdot \left\{ \sin(\eta + \tfrac{1}{4}\pi) \left[\sum_{m=0}^{M} (-1)^m c_{2m} \eta^{-2m} + O(\eta^{-2M-2}) \right] \right.$$

$$\left. - \cos(\eta + \tfrac{1}{4}\pi) \left[\sum_{m=0}^{M} (-1)^m c_{2m+1} \eta^{-2m-1} + O(\eta^{-2M-3}) \right] \right\},$$

$$|\arg y| < \tfrac{2}{3}\pi,$$

(12) $\quad \mathrm{Bi}(y) = \pi^{-1/2} y^{-1/4} \exp(\eta) \left[\sum_{m=0}^{M} c_m \eta^{-m} + O(\eta^{-M-1}) \right], \quad |\arg y| < \tfrac{1}{3}\pi,$

(13) $\quad \mathrm{Bi}(-y) = \pi^{-1/2} y^{-1/4}$

$$\cdot \left\{ \cos(\eta + \tfrac{1}{4}\pi) \left[\sum_{m=0}^{M} (-1)^m c_{2m} \eta^{-2m} + O(\eta^{-2M-2}) \right] \right.$$

$$\left. + \sin(\eta + \tfrac{1}{4}\pi) \left[\sum_{m=0}^{M} (-1)^m c_{2m+1} \eta^{-2m-1} + O(\eta^{-2M-3}) \right] \right\},$$

$$|\arg y| < \tfrac{2}{3}\pi,$$

the coefficients c_m being defined as follows:

(14) $\quad c_0 = 1,$
(15a) $\quad c_m = \Gamma(3m + \tfrac{1}{2}) / [54^m m! \Gamma(m + \tfrac{1}{2})], \quad m = 1, 2, 3, \ldots,$
(15b) $\quad c_m = (2m+1)(2m+3) \cdots (6m-1) / (216^m m!), \quad m = 1, 2, 3, \ldots.$

We also introduce for convenience the function $\mathrm{Ei}(y)$ (not to be confused with the error integral!), defined by

(16) $\quad \text{Ei}(y) = \text{Bi}(y) - i\text{Ai}(y).$

Note that the formulae given above imply the following properties:

(17) $\quad \text{Ei}'(y)\text{Ai}(y) - \text{Ei}(y)\text{Ai}'(y) = 1/\pi,$

(18) $\quad \text{Ai}(y) \to \tfrac{1}{2}\pi^{-1/2} y^{-1/4} \exp(-\tfrac{2}{3} y^{3/2}) \quad \text{as } y \to +\infty,$

(19) $\quad \text{Ai}(y) \to \pi^{-1/2}(-y)^{-1/4} \sin\!\left[\tfrac{2}{3}(-y)^{3/2} + \tfrac{1}{4}\pi\right] \quad \text{as } y \to -\infty,$

(20) $\quad \text{Ei}(y) \to \pi^{-1/2} y^{-1/4} \exp(\tfrac{2}{3} y^{3/2}) \quad \text{as } y \to +\infty,$

(21) $\quad \text{Ei}(y) \to \pi^{-1/2}(-y)^{-1/4} \exp\!\left\{-i\!\left[\tfrac{2}{3}(-y)^{3/2} + \tfrac{1}{4}\pi\right]\right\} \quad \text{as } y \to -\infty,$

(22) $\quad |\text{Ai}(y)\,\text{Ei}(y)| = O(|y|^{-1/2}), \quad |y| \to \infty, \; -\tfrac{1}{3}\pi \leq \arg y \leq \pi.$

We now identify two solutions of (1), $\Phi(x,z)$ and $F(x,z)$, via the asymptotic conditions

(23) $\quad \lim_{x \to +\infty} [\Phi(x,z)/\text{Ai}(x-z)] = 1,$

(24) $\quad \lim_{x \to -\infty} [F(x,z)/\text{Ei}(x-z)] = 1,$

and we define the (Jost) function $f(z)$ by the formula

(25) $\quad f(z) = \pi[F_x(x,z)\Phi(x,z) - F(x,z)\Phi_x(x,z)].$

Thus, for $u(x) = 0$, $f(z) = 1$ (see (17)).

We assume hereafter, for simplicity, that $u(x)$ is real. It is then easily seen that (25) implies

(26) $\quad \Phi(x,z) = (2i)^{-1}[f(z)F^*(x,z^*) - f^*(z^*)F(x,z)],$

and this, together with (21), yields

(27) $\quad \Phi(x,z) \to (2i)^{-1}\pi^{-1/2}(-x)^{-1/4}\!\left[f(z)\exp\!\left\{i\!\left[\tfrac{2}{3}(z-x)^{3/2} + \tfrac{1}{4}\pi\right]\right\}\right.$
$\left. - f^*(z^*)\exp\!\left\{-i\!\left[\tfrac{2}{3}(z-x)^{3/2} + \tfrac{1}{4}\pi\right]\right\}\right] \quad \text{as } x \to -\infty.$

(Note that, for the sake of generality, and in view of future utilization, we have written these equations for the general case of complex z; to derive them we have used the fact that $F(x,z)$, $F^*(x,z^*)$, $\Phi(x,z)$ and $\Phi^*(x,z^*)$ all satisfy the same equation, (1), and moreover $\Phi(x, z) = \Phi^*(x, z^*)$ due to (23)).

The direct spectral problem is the calculation of $f(z)$, given $u(x)$; clearly the formulae given above, in particular (1), (23), (24) and (25), indicate that it has a unique solution. The inverse spectral problem consists in the calculation of $u(x)$ given $f(z)$, or rather, given $|f(z)|$, because, as we show below, knowledge of the modulus of the Jost function $f(z)$ is sufficient to determine $u(x)$ uniquely. Of course, in order that a *bona fide* $u(x)$ correspond to a given $|f(z)|$, this function must have the properties that characterize the modulus of a Jost function $f(z)$ produced by the solution of the direct spectral problem described above. Thus, before proceeding with the solution of the inverse spectral problem, let us report here some other properties obtained in the framework of the direct spectral problem.

The *integral equation*

$$(28) \quad \Phi(x, z) = \text{Ai}(x-z) + \pi \int_x^\infty dy \left[\text{Ai}(x-z)\text{Bi}(y-z) \right.$$
$$\left. - \text{Ai}(y-z)\text{Bi}(x-z) \right] u(y) \Phi(y, z),$$

corresponds to (1), (4), (5), (6) and (23). It yields, through (25), the *integral representation*

$$(29) \quad f(z) = 1 + \pi \int_{-\infty}^{+\infty} dx \, \text{Ei}(x-z) \Phi(x, z) u(x).$$

It is now easy to identify, from this equation, the requirement on the asymptotic rate of vanishing of $u(x)$ that characterizes the class of potentials under consideration. Since (20), (21), (23), and (27) imply

(30a) $\quad |\text{Ei}(x-z)\Phi(x, z)| = O(x^{-1/2}) \quad$ as $x \to +\infty$,

(30b) $\quad |\text{Ei}(x-z)\Phi(x, z)| \to (2\pi)^{-1}|f(z)||x|^{-1/2}\exp(2|x|^{1/2}\text{Im } z)$

as $x \to -\infty$,

the condition that the integral in the r.h.s. of (29) converge (for real z) corresponds to the requirement that $u(x)$ vanish asymptotically faster than $|x|^{-1/2}$, or more precisely, to the condition

$$(31) \quad \int_{-\infty}^{+\infty} dx \, |u(x)|/(1+|x|)^{1/2} < \infty.$$

Potentials that violate this condition marginally are also considered below.

Properties of the Jost function $f(z)$. If (31) holds, $f(z)$ exists for real z, being defined either by (25) or by (29); it can moreover be analytically continued, off the real axis, into the half-plane $\operatorname{Im} z \leq 0$, and in this half-plane it is holomorphic and without zeros. The first property is a consequence of (28), (29) and (30b); the second property is obvious for $\operatorname{Im} z = 0$ (see (23), (24), and (25)) and, for $\operatorname{Im} z < 0$, can be proved per absurdum: assume $f(\bar{z}) = 0$ with $\operatorname{Im} \bar{z} < 0$, integrate from x to ∞ the derivative of the wronskian of $\Phi(x, \bar{z})$ and $\Phi^*(x, \bar{z})$, use (1), take the limit as $x \to -\infty$ (using (27)); and in this manner obtain

$$(32) \qquad 0 = \operatorname{Im} \bar{z} \int_{-\infty}^{+\infty} dx |\Phi(x, \bar{z})|^2, \quad (f(\bar{z}) = 0, \quad \operatorname{Im} \bar{z} < 0)$$

(note that, for $f(\bar{z}) = 0$ and $\operatorname{Im} \bar{z} < 0$, the integral in the r.h.s. of this equation is in fact convergent).

Another important characteristic of $f(z)$ is its *asymptotic behaviour* as $z \to -\infty$. It can be shown that it is given by the following expansion,

$$(33) \qquad f(z) = 1 + \sum_{n=1}^{N} f_n (-4z)^{-n/2} + O(z^{-(N+1)/2}),$$

with the coefficients f_n expressed in terms of the potential $u(x)$ by the formulae

$$(34) \qquad f_{2m+1} = \int_{-\infty}^{+\infty} dx\, u(x) \left[v_{2m+1}(x) + \sum_{j=1}^{m} M^{m-j} N v_{2j-1}(x) \right],$$

$$m = 0, 1, 2, \ldots,$$

$$(35) \qquad f_{2m} = \int_{-\infty}^{+\infty} dx\, u(x) M^{m-1} \int_{x}^{+\infty} dy\, u(y), \quad m = 1, 2, 3, \ldots.$$

In these equations M and N are two integrodifferential operators, defined by the following formulae that specify their action on a generic function $g(x)$ (vanishing as $x \to +\infty$, in the case of M):

$$(36) \qquad M \cdot g(x) = g_{xx}(x) - 4x g(x) - 2 \int_{x}^{+\infty} dy\, g(y) + N \cdot g(x),$$

$$(37) \qquad N \cdot g(x) = -u(x) g(x) + \int_{x}^{+\infty} dy\, u(y) g_y(y)$$

$$+ \int_{x}^{+\infty} dy\, u(y) \int_{y}^{+\infty} dz\, u(z) g(z);$$

while the functions $v_{2m+1}(x)$ are polynomials of degree m containing $1+[[\frac{1}{3}m]]$ terms ($[[q]]\equiv$ integral part of q),

$$\tag{38} v_{2m+1}(x) = \sum_{j=0}^{[[m/3]]} v_{mj} x^{m-3j},$$

$$\tag{39} v_{mj} = (-1)^{m+j} 2^{2(m-j)} 3^{2j} \Gamma(m+\tfrac{1}{2})\left[(m-3j)!\,\Gamma(3j+\tfrac{1}{2})\right]^{-1}$$
$$\cdot \sum_{k=0}^{2j} (-1)^k c_k c_{2j-k}.$$

The numerical coefficients c_m in the last formula are those defined above, see (14) and (15).

The explicit expressions of the first 4 polynomials v_{2m+1} are:

$$\tag{40} v_1 = 1, \quad v_3 = -2x, \quad v_5 = 6x^2, \quad v_7 = -20x^3 + 10.$$

The explicit expressions of the first 4 coefficients f_m are:

$$\tag{41a} f_1 = \int_{-\infty}^{+\infty} dx\, u(x),$$

$$\tag{41b} f_2 = \tfrac{1}{2} f_1^2,$$

$$\tag{41c} f_3 = \tfrac{1}{6} f_1^3 - \int_{-\infty}^{+\infty} dx\, u(x)[2x + u(x)],$$

$$\tag{41d} f_4 = f_1 f_3 - \tfrac{1}{8} f_1^4.$$

Clearly it appears that every coefficient of even order can be expressed in terms of coefficients f_m of lower (odd) order (thus the apparently neater structure of (35) as compared to (34) is somewhat deceptive).

These formulae for the asymptotic behaviour as $z \to -\infty$ of $f(z)$ will be used, in section 6.3.3, to study the conservation laws associated with the nonlinear evolution equation (6.3.2.-1).

Let us proceed now to describe the *inverse spectral problem*; again we report here the results, referring for their derivation to the literature (see Notes).

Given $|f(z)|$, construct the "spectral function" $\rho(z)$ by the simple formula

$$\tag{42} \rho(z) = |f(z)|^{-2} - 1;$$

introduce next the kernel $M(x, y)$ by the definition

(43) $\quad M(x, y) = \int_{-\infty}^{+\infty} dz\, \rho(z) \mathrm{Ai}(x-z) \mathrm{Ai}(y-z);$

set up finally the "Gel'fand–Levitan equation"

(44) $\quad K(x, y) + M(x, y) + \int_{x}^{\infty} dz\, K(x, z) M(z, y) = 0, \quad x < y.$

This is a Fredholm integral equation for the, a priori unknown, function $K(x, y)$; note that the integral equation refers to the dependence of this function on its second argument, while the dependence of $K(x, y)$ on the first variable, x, results, as it were, parametrically, from the dependence of $M(x, y)$ on x and from the presence of x as lower limit of integration. This Fredholm equation identifies the function $K(x, y)$ uniquely; and this function determines $u(x)$ through the formulae

(45) $\quad w(x) = 2 \lim_{\varepsilon \to 0} K(x, x + |\varepsilon|),$

(46a) $\quad w(x) = \int_{x}^{\infty} dy\, u(y),$

(46b) $\quad u(x) = -w_x(x).$

This completes the description of the procedure to solve the inverse spectral problem, namely to reconstruct $u(x)$ from $f(z)|$. Let us also report here three other important formulae involving the spectral function $\rho(z)$ and the kernel $K(x, y)$, together with the "regular solution", $\Phi(x, z)$, of the Schroedinger equation (1) (see (23)):

(47) $\quad \int_{-\infty}^{+\infty} dz\, [1 + \rho(z)] \Phi(x, z) \Phi(y, z) = \delta(x - y),$

(48) $\quad \Phi(x, z) = \mathrm{Ai}(x - z) + \int_{x}^{\infty} dy\, K(x, y) \mathrm{Ai}(y - z),$

(49) $\quad K(x, y) = -\int_{-\infty}^{+\infty} dz\, \rho(z) \Phi(x, z) \mathrm{Ai}(y - z).$

Let us pause and summarize our findings. We are now in the position to identify the spectral transform of a function $u(x)$ (regular for all x and vanishing at infinity; see (31)) as the function $|f(z)|$, or equivalently $\rho(z)$ (see (42)). The *direct spectral problem*, namely the calculation of $|f(z)|$ given $u(x)$, is characterized by (1), (23), (24) and (25) (or, equivalently, by (28) and (29)); the *inverse spectral problem*, namely the calculation of $u(x)$ given $|f(z)|$, proceeds through (42), (43), (44), (45) and (46). Moreover, the coefficients of the asymptotic expansion of $f(z)$ as $z \to -\infty$, see (33), are directly related to integrals over all space of nonlinear combinations of $u(x)$ and its derivatives, and polynomials in x (see (34), (35) and (41)).

As is clear from the treatment given thus far, and as is also obvious on "physical" grounds, the Schroedinger operator

(50) $\qquad -\partial^2/\partial x^2 + x + u(x)$

does not possess discrete eigenvalues. Thus, when the spectral transform introduced here is used to solve nonlinear evolution equations (see next section), no solitons will emerge. There is, however, a potential that may be considered to constitute, in some sense, the analog of the potential possessing, in the more standard Schroedinger spectral problem (without the additional linear potential), just one discrete eigenvalue and a vanishing reflection coefficient. This potential, however, violates marginally the condition (31); and indeed the corresponding Jost function $f(z)$ is quite peculiar. It results from

(51) $\qquad \rho(z) = \bar{\rho}\,\delta(z - \bar{z}),$

with $\bar{\rho}$ and $\bar{z}$ two real constants. Insertion of this formula in (43) yields

(52) $\qquad M(x, y) = \bar{\rho}\,\text{Ai}(x - \bar{z})\,\text{Ai}(y - \bar{z}),$

and with this kernel the Fredholm equation (44) becomes separable and can therefore be easily solved. There obtains the potential

(53) $\qquad u(x) = \bar{u}(x - \bar{z}, \bar{\rho})$

with

(54) $\qquad \bar{u}(y, \bar{\rho}) = 2\bar{\rho}\bigl[2\,\text{Ai}'(y)\,\text{Ai}(y) + \bar{\rho}[\text{Ai}(y)]^4 G(\bar{\rho}, y)\bigr] G(\bar{\rho}, y),$

(55) $\qquad G(\bar{\rho}, y) = \bigl\{1 + \bar{\rho}[\text{Ai}'(y)]^2 - \bar{\rho}y[\text{Ai}(y)]^2\bigr\}^{-1}.$

In the last two formulae the primes indicate of course differentiation.

The function $\bar{u}(y)$ is (real and) regular for $-\infty < x < \infty$, and has moreover the following properties:

(56a) $\quad \bar{u}(y,\bar{\rho}) = -(\bar{\rho}/\pi)\exp(-\frac{4}{3}y^{3/2})[1+O(y^{-3/2})], \quad y \to +\infty,$

(56b) $\quad \bar{u}(y,\bar{\rho}) = -2(-y)^{-1/2}\cos[\frac{4}{3}(-y)^{3/2}][1+O(|y|^{-3/2})], \quad y \to -\infty,$

(57) $\quad \int_{-\infty}^{+\infty} dy\, \bar{u}(y,\bar{\rho}) = 0.$

(Note however that the integral in the last formula does not converge absolutely).

A graph of $\bar{u}(y)$ for $\bar{\rho}=1$ is displayed in the following figure (computer-produced by C. Poppe):

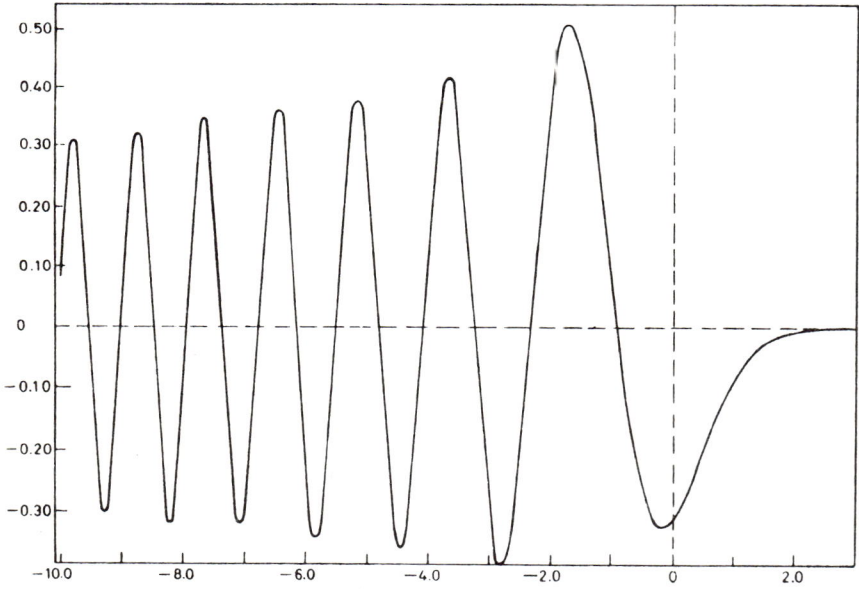

The corresponding expressions of $K(x,y)$ and $\Phi(x,y)$ read

(58) $\quad K(x,y) = -\bar{\rho}G(\bar{\rho}, x-\bar{z})\operatorname{Ai}(x-\bar{z})\operatorname{Ai}(y-\bar{z}),$

(59) $\quad \Phi(x,z) = G(\bar{\rho}, x-\bar{z})\Big\{ \operatorname{Ai}(x-\bar{z}) + \bar{\rho}\int_x^\infty dy\, \operatorname{Ai}(y-\bar{z})$

$\qquad \cdot [\operatorname{Ai}(y-\bar{z})\operatorname{Ai}(x-z) - \operatorname{Ai}(x-\bar{z})\operatorname{Ai}(y-z)]\Big\},$

6.3.1 Addition of a linear potential

implying

(60) $\quad f(z) = 1 \quad \text{for } z \neq \bar{z}, \quad f(\bar{z}) = 0;$

note the consistency of this last formula with (51) and (42).

It is of course easy to construct analogous potentials corresponding to a spectral function $\rho(z)$ that is the sum of a finite number of terms such as (51); these potentials always satisfy (57), and their asymptotic behaviour is directly given by a sum of terms such as (56). Clearly they provide the analogs of the pure multisoliton solutions, in the same sense as the function (53) provides the analog of the single-soliton solution (see following section).

Finally let us turn to the wronskian relations that can be established in the framework of the spectral problem treated in this section, and that will provide the starting point to relate the spectral transform introduced here to certain nonlinear evolution equations (see next section). We report here only those wronskian relations that are needed for these developments.

The standard wronskian relation yields

(61) $\quad f^{(1)}(z) - f^{(2)}(z) = \pi \int_{-\infty}^{+\infty} dx \, F^{(1)}(x,z) \, \Phi^{(2)}(x,z)$
$$\cdot \left[u^{(1)}(x) - u^{(2)}(x) \right].$$

Of course an analogous relation holds with the superscripts 1 and 2 exchanged.

Two other relations, referring now to a single potential, read:

(62) $\quad f_z(z) = \pi \int_{-\infty}^{+\infty} dx \, F(x,z) \, \Phi(x,z) u_x(x),$

(63) $\quad -4zf_z(z) = \pi \int_{-\infty}^{+\infty} dx \, F(x,z) \, \Phi(x,z) L_1 \cdot u_x(x),$

with the integrodifferential operator L_1 defined by the formula

(64) $\quad L_1 \cdot g(x) = g_{xx}(x) - 4[x + u(x)] g(x) + 2[1 + u_x(x)] \int_x^{\infty} dy \, g(y),$

that specifies its action on a generic function $g(x)$ (vanishing as $x \to +\infty$). The interested reader will find the proofs of these formulae in the literature (see Notes).

6.3.2. Solution of a nonlinear evolution equation including as a special case the cylindrical KdV equation

The nonlinear evolution equation whose solution is discussed in this section reads

(1) $$u_t = \alpha_0(t) u_x + \alpha_1(t) [u_{xxx} - 6u_x u - 4xu_x - 2u],$$
$$u \equiv u(x, t), \quad -\infty < x < \infty.$$

Let us note that, in the special case

(2) $$\alpha_0(t) = 0, \quad \alpha_1(t) = -(12t)^{-1},$$

it yields the cylindrical KdV equation

(3) $$q_t + q_{yyy} - 6q_y q + (2t)^{-1} q = 0, \quad q \equiv q(y, t), \quad -\infty < y < \infty,$$

via the change of variables

(4) $$q(y, t) = (12t)^{-2/3} u(x, t), \quad y = (12t)^{1/3} x$$

(see subsection 6.3.4). Note that throughout this subsection, and the following two as well, we restrict attention, whenever considering the cKdV (3), to positive values of t, $t > 0$ (the extension of the results to negative values of t can be trivially effected noting the invariance of (3) under the transformation $t \to -t$, $y \to -y$).

Let us now introduce the Jost function $f(z, t)$ associated with $u(x, t)$ via the spectral problem treated in the preceding section, 6.3.1. The wronskian relations written at the end of that section, (6.3.1.-61, 62, 63), yield now the formulae

(5) $$f_t(z, t) = \pi \int_{-\infty}^{+\infty} dx\, F(x, z, t) \Phi(x, z, t) u_t(x, t),$$

(6) $$f_z(z, t) = \pi \int_{-\infty}^{+\infty} dx\, F(x, z, t) \Phi(x, z, t) u_x(x, t),$$

(7) $$-4z f_z(z, t) = \pi \int_{-\infty}^{+\infty} dx\, F(x, z, t) \Phi(x, z, t) [u_{xxx}(x, t)$$
$$- 6u_x(x, t) u(x, t) - 4xu_x(x, t) - 2u(x, t)];$$

and these formulae imply that, if $u(x, t)$ evolves in time according to (1), $f(z, t)$ evolves according to the linear partial differential equation

(8) $$f_t(z, t) = [\alpha_0(t) - 4z\alpha_1(t)] f_z(z, t).$$

6.3.2 An equation including as a special case the cKdV equation

The Cauchy problem for this last equation is explicitly solved by the formula

(9) $$f(z,t) = f_0 \left(z \exp\left[-4 \int_{t_0}^t dt' \, \alpha_1(t') \right] + \int_{t_0}^t dt' \, \alpha_0(t') \exp\left[-4 \int_{t_0}^{t'} dt'' \, \alpha_1(t'') \right] \right),$$

where of course $f_0(z)$ is the (initial) datum at $t = t_0$,

(10) $$f_0(z) \equiv f(z, t_0).$$

The Cauchy problem for (1), with (initial) datum $u_0(x) = u(x, t_0)$, is therefore also solved: via the direct spectral transform, $u_0(x)$ yields $f_0(z)$; from $f_0(z)$, $f(z, t)$ is obtained via (9); and finally from $f(z, t)$ (or rather from its modulus), $u(x, t)$ is retrieved by the inverse spectral transformation, as explained in the preceding section.

Note that, in the special case of the cKdV equation, (9) takes the simple form

(11) $$f(z, t) = f_0 \left[z(t/t_0)^{1/3} \right].$$

Also note that the relation (6.3.1.-42), together with (9), implies that the spectral function $\rho(z, t)$ associated with $u(x, t)$ evolves according to the explicit formula

(12) $$\rho(z, t) = \rho_0 \left(z \exp\left[-4 \int_{t_0}^t dt' \, \alpha_1(t') \right] + \int_{t_0}^t dt' \, \alpha_0(t') \exp\left[-4 \int_{t_0}^{t'} dt'' \, \alpha_1(t'') \right] \right),$$

where of course

(13) $$\rho_0(z) \equiv \rho(z, t_0);$$

and again in the special case of the cKdV equation, (12) takes the simple form

(14) $$\rho(z, t) = \rho_0 \left[z(t/t_0)^{1/3} \right].$$

We end this section by noting that the results of the previous section (see in particular (6.3.1.-51)) imply that a special solution of (1) is given by the formula

(15) $\quad u(x,t)=\bar{u}[x-\bar{z}(t),\bar{\rho}(t)]$,

where $\bar{u}(y,\bar{\rho})$ is defined by (6.3.1.-54) and (6.3.1.-55) with

(16) $\quad \bar{z}(t)=\bar{z}_0\exp\left[4\int_{t_0}^t dt'\,\alpha_1(t')\right]-\int_{t_0}^t dt'\,\alpha_0(t')\exp\left[4\int_{t'}^t dt''\,\alpha_1(t'')\right]$,

(17) $\quad \bar{\rho}(t)=\bar{\rho}_0\exp\left[4\int_{t_0}^t dt'\,\alpha_1(t')\right]$,

$\bar{z}_0$ and $\bar{\rho}_0$ being two arbitrary (real) constants. In the special case of the cKdV equation (16) and (17) simplify:

(18) $\quad \bar{z}(t)=\bar{z}_0(t/t_0)^{-1/3}$,

(19) $\quad \bar{\rho}(t)=\bar{\rho}_0(t/t_0)^{-1/3}$.

Clearly the solution (15) has some claim to be considered the analog, for the nonlinear evolution equation (1) (and, after the appropriate changes of variables, also for the cKdV equation (3)), of the single-soliton solution that is possessed by any equation of the class, say, (3.1.-28) (and in particular by the KdV equation (3.1.-33)). However neither is it a solitary wave (its shape changes through the time evolution), nor is it localized, since it has a wiggling tail extending to the left whose envelop vanishes, as $x\to-\infty$, proportionally only to $|x|^{-1/2}$ (see (6.3.1.-56b), as well as the graph displayed in the preceding section).

It is also clear how the analogs of multisoliton solutions can be explicitly exhibited (see preceding section).

Let us end this section by emphasizing that the results we have found so far indicate that, in contrast to what might have been naively expected, the long-time behaviour of solutions of the cKdV equation (3) does not reproduce the behaviour of solutions of the KdV equation (that is for instance characterized, for a large class of initial data, by the presence of solitons of the ordinary, form-preserving and uniformly moving, type). Additional support for this finding is provided by the results of next section.

6.3.3. Conservation laws

In this section we discuss the conservation laws associated with the nonlinear evolution equation (6.3.2.-1), as well as with the cKdV equation (6.3.2.-3).

6.3.3 Conservation laws

The derivation is actually quite straightforward, being based on the asymptotic expansion as $z \to -\infty$ of the Jost function $f(z,t)$ associated with $u(x,t)$. This expansion reads (see (6.3.1.-33)):

(1) $$f(z,t) = 1 + \sum_{n=1}^{N} f_n(t)(-4z)^{-n/2} + O(z^{-(N+1)/2}).$$

Now inserting this formula in (6.3.2.-9) one immediately gets the formula

(2) $$f_n(t) = \sum_{m=0}^{[[(n-1)/2]]} f_{n-2m}(t_0) \exp\left[2(n-2m)\int_{t_0}^{t} dt'\, \alpha_1(t')\right]$$
$$\cdot [A_0(t_0,t)]^m (n-2)!! / [2^m m! (n-2m-2)!!].$$

In writing this equation we have used the symbol $[[\tfrac{1}{2}(n-1)]]$ for the integral part of $\tfrac{1}{2}(n-1)$, and the usual convention $(-1)!! = (0)!! = 1$; while the function $A_0(t_0,t)$ is defined by

(3) $$A_0(t_0,t) = 4\int_{t_0}^{t} dt'\, \alpha_0(t') \exp\left[4\int_{t'}^{t} dt''\, \alpha_1(t'')\right].$$

The formula (2) is, of course, particularly simple if, in (6.3.2.-1), $\alpha_0 = 0$; since in that case it reads simply

(4) $$f_n(t) = f_n(t_0) \exp\left[2n \int_{t_0}^{t} dt'\, \alpha_1(t')\right].$$

This is indeed the case for the cKdV equation, when (4) becomes simply

(5) $$f_n(t) = f_n(t_0)(t/t_0)^{-n/6}.$$

The formula (2) (and, when applicable, (4) or (5)) displays explicitly the time-evolution of the coefficients $f_n(t)$ of the asymptotic ($z \to -\infty$) expansion of $f(z,t)$; on the other hand these coefficients are also explicitly given in terms of $u(x,t)$ by the formulae given in section 6.3.1 (see in particular (6.3.1.-34, 35) and (6.3.1.-41)). Thus (2) provides the explicit time evolution of an endless sequence of integrals, extended over all space, of appropriate polynomials of x, u and the x-derivatives of u. This provides, of course, useful information on the time evolution of $u(x,t)$; and clearly to get this information it is sufficient to use only the f_n's with n odd, since the f_n's with n even are merely combinations of these.

On the other hand the formulae (2), (4) and (5) remain evidently valid if the rôles of t and t_0 are exchanged; and they then provide an endless sequence of explicit constants of the motion associated with the flow (6.3.2.-1) (or, in the case of (5), with the cKdV equation (6.3.2.-3); after the appropriate changes of variables have been performed); again, however, only the $f_n(t_0)$'s with odd n need be considered. Thus the constants of the motion associated with (6.3.2.-1) can be written as

(6) $\quad V_n = f_{2n+1}(t_0), \quad n = 0, 1, 2, \ldots,$

(with the understanding that $f_{2n+1}(t_0)$ be given by (2), with t and t_0 exchanged; and the quantities $f_{2m+1}(t)$ appearing, after such an exchange, in the r.h.s. of (2), be expressed in terms of $u(x, t)$ via (6.3.1.-34)). For instance the first 2 of these conserved quantities have the explicit expressions

(7a) $\quad V_0 = \exp\left[-2\int_{t_0}^{t} dt'\, \alpha_1(t')\right] \int_{-\infty}^{+\infty} dx\, u(x, t),$

(7b) $\quad V_1 = \tfrac{1}{6} V_0^3 - 2V_0 \int_{t_0}^{t} dt'\, \alpha_0(t') \exp\left[4\int_{t'}^{t_0} dt''\, \alpha_1(t'')\right]$

$\quad\quad - \exp\left[-6\int_{t_0}^{t} dt'\, \alpha_1(t')\right] \int_{-\infty}^{+\infty} dx\, u(x, t)[2x + u(x, t)].$

In the special case of the cKdV equation (6.3.2.-3) the corresponding first two constants of motion read

(8a) $\quad q_0 = t^{1/2} \int_{-\infty}^{+\infty} dy\, q(y, t),$

(8b) $\quad q_1 = t^{1/2} \int_{-\infty}^{+\infty} dy\, q(y, t)[y + 6tq(y, t)].$

In addition to the infinite sequence of conserved quantities that we have just described, the cKdV equation possesses, as can be readily verified by using (6.3.2.-3), one additional constant of motion (as the KdV equation; see (5.-58)), that reads

(9) $\quad q = t \int_{-\infty}^{+\infty} dy\, q^2(y, t).$

Using this last formula, in conjunction with (8), one easily obtains that the centre-of-mass,

(10) $\quad Y(t) \equiv \int_{-\infty}^{+\infty} dy\, y q(y, t) \Big/ \int_{-\infty}^{+\infty} dy\, q(y, t),$

of the generic solution $q(y, t)$ of the cKdV equation (6.3.2.-3) moves

6.3.3 Conservation laws

according to the simple law

(11) $\quad Y(t) = Y_0 + Y_1 t^{1/2}$

with Y_0 and Y_1 constant,

(12a) $\quad Y_0 = q_1/q_0.$
(12b) $\quad Y_1 = 6q/q_0.$

A comparison of (11) with the analogous formula, (5.-62), valid for the ordinary KdV equation, underscores once more the difference between the behaviours of solutions of the ordinary and cylindrical KdV equations; note that the difference between (11) and (5.-62) becomes particularly dramatic as $t \to \infty$, namely precisely when, on the basis of a superficial inspection of the difference between ordinary and cylindrical KdVs, one might have expected an identical behaviour.

Before closing this section we give below an alternative derivation of the conserved quantities (6) and (9), that does not rely on the spectral problem treated in section 6.3.1., but instead exploits directly the connection with (linearly diverging solutions of) the KdV equation. In fact the starting point here is the remark that if $u(x,t)$ satisfies (6.3.2.-1), then

(13) $\quad \tilde{u}(x,t) = a(t)x + b(t) + [a(t)]^{2/3} u\left([a(t)]^{1/3} x, t\right)$

satisfies the evolution equation

(14) $\quad \tilde{u}_t = \tilde{a}_0(t)\tilde{u}_x + \tilde{a}_1(t)[\tilde{u}_{xxx} - 6\tilde{u}_x \tilde{u}], \quad \tilde{u} \equiv \tilde{u}(x,t),$

with

(15a) $\quad a(t) = \exp\left[-6\int_{t_0}^{t} dt'\, \alpha_1(t')\right],$

(15b) $\quad b(t) = \int_{t_0}^{t} dt'\, \alpha_0(t') \exp\left[-4\int_{t_0}^{t'} dt''\, \alpha_1(t'')\right],$

(16a) $\quad \tilde{a}_0(t) = \alpha_0(t)[a(t)]^{-1/3} + 6\alpha_1(t)b(t)/a(t),$
(16b) $\quad \tilde{a}_1(t) = \alpha_1(t)/a(t),$

where $\alpha_0(t)$ and $\alpha_1(t)$ are of course the two coefficients characterizing the evolution equation (6.3.2.-1). It is thus seen that the evolution equation (6.3.2.-1) can be transformed into the equation (14), that is merely a special case (namely, with $\alpha(z,t) = \tilde{a}_0(t) + \tilde{a}_1(t)z$) of (5.-1), whose constants of the motion C_m's have been shown in chapter 5 to be integrals over the whole x-axis of polynomial combinations of the solution of (5.-1) and its x-derivatives (see, e.g., the expression (1.7.3.-12) of the first three constants of the motion).

The problem here is that the solution $\tilde{u}(x,t)$ of (14) diverges (linearly) in x as $x \to \pm\infty$, since the solution $u(x,t)$ of (6.3.2.-1) is assumed to vanish asymptotically, i.e.

(17) $\quad u(\pm\infty,t)=0$

(see (13)); therefore the integral expressions for the constants of the motion C_m associated with (14) all diverge for the (asymptotically diverging) solution (13). This difficulty can be easily by-passed by noticing that

(18) $\quad u^{(a)}(x,t)=a(t)x+b(t)$

is itself a solution of (14), with (15) and (16). This suggests a simple way to regularize the diverging integral defining C_m, by subtracting the diverging contribution due to the solution (18). To work this out we consider first the local conservation law

(19) $\quad \tilde{g}_t^{(m)}(x,t) = \tilde{\gamma}_x^{(m)}(x,t), \quad m=0,1,2,\ldots,$

that obtains by writing down the local conservation law (5.-25) with $u(x,t)$ replaced by the solution (13), $\tilde{u}(x,t)$, of (14). This can be certainly done since both $g^{(m)}$ and $\gamma^{(m)}$ are local quantities (namely, they are expressed in terms of the solution of (5.-1), and of its derivatives, evaluated at x and t only; see (5.-27), (3.1.-32) and appendix A.9, and (5.-29)). Consider next the same conservation law (5.-25), associated again with (14), but for the solution (18), namely, with obvious notation,

(20) $\quad g_t^{(a)(m)}(x,t) = \gamma_x^{(a)(m)}(x,t), \quad m=0,1,2,\ldots.$

Subtracting now (20) from (19) yields the conservation law

(21) $\quad \rho_t^{(m)}(x,t) = J_x^{(m)}(x,t), \quad m=0,1,2,\ldots,$

where

(22a) $\quad \rho^{(m)}(x,t) \equiv \tilde{g}^{(m)}(x,t) - g^{(a)(m)}(x,t),$

(22b) $\quad J^{(m)}(x,t) \equiv \tilde{\gamma}^{(m)}(x,t) - \gamma^{(a)(m)}(x,t).$

It is evident that the "density" and the "current" in (21) satisfy the asymptotic conditions

(23) $\quad \rho^{(m)}(\pm\infty, t) = 0, \quad J^{(m)}(\pm\infty, t) = 0,$

that guarantee the time-independence of the quantities

(24) $\quad \Gamma_m = (-1)^m (2m+1)^{-1} \int_{-\infty}^{+\infty} dx\, \rho^{(m)}(x, t), \quad m = 0, 1, 2, \ldots\,.$

This set of conserved quantities is equivalent to the sequence (6) obtained above, and, for instance, the first two constants of the motion (24) read

(25a) $\quad \Gamma_0 = \exp\left[-2\int_{t_0}^{t} dt'\, a_1(t')\right] \int_{-\infty}^{+\infty} dx\, u(x, t),$

(25b) $\quad \Gamma_1 = \exp\left[-6\int_{t_0}^{t} dt'\, a_1(t')\right] \int_{-\infty}^{+\infty} dx\, \{u(x,t)[2x + u(x,t)]\}$
$\quad\quad + 2\int_{t_0}^{t} dt'\, a_0(t') \exp\left[4\int_{t'}^{t_0} dt''\, a_1(t'')\right] \Gamma_0;$

their relation to (7) is

(26a) $\quad \Gamma_0 = V_0,$
(26b) $\quad \Gamma_1 = -V_1 + \tfrac{1}{6} V_0^3.$

The expressions (25) are easily evaluated by using (24), (22a), (A.9.-100), (13), (18) and (15).

In the same way one may deal with the conserved quantity that is related to the motion of the center of mass associated with the solution of (6.3.2.-1). In this case one starts from the local conservation law (5.-73) (with $a_0(t)$ and $a_1(t)$ replaced by $\tilde{a}_0(t)$ and $\tilde{a}_1(t)$, and $a_m(t) = 0$ for $m > 1$); then, by applying the subtraction technique (an exercise that is left to the reader), one ends up with the additional constant of the motion

(27) $\quad \Gamma = \int_{-\infty}^{+\infty} dx\, xu(x,t) + \Gamma_0 \int_{t_0}^{t} dt'\, \tilde{a}_0(t') - 3\Gamma_1 \int_{t_0}^{t} dt'\, \tilde{a}_1(t'),$

where $\tilde{a}_0(t)$ and $\tilde{a}_1(t)$ are given in terms of $a_0(t)$ and $a_1(t)$ by (16) and (15). This expression finally implies that the center of mass, defined in terms of the solution $u(x,t)$ of (6.3.2.-1) by the formula

(28) $$X(t) = \int_{-\infty}^{+\infty} dx\, xu(x,t) \bigg/ \int_{-\infty}^{+\infty} dx\, u(x,t),$$

moves according to the explicit equation

(29) $$X(t) = \left[X_0 - \int_{t_0}^{t} dt'\, \tilde{\alpha}_0(t') + F \int_{t_0}^{t} dt'\, \tilde{\alpha}_1(t') \right] \exp\left[-2 \int_{t_0}^{t} dt'\, \alpha_1(t') \right],$$

where we have set

(30a) $X_0 = \Gamma/\Gamma_0$,
(30b) $F = 3\Gamma_1/\Gamma_0$

(see also (16) and (15)). As already emphasized, these formulae hold for a generic solution $u(x,t)$ of (6.3.2.-1) that vanishes (sufficiently fast) as $x \to \pm\infty$ (see (17)). Of course, in the particular case of the cKdV equation (6.3.2.-3), (29) reproduces (11) (via (6.3.2.-2,4)), with the relationships

(31a) $Y_0 = (\tfrac{1}{6}F)(12t_0)^{1/3}$,
(31b) $Y_1 = (12)^{1/3} t_0^{-1/6}(X_0 - \tfrac{1}{6}F)$.

6.3.4. The cylindrical KdV equation

Most of the results on the (so-called) cylindrical KdV equation have already been reported in the preceding subsections; here we review them tersely. As it was already noted, the cylindrical KdV equation,

(1) $$q_t + q_{yyy} - 6q_y q + (2t)^{-1} q = 0, \quad q \equiv q(y,t),$$

is obtained from (6.3.2.-1),

(2) $$u_t = \alpha_0(t) u_x + \alpha_1(t)[u_{xxx} - 6u_x u - 4xu_x - 2u], \quad u \equiv u(x,t),$$

by setting

(3) $\alpha_0(t) = 0, \quad \alpha_1(t) = -(12t)^{-1}$,
(4) $q(y,t) = (12t)^{-2/3} u(x,t), \quad y = (12t)^{1/3} x$.

The solvability of (2) via the spectral transform introduced in subsection 6.3.1 implies therefore that the cKdV equation (1) can also be solved. Let us emphasize that one is referring here to the Cauchy problem, in which q is assigned at some given positive time t_0,

(5) $\qquad q(y, t_0) = q_0(y),$

and is to be determined for all positive time t (note that (1) is singular at $t=0$; the converse problem, to determine $q(x, t)$ for all *negative* time t from $q(x, t_0)$ assigned at a given *negative* time t_0 is trivially related to that discussed here). Attention is moreover focussed on the class of asymptotically vanishing solutions,

(6) $\qquad q(\pm\infty, t) = 0$

(we refer to the literature, see Notes, for a discussion of the required rate of vanishing; that this may be quite slow, at least at one end, is implied by some of the examples given above, see (6.3.2.-15, 18, 19)). The results described above and below imply of course that it is sufficient that (6) hold at the "initial" (positive) time t_0, in order that it hold for all (positive) time.

In fact a more general class of nonlinear evolution equations, that includes (1) as a special case, can be similarly treated. This class reads

(7) $\qquad q_t = a q_{yyy} + A q_y + B q_y q + (C_t/C) y q_y + (D_y/D) q, \quad q \equiv q(y, t).$

Here

(8) $\qquad A \equiv A(t), \quad B \equiv B(t), \quad C \equiv C(t), \quad D \equiv D(t)$

are arbitrary functions of time, while a is related to these functions by the formula

(9) $\qquad a(t) = -B(t) D(t) E(t) / 6 C^2(t),$

where

(10) $\qquad E(t) \equiv \int_0^t dt'\, B(t') C(t') D(t').$

The cKdV equation is easily seen to correspond to

(11) $\qquad A = 0, \quad B = 6, \quad C = 12 c^2, \quad D = c t^{-1/2},$

c being an arbitrary constant.

The solvability of (7) via the spectral transform introduced in 6.3.1 is indeed guaranteed by its relation to (2) via the following change of dependent and independent variables:

(12) $\quad q(y,t) = \beta(t)u(x,t),$

(13) $\quad y = \gamma(t)x,$

(14) $\quad \beta(t) = D(t)[E(t)]^{-1/3},$

(15) $\quad \gamma(t) = [E(t)]^{2/3}/C(t),$

(16) $\quad \alpha_0(t) = A(t)C(t)[E(t)]^{-2/3},$

(17) $\quad \alpha_1(t) = -B(t)C(t)D(t)/6E(t) = -\frac{1}{6}[\ln E(t)]_t.$

Again, in connection with (7), the (Cauchy) problem under consideration is the determination of $q(y,t)$ from given $q(y,t_0)$, see (5), in the class of asymptotically vanishing functions, see (6).

It is plain that all the results derived in the preceding two subsections can be rewritten for the evolution equation (7) using the transformation formulae (12–17). For instance the "single-soliton" solution easily obtains by transforming the expression (6.3.2.-15) (its specialization to the cKdV (1) has been already given there; see (6.3.2.-18, 19)). In addition, an infinite sequence of conserved quantities associated with the flow (7) (and in particular, with the cKdV (1)) can be easily obtained from those derived in 6.3.3 for the evolution equation (2) (see (6.3.3.-6) or, equivalently, (6.3.3.-25)). It is easily seen that these conserved quantities are independent from each other, and have the form of integrals over all space of nonlinear combinations of q and its derivatives, with explicitly y- and t-dependent coefficients. The first two of this sequence (obtained from (6.3.3.-25)) read

(18a) $\quad Q_0 = [E(t_0)]^{-1/3}[C(t)/D(t)]\int_{-\infty}^{+\infty} dy\, q(y,t),$

(18b) $\quad Q_1 = \{C(t)/[E(t_0)D(t)]\}\left\{2\left[\int_{t_0}^{t} dt'\, A(t')C(t')\right]\int_{-\infty}^{+\infty} dy\, q(y,t)\right.$

$\quad\quad \left. + \int_{-\infty}^{+\infty} dy\, q(y,t)\{2yC(t) + [E(t)/D(t)]q(y,t)\}\right\}.$

These expressions refer to solutions of the evolution equation (7); for the (special case of (7) corresponding to the) cKdV equation (1), the explicit form of the corresponding constants of motion has been displayed above, see (6.3.3.-8).

6.3.4 The cylindrical KdV equation

Another result that obtains directly from the preceding subsection is the formula detailing the motion of the center of mass,

(19) $$Y(t) = \int_{-\infty}^{+\infty} dy\, y q(y,t) \bigg/ \int_{-\infty}^{+\infty} dy\, q(y,t),$$

associated with a generic solution of (7); it reads

(20) $$Y(t) = [E(t)/C(t)]\overline{Y}$$
$$- [C(t)]^{-1} \int_{t_0}^{t} dt'\, C(t')[A(t') + GB(t')D(t')].$$

Here the two t-independent quantities $\overline{Y}$ and G are defined by

(21a) $$\overline{Y} = [E(t_0)]^{-1/3} Q/Q_0,$$

(21b) $$G = Q_1 / \{2 Q_0 [E(t_0)]^{1/3}\},$$

where Q_0 and Q_1 are given by (18) and Q is the additional conserved quantity,

(22) $$Q = \{C(t)/[E(t_0) D(t)]\} \left\{ \left[\int_{t_0}^{t} dt'\, A(t') C(t') \right] \int_{-\infty}^{+\infty} dy\, q(y,t) \right.$$
$$+ C(t) \int_{-\infty}^{+\infty} dy\, y q(y,t)$$
$$\left. + \left\{ \left[\int_{t_0}^{t} dt'\, B(t') C(t') D(t') \right] / [2 D(t)] \right\} \int_{-\infty}^{+\infty} dy\, [q(y,t)]^2 \right\},$$

corresponding to (6.3.3.-27). In the case of the cKdV equation, (20) reduces of course to (6.3.3.-11) with

(23a) $$Y_0 = 12 c G t_0^{1/2},$$
(23b) $$Y_1 = 12 c (\overline{Y} - G),$$

c being the arbitrary constant introduced by (11).

As for the relation of the cKdV equation to the KdV equation (6.3.-1), we have already displayed the change of variables (6.3.-7,8,9,3,2) that transforms one into the other. However, in the context of the application of the spectral transform method, this transformation is hardly of use since it maps a solution $q(y,t)$ of the cKdV equation satisfying (6) into a solution of the KdV equation that is out of the class of *bona fide* potentials. More generally, in fact, we have seen that a solution of the more general equation (2) may be transformed into the solution (6.3.3.-13) of the evolution equation (6.3.3.-14), that belongs to the class of evolution equations that has been extensively investigated in the previous chapters (and it is also clear that it is similarly possible to map, via (12–17), (7) into (6.3.3.-14)); but again, this transformation takes a solution $u(x,t)$ of (2), that vanishes as $x \to \pm \infty$, into a solution of (6.3.3.-14) that linearly diverges at infinity. On the other hand, it is worth noting that, if the numerical coefficient 2 in the last term in the r.h.s. of (2) were replaced by 8 (without changing the other coefficients, in particular the 4 in the next-to-last term), then (2) would belong to the class (6.2.-52), and would therefore be reducible by the transformations

(24a) $\quad \tilde{u}(x,t) = [\gamma(t)]^2 u(\gamma(t)x, t),$

(24b) $\quad \tilde{\alpha}_0(t) = \alpha_0(t)/\gamma(t), \qquad \tilde{\alpha}_1(t) = \alpha_1(t)/[\gamma(t)]^3,$

(24c) $\quad \gamma(t) = \exp\left[4 \int_{t_0}^{t} dt' \, \alpha_1(t')\right],$

to the usual equation (6.3.3.-14) (see also the formulae following (6.2.-52)). The difference with respect to the other transformation (6.3.3.-13) is of course that here, if $u(x,t)$ is a *bona fide* potential, so is $\tilde{u}(x,t)$ given by (24a).

Let us end this subsection by reemphasizing that the possibility to solve by the spectral transform technique the cKdV equation (1) indicates that the scope of this technique extends beyond what one might a priori have expected. For instance, the cKdV equation (1) (or, for this matter, the equation (2) as well as (7)) does not possess solitary wave solutions; nor are there solitons (unless one is prepared to extend somewhat the notion of what a soliton is, to include the solution

(25) $\quad \bar{q}(y,t) = (12t)^{-2/3} \bar{u}\left[(12t)^{-1/3} y - \bar{z}(t), \bar{p}(t)\right]$

with (6.3.2.-18, 19) and (6.3.1.-54, 55)). In this connection an interesting open problem is the status of the so-called *spherical KdV equation*,

(26) $\quad q_t + q_{yyy} - 6q_y q + t^{-1} q = 0, \quad q \equiv q(y,t),$

for which no technique of exact solution is so far known.

6.4. Solution of the KdV equation with one real double pole

It is trivial to verify that the Korteweg–de Vries (KdV) equation

(1) $\quad u_t + u_{xxx} - 6u_x u = 0, \quad u \equiv u(x,t)$

admits the (time-independent) solution

(2) $\quad \bar{u} = 2(x-\bar{x})^{-2},$

with $\bar{x}$ an (arbitrary) constant. Of course $\bar{x}$ must be real in order that $\bar{u}$ be real (for real x); but then $\bar{u}$ has a double pole at $x=\bar{x}$, and therefore it does not belong to the class of *bona fide* potentials. This solution is the simplest instance of the rational solutions of the KdV equation that are tersely treated in section 3.2.4.1.

More generally, one can show that

(3) $\quad u^{(n)} = n(n+1)(x-\bar{x})^{-2}, \quad n=1,2,3,\ldots.$

is a solution of the evolution equation

(4) $\quad u_t = \sum_{m=n}^{M} \alpha_m(t) L^m u_x, \quad u \equiv u(x,t), \, M \geq n,$

where $\alpha_m(t)$ is an arbitrary function of time and L is the usual integrodifferential operator (1.2.-2). Thus (2) is actually a solution for the whole class

(5) $\quad u_t = \alpha(L,t) u_x, \quad u \equiv u(x,t),$

of evolution equations, with the single restriction that $\alpha(z,t)$ be a polynomial in z that vanishes at the origin,

(6) $\quad \alpha(z,t) = \sum_{m=1}^{M} \alpha_m(t) z^m, \quad M \geq 1,$

since (2) corresponds to (3) with $n=1$.

In this section we confine our interest to study a class of solutions of the KdV equation (1) that contains a double pole, leaving the extension of this investigation to the whole class (5) with (6) as an exercise for the interested reader. Of course, a solution with a double pole satisfies (1) everywhere except at the pole, and is associated with a "natural" prescription to connect the solutions on the two sides of the pole (see below). The treatment shall be quite terse; only the more interesting results will be presented, without any proof. The interested reader may seek more details in the literature (see Notes).

Incidentally, it is easy to verify quite generally that, if $u(x,t)$ satisfies (1) and has, as a function of x, a polar singularity, then in the neighborhood of that singularity its Laurent expansion reads

$$(7) \qquad u(x,t) = 2(x-\xi)^{-2} + v_0 + \sum_{m=2}^{\infty} v_m (x-\xi)^m,$$

where of course the quantities ξ and v_m, $m = 0, 2, 3, 4, \ldots$, are generally time-dependent. The fact that the sum in the r.h.s. of this equation starts from $m=2$, as well as the lack of the simple pole term, result from the requirement that the Laurent expansion be compatible with the time evolution described by (1). Of course this formula is required to hold *almost always*; at some specific value of t it could be violated, consistently with the possibility in principle to chose arbitrarily the form of the solution at any specific time (as an initial condition for the Cauchy problem). Indeed an explicit instance of such a violation is given below.

An approach to the problem at hand is suggested by the treatment of section 6.3; (see (6.3.-3)). The analogous ansatz

$$(8) \qquad u(x,t) = \bar{u}(x) + q(x,t),$$

with $\bar{u}$ given by (2) and the requirement that $u(x,t)$ satisfy (1), yields for $q(x,t)$ the evolution equation

$$(9) \qquad q_t + q_{xxx} - 6 q_x q - 12 \left[(x-\bar{x})^{-2} q \right]_x = 0, \quad q \equiv q(x,t).$$

As a consequence of the fact that $\bar{u}(x)$ in (8) is itself a solution of (1), this equation has the nice property to admit $q=0$ as a (trivial!) solution. On the other hand, it does not display conveniently the property of the class of solutions under consideration, to have just one double pole at $x = \xi(t)$ (see (7)). In fact, comparing (8) with (7), one sees that the solution q of (9) must have two double poles, one at $x = \bar{x}$ to cancel out the double pole of $\bar{u}$ (see (2) and (8)), and one at $x = \xi$, consistently with (7). The point here is that the pole of $u(x,t)$ does in general move with time (except, of course, for the particular solution (2)), so that a more convenient ansatz for u than (8) reads

$$(10) \qquad u(x,t) = 2 [x-\xi(t)]^{-2} + v(x,t),$$

where $v(x, t)$ is a regular function (that may belong to the class of *bona fide* potentials), satisfying moreover the condition

(11) $\quad v_x[\zeta(t), t] = 0,$

as implied by (7). Insertion of (10) in (1) yields two coupled evolution equations for the pole position, $\zeta(t)$, and for the function $v(x, t)$:

(12) $\quad \dot{\zeta}(t) = -6v[\zeta(t), t],$

(13) $\quad v_t(x, t) + v_{xxx}(x, t) - 6v_x(x, t) v(x, t)$

$\qquad - 12 \{ [x - \zeta(t)]^{-2} \{ v(x, t) - v[\zeta(t), t] \} \}_x = 0,$

whose consistency with (11), and the regularity of $v(x, t)$, is easily verified. Of course in these equations the superimposed dot denotes time-differentiation.

The splitting (8) of $u(x, t)$, with (2), is on the other hand convenient to set up the appropriate spectral transform technique, to solve the evolution equations (12) and (13) (or, equivalently, (9); or (1) within the class of solutions characterized by (7)). This is based on the direct and inverse spectral problem associated with the ODE

(14) $\quad -\psi_{xx} + [2(x - \bar{x})^{-2} + q(x, t)] \psi = k^2 \psi, \quad \psi \equiv \psi(x, k; t),$

where of course the variable t plays the rôle of an external parameter. Although the treatment of the spectral transform method in this case follows closely that given in chapter 2 (and appendix A.5), some care must be exercised because of the presence of the double pole at $x = \zeta$ (see (8), (2), (10) and (7)), this point being a singular (but regular) point for the ODE (14). A "natural" way to extend the solution ψ of (14) across the double pole comes from the requirement that ψ be monodromic (in x) in a neighborhood of that pole. Indeed, it is remarkable that this requirement on ψ leads precisely to the same expansion (7) that was obtained above from the time-evolution described by (1).

Thus one may conclude that a formalism based on the spectral transform technique exists to solve the evolution equations (12) and (13), this being equivalent to solving the KdV equation (1) for the class (10) (with (11)) of solutions. Of course the initial data in the Cauchy problem include both the position, say $\zeta(0) \equiv \zeta_0$, of the pole and the function $v(x, 0) \equiv v_0(x)$, and must be consistent with (11) (the time evolutions (12) and (13) then guarantee automatically the validity of (11) for all time). Moreover, the technique of solution via the spectral transform yields all the additional results that characterize the nonlinear evolution equations discussed in this book (and in particular the KdV equation (1)): Bäcklund transformations, conservation laws, explicit solutions, etc. We refer for all these results (as well as the description of the appropriate spectral transform to solve (12) and (13)) to the literature (see Notes). Here we report only the explicit expression of the "N-soliton + 1-pole" solution of the KdV equation (1), that reads

(15) $\quad u(x,t) = -2\{\ln(x \det[I + J(x,t)])\}_{xx}$,

with I the unit matrix of order N and $J(x,t)$ the matrix of order N with matrix elements

(16) $\quad J_{mn}(x,t) = c_m(t) c_n(t) \left[(p_m + p_n)^{-1} + (p_m p_n x)^{-1} \right]$

$\qquad \cdot \exp[-(p_m + p_n)x]$.

Here the N positive quantities p_m are time-independent, while the N positive quantities $c_m(t)$ evolves according to the simple formula

(17) $\quad c_m(t) = c_m(0) \exp(4 p_m^3 t), \quad m = 1, 2, \ldots, N$.

Note that (15) provides a $2N$-parameter family of solutions, since the N quantities p_m and the N quantities $c_m(0)$ can be assigned arbitrarily (except for the requirement that they be all positive). Note also that this solution has been obtained setting $\bar{x} = 0$ in (14) for simplicity.

The formula (15) provides of course also a solution of (12) and (13), via (10); but an explicit expression of $\xi(t)$ cannot be exhibited.

For $N = 1$ the ("1-pole + 1-soliton") solution (15) takes the explicit form

(18) $\quad u(x,t) = -2p^2 \left[\cosh^{-2}\{p[x - \xi(t)]\} \right.$

$\qquad \left. - (1+px)^{-2} \operatorname{tgh}^2\{p[x - \xi(t)]\} \right]$

$\qquad \cdot \left[1 - (1+px)^{-1} \operatorname{tgh}\{p[x - \xi(t)]\} \right]^{-2}$,

where

(19) $\quad p = p_1$,

(20) $\quad \xi(t) = (2p)^{-1} \ln[c_1^2(t) / 2p]$.

6.4 Solutions with one real double pole

The first of these quantities, p, is of course time-independent, while the second evolves linearly,

(21) $\quad \xi(t) = \xi(0) + 4p^2 t,$

as implied by (17), (19) and (20).

The solution (18) describes the collision between a pole and a soliton. This collision process can be described as follows. The soliton moves asymptotically, as $t \to \pm\infty$, with the constant speed $v = 4p^2$, and it has the standard sech² shape; thus, as $t \to \pm\infty$, the solution factors into the sum of a standard sech² bump and a double pole (that deforms the soliton through the addition of its x^{-2} tail); of course the soliton can be precisely localized only in the asymptotic limit, as $t \to \pm\infty$ (see below). As for the double pole, located at the (real) position $\mathcal{S}(t)$, it moves, as the time spans the interval from $-\infty$ to $+\infty$, over the *finite* interval

(22a) $\quad \mathcal{S}(-\infty) > \mathcal{S}(t) > \mathcal{S}(+\infty),$
(22b) $\quad \mathcal{S}(-\infty) = 0, \qquad \mathcal{S}(+\infty) = -2/p,$

with the negative (time-dependent) velocity

(23) $\quad \dot{\mathcal{S}}(t) = -4p^2 \sinh^{-2}\{p[\mathcal{S}(t) - \xi(t)]\}.$

These results easily follow from the equation defining the zero of the denominator in the expression (18), namely

(24) $\quad p\mathcal{S}(t) = \tgh\{p[\mathcal{S}(t) - \xi(t)]\} - 1.$

Note that the soliton goes through the pole. The collision is most "violent" at the time

(25) $\quad \bar{t} = -[\xi(0) + 1/p]/4p^2,$

when the pole crosses the middle point of the interval (22),

(26) $\quad \mathcal{S}(\bar{t}) = \xi(\bar{t}) = -1/p,$

with infinite speed, as shown by the behaviour

(27) $\quad \mathcal{S}(t) \approx [12(t - \bar{t})]^{1/3}, \quad \text{as } t \to \bar{t}.$

Note that, at this particular time, $t = \bar{t}$, the solution $u(x, \bar{t})$ has a double pole at $x = -1/p$ with the "wrong" coefficient

(28) $$\lim_{x \to -1/p} \left[(x + 1/p)^2 u(x, \bar{t}) \right] = 6;$$

in this case the ordinary analysis of the solution of (1) at fixed time breaks

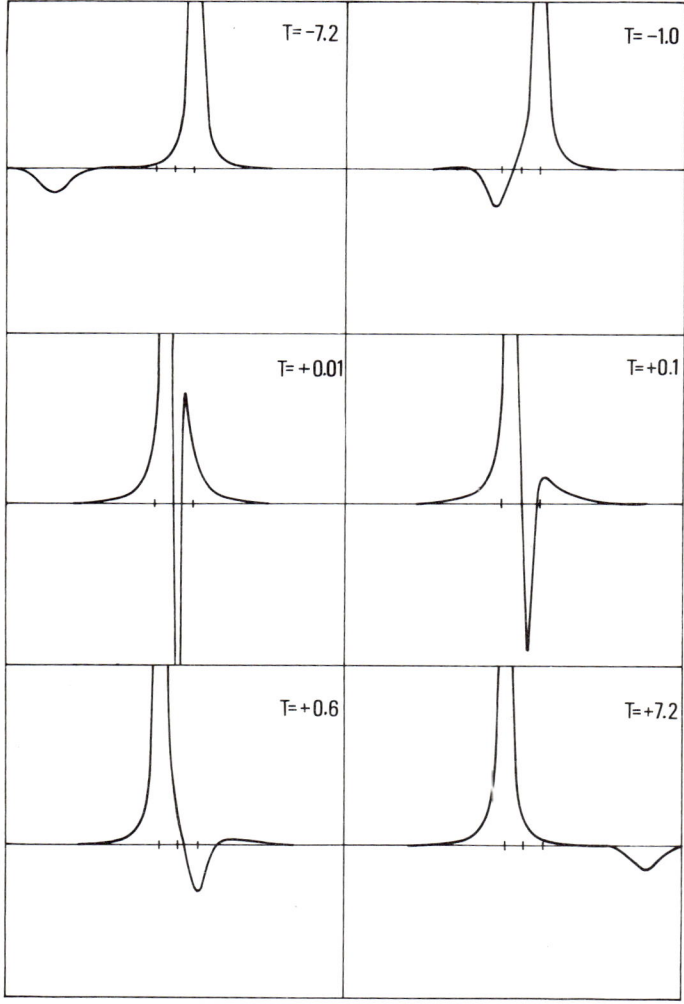

Figure 1. These plots show the x-dependence of the 1-pole-1-soliton solution (18) with $p = 0.6$ and $\bar{t} = 0$, at six different values of time. The pole and the soliton "cross each other" at the middle point of the interval shown, whose length is $2/p$ (see (22)).

down because of the infinite velocity of the pole itself. Moreover the value of the solution (18) at $x=-1/p$ and at $t=\bar{t}$ is not well defined as $u(x,t)$ can take any real value there, depending on the direction taken in the (x,t) plane to approach the singular point $x=-1/p$, $t=\bar{t}$.

An explicit example of this time evolution is displayed in figure 1.

6.5. Evolution equations associated with the spectral problem based on the ODE $-\psi_{xx}(x)+u(x)\psi(x)=k^2\rho^2(x)\psi(x)$

In this section we discuss the class of evolution equations associated with the spectral problem based on the second-order linear ODE

$$\text{(1)} \qquad \psi_{xx}(x,k) = \{u(x) - k^2[\rho(x)]^2\}\psi(x,k).$$

For $\rho(x)=1$, this reduces of course to the standard Schroedinger equation. If $\rho(x)\neq 1$, (1) can be transformed back to the standard Schroedinger form,

$$\text{(2)} \qquad \psi'_{x'x'}(x',k) = [u'(x') - k^2]\psi'(x',k),$$

by the change of variables

$$\text{(3)} \qquad x' = x + \int_x^{+\infty} dy[1-\rho(y)],$$

$$\text{(4)} \qquad u'(x') = [\rho(x)]^{-2}\Big[u(x) - [\rho(x)]^{1/2}\{[\rho(x)]^{-1/2}\}_{xx}\Big],$$

$$\text{(5)} \qquad \psi'(x',k) = [\rho(x)]^{1/2}\psi(x,k).$$

Thus all the results of this section could be obtained from the results given elsewhere in this book, by an appropriate change of variables. We deem nevertheless the techniques and results reported below to be of sufficient interest to justify this presentation, that shall however be quite terse.

Throughout this section we assume $u(x)$, $1-\rho(x)$ and $1-[\rho(x)]^{-1}$ to belong to the class of *bona fide* potentials; we are thus, in particular, assuming $\rho(x)$ not to vanish for real x and to tend to unity (sufficiently fast!) as $x \to \pm\infty$:

(6a) $\qquad \lim_{x\to\pm\infty} [\rho(x)] = 1,$

(6b) $\qquad \rho(x) \neq 0.$

The reflection and transmission coefficients for the spectral problems (1) and (2) are defined by the standard formulae:

(7a) $\quad \psi(x,k) \to T(k)\exp(-ikx), \quad x \to -\infty$

(7b) $\quad \psi(x,k) \to \exp(-ikx) + R(k)\exp(ikx), \quad x \to +\infty;$

(8a) $\quad \psi'(x,k) \to T'(k)\exp(-ikx), \quad x \to -\infty$

(8b) $\quad \psi'(x,k) \to \exp(-ikx) + R'(k)\exp(ikx), \quad x \to +\infty.$

Thus (see (3), (5) and (6)) they are related by the formulae

(9a) $\quad T'(k) = T(k)\exp\left\{ ik\int_{-\infty}^{+\infty} dx[1-\rho(x)] \right\},$

(9b) $\quad R'(k) = R(k).$

For simplicity we omit hereafter to consider the discrete eigenvalues, leaving their treatment as a (rather trivial) exercise for the diligent reader.

The basic formula of wronskian type (for infinitesimal variations only) that provides a convenient starting point for our treatment based on (1) reads

(10) $\quad 2ikg(-4k^2)\,\delta R(k) = \int_{-\infty}^{+\infty} dx\,[\psi(x,k)]^2 g(\mathcal{L})\{\delta u(x)$
$\qquad\qquad + \tfrac{1}{4}\mathcal{L}\delta[\rho(x)]^2\},$

with the integrodifferential operator $\mathcal{L}$ defined by the formula

(11) $\quad \mathcal{L} = -\boldsymbol{LDrIr}.$

Here we have introduced the convenient notation

(12) $\quad r(x) \equiv 1/\rho(x),$

and the symbol r indicates the corresponding multiplicative operator,

(13) $\quad rf(x) \equiv r(x)f(x),$

while the operators $\boldsymbol{L}, \boldsymbol{D}$ and $\boldsymbol{I}$ are defined as in appendix A.9:

(14) $\quad \boldsymbol{D}f(x) \equiv f_x(x),$

(15) $\quad \boldsymbol{I}f(x) \equiv \int_x^{+\infty} dy\, f(y),$

(16a) $\quad \boldsymbol{L}f(x) \equiv f_{xx}(x) - 4u(x)f(x) + 2u_x(x)\int_x^{+\infty} dy\, f(y),$

(16b) $\quad \boldsymbol{L} = \boldsymbol{D}^2 - 2u + 2\boldsymbol{D}u\boldsymbol{I}.$

6.5 Evolution equations associated with the spectral problem

Clearly the integrodifferential operator $\mathcal{L}$ is characterized by the property

(17) $\quad \int_{-\infty}^{+\infty} dx\, [\psi(x,k)]^2 \mathcal{L} f(x) = (2ik)^2 \int_{-\infty}^{+\infty} dx\, [\psi(x,k)]^2 f(x),$

and it reduces to the operator L if $\rho(x) = r(x) = 1$.

Let us now assume that u and ρ, and therefore R, T and ψ, are time-dependent. Then, as a special case of (10), we write

(18) $\quad 2ikg(-4k^2, t)R_t(k, t) = \int_{-\infty}^{+\infty} dx\, [\psi(x,k,t)]^2 g(\mathcal{L},t)\{u_t(x,t)$

$+ \tfrac{1}{2}\mathcal{L}[\rho_t(x,t)\rho(x,t)]\}.$

Moreover, by considering the special cases of (10) corresponding to infinitesimal translations and to infinitesimal scale transformations (see section 2.1.1), we also get

(19) $\quad (2ik)^2 g(-4k^2, t)R(k,t) = \int_{-\infty}^{+\infty} dx\, [\psi(x,k,t)]^2 g(\mathcal{L},t)$

$\cdot \{u_x(x,t) + \tfrac{1}{2}\mathcal{L}[\rho_x(x,t)\rho(x,t)]\},$

(20) $\quad -2ik^2 g(-4k^2, t)R_k(k,t) = \int_{-\infty}^{+\infty} dx\, [\psi(x,k,t)]^2 g(\mathcal{L},t)$

$\cdot \{xu_x(x,t) + 2u(x,t) + \tfrac{1}{2}\mathcal{L}[x\rho_x(x,t)\rho(x,t)]\}.$

These 3 formulae are the starting point to identify and discuss evolution equations, for they clearly imply that, if $u(x,t)$ and $\rho(x,t)$ evolve according to the nonlinear equation

(21) $\quad g(\mathcal{L},t)[u_t + \tfrac{1}{2}\mathcal{L}(\rho_t\rho)] + h(\mathcal{L},t)[u_x + \tfrac{1}{2}\mathcal{L}(\rho_x\rho)]$

$+ f(\mathcal{L},t)[xu_x + 2u + \tfrac{1}{2}\mathcal{L}(x\rho_x\rho)] = 0,$

the quantity $R(k,t)$ evolves according to the *linear* equation

(22) $\quad g(-4k^2, t)R_t + (2ik)h(-4k^2, t)R - kf(-4k^2, t)R_k = 0.$

However for simplicity in the following we concentrate mainly on some simpler instances of the equations contained in the class (21), corresponding to $g(z,t) = 1$ and $f(z,t) = 0$. We also set, for consistency with previous notation, $h(z,t) = -\alpha(z,t)$, so that (21) specializes into

(23) $\quad u_t + \tfrac{1}{4}\mathcal{L}(\rho^2)_t = \alpha(\mathcal{L},t)[u_x + \tfrac{1}{4}\mathcal{L}(\rho^2)_x],$

$u \equiv u(x,t), \quad \rho \equiv \rho(x,t),$

and correspondingly

(24a) $\quad R_t = 2ik\alpha(-4k^2, t)R, \quad R \equiv R(k, t),$

implying

(24b) $\quad R(k, t) = R(k, 0) \exp\left[2ik \int_0^t dt' \alpha(-4k^2, t')\right].$

Extension of the discussion given below to the more general class (21) (rather than (23)) is an instructive exercise, that may be usefully undertaken by the reader who has absorbed the material presented in this section (above and below) and in section 6.2 (and, if need be, in appendix A.18).

Let us emphasize that the evolution equation (23) involves *two* fields, $u(x, t)$ and $\rho(x, t)$, while (24) involves only the reflection coefficient $R(k, t)$; and of course, while $u(x, t)$ and $\rho(x, t)$ together determine $R(k, t)$, knowledge of $R(k, t)$ does not determine separately $u(x, t)$ and $\rho(x, t)$, but merely the combination (see (4))

(25) $\quad u'(x', t) = [\rho(x, t)]^{-2} \Big[u(x, t) - [\rho(x, t)]^{1/2} \{[\rho(x, t)]^{-1/2}\}_{xx} \Big],$

with x' and x related (see (3)) by the time-dependent relation

(26) $\quad x' = x + \int_x^\infty dy\, [1 - \rho(y, t)].$

Indeed clearly the evolution equation (23) implies that $u'(x, t)$ evolves according to the (higher KdV) equation

(27) $\quad u'_t(x, t) = \alpha(L', t) u'_x(x, t),$

where of course L' is the operator L, see (16), with $u(x)$ replaced by $u'(x, t)$ (this is implied by (24) and (9b)).

As a consequence, the usual infinite sequence of conserved quantities is associated to these flows (see chapter 5). Through (25) and (26), which also implies

(28) $\quad dx'/dx = \rho(x, t),$

these conserved quantities can be rewritten in term of the fields $u(x, t)$ and

$\rho(x, t)$. For instance the first two constants of the motion, C_0 and C_1, read

(29a) $\quad C_0 = \int_{-\infty}^{+\infty} dx \left\{ [u(x,t)/\rho(x,t)] + \left[([\rho(x,t)]^{-1/2})_x \right]^2 \right\}$,

(29b) $\quad C_1 = \int_{-\infty}^{+\infty} dx [\rho(x,t)]^{-3}$

$\cdot \left\{ u(x,t) - [\rho(x,t)]^{1/2} ([\rho(x,t)]^{-1/2})_{xx} \right\}^2$

(note that, to obtain the first of these formulae, an integration by parts has been performed, taking advantage of the boundary conditions (6a)).

The fact that the evolution equation (23) involves two fields leaves a certain freedom, of which advantage may be taken to explore various possibilities. One of the two fields, $u(x, t)$ or $\rho(x, t)$, may for instance be taken as given; then (23) determines the time evolution of the other. Or a relationship between u and ρ may be imposed, obtaining thereby again an equation that determines the evolution of a single field. Or two separate equations may be considered, that determine the evolution of both fields, and that together imply (23). These possibilities are now illustrated by treating tersely a few examples, rather than by presenting an exhaustive analysis.

Let us assume first of all that $\rho(x, t)$ is a given, known function of x and t. One may then consider the class of evolution equations

(30) $\quad u_t = \tfrac{1}{2} \sigma_{xxx} - 2\sigma_x u - \sigma u_x + \alpha(\mathcal{L}, t) \left[u_x - \tfrac{1}{2} r_{xxx} + 2 r_x u + (r-1) u_x \right]$.

To write this equation in more compact form we have introduced the convenient notation

(31) $\quad \sigma(x, t) = [\rho(x,t)]^{-1} \int_x^{\infty} dy \, \rho_t(y, t)$

in addition to (12); of course σ vanishes identically if ρ is time-independent.

We may therefore assert that the class of nonlinear evolution equations (30) for the dependent variable $u(x, t)$ can be integrated via the spectral transform technique. Note that indeed, if $\rho(x, t)$ is known, $u(x, t)$ is obtainable, with one integration, from (25) and (26), once $u'(x, t)$ is known; thus the Cauchy problem (given $u(x, 0)$, to compute $u(x, t)$) is solvable, via the spectral transform technique, for the class of evolution equations (30). The explicit form of the simplest nonlinear evolution equation of this class (corresponding to $\alpha(z,t) = -z$) is immediately obtained noting that

$$(32) \quad \mathcal{L}f(x) = -\left[r(x,t)\int_x^\infty dy\, r(y,t)f(y)\right]_{xxx}$$
$$+ 4\left[u(x,t)r(x,t)\int_x^\infty dy\, r(y,t)f(y)\right]_x$$
$$- 2u_x(x,t)r(x,t)\int_x^\infty dy\, r(y,t)f(y).$$

It is easily seen that it reduces to the KdV equation if $\rho(x,t) = r(x,t) = 1$; of course it corresponds to the KdV equation via the change of dependent and independent variables (25) and (26).

Another class of nonlinear evolution equations obtains from (21) if one assumes $u(x,t)$ to be a given, known, function, and considers $\rho(x,t)$ (or equivalently $r(x,t)$; see (12)) as the dependent variable. For instance, to the simple choice $g(z,t) = 1/z$, $h(z,t) = -\beta(z,t)$, $f(z,t) = 0$ and $u(x,t) = 0$ there correspond the class of evolution equations

$$(33a) \quad \rho_t = r\beta(-D^3 r\mathbf{I}r, t)\rho_x\rho, \quad r \equiv r(x,t) = 1/\rho(x,t),$$
$$(33b) \quad (\rho^2)_t = \beta(-D^3 r\mathbf{I}r, t)(\rho^2)_x,$$

that we write here in a mixed notation based on (12)÷(15). The simplest nonlinear equation of this class, corresponding to $\beta(z,t) = z$, is the so-called Harry Dym equation:

$$(34a) \quad r_t(x,t) = [r(x,t)]^3 r_{xxx}(x,t), \quad r(x,t) \equiv 1/\rho(x,t),$$
$$(34b) \quad \eta_t(x,t) = -2\{[\eta(x,t)]^{-1/2}\}_{xxx},$$
$$\eta(x,t) = [\rho(x,t)]^2 = [r(x,t)]^{-2}.$$

The next simplest equation, corresponding to $\beta(z,t) = z^2$ reads

$$(35) \quad r_t = r^3\left[r\left(rr_{xx} - \tfrac{1}{2}r_x^2\right)\right]_{xxx}, \quad r \equiv r(x,t) = 1/\rho(x,t).$$

Note however that, to solve the Cauchy problem (given $r(x,0)$, to compute $r(x,t)$), it is not quite sufficient to use the spectral transform technique, since this only yields (see (25) and (12)) the field $u'(x,t)$, related to $r(x, t)$ by the formulae

(36) $$u'(x', t) = -[r(x, t)]^{3/2} \{[r(x, t)]^{1/2}\}_{xx},$$

(37) $$x' = x + \int_x^\infty dy \{1 - [r(y, t)]^{-1}\};$$

and clearly, even when $u'(x, t)$ is known, the task to evince $r(x, t)$ from these equations is far from trivial. On the other hand it is plain how an explicit infinite sequence of conserved quantities obtains; for instance the first two constants read (see (29))

(38a) $$C_0 = \int_{-\infty}^{+\infty} dx \left[\{[r(x, t)]^{1/2}\}_x\right]^2,$$

(38b) $$C_1 = \int_{-\infty}^{+\infty} dx [r(x, t)]^2 \left[\{[r(x, t)]^{1/2}\}_{xx}\right]^2.$$

A more general technique, to extract an evolution equation for a *single* function from (23) (that involves the *two* functions $u(x, t)$ and $\rho(x, t)$), is to postulate a relationship between these two functions (rather than merely assume one of the two is a given, known, function). There is a large arbitrariness in the way this can be done. We limit again our analysis to the display of one example.

Let us assume that ρ and u are related by the formula

(39) $$\rho(x, t) = F[u'(x', t), t],$$

where $u'(x', t)$ is defined by (25) and (26) and $F(z, t)$ is an easily invertible function of its first argument (with the property $F(0, t) = 1$, to maintain consistency with the asymptotic properties of u and ρ). For instance one could take

(40) $$F(z, t) = 1 + z,$$

so that (39) would be rewritten in the form

(41) $$u(x, t) = [\rho(x, t)]^{1/2} \{[\rho(x, t)]^{-1/2}\}_{xx} + [\rho(x, t)]^2 [\rho(x, t) - 1].$$

The insertion of such an explicit, if complicated, expression of $u(x, t)$ in (23) yields a class of evolution equations for the single field $\rho(x, t)$, whose

Cauchy problem can then be solved via the spectral transform technique outlined above (note however the nontrivial nature of the final step that is necessary in order to evaluate $\rho(x, t)$ from the function $u'(x', t)$ via (39) and (26)). It is also easy to write in quite explicit form the endless sequence of conserved quantities that are associated to such a class of evolution equations. For instance, for the choice (40), the first two conserved quantities read (see (39) and (28))

$$\text{(42a)} \quad C_0 = \int_{-\infty}^{+\infty} dx\, \rho(x, t) [\rho(x, t) - 1],$$

$$\text{(42b)} \quad C_1 = \int_{-\infty}^{+\infty} dx\, \rho(x, t) [\rho(x, t) - 1]^2.$$

But unfortunately all the equations of this class are too complicated to motivate much interest in their study.

Let us next proceed to consider classes of (coupled) evolution equations for both fields $u(x, t)$ and $\rho(x, t)$. A simple choice directly suggested by (23) reads

$$\text{(43a)} \quad u_t(x, t) = \alpha(\mathcal{L}, t) u_x(x, t),$$

$$\text{(43b)} \quad [\rho^2(x, t)]_t = \alpha(\mathcal{L}, t) [\rho^2(x, t)]_x.$$

Then, given $u(x, 0)$ and $\rho(x, 0)$, $u(x, t)$ and $\rho(x, t)$ are determined; however, only their combination (25) (with (26)) can be obtained by the spectral transform technique (i.e., from (24b)). The main motivation for considering this class of coupled evolution equations is that, when $\alpha(z)$ is a low-order polynomial, these equations are not too complicated. For instance for $\alpha(z) = -z$ they can be cast in the pure differential form

$$\text{(44a)} \quad u_t + v_{xxx} - 4uv_x - 2vu_x = 0,$$

$$\text{(44b)} \quad r_t + r^3[r_{xxx} - 4r_x u - 2(r-1)u_x] = 0,$$

$$\text{(44c)} \quad ru_x = (v/r)_x.$$

Here of course $u \equiv u(x, t)$, $v \equiv v(x, t)$, $r \equiv r(x, t) = 1/\rho(x, t)$, and we assume $u(\pm\infty, t) = v(+\infty, t) = 0$, $r(\pm\infty, t) = \rho(\pm\infty, t) = 1$. Of course for this class of coupled evolution equations there exist the infinite sequence of (explicitly known) conserved quantities, of which (29a) and (29b) are the first two instances.

Let us finally consider another simple case, that obtains setting

(45) $\quad \alpha(z,t)=0$

in (23), and

(46a) $\quad u_t + LDU[u(x,t)] = 0.$

Here L and D are the operators defined by (14) and (16), $U(z)$ is an arbitrary (given) function (except for the condition $U(0)=0$; it is left as an exercise to consider the more general case when U depends explicitly on x and/or t as well; and also the case when $\alpha(z,t)$ does not vanish identically but is instead a low-order polynomial in z).

Of course (45) implies

(47) $\quad u'(x,t) = u'(x,0)$

(see (24b)), where u' is related to u and ρ by (25) and (26). Thus, if we take the initial conditions

(48a) $\quad \rho(x,0) = 1,$
(48b) $\quad u(x,0) = u_0(x),$

from (25), (26) and (47) we get

(49) $\quad u(x,t) = [\rho(x,t)]^{1/2} \{[\rho(x,t)]^{-1/2}\}_{xx}$
$\qquad + [\rho(x,t)]^2 u_0\!\left(x + \int_x^\infty dy\,[1-\rho(y,t)]\right)$

(to obtain this formula note that (25), (26) and (48a) imply, at $t=0$, $u'(x,0) = u(x,0)$).

On the other hand (45), (46a), (23), (11) and (12) yield

(50a) $\quad \rho_t(x,t) = 2\{\rho(x,t)U[u(x,t)]\}_x.$

Before summarizing our findings, let us rewrite (46a) in the more explicit form

(46b) $\quad u_t(x,t) + U_{xxx}[u(x,t)] - 4u(x,t)U_x[u(x,t)]$
$\qquad - 2u_x(x,t)U[u(x,t)] = 0,$

and let us note that (50a) and (49) imply

(50b) $$\rho_t(x,t) = 2\left\{\rho(x,t)U\left[[\rho(x,t)]^{1/2}\left\{[\rho(x,t)]^{-1/2}\right\}_{xx}\right.\right.$$
$$\left.\left. + [\rho(x,t)]^2 u_0\left(x + \int_x^\infty dy[1-\rho(y,t)]\right)\right]\right\}_x.$$

It is thus seen that *the task to solve the Cauchy problem, with the initial condition* (48b), *for the evolution equation* (46b) *(where $U(z)$ is an arbitrary given function), is equivalent (via* (49)) *to the task of solving the Cauchy problem for the evolution equation* (50b), *with the initial condition* (48a). Note that, for the special choice $U(z) = z$, (46b) becomes the KdV equation

(51) $$u_t(x,t) + u_{xxx}(x,t) - 6u_x(x,t)u(x,t) = 0,$$

while (50b) becomes

(52) $$\rho_t(x,t) = 2\left\{[\rho(x,t)]^{3/2}\left([\rho(x,t)]^{-1/2}\right)_{xx}\right\}_x$$
$$+ 2\left\{[\rho(x,t)]^3 u_0\left(x + \int_x^\infty dy[1-\rho(y,t)]\right)\right\}_x.$$

This finding may of course be looked at both ways; as an intriguing alternative formulation of the Cauchy problem for the evolution equation (46b) (and in particular for the KdV equation (51)); or as a technique to solve (50b), with $u_0(x)$ an arbitrary function (say, in the class of *bona fide* potentials), wherever (46b) is solvable (as is for instance the case in the KdV case (51); thus (52) may be considered a solvable equation).

Let us finally note that these results imply that there is an infinite sequence of conserved quantities associated with the evolution equation (46b) (of which (29) provide the first two instances); but these quantities are given in terms of *both* fields $u(x,t)$ and $\rho(x,t)$, the latter field being determined, in terms of $u(x,t)$, by the *linear first order* PDE (50a).

6.N. Notes to chapter 6

The idea to generalize the spectral transform of $u(x)$ by introducing a reference potential in the Schroedinger spectral problem, as in (6.-1), appears, in the context of the investigation of nonlinear evolution equations, in [KMi1974]. This approach is particularly fruitful if the reference potential is itself a solution of the evolution equation under consideration.

In addition to those treated in this chapter, many other extensions of the spectral transform method to solve nonlinear evolution equations have been investigated. Many of them originate from the natural idea to consider another spectral problem than that based on the Schroedinger equation; the associated spectral transform is therefore different from that discussed in chapter 2, and generally involves more than one function of x. Some such extensions, mainly based on replacing scalar functions by matrix valued functions, are detailed in volume II.

6.1. The first suggestion to extend in this manner the approach to problems in several space dimensions can be found in [C1975]; but the idea is so simple that it must have been known independently to several researchers at the same time or even earlier. No in-depth analysis of the behaviour of solutions (even just of soliton solutions) of this class of equations appears to have been made, perhaps because no applications have so far been suggested.

6.2. The new wronskian relation (6.2.-3), with $f(z,t)=1$, is known in the quantum theory of scattering as the "virial theorem". The extended class (6.2.-4) of nonlinear evolution equations has been independently pointed out in [N1977] and [CD1978a]; see also [N1980], [CD1978b], [CD1978c], [CD1978d], [CD1978e], [CD1978f]. The Lax approach to this class is due to [BR1980b].

6.3. This section reproduces mainly the material first reported in [CD1978g], [CD1978h], [CD1978i] and [D1979a]. The inverse problem of subsection 6.3.1 was independently solved in [LYS1981]. A (singular) solution of the cylindrical KdV has been first reported in [Dr1976]. Additional solutions (of multisoliton type, but still singular) of this equation have been given in [BM1979]. The change of variables (6.3.-7,8,9) with (6.3.-3,2), that connects the cylindrical KdV equation and the KdV equation, has been introduced in [LL1964]. For an investigation of the cylindrical KdV equation by means of the prolongation structure technique of differential geometry see [LLSM1982]; for the Bäcklund transformations see also [Na1980].

6.4. The 1-pole–N-soliton solution of the KdV equation has been given in [AC1979]. The spectral transform technique for the class of solutions of the

KdV equation, with one double pole, has been introduced in [D1979a]. The plots of figure 1 have been produced at UMIST (Manchester) by Dr. E. Abraham.

6.5. The equation (34) was discovered by Harry Dym and Martin Kruskal, in a joint exploration of evolution equations solvable using the spectral problem based on the string equation rather than the Schroedinger equation (i.e., on the ODE (6.5.-1) with $u(x)=0$, rather than with $\rho(x)=1$); but these results were never published, and the "Harry Dym equation" (34b) was only mentioned in [K1974]. This equation, and its generalizations, were then independently rediscovered by P. C. Sabatier [Sab1979a], [Sab1979b] and [Sab1979c], by Li Yi Shen [LYS1982] and by one of us (FC), who introduced two young Dutchmen to the study of this problem [DD1979]. The treatment given here is rather comprehensive and general, yet some additional results (especially concerning special solutions) can be found in [Sab1979a], [Sab1979b], [Sab1979c] and especially in [LYS1982]; moreover the latter paper uses a local approach (á la AKNS; see appendix A.20), and thereby extends the validity of the results (for instance to functions ρ that do not satisfy the boundary conditions (6.5.-6), tending to different limits at the two ends, i.e. as $x \to +\infty$ and as $x \to -\infty$).

APPENDICES

A.1. On the number of discrete eigenvalues of the Schroedinger spectral problem on the whole line

In this appendix we report and prove some results on the number N of discrete eigenvalues $-p_n^2$ for the Schroedinger spectral problem characterized by the equation

(1) $\quad -\psi_{xx}(x) + u(x)\psi(x) = -p^2 \psi(x), \quad -\infty < x < +\infty.$

At least some of these results are probably new. For analogous results in the context of the Schroedinger problem on the half-line see [C1967].

We always assume the function $u(x)$ to be real and regular for all (real) values of x and to vanish as $x \to \pm\infty$; indeed it is generally convenient to assume that $u(x)$ has compact support, namely that it vanishes identically for sufficiently large real values of $|x|$, say $|x| > x_\infty$, although the final results hold clearly more generally. It is of course well-known that all eigenvalues are nondegenerate and that they require that p be real; we conventionally assume $p > 0$.

Lemma 1. *The number of discrete eigenvalues of* (1) *coincides with the number of (real) zeros* ξ_j *of the function* $\varphi(x)$ *defined by the equation*

(2) $\quad \varphi_{xx}(x) = u(x)\varphi(x)$

and by the boundary conditions

(3) $\quad \varphi(-\infty) = 1, \quad \varphi_x(-\infty) = 0,$

namely

(4) $\quad \varphi(\xi_j) = 0, \quad \xi_j < \xi_{j+1}, \quad j = 1, 2, \ldots, N.$

Proof. This is a well-known result; it may be established by noting that a way to compute, and therefore also count, the discrete eigenvalues, is to follow the zeros $\xi_j(p)$ of the solution $\varphi(x, p)$ of (1) characterized by the boundary condition

(5) $$\lim_{x \to -\infty} [\exp(-px)\varphi(x, p)] = 1$$

for progressively larger values of p (starting from $p \approx 0$), and to note that, to every value of p at which the number of these zeros decreases by one (due to the exit, as it were, at ∞, of the rightmost zero), there corresponds a discrete eigenvalue.

Remark. Here, and hereafter, we exclude for simplicity from consideration the marginal case when $\varphi(\infty)$ has a finite value, corresponding in some sense to the occurrence of a discrete eigenvalue at $p=0$. This incidentally accounts for the fact that some of the results given below appear to be contradicted by the trivial case $u(x)=0$, implying of course $N=0$.

Lemma 2. *Define the function $\theta(x)$ via the nonlinear first-order differential equation*

(6) $$\theta_x(x) = -yu(x)\cos^2\theta(x) + y^{-1}\sin^2\theta(x)$$

and the boundary condition

(7) $$\theta(-\infty) = 0,$$

y being an arbitrary positive constant. Then the number of discrete eigenvalues N is given by the formula

(8) $$N = (1/\pi)\theta(+\infty).$$

Proof. Set

(9) $$\varphi_x(x)/\varphi(x) = -y^{-1}\operatorname{tg}\theta(x).$$

Then (2) and (3) imply (6) and (7). Moreover from (4) there clearly follows

(10) $$\theta(\xi_j) = (j - \tfrac{1}{2})\pi;$$

note that the ordering of the ξ_j's, as given by (4), is consistent with (10), as implied by

(11) $\quad \theta_x(\xi_j) = y^{-1} > 0.$

Finally it is clear from (6) that the asymptotic value $\theta(+\infty)$ must coincide with an integral multiple of π; and this, together with (10) and (4), implies (8). Q.E.D.

Lemma 3. *Define the function $\mu(x)$ via the nonlinear first-order differential equation*

(12) $\quad \mu_x(x) = -yu(x)\{[(x-z)/y]\sin\mu(x) - \cos\mu(x)\}^2$

and the boundary condition

(13) $\quad \mu(-\infty) = 0,$

y and z being arbitrary constants (but y>0). Then the number of discrete eigenvalues N is given by the formula

(14) $\quad N = $ *integral part of* $\left[1 + \mu(\infty)/\pi\right]$

(of course, provided the r.h.s. is positive; otherwise $N=0$).

Proof. Set

(15) $\quad \cotg\mu(x) = (x-z)/y + \cotg\theta(x).$

Then (12) follows immediately from (6), and (13) obtains easily from (7). To prove (14) it is convenient to introduce the coordinates η_j, at which φ_x, rather than φ, vanishes. Then

(16) $\quad \theta(\eta_j) = j\pi$

and clearly (10) and (11) imply

(17) $\quad \xi_j < \eta_j < \xi_{j+1}$

(note that there might possibly be, if $u(x)$ changes sign, more than one η_j for each value of j; in such a case there should of course be an odd number of them, and the argument given here applies equally by considering only the rightmost of them). On the other hand (15) implies

(18) $\quad \mu(\eta_j)=j\pi, \quad j=1,2,\ldots,N-1.$

Assume now that $u(x)=0$ for $x>x_\infty$, and that $j\pi<\mu(x_\infty)=\mu(\infty)\leq(j+1)\pi$. Since $\theta_x(\eta_j)$ has the same sign as $\mu_x(\eta_j)$, this implies $j\pi<\theta(x_\infty)\leq(j+1)\pi$ and therefore (see (6)) $\theta(\infty)=(j+1)\pi$. And this, together with (8), completes the argument.

Lemma 4. *Let $N^{(m)}$ be the number of discrete eigenvalues associated, via (1), to $u^{(m)}(x)$, $m=1,2$, and assume*

(19) $\quad u^{(1)}(x)\leq u^{(2)}(x), \quad -\infty<x<\infty.$

Then

(20) $\quad N^{(1)}\geq N^{(2)}.$

This result is too trivial and well-known to require a proof.

Let us proceed now to prove a lower bound for the number N of discrete eigenvalues. Other bounds can be obtained using the results given above and following, for instance, the treatment of [C1967].

Theorem 1 (lower bound). *The number N of discrete eigenvalues of (1) is bounded below by the formula*

(21) $\quad N\geq \text{integral part of } \left\{1+\pi^{-1}\int_{-\infty}^{+\infty}dx\min[y^{-1},-yu(x)]\right\},$

where y is a positive, but otherwise arbitrary, constant and by definition

(22a) $\quad \min(a,b)=a \quad \text{if } a\leq b,$

(22b) $\quad \min(a,b)=b \quad \text{if } a\geq b.$

Proof. Let $u(x)=0$ for $x>x_\infty$. Then (6) implies that, if $n\pi<\theta(x_\infty)\leq(n+1)\pi$, $\theta(\infty)=(n+1)\pi$. This, together with (8), implies the formula

(23) $\quad N=\text{integral part of }[1+\theta(x_\infty)/\pi]$

(ignoring the marginal case when $\theta(x_\infty)$ is an integral multiple of π). On the other hand (6) implies

(24) $\quad \theta_x(x)\geq\min[y^{-1},-yu(x)],$

since the minimum value of $a\cos^2 x + b\sin^2 x$ over all (real) values of x is $\min[a, b]$; and this, together with (7), yields

(25a) $$\theta(x_\infty) \geq \int_{-\infty}^{x_\infty} dx \min[y^{-1}, -yu(x)]$$

or equivalently (since $\min[a, 0] = 0$ if a is positive)

(25b) $$\theta(x_\infty) \geq \int_{-\infty}^{+\infty} dx \min[y^{-1}, -yu(x)].$$

This formula, together with (23), yields (21). Q.E.D.

Corollary 1.1. If $u(x) \leq 0$ for $-\infty < x < \infty$, then $N \geq 1$.

Proof. This follows trivially from (21). (Recall that we are excluding from consideration the trivial case when $u(x)$ vanishes identically, in which case obviously $N = 0$).

Corollary 1.2

(26) $$N \geq \text{integral part of } \left\{ 1 - (\pi g)^{-1} \int_{-\infty}^{+\infty} dx\, u(x) \right\}, \quad g > 0,$$

provided

(27) $$u(x) \geq -g^2, \quad -\infty < x < \infty.$$

Proof. This follows from (21), choosing $y = g^{-1}$.

We proceed next to derive upper bounds to the number N of discrete eigenvalues associated with $u(x)$. It is actually convenient to establish such bounds for the (nonpositive) potential $u_-(x)$ related to $u(x)$ by the definition

(28a) $\quad u_-(x) = u(x) \quad$ wherever $u(x) < 0$,
(28b) $\quad u_-(x) = 0 \quad\quad\;\;$ wherever $u(x) \geq 0$.

It is then guaranteed by Lemma 4 that such bounds hold *a fortiori* for $u(x)$.

These bounds are conveniently expressed in terms of the integrals

(29) $$I_m \equiv \int_{-\infty}^{+\infty} dx\, |x - z|^m |u_-(x)|, \quad m = 0, 1, 2.$$

In this formula, and always below, z is an arbitrary (real) constant.

Theorem 2 (upper bound). *The number N of discrete eigenvalues of* (1) *is bounded above by the formula*

(30) $\quad N \leqslant 1 + (2/\pi)(I_0 I_2)^{1/2}.$

Proof. The starting point is (12), that can be rewritten (with $u(x)$ replaced by $u_-(x)$) in the form

(31) $\quad \mu_x(x) = |u_-(x)| [y + (x-z)^2/y] \sin^2[\mu(x) + \chi(x)]$

where of course

(32) $\quad \cotg \chi(x) = (z-x)/y.$

From (13), (29) and (31) there immediately follows

(33) $\quad \mu(\infty) \leqslant y I_0 + y^{-1} I_2.$

The (optimal) choice $y = (I_2/I_0)^{1/2}$, together with (14), yields (30). Q.E.D.

Theorem 3 (upper bound à la Bargmann). *The number N of discrete eigenvalues of* (1) *is bounded above by the formula*

(34) $\quad N \leqslant 1 + I_1.$

Proof. Consider the function $\mu(x)$ defined by (31) and (13); it is clearly a nondecreasing function of x. Define the positions η_j and λ_j by the formulae

(35a) $\quad \mu(\lambda_j) = (j - \tfrac{1}{2})\pi, \quad j = 1, 2, \ldots, N-1,$
(35b) $\quad \mu(\eta_j) = j\pi, \quad j = 1, 2, \ldots, N-1.$

Note the consistency of these formulae with (14) (recall that we neglect the marginal case when $\mu(\infty) = N\pi$); of course there holds the ordering rule

(36) $\quad \lambda_j < \eta_j < \lambda_{j+1} < \eta_{j+1}.$

Now define the function $\nu(x)$ setting

(37) $\quad \tg \nu(x) = (|x-z|/y) \tg \mu(x).$

Then (31) yields

(38) $\quad \nu_x(x) = y^{-1} \text{sign}(x-z) \tg \mu(x) \cos^2 \nu(x)$
$\quad\quad\quad + 2|x-z| |u_-(x)| \sin^2[\nu(x) - \tfrac{1}{4}\pi \text{sign}(x-z)],$

while (13) and (35) yield

(39a) $\quad \nu(\eta_{j-1}) = (j-1)\pi, \quad j = 1, 2, \ldots, N,$

(39b) $\quad \nu(\lambda_j) = (j - \tfrac{1}{2})\pi, \quad j = 1, 2, \ldots, N,$

with the convention

(40) $\quad \eta_0 = -\infty, \quad \lambda_N = +\infty.$

It is now convenient to define the $2N+1$ quantities a_j and b_j as follows:

(41a) $\quad a_j = \int_{\eta_{j-1}}^{\lambda_j} dx\, |x - z| |u_-(x)|, \quad j = 1, 2, \ldots, N,$

(41b) $\quad b_j = \int_{\lambda_j}^{\eta_j} dx\, |x - z| |u_-(x)|, \quad j = 1, 2, \ldots, N-1.$

Then clearly

(42) $\quad I_1 = \sum_{j=1}^{N} a_j + \sum_{j=1}^{N-1} b_j,$

this formula corresponding to the partitioning of the integration range $-\infty < x < +\infty$ in (29) into the $2N-1$ intervals $-\infty = \eta_0$ to λ_1, λ_1 to η_1, η_1 to λ_2, etc., η_{N-1} to $\lambda_N = +\infty$.

We now prove that,

(43a) $\quad$ if $z \geq \lambda_j$, then $a_j \geq 1$,

(43b) $\quad$ if $z \leq \lambda_j$, then $b_j \geq 1$.

This implies that, whichever is the value of z, in the sum of $2N-1$ positive terms in the r.h.s. of (42) at least $N-1$ are not less than unity; and this implies (34). Q.E.D.

There remains to prove the inequalities (43). Consider first the case $z \geq \lambda_j$. Then, for $\eta_{j-1} \leq x \leq \lambda_j$, (38) and (35) (implying $\operatorname{tg} \mu(x) \geq 0$) yield

(44a) $\quad \nu_x(x) / \sin^2[\nu(x) + \tfrac{1}{4}\pi] \leq 2|x - z| |u_-(x)|.$

Integration from η_{j-1} to λ_j yields, using (39), $a_j \geq 1$; (43a) is thus proven. Consider next the case $z \leq \lambda_j$. Then, for $\lambda_j \leq x \leq \eta_j$, (38) and (35) (implying now $\operatorname{tg} \mu(x) \leq 0$) yield

(44b) $\quad \nu_x(x) / \sin^2[\nu(x) - \tfrac{1}{4}\pi] \leq 2|x - z| |u_-(x)|.$

Integration from λ_j to η_j, using (39), yields now $b_j \geq 1$; (43b) is thus also proven.

Note incidentally that, from (34) and the obvious inequality

(45) $\quad I_1 \leq (I_0 I_2)^{1/2}$,

there follows an inequality analogous to, but less stringent than, (30) (since $2/\pi < 1$).

It is easily seen that the upper and lower bounds of Theorems 1, 2 and 3 are all *best possible*; there are indeed special functions $u(x)$ that saturate them. For the bound of Theorem 1 and arbitrary N, the saturating function is a square well, $u(x) = -g^2$ for $|x-z| < a$, $u(x) = 0$ for $|x-z| > a$, with $ga = \frac{1}{2}\pi(N-1) + \varepsilon$, $\varepsilon > 0$; for the bounds of Theorems 1, 2 and 3, with $N=1$, the saturating function is $u(x) = -g^2 \delta(x-z)$.

It is also clear that the flexibility associated with the (arbitrary) constants y and z is merely a reflection of the invariance of N under scale transformations and translations (see section 2.1.1).

A.2. Orthogonality and completeness relations for the Schroedinger spectral problem on the whole line

For the Jost solutions of (2.1.-1) characterized by (2.1.-4) the orthogonality relations read:

(1) $\quad (2\pi)^{-1} \int_{-\infty}^{+\infty} dx\, f^{(\pm)}(x,k) f^{(\pm)}(x,q) = [T(k)T(-k)]^{-1} \delta(k+q)$
$\quad \mp R(\mp k)[T(k)T(\mp k)]^{-1} \delta(k-q)$,

(2) $\quad (2\pi)^{-1} \int_{-\infty}^{+\infty} dx\, f^{(\pm)}(x,k) f^{(\mp)}(x,q) = [T(k)]^{-1} \delta(k-q)$.

Here k and q are real, namely these formulae refer to the continuous part of the spectrum. Let us recall that to each eigenvalue k^2 there correspond two independent eigenfunctions, for instance $f^{(+)}(x,k)$ and $f^{(-)}(x,k)$; other eigenfunctions that correspond to the same eigenvalue, and that can of course be expressed as linear combinations of $f^{(+)}(x,k)$ and $f^{(-)}(x,k)$, are $f^{(\pm)}(x,-k)$ and $\psi(x, \pm k)$ (see (2.1.-5)). If the potential $u(x)$ vanishes, then $f^{(\pm)}(x,k) = \exp(\pm ikx)$, $R(k) = 0$, $T(k) = 1$ and the relations (1) and (2) are usually written for the eigenfunctions

(3) $\quad \varphi^{(0)}(x,k) = (2\pi)^{-1/2} \exp(ikx)$

that provide the basis for the Fourier expansion, namely

(4) $$\int_{-\infty}^{+\infty} dx\, \varphi^{(0)}(x,k)\varphi^{(0)}(x,q) = \delta(k+q).$$

Also in the general case ($u(x) \neq 0$), it is of course possible to choose, for each eigenvalue k^2, two eigenfunctions, say $\varphi(x,k)$ and $\varphi(x,-k)$, that satisfy the same relation (4), i.e.

(5) $$\int_{-\infty}^{+\infty} dx\, \varphi(x,k)\varphi(x,q) = \delta(k+q).$$

Indeed for instance

(6) $$\varphi(x,k) = (2\pi)^{-1/2}\{a(k)T(-k)f^{(+)}(x,k)$$
$$+ [2a(k)]^{-1}[T(k)T(-k)]^{1/2}R(-k)f^{(-)}(x,k)\}$$

together with

(7) $$a(k) = 2^{-1/2}\{1 + [T(k)T(-k)]^{1/2}\}^{1/2},$$

clearly imply (5).

These eigenfunctions of the continuum are of course orthogonal to the eigenfunctions corresponding to the discrete eigenvalues $-p_n^2$ (if any):

(8a) $$\int_{-\infty}^{+\infty} dx\, f^{(\pm)}(x,k)\varphi_n(x) = 0,$$

(8b) $$\int_{-\infty}^{+\infty} dx\, \varphi(x,k)\varphi_n(x) = 0;$$

and the eigenfunctions corresponding to the discrete eigenvalues are of course orthogonal among themselves

(9) $$\int_{-\infty}^{+\infty} dx\, \varphi_n(x)\varphi_m(x) = \delta_{nm}.$$

Here for the eigenfunctions of the discrete spectrum we use the same notation as in section 2.1, see (2.1.-21).

The derivation of these formulae is plain. For instance, to get (1), one may start from the formula

(10) $$\int_{x_1}^{x_2} dx\, f^{(\pm)}(x,k)f^{(\pm)}(x,q)$$
$$= (k^2 - q^2)^{-1} W[f^{(\pm)}(x,k), f^{(\pm)}(x,q)]\Big|_{x=x_1}^{x=x_2},$$

which is an immediate consequence of (the wronskian theorem applied to) (2.1.-1), and then evaluate the limit as $x_1 \to -\infty$ and $x_2 \to +\infty$ using (2.1.-7), (2.1.-4) and the well-known formula of distribution theory

(11) $\quad P[k^{-1}\exp(ikx)] = i\pi\delta(k) \quad$ as $x \to \infty$.

The other proofs are analogous, and even more trivial.

For the validity of (1), (2), (8) and (9) it is not required that $u(x)$ be real; if however $u(x)$ is real, then these equations can be rewritten in other forms using the reflection properties

(12a) $\quad [f^{(\pm)}(x, k)]^* = f^{(\pm)}(x, -k), \quad \text{Im } k = 0,$
(12b) $\quad [\varphi(x, k)]^* = \varphi(x, -k), \quad \text{Im } k = 0,$

as well as (2.1.-12) and the fact that $\varphi_n(x)$ is real. In this case the orthonormality relations take the usual expression (in a self-explanatory notation)

(13) $\quad \langle \varphi(k), \varphi(q) \rangle = \delta(k-q), \quad \langle \varphi(k), \varphi_n \rangle = 0, \quad \langle \varphi_n, \varphi_m \rangle = \delta_{nm},$

by introducing the positive definite scalar product

(14) $\quad \langle f, g \rangle \equiv \int_{-\infty}^{+\infty} dx [f(x)]^* g(x).$

The corresponding completeness (or closure) relation reads

(15) $\quad \sum_{n=1}^{N} \varphi_n(x)\varphi_n(y) + \int_{-\infty}^{+\infty} dk\, \varphi(x, k)[\varphi(y, k)]^* = \delta(x-y).$

These formulae imply that any function $f(x)$, whose square modulus is integrable, can be expanded in the form

(16) $\quad f(x) = \sum_{n=1}^{N} \hat{f}_n \varphi_n(x) + \int_{-\infty}^{+\infty} dk\, \hat{f}(k)\varphi(x, k)$

where

(17) $\quad \hat{f}_n = \langle \varphi_n, f \rangle, \quad \hat{f}(k) = \langle \varphi(k), f \rangle.$

A.3. Asymptotic behaviour (in k) of the transmission and reflection coefficients

In this appendix we indicate tersely how the formulae (2.1.-39), namely

(1a) $\qquad \lim\limits_{k \to \pm \infty} \left[|k|^M (1 - |T(k)|) \right] = 0, \quad M < \infty, \ \mathrm{Im}\, k = 0,$

(1b) $\qquad \lim\limits_{k \to \pm \infty} \left[|k|^M R(k) \right] = 0, \quad M < \infty, \ \mathrm{Im}\, k = 0,$

are proved. We also indicate how (2.1.-40) with (2.1.-41), namely

(2) $\qquad \theta(k) = \sum\limits_{m=0}^{M} \theta_m (2k)^{-2m-1} + O(k^{-2M-3}),$

(3) $\qquad \theta_m = (-1)^{m+1} (2m+1)^{-1} \int_{-\infty}^{+\infty} dx\, L^m [x u_x(x) + 2u(x)],$

$$m = 0, 1, 2, \ldots,$$

are obtained; here of course $\theta(k)$ is the phase of $T(k)$ if the potential is real, and more generally is defined by

(4) $\qquad T(k) = T(-k) \exp[2i\theta(k)], \quad \mathrm{Im}\, k = 0$

(see (2.1.-15) and (2.1.-16)), and L is the usual integrodifferential operator defined, for instance, by (2.1.-42).

There are of course many ways to prove these formulae. We believe the route followed below is the most direct one; it makes the validity of these results immediately evident. Note however that we do not present here detailed proofs, but merely indicate how the arguments go. The filling in of the technical details required in order to produce rigorous proofs is left as an easy exercise for the diligent reader.

For simplicity, we restrict attention to the case of real $u(x)$ and to real values of k, although the following arguments are easily extended to more general cases. We also keep in mind the formula

(5) $\qquad \lim\limits_{k \to \pm \infty} [\psi(x, k) \exp(ikx)] = 1,$

which is an immediate consequence of (2.1.-1) and (2.1.-3) (or, equivalently, of (2.1.-10) and (2.1.-11)). Of course we also assume $u(x)$ to be infinitely differentiable for all real values of x (this assumption is essential for the validity of these results).

Equation (1b) follows then immediately from the formula

(6) $$(-4k^2)^{m+1} R(k) = \int_{-\infty}^{+\infty} dx \, [\psi(x,k)]^2 L^m u_x(x), \quad m=0,1,2,\ldots,$$

which is merely a special case of (2.4.1.-12b), since the r.h.s. of this formula is obviously finite, as $k \to \pm\infty$, for any (positive integral) value of m.

Similarly (1a) follows from

(7) $$(-4k^2)^{m+1}[1-|T(k)|^2] = \int_{-\infty}^{+\infty} dx \, \psi(x,k) \psi(x,-k) L^m u_x(x),$$

$$m = 0, 1, 2, \ldots,$$

which is a special case of (2.4.1.-26) (using (2.1.-12) and (2.1.-13)).

The starting point to prove (3) is the formula

(8) $$\delta\theta_m = (-1)^{m+1} \int_{-\infty}^{+\infty} dx \, L^m \, \delta u(x)$$

which relates any infinitesimal change $\delta u(x)$ of the (*bona fide*) potential $u(x)$ in the Schroedinger equation (2.1.-1) to the corresponding changes $\delta\theta_m$ of the coefficients of the asymptotic expansion (2) of the phase $\theta(k)$ of the transmission coefficient $T(k)$. This formula obtains analogously to the results described above, but it is sufficiently important to deserve a separate treatment; see therefore appendix A.8 for its proof (and for the display of additional results it implies).

Once (8) is established, the derivation of (3) is trivial, since it is merely the special case of (8) corresponding to an infinitesimal scale transformation: indeed (2.1.1.-4) and (2.1.1.-4c), together with (4), imply that to the infinitesimal change

(9) $$\delta u(x) = [xu_x(x) + 2u(x)] \varepsilon$$

there corresponds

(10) $$\delta\theta(k) = -k[d\theta(k)/dk]\varepsilon;$$

and this last formula, together with (2), implies

(11) $$\delta\theta_m = (2m+1)\theta_m \varepsilon.$$

Insertion of (9) and (11) in (8) yields (3).

A.4. Dispersion relations for the transmission coefficient

In this appendix we outline the (standard) derivation of (2.1.-43), (2.1.-44) and (2.1.-45).

Define the function

$$(1) \qquad G(k) = [T(k)]^{-1} \prod_{n=1}^{N} [(k+ip_n)/(k-ip_n)].$$

Here $T(k)$ is the transmission coefficient and the *positive* quantities p_n identify the discrete eigenvalues. The results of section 2.1 imply that this function is holomorphic in the upper half of the complex k-plane, Im $k \geq 0$, since $[T(k)]^{-1}$ is analytic there (as implied by (2.1.-6a) and by the analyticity of the Jost solutions $f^{(\pm)}(x,k)$) and it has simple zeros at $k = ip_n$ (these are the simple poles of $T(k)$ that correspond to the discrete eigenvalues). Moreover $G(k)$ has no zeros in Im $k \geq 0$, since $T(k)$ is analytic there except for the single poles at $k = ip_n$ that clearly yield no zeros of $G(k)$. Finally

$$(2) \qquad \lim_{|k| \to \infty} [G(k)] = 1, \quad \text{Im } k \geq 0.$$

Thus the function

$$(3) \qquad F(k) = \ln G(k)$$

is itself holomorphic in Im $k \geq 0$, and it vanishes asymptotically,

$$(4) \qquad \lim_{|k| \to \infty} [F(k)] = 0, \quad \text{Im } k \geq 0.$$

A standard application of the Cauchy theorem to a closed contour composed of the real k-axis and of a semicircle at infinity in the upper half k-plane yields therefore the dispersion relations

$$(5a) \qquad F(k) = (2\pi i)^{-1} \int_{-\infty}^{+\infty} dq\, F(q)/(q-k), \quad \text{Im } k > 0,$$

$$(5b) \qquad F(k) = (\pi i)^{-1} P \int_{-\infty}^{+\infty} dq\, F(q)/(q-k), \quad \text{Im } k = 0,$$

where P denotes of course the (Cauchy) principal value.

The imaginary part of (5b) then yields, for real k,

$$(6) \qquad \theta(k) = 2 \sum_{n=1}^{N} \text{arctg}(p_n/k) - \pi^{-1} P \int_{-\infty}^{+\infty} dq\,(q-k)^{-1} \ln|T(q)|,$$

where we are of course using the definition

(7) $\quad T(k)=|T(k)|\exp[i\theta(k)], \quad \text{Im } k=0$

(see (2.1.-15a)), together with (3) and (1). The symmetry properties (2.1.-14a) and (2.1.-16a) imply moreover

(8) $\quad \theta(k)=2\sum_{n=1}^{N}\text{Arctg}(p_n/k)-(k/\pi)\text{P}\int_{-\infty}^{+\infty}dq(q^2-k^2)^{-1}\ln|T(q)|,$

$$\text{Im } k=0,$$

which is precisely (2.1.-43b).

From this formula the Levinson theorem (2.1.-44) immediately follows. Similarly, from the real part of (5b), there follows

(9) $\quad |T(k)|=\prod_{n=1}^{N}\left[1+(p_n/k)^2\right]^{-1}$

$$\cdot\exp\left[\pi^{-1}\text{P}\int_{-\infty}^{+\infty}dq(q^2-k^2)^{-1}q\theta(q)\right], \quad \text{Im } k=0,$$

which is precisely (2.1.-43a); and (2.1.-46) follows from (5a).

If $u(x)$ is infinitely differentiable, then as $k\to\pm\infty$, $1-|T(k)|$ vanishes faster than any inverse power of k (see (2.1.-39a) and appendix A.3). Thus in this case there follow from (9) an infinite sequence of sum rules involving the discrete eigenvalues $-p_n^2$ and the momenta of the phase $\theta(k)$,

(10) $\quad \mu_m = \makebox[0pt][l]{$\int$}{\,\text{---}}_{-\infty}^{+\infty} dk\, k^{2m-1}\theta(k), \quad m=1,2,\ldots .$

Here the regularized integral $\makebox[0pt][l]{$\int$}{\,\text{---}}_{-\infty}^{+\infty} dk$ is defined by the prescription

(11) $\quad \makebox[0pt][l]{$\int$}{\,\text{---}}_{-\infty}^{+\infty} dk\, k^{2m-1}\theta(k) = -[(2m-1)!]^{-1}\int_{-\infty}^{+\infty} dk\, k^{2m-1}$

$$\cdot\frac{d^{2m-1}}{dk^{2m-1}}\left[k^{2m-1}\theta(k)\right];$$

it is easily seen that the asymptotic behavior of $\theta(k)$, see (2.1.-40), implies that the integral in the r.h.s. of this equation is convergent (while (11) would clearly be an identity if the integral on the l.h.s. were convergent without regularization).

These sum rules are given in explicit form by the neat formula

(12a) $$\sum_{n=1}^{N} p_n^{2m} = (-1)^m (m/\pi)\mu_m, \quad m=1,2,3,\ldots, \quad (N>0),$$

which obtains easily by noticing that (9) yields an asymptotic expansion of $1-|T(k)|$ at large k in inverse powers of k^2, whose coefficients must all vanish; and by then looking at the asymptotic expansion in inverse powers of k of $\ln|T(k)|$, with $|T(k)|$ given by (9), whose coefficients must also vanish. Note that (12a) is appropriate to the case with discrete eigenvalues, $(N>0)$; if there are no discrete eigenvalues, (12a) is replaced by the condition that all the momenta μ_m vanish,

(12b) $\quad \mu_m = 0, \quad m=1,2,3,\ldots, \quad (N=0).$

In the case of the reflectionless potentials (2.2.-20) the function $F(k)$ defined by (3) and (1) vanishes identically (see (2.2.-23)) and the sum rules (12a) reduce to the elementary identity

(13) $$p^{2m} = (-1)^m (2m/\pi) \int_{-\infty}^{+\infty} dk \, k^{2m-1} \operatorname{Arctg}(p/k),$$

$$p>0, \quad m=1,2,\ldots$$

Let us end this appendix by reporting the explicit expression (see (2.1.-45)):

(14) $$\theta_m = (-1)^m (m+\tfrac{1}{2})^{-1} \sum_{n=1}^{N} (2p_n)^{2m+1}$$
$$+ (2/\pi) \int_{-\infty}^{+\infty} dk \, (2k)^{2m} \ln|T(k)|, \quad m=0,1,2,\ldots$$

of the coefficients θ_m of the asymptotic expansion at large (real) k of the phase $\theta(k)$ of the transmission coefficient $T(k)$, a formula that is easily obtained by expanding (8) at large k (and using the definition, (2.1.-40) or (A.3-2), of the coefficients θ_m).

A.5. The inverse spectral Schroedinger problem on the whole line

Several approaches to the inverse spectral problem have been given; at various levels of generality and mathematical rigour, the methods applied range from those of the spectral theory of differential operators to the contour-integral techniques in the complex plane of the spectral parameter, as borrowed from the scattering theory of waves. In any case, all these methods eventually reduce the inverse problem to solving a Fredholm integral equation, whose theory is well established. In those cases where more than one method could apply, the choice of that one which is most suited to the problem of interest may be a matter of personal taste; however, in some cases one particular approach could turn out to be much more cumbersome than another (an instance of such a case is mentioned below).

Here we present an elementary and partially heuristic approach to the inverse spectral problem, that should be understandable to readers with minimal mathematical background and expertise in this field. We focus on the underlying ideas and computational techniques rather than trying to present a rigorous derivation of the all important Gel'fand–Levitan–Marchenko equation. To be specific, we treat the scalar Schroedinger problem of chapter 2 (Schroedinger equation on the whole line, with asymptotically vanishing potential), though other spectral problems, such as the Schroedinger problem with an additional linear potential (see subsection 6.3.1) or the generalized Zakharov-Shabat scattering problem (see volume II), can be similarly analysed.

As an introductory step, consider first the differential equation

(1) $\quad f_{xx}^{(\pm)} = [u(x) - k^2] f^{(\pm)}, \quad f^{(\pm)} \equiv f^{(\pm)}(x, k),$

and its Jost-solutions $f^{(\pm)}(x, k)$, which are identified by the asymptotic condition

(2) $\quad \lim_{x \to \pm \infty} [\exp(\mp ikx) f^{(\pm)}(x, k)] = 1;$

in order to simplify the following discussion, we assume that the function $u(x)$ in the r.h.s. of (1) is real, bounded and of compact support, namely

(3a) $\quad u(x) = u^*(x),$
(3b) $\quad u(x) = 0, \quad |x| > x_\infty, \quad 0 < x_\infty < \infty.$

The parameter k is a complex number and the reality of u implies that

(4) $\quad f^{(\pm)}(x, k) = [f^{(\pm)}(x, -k^*)]^*;$

moreover since the transformation $k \to -k$ leaves the equation (1) invariant, the two functions $f^{(+)}(x, k)$ and $f^{(+)}(x, -k)$, as well as $f^{(-)}(x, k)$ and

$f^{(-)}(x, -k)$, are (for $k \neq 0$) linearly independent solutions of (1), as it is implied by the boundary condition (2). Therefore we introduce two complex functions, $A(k)$ and $B(k)$, of the complex variable k, which are uniquely defined by the linear combination

$$f^{(+)}(x, k) = A(k) f^{(-)}(x, k) + B(k) f^{(-)}(x, -k). \tag{5}$$

Note that $A(k)$ and $B(k)$ characterize completely the asymptotic behaviour of $f^{(+)}(x, k)$ as x becomes large and negative, since (2) and (5) imply that

$$f^{(+)}(x, k) \underset{x \to -\infty}{\to} A(k) \exp(-ikx) + B(k) \exp(ikx). \tag{6}$$

This suggests that the functions $A(k)$ and $B(k)$ can obtain by integrating (1), starting from $x > x_\infty$ where $f^{(+)}(x, k) = \exp(ikx)$, as implied by (2) and (3b), and then, after the integration has been carried out up to $x < -x_\infty$, by extracting the coefficients of the exponential functions from the asymptotic behaviour (6) (actually if (3b) holds, for $x < -x_\infty$ (6) becomes a strict equality). As a result of the integration of (1) through the whole x-axis, the asymptotic coefficients $A(k)$ and $B(k)$ for each value of k, have memorized some information on the function $u(x)$. Thus, knowledge of these coefficients for all real values of k may be sufficient to reconstruct $u(x)$. In fact, it may be necessary to have also some additional information for complex k. The information actually needed for the reconstruction of $u(x)$, and the procedure to accomplish this task, is the main topic of this appendix.

Consider now two different functions $u^{(1)}(x)$ and $u^{(2)}(x)$ satisfying (3). All the quantities which have been previously introduced through the Schroedinger differential equation (1), with $u(x) = u^{(j)}(x)$, $j = 1, 2$, such as the Jost solutions or the asymptotic coefficients, will be represented by the same symbol, with the index $j = 1, 2$ that specifies which one of the two functions $u^{(j)}(x)$ they correspond to. With the aim of constructing a transformation relating to each other the Jost solutions of these two differential equations, we consider first of all an operator Ω satisfying the equation

$$H^{(1)} \Omega = \Omega H^{(2)}, \qquad H^{(j)} \equiv -\partial_x^2 + u^{(j)}(x). \tag{7}$$

It is evident that Ω transforms a solution $\psi^{(2)}$ of the equation $H^{(2)} \psi^{(2)} = k^2 \psi^{(2)}$ into a solution $\psi^{(1)} = \Omega \psi^{(2)}$ of the equation $H^{(1)} \psi^{(1)} = k^2 \psi^{(1)}$. Moreover, we impose on the operator Ω the further condition of connecting Jost solutions satisfying the same boundary condition, to say that

(8) $$f^{(+)(1)}(x,k) = \Omega[f^{(+)(2)}(x,k)]$$
$$\equiv f^{(+)(2)}(x,k) + \int_{-\infty}^{+\infty} dy\, Z(x,y) f^{(+)(2)}(y,k),$$

where the kernel function $Z(x,y)$ vanishes if $u^{(1)} = u^{(2)}$, and is such that

(9) $$Z(x, \pm\infty) = Z_y(x, \pm\infty) = 0.$$

Note that the kernel Z (as well as the operator Ω) should be equipped with the symbol $(+)$ and the indices 1 and 2, say $Z \equiv Z^{(+)(12)}$, since its actual expression depends on the two functions $u^{(1)}$ and $u^{(2)}$ and the choice of the Jost solutions $f^{(+)(j)}$; we have omitted all the indeces for notational simplicity. In self-explanatory abstract notation

(10) $$\Omega = 1 + Z,$$

and (7) reads

(11) $$u^{(1)} - u^{(2)} = [D^2, Z] - u^{(1)} Z + Z u^{(2)}, \quad D \equiv \partial_x.$$

A more explicit version of this equation, in terms of kernels of integral operators, is

(12) $$[u^{(1)}(x) - u^{(2)}(x)] \delta(y-x) = Z_{xx}(x,y) - Z_{yy}(x,y)$$
$$- [u^{(1)}(x) - u^{(2)}(y)] Z(x,y),$$

where $\delta(t)$ is the Dirac distribution and, in the integrations by parts, (9) has been taken into account.

It is then easily seen, that a function $Z(x,y)$ consistent with (8) and (2), and yielding the Dirac distribution appearing in the l.h.s. of (12), must have the triangular form

(13) $$Z(x,y) = K(x,y) \theta(y-x),$$

where $\theta(t)$ is the unit step (Heaviside) function, namely $\theta(t) = 1$ for $t > 0$ and $\theta(t) = 0$ for $t < 0$. In fact, it is easily seen that (12) is satisfied if the function $K(x,y)$, introduced through (13), obeys the following hyperbolic second order partial differential equation

(14) $$K_{xx}(x,y) - K_{yy}(x,y) - [u^{(1)}(x) - u^{(2)}(y)] K(x,y) = 0, \quad y \geq x,$$

together with the boundary condition

(15) $\quad K(x,x) = \frac{1}{2}\int_{x}^{+\infty} dy\left[u^{(1)}(y) - u^{(2)}(y)\right].$

We summarize these findings in the following proposition:

Proposition. *If $f^{(+)(j)}(x,k)$ satisfies the second order differential equation*

(16) $\quad f_{xx}^{(+)(j)} = \left[u^{(j)}(x) - k^2\right]f^{(+)(j)}, \qquad f^{(+)(j)} \equiv f^{(+)(j)}(x,k), \quad j=1,2,$

with the boundary condition

(17) $\quad \lim_{x\to +\infty}\left[\exp(-ikx)f^{(+)(j)}(x,k)\right] = 1, \quad j=1,2,$

and $u^{(j)}(x)$ satisfies (3), and if $K(x,y)$ satisfies (14), in the half-plane $y - x \geq 0$, with (15) and

(18) $\quad K(x, +\infty) = K_y(x, +\infty) = 0,$

then $f^{(+)(1)}(x,k)$ is related to $f^{(+)(2)}(x,k)$ by the transformation

(19) $\quad f^{(+)(1)}(x,k) = f^{(+)(2)}(x,k) + \int_{x}^{+\infty} dy\, K(x,y)f^{(+)(2)}(y,k).$

Note that the conditions (3) on $u^{(j)}(x)$, $j=1,2$, imply the following properties of $K(x,y)$:

(20) $\quad K(x,y) = K^*(x,y), \qquad K(x,y) = 0 \quad \text{for } x_\infty < x \leq y.$

These properties are easily inferred from (4) and from the fact that, for $x > x_\infty$, $f^{(+)(j)}(x,k) = \exp(ikx)$ so that (19) implies

(21) $\quad \int_{x}^{+\infty} dy\, K(x,y)\exp(iky) = 0, \quad x_\infty < x,$

for any k. The k-independence of the kernel $K(x,y)$ should be emphasized as a crucial feature of the transformation (19).

Of course, the validity of (19) depends on the existence of the solution of (14) with (15) and (20). We now show that the actual existence and uniqueness of the function $K(x,y)$ follows from a theorem by Goursat, which establishes the existence and uniqueness of the solution of the so-called Goursat problem. In our particular case, this is the problem of finding the solution of (14), whose values on two intersecting characteristics have been arbitrarily given, i.e.

(22) $\quad K(x, \pm x + y_0 \mp x_0) = \chi_\pm(x);$

(x_0, y_0) is the point of the (x, y) plane where the straight lines ("characteristics") $y - y_0 = \pm(x - x_0)$ cross each other. The Goursat theorem states that the solution $K(x, y)$ of (14) is completely determined *inside* a parallelogram whose sides run along the characteristics, if its values on two intersecting sides of the parallelogram are assigned; and in particular $K(x, y)$ vanishes identically inside any such parallelogram if it vanishes on two intersecting boundaries. Note incidentally that in our case the characteristics intersect at right angles, so that the parallelograms turn out to be rectangles. Apply now this theorem to (14) and (22), firstly to the square with vertices at the points (x_∞, x_∞), $(x_\infty + d, x_\infty + d)$, $(x_\infty, x_\infty + 2d)$ and $(x_\infty - d, x_\infty + d)$ and note that, by choosing in (22) $y_0 = x_0 = x_\infty + d$ $(d > 0)$, $K(x, y)$ vanishes on the two sides that lie in the half-plane $x \geqslant x_\infty$ (see (20)). There follows, by letting d go to infinity, that

(23) $\quad K(x, y) = 0 \quad \text{for } y + x \geqslant 2x_\infty.$

Next consider the two characteristics $y = x$ and $y = -x + 2x_\infty$, and note that $K(x, y)$ is given on them by (15) and (23); therefore $K(x, y)$, in the domain $y - x \geqslant 0$ and $y + x \geqslant 2x_\infty$, is the unique solution of the Goursat problem corresponding to our equation (14), and this finally proves the important formula (19).

It is evident that these results provide another route to compute the Jost solution $f^{(+)(1)}(x, k)$, that may replace the direct integration of the Schroedinger equation for each value of k. In fact, if we assume that $f^{(+)(2)}(x, k)$ is already known, for given $u^{(1)}(x)$ the solution $K(x, y)$ of (14) yields, via the integral relation (19), the solution $f^{(+)(1)}(x, k)$ of the Schroedinger equation corresponding to $u^{(1)}(x)$. It should be noted that since $f^{(+)(1)}(x, k)$ is a function of two variables, its computation requires either the integration of infinitely many ordinary differential equations, i.e. a Schroedinger equation for each value of k, or the solution of the partial differential equation (14) in two independent variables.

For the sake of completeness, we report here the integral equation

(24) $\quad K(x, y) = \tfrac{1}{2}\left[w^{(1)}\left(\tfrac{1}{2}(x+y)\right) - w^{(2)}\left(\tfrac{1}{2}(x+y)\right)\right] + \int_0^{(y-x)/2} \mathrm{d}\xi$
$\cdot \int_{(y+x)/2}^{x_\infty} \mathrm{d}\eta \left[u^{(1)}(\eta - \xi) - u^{(2)}(\eta + \xi)\right] K(\eta - \xi, \eta + \xi)$

where

(25) $$w^{(j)}(x) \equiv \int_x^{+\infty} dy\, u^{(j)}(y), \quad j=1,2,$$

which is satisfied by the solution $K(x, y)$ of (14) with the conditions (15) and (20). This is obtained by integrating (14), and by taking into account the boundary conditions. Moreover, by differentiating (24) with respect to y and setting $x=y$, one obtains the expression

(26) $$K_y(x, x) = \tfrac{1}{4}\left\{w_x^{(1)}(x) - w_x^{(2)}(x) + \tfrac{1}{2}\left[w^{(1)}(x) - w^{(2)}(x)\right]^2\right\},$$

which implies, together with (15),

(27) $$K_x(x, x) = \tfrac{1}{4}\left\{w_x^{(1)}(x) - w_x^{(2)}(x) - \tfrac{1}{2}\left[w^{(1)}(x) - w^{(2)}(x)\right]^2\right\}.$$

(Here of course $K_y(x, x) \equiv K_y(x, y)|_{y=x}$ and $K_x(x, x) \equiv K_x(x, y)|_{y=x}$.)

So far the function $K(x, y)$ has been introduced and investigated in the framework of the direct problem, namely that of computing the functions $A^{(1)}(k)$ and $B^{(1)}(k)$ through the asymptotic behaviour of the solution $f^{(+)(1)}(x, k)$ corresponding to the function $u^{(1)}(x)$ (the direct problem related to $u^{(2)}(x)$ is considered as solved; the simplest instance, to which consideration was confined in section 2.1, corresponds to $u^{(2)}(x)=0$, implying of course $f^{(+)(2)}(x, k) = \exp(ikx)$).

However, the function $K(x, y)$ plays more of a rôle in the inverse problem than in the direct problem, as it will be eventually clear. Let us now proceed to establish an important integral representation for this kernel. Notice that the special structure of the partial differential equation (14) suggests that any bilinear expression of the form $g^{(1)}(x, k)g^{(2)}(y, k)$ satisfies (14), provided $g^{(j)}(x, k)$ is a solution of the Schroedinger equation (16) corresponding to $u^{(j)}(x)$ and k is an arbitrary complex number. Because of the linearity of (14), the following *ansatz* is naturally suggested:

(28) $$K(x, y) = \int_C dk\, \rho(k) f^{(+)(1)}(x, k) f^{(+)(2)}(y, k), \quad y \geq x.$$

Here the integration runs on a curve C in the complex k-plane (this curve should always be understood in a generalized sense, as part of it could reduce to a discrete set of points). Of course, the integral (28), if convergent, yields for any C and $\rho(k)$ a solution of (14). The important point that must still be considered is the assumption that a curve C and a function $\rho(k)$

exist such that the integral (28) provides a representation of the solution of the equation (14) satisfying the conditions (20), (15) and (18).

First of all, because of the condition (18), the variable k in the integration (28) cannot take values with negative imaginary part; in fact, the solution $f^{(+)(2)}(y,k)$ diverges exponentially as $y \to +\infty$ if k lies in the lower half-plane. Therefore, the curve C is required to be in the upper half-plane Im $k \geq 0$. Next we note that, since the solutions $f^{(+)(j)}(x,k)$ are bounded for real k, the curve C would in general include the real axis of the k-plane, provided the function $\rho(k)$ vanishes as $|k|$ increases,

(29) $\qquad \lim_{k \to \pm \infty} \rho(k) = 0, \quad \text{Im } k = 0.$

The asymptotic rate of vanishing of $\rho(k)$ should be sufficiently fast to insure the convergence of the integral in (28); a condition that is easily achieved, since the solutions $f^{(+)(j)}(x,k)$ are not only bounded but in fact oscillating as $k \to \pm \infty$, being asymptotically proportional to $\exp(ikx)$.

As for the contributions to $K(x,y)$ coming, in the integral (28), from values of k with Im $k > 0$, it should be first noticed that, for a generic k with Im $k > 0$, the Jost solution $f^{(+)(j)}(x,k)$ would exponentially diverge as $x \to -\infty$. On the other hand, this divergence is not compatible with the relation (15) which, because of our assumptions on the functions $u^{(j)}(x)$, $j=1,2$, demands that $K(x,x)$ remain finite and constant as $x \to -\infty$. Therefore, only those values of k in the upper half-plane are allowed such that the product $f^{(+)(1)}(x,k)f^{(+)(2)}(x,k)$ converge to a constant as $x \to -\infty$. This can happen only if (at least) one of the two Jost solutions asymptotically vanishes in the limit $x \to -\infty$. We conclude then, looking at the asymptotic behaviour (6), that in the upper half-plane the curve C degenerates to the set of points which are either the zeros of $B^{(2)}(k)$ (if any),

(30) $\qquad B^{(2)}(k_n^{(2)}) = 0, \quad \text{Im } k_n^{(2)} > 0, \qquad n = 1, 2, \ldots, N^{(2)},$

or the zeros of $B^{(1)}(k)$ (if any),

(31) $\qquad B^{(1)}(k_n^{(1)}) = 0, \quad \text{Im } k_n^{(1)} > 0, \qquad n = 1, 2, \ldots, N^{(1)}.$

It is easily seen (see (5)) that these values correspond respectively to the discrete eigenvalues of the Schroedinger problems with potentials $u^{(2)}(x)$ respectively $u^{(1)}(x)$. Thus (see section 2.1 and appendix A.1) all these values $k_n^{(2)}$ and $k_n^{(1)}$ lie on the upper imaginary axis, and there is a *finite* number of them (possibly none):

(32) $\qquad k_n^{(j)} = i p_n^{(j)}, \quad p_n^{(j)} > 0, \qquad n = 1, 2, \ldots, N^{(j)}, \quad j = 1, 2.$

A.5 The inverse spectral Schroedinger problem

In conclusion, the most general curve C is the union of the real axis $\operatorname{Im} k = 0$ and of a finite number of discrete points on the imaginary axis. On C the measure $dk\,\rho(k)$ is characterized by a function of the real variable k, satisfying (29) and the reality condition

(33) $\quad \rho(k) = \rho^*(-k), \quad \operatorname{Im} k = 0,$

and by $N^{(1)} + N^{(2)}$ real numbers corresponding to the discrete set (32). The integral representation (28) then reads

(34) $\quad K(x,y) = \int_{-\infty}^{+\infty} dk\,\rho(k) f^{(+)(1)}(x,k) f^{(+)(2)}(y,k)$

$\qquad - \sum_{n=1}^{N^{(1)}} \rho_n^{(1)} f^{(+)(1)}(x, i p_n^{(1)}) f^{(+)(2)}(y, i p_n^{(1)})$

$\qquad + \sum_{n=1}^{N^{(2)}} \rho_n^{(2)} f^{(+)(1)}(x, i p_n^{(2)}) f^{(+)(2)}(y, i p_n^{(2)}).$

The condition (33) on $\rho(k)$ easily follows from the reality of $K(x,y)$, see (20), and from the property (4) of the Jost solutions. The sign in front of the two sums has been chosen so that the real numbers $\rho_n^{(1)}$ and $\rho_n^{(2)}$ are positive,

(35) $\quad \rho_n^{(j)} > 0, \quad n = 1, 2, \ldots, N^{(j)}, \quad j = 1, 2,$

as is proved below.

In the framework of the direct spectral problem, $u^{(1)}(x)$ and $u^{(2)}(x)$ uniquely determine the function $K(x,y)$ (via the Goursat problem, see above); thus, in this framework the representation (34) should be completed by the specification of a procedure to evaluate the function $\rho(k)$ and the numbers $\rho_n^{(j)}$ from $u^{(1)}(x)$ and $u^{(2)}(x)$. To this end, we start by rewriting the relationship (19) with $K(x,y)$ represented by (34), with the assumption that the order of the integrations involved can be exchanged (of course, in the context of distribution theory; see (37) below):

(36) $\quad f^{(+)(1)}(x,k) = f^{(+)(2)}(x,k) + \int_{-\infty}^{+\infty} dk'\,\rho(k') f^{(+)(1)}(x,k')$

$\qquad \cdot \int_x^{+\infty} dy\, f^{(+)(2)}(y,k') f^{(+)(2)}(y,k)$

$\qquad - \sum_{n=1}^{N^{(1)}} \rho_n^{(1)} f^{(+)(1)}(x, i p_n^{(1)}) \int_x^{+\infty} dy\, f^{(+)(2)}(y, i p_n^{(1)}) f^{(+)(2)}(y,k)$

$\qquad + \sum_{n=1}^{N^{(2)}} \rho_n^{(2)} f^{(+)(1)}(x, i p_n^{(2)}) \int_x^{+\infty} dy\, f^{(+)(2)}(y, i p_n^{(2)}) f^{(+)(2)}(y,k).$

We first investigate this formula and its implications by taking k real and making use of the wronskian relations, in particular

$$(37) \quad \int_x^{+\infty} dy \, f^{(+)(2)}(y,k') f^{(+)(2)}(y,k)$$
$$= (k'^2 - k^2)^{-1} W[f^{(+)(2)}(x,k), f^{(+)(2)}(x,k')] + \pi \delta(k'+k),$$
$$\text{Im } k = 0, \quad \text{Im } k' = 0,$$

where $W[f,g] \equiv fg_x - gf_x$ and the formula

$$(38) \quad \lim_{x \to +\infty} P\left(\frac{1}{q} \exp(iqx)\right) = i\pi\delta(q)$$

has been used (see appendix A.2). *After* all the integrations in the variable y have been performed in (36) through the wronskian relations, we consider the limit of this equation as $x \to -\infty$ and equate the coefficients of the exponential functions $\exp(\pm ikx)$ which arise from the asymptotic behaviour (6). After a lengthy but trivial calculation, one obtains the two equations

$$(39) \quad A^{(1)}(k) = 2\pi\rho(-k) B^{(1)}(-k) + A^{(2)}(k) Y(k), \quad \text{Im } k = 0$$
$$(40) \quad B^{(1)}(k) = B^{(2)}(k) Y(-k), \quad \text{Im } k = 0,$$

where we have defined

$$(41) \quad Y(k) \equiv 1 + b(k) + (i/\pi) P \int_{-\infty}^{+\infty} dk' \, b(k')/(k'-k)$$
$$+ i \sum_{n=1}^{N^{(1)}} \rho_n^{(1)} \gamma_n^{(1)} B^{(2)}(ip_n^{(1)})/(k - ip_n^{(1)})$$
$$- i \sum_{n=1}^{N^{(2)}} \rho_n^{(2)} \gamma_n^{(2)} B^{(1)}(ip_n^{(2)})/(k - ip_n^{(2)}), \quad \text{Im } k = 0,$$

and

$$(42) \quad b(k) \equiv \pi[\rho(k) B^{(2)}(k) A^{(1)}(k) + \rho(-k) A^{(2)}(-k) B^{(1)}(-k)];$$

the constants $\gamma_n^{(j)}$'s are defined by the asymptotic decay at negative x of the Jost solutions at the points $k_n^{(j)} = ip_n^{(j)}$, namely

$$(43) \quad f^{(+)(j)}(x, ip_n^{(j)}) = \gamma_n^{(j)} f^{(-)(j)}(x, ip_n^{(j)}), \quad n = 1, 2, \ldots, N^{(j)}, \quad j = 1, 2,$$

or, in the notation of section 2.1 (see (2.1.-28)),

(44) $\quad \gamma_n^{(j)} = c_n^{(-)(j)}/c_n^{(+)(j)}.$

Note that, in order to derive (39) and (40), the so called unitarity relation

(45) $\quad B(k)B(-k) = 1 + A(k)A(-k)$

has been used; this follows from (5) and from the trivial relation

(46) $\quad W[f^{(+)}(x,-k), f^{(+)}(x,k)] \equiv f^{(+)}(x,-k)f_x^{(+)}(x,k)$
$\quad\quad\quad -f_x^{(+)}(x,-k)f^{(+)}(x,k) = 2ik$

implied by the x-independence of the wronskian and by (2). By substituting $Y(k)$ in (39) with its expression derived from (40), we obtain the function $\rho(k)$ in terms of the asymptotic coefficients,

(47) $\quad \rho(k) = (2\pi)^{-1}\{[A^{(1)}(-k)/B^{(1)}(k)] - [A^{(2)}(-k)/B^{(2)}(k)]\}.$

As for the expression of the positive numbers $\rho_n^{(j)}$, we start again from (36), but we now take k in the upper half-plane, i.e. Im $k > 0$. Again we eliminate the integration in the variable y via the standard wronskian expression and investigate the asymptotic behaviour of the l. and r.h.s. of (36) as $x \to -\infty$. For arbitrary k, both terms of (36) exponentially diverge, since Im $k > 0$ and their equality implies, for the coefficients of the leading exponential exp(ikx) (see (6)),

(48) $\quad [B^{(1)}(k)/B^{(2)}(k)] = 1 + (2i\pi)^{-1}\int_{-\infty}^{+\infty} dk'$
$\quad\quad \cdot \{[B^{(1)}(k')/B^{(2)}(k')] - 1\}/(k'-k)$
$\quad\quad + (2i\pi)^{-1}\int_{-\infty}^{+\infty} dk' \{[B^{(2)}(k')/B^{(1)}(k')] - 1\}/(k'+k)$
$\quad\quad - i \sum_{n=1}^{N^{(1)}} \rho_n^{(1)} \gamma_n^{(1)} B^{(2)}(ip_n^{(1)})/(k + ip_n^{(1)})$
$\quad\quad + i \sum_{n=1}^{N^{(2)}} \rho_n^{(2)} \gamma_n^{(2)} B^{(1)}(ip_n^{(2)})/(k - ip_n^{(2)}),\quad$ Im $k > 0$;

this equation obtains from (40) by using the previous result (47) to reexpress the function $b(k)$, defined by (42), in the form

(49) $\quad b(k) = \frac{1}{2}\{[B^{(1)}(-k)/B^{(2)}(-k)] - [B^{(2)}(k)/B^{(1)}(k)]\}, \quad \text{Im } k = 0.$

Note, moreover, that the convergence of the integrals in (48) follows from the limiting behaviour

(50) $\quad \lim_{k \to \infty} [B^{(j)}(k) - 1] = 0, \quad \text{Im } k \geq 0, \quad j = 1, 2$

(this property is rather obvious from the very definition (5); for a proof, see (57) below and appendix A.3). Since the coefficient $B^{(j)}(k)$, if the conditions (3) hold, is analytic in the upper half of the complex k-plane (see e.g. (57) below and section 2.1), the integrals in the r.h.s. of (48) can be easily evaluated by the standard technique of closing the path of integration in the upper half-plane along a semicircle of infinite radius, and using the Cauchy theorem. Since the zeros of the function $B^{(j)}(k)$ are simple (as it is proved below; see (63)), the final result reads

(51) $\quad \sum_{n=1}^{N^{(1)}} B^{(2)}(i p_n^{(1)}) \{[B_k^{(1)}(i p_n^{(1)})]^{-1} - i \rho_n^{(1)} \gamma_n^{(1)}\} / (k + i p_n^{(1)})$

$\quad - \sum_{n=1}^{N^{(2)}} B^{(1)}(i p_n^{(2)}) \{[B_k^{(2)}(i p_n^{(2)})]^{-1} - i \rho_n^{(2)} \gamma_n^{(2)}\} / (k - i p_n^{(2)}) = 0,$

where $B_k(k) \equiv d B(k)/dk$. Clearly, this equation implies that

(52) $\quad \rho_n^{(j)} = [i \gamma_n^{(j)} B_k^{(j)}(i p_n^{(j)})]^{-1}, \quad n = 1, 2, \ldots, N^{(j)}, \quad j = 1, 2.$

The formulae (47) and (52) are the main results of these calculations and show that $\rho(k)$ and $\rho_n^{(j)}$ can be directly computed from the functions $u^{(1)}(x)$ and $u^{(2)}(x)$, through the Jost solutions $f^{(+)(j)}(x, k)$ which obtain by integrating the Schroedinger equation (16) with (17).

Before proceeding further it is convenient to display the connection with the formalism of section (2.1) (but we try to be self-consistent throughout this appendix, even at the cost of rewriting here some of the formulae of section 2.1). Let $\psi(x, k)$ be the solution, for real k, of the Schroedinger equation

(53) $\quad \psi_{xx} = [u(x) - k^2] \psi, \quad \psi \equiv \psi(x, k),$

with the asymptotic conditions

(54a) $\psi(x,k) \underset{x\to+\infty}{\to} \exp(-ikx)+R(k)\exp(ikx),$

(54b) $\psi(x,k) \underset{x\to-\infty}{\to} T(k)\exp(-ikx),$

which define the transmission and reflection coefficients $T(k)$ and $R(k)$. These formulae refer to the continuous part of the spectrum. The discrete spectrum (if any) is defined by the formulae

(55) $\varphi_{nxx} = [u(x)+p_n^2]\varphi_n, \quad \int_{-\infty}^{+\infty} dx [\varphi_n(x)]^2 = 1, \quad n=1,2,\ldots,N.$

The asymptotic conditions (54), together with (2), imply

(56) $T(k)f^{(-)}(x,k) = \psi(x,k) = f^{(+)}(x,-k) + R(k)f^{(+)}(x,k);$

substituting (5) into this last equality yields

(57) $A(k) = -R(-k)/T(-k), \quad B(k) = 1/T(k).$

On the discrete spectrum, we have

(58) $\varphi_n(x) = c_n f^{(+)}(x,ip_n), \quad c_n^2 = \left\{ \int_{-\infty}^{+\infty} dx [f^{(+)}(x,ip_n)]^2 \right\}^{-1},$

$$n=1,2,\ldots,N;$$

a simple connection between the constants c_n and the parameter $\rho_n = [i\gamma_n B_k(ip_n)]^{-1}$ (see (52) and (34)),

(59) $\rho_n = c_n^2,$

is provided by the following theorem which also proves the simplicity of the zeros of $B(k)$ in the upper half-plane (and the positivity of ρ_n). Consider the wronskian type relation

(60) $f^{(+)}(x,k)f^{(+)}_{kxx}(x,k) - f^{(+)}_x(x,k)f^{(+)}_k(x,k)$

$$= 2k \int_x^{+\infty} dy [f^{(+)}(y,k)]^2, \quad \operatorname{Im} k > 0,$$

and let $k = ip_n$ be a zero of $B(k)$, i.e.

(61) $B(ip_n) = 0;$

then, from (43), (6) and (61), there follow the asymptotic formulae

(62a) $\quad f^{(+)}(x,ip_n) \underset{x\to-\infty}{\to} \gamma_n \exp(p_n x),$

(62b) $\quad f_k^{(+)}(x,ip_n) \underset{x\to-\infty}{\to} B_k(ip_n)\exp(-p_n x),$

which, inserted in (60), yield

(63) $\quad i\gamma_n B_k(ip_n) = \int_{-\infty}^{+\infty} dx \left[f^{(+)}(x,ip_n) \right]^2.$

Because of the reality of $f^{(+)}(x,ip_n)$ implied by (4), this formula shows that $B_k(ip_n)$ cannot vanish and, furthermore, it provides the alternative expression for the parameters (52),

(64) $\quad \rho_n^{(j)} = \left\{ \int_{-\infty}^{+\infty} dx \left[f^{(+)(j)}(x,ip_n^{(j)}) \right]^2 \right\}^{-1} = \left[c_n^{(j)} \right]^2,$

where we have inserted again the index j.

If we associate, through the Schroedinger spectral problem, to a function $u(x)$ satisfying (3) the quantities

(65) $\quad S[u] = \{R(k), p_n, \rho_n\},$

known as the *spectral transform* of $u(x)$, we conclude, from (34), (47) and (57) that the kernel $K(x, y)$ takes the final expression

(66) $\quad K(x,y) = -(2\pi)^{-1} \int_{-\infty}^{+\infty} dk\, f^{(+)(1)}(x,k) f^{(+)(2)}(y,k)$

$\cdot \left[R^{(1)}(k) - R^{(2)}(k) \right] - \sum_{n=1}^{N^{(1)}} f^{(+)(1)}(x,ip_n^{(1)}) f^{(+)(2)}(y,ip_n^{(1)}) \rho_n^{(1)}$

$+ \sum_{n=1}^{N^{(2)}} f^{(+)(1)}(x,ip_n^{(2)}) f^{(+)(2)}(y,ip_n^{(2)}) \rho_n^{(2)},$

in terms of the Jost solutions and of the spectral transforms of the functions $u^{(1)}(x)$ and $u^{(2)}(x)$. By itself, this formula is not useful in the context of the direct problem, since it requires knowledge of the solutions of the spectral problem, both for $u^{(1)}(x)$ and for $u^{(2)}(x)$. On the other hand, it is useful to formulate the inverse problem and to find its basic equation. In fact, by writing in the expression (66) the Jost solution $f^{(+)(1)}(x,k)$ in terms of

$f^{(+)(2)}(x,k)$ via the transformation (19), we obtain for $K(x, y)$ the Gel'fand–Levitan–Marchenko equation

(67) $\quad K(x, y) + M(x, y) + \int_{x}^{+\infty} dz\, K(x, z) M(z, y) = 0, \quad x \leqslant y,$

where the kernel $M(x, y)$ is defined as follows:

(68) $\quad M(x, y) = (2\pi)^{-1} \int_{-\infty}^{+\infty} dk\, f^{(+)(2)}(x, k) f^{(+)(2)}(y, k)$

$\cdot \left[R^{(1)}(k) - R^{(2)}(k) \right] + \sum_{n=1}^{N^{(1)}} f^{(+)(2)}(x, i p_n^{(1)}) f^{(+)(2)}(y, i p_n^{(1)}) \rho_n^{(1)}$

$- \sum_{n=1}^{N^{(2)}} f^{(+)(2)}(x, i p_n^{(2)}) f^{(+)(2)}(y, i p_n^{(2)}) \rho_n^{(2)}.$

Note that the variable x in (67) plays the role of a parameter and that, for each fixed value of x, (67) is a Fredholm integral equation, whose solution exists and is unique. In particular, it follows that, for a given function $R(k)$ (within an appropriate class) and a given set of positive numbers, p_n and ρ_n, one and only one function $u(x)$ exists, such that, through the Schroedinger spectral problem, (65) holds. To show this, let us choose

(69) $\quad u^{(2)}(x) = 0, \quad u^{(1)}(x) = u(x),$

so that the integral equation (67) now specializes to

(70) $\quad K(x, y) + M(x+y) + \int_{x}^{+\infty} dz\, K(x, z) M(z+y) = 0, \quad x \leqslant y,$

with

(71) $\quad M(z) = (2\pi)^{-1} \int_{-\infty}^{+\infty} dk\, \exp(ikz)\, R(k) + \sum_{n=1}^{N} \exp(-p_n z)\, \rho_n.$

Note that, given $R(k)$, p_n and ρ_n, the kernel of the integral equation (70) is known, and explicitly given by (71). Then $K(x, y)$ is obtained uniquely, by solving (70). Finally the function $u(x)$ is easily recovered by the relation (15), with (69), namely

(72) $\quad u(x) = -2\, dK(x, x)/dx.$

Thus, the inverse problem, which is the problem of reconstructing the function $u(x)$ from its spectral transform (65), is reduced to the problem of solving the integral equation (70). Actually, the solution $K(x, y)$ of (70) yields not only the function $u(x)$, but also the corresponding Jost solution of the Schroedinger equation, through the relation (19) that now reads

(73) $\quad f^{(+)}(x, k) = \exp(ikx) + \int_{x}^{+\infty} dy\, K(x, y) \exp(iky).$

We summarize the main content of this appendix by graphically displaying the connections between the various quantities which are relevant both to the direct and to the inverse spectral problems:

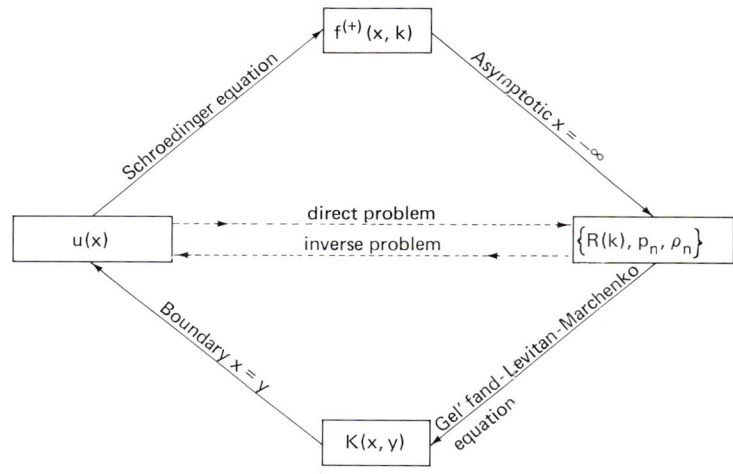

The solution of the direct problem, indicated by the dotted line going from $u(x)$ to $\{R(k), p_n, \rho_n\}$, requires the integration of the Schroedinger equation (1), for instance with the boundary conditions (2); this is a linear problem, whose solution, say $f^{(+)}(x, k)$, is a function of two variables, since, in addition to x, it depends also on the spectral parameter k. From this Jost solution, the spectral transform (65) of $u(x)$ is obtained from the asymptotic coefficients $A(k)$ and $B(k)$, defined by (6), according to (57), i.e.

(74) $\quad R(k) = -A(-k)/B(k), \quad \operatorname{Im} k = 0,$

and from (61) and (64). Alternatively (see section 2.1), the formulae (53),

(54) and (55) could be used, together with

(75) $$\lim_{x \to +\infty} \left[\exp(2p_n x) \varphi_n^2(x) \right] = \rho_n, \quad n = 1, 2, \ldots, N,$$

as implied by (58) and (59).

As for the inverse spectral problem, namely the reconstruction of $u(x)$ from $\{R(k), p_n, \rho_n\}$, the first step consists in solving the Gel'fand–Levitan–Marchenko equation (70), whose kernel is given by (71); again, this is a linear problem, whose solution is the function $K(x, y)$ of two (real) variables. Then the function $u(x)$ is easily recovered from the boundary function $K(x, x)$, see (72).

It should be finally emphasized that these results, and therefore the spectral transformation shown by the above diagram, apply to a larger class of functions $u(x)$ than has been considered here; in fact, the conditions (3), which have been assumed to make the arguments as simple as possible, can be weakened by requiring the function $u(x)$ to vanish asymptotically, say faster than $|x|^{-2}$ so that

(76) $$\int_{-\infty}^{+\infty} dx (1 + |x|) |u(x)| < \infty.$$

Of course, the formalism with two functions, $u^{(1)}(x)$ and $u^{(2)}(x)$, is more general and, indeed, it yields some of the equations displayed in section 2.4 and used extensively in chapters 3 and 4. For instance the important formula

(77) $$w^{(1)}(x) - w^{(2)}(x)$$

$$= -\pi^{-1} \int_{-\infty}^{+\infty} dk \, f^{(+)(1)}(x, k) f^{(+)(2)}(x, k) \left[R^{(1)}(k) - R^{(2)}(k) \right]$$

$$- 2 \sum_{n=1}^{N^{(1)}} f^{(+)(1)}(x, i p_n^{(1)}) f^{(+)(2)}(x, i p_n^{(1)}) \rho_n^{(1)}$$

$$+ 2 \sum_{n=1}^{N^{(2)}} f^{(+)(1)}(x, i p_n^{(2)}) f^{(+)(2)}(x, i p_n^{(2)}) \rho_n^{(2)}$$

is obtained directly from (15), (25) and (66).

The approach to the inverse problem described here applies as well to many other spectral problems, such as the generalized Zakharov–Shabat and the matrix Schroedinger problems, with few changes requiring (at most) some technical skill but no new ideas. However, for these problems (as well as for the simpler one treated here), another, possibly more convenient, procedure to derive the Gel'fand–Levitan–Marchenko equation and all related results may be based on the nice analyticity properties of the Jost solutions in the complex k-plane, and the use of standard contour-integral techniques. On the other hand, there are other cases when the analyticity properties of the Jost solutions are not so simple, and in these cases it is rather an approach along the lines followed here that is preferable. One such case is the Schroedinger problem with an additional linear potential (see subsection 6.3.1).

A.5.N. Notes to appendix A.5

Some comments and references on the inverse problem are given in section 2.N. In addition to those, we quote here the original papers [M1950] and [GL1951], that refer however to the spectral problem on the semiline or a finite interval. For the treatment of the inverse problem on the whole line (one-dimensional Schroedinger problem) see [Kay1955], and also [B1953], [KayMo1956], [KayMo1957] and [F1958]; see also [Mos1979]. A more general inverse problem, associated with the Schroedinger equation with a potential depending linearly on the spectral parameter k, has been treated in [JJ1975].

As for the explicit solutions of the inverse problem, in addition to those corresponding to a purely discrete spectrum (yielding reflectionless potentials, see section 2.2. and appendix A.19), those corresponding to a reflection coefficient that is a rational function of the spectral variable k have been studied by [Mos1956], [Kay1960], and [Mos1964]; see also [P1967] and [SCYA1976]. The construction of these solutions is detailed in appendix A.19.

A.6. Wronskian integral relations: proofs

In this appendix we indicate how the wronskian integral relations of subsection 2.4.1 are proved.

A.6 Wronskian integral relations: proofs

The basic formula, from which all others are easily derived, reads

(1) $$4k^2 \int_{x_1}^{x_2} dx\, \psi^{(1)}(x)\psi^{(2)}(x)F_x(x) = \int_{x_1}^{x_2} dx\, \psi^{(1)}(x)\psi^{(2)}(x)$$
$$\cdot \Big\{ -F_{xxx}(x) + 2[u^{(1)}(x)+u^{(2)}(x)]F_x(x) + [u_x^{(1)}(x)+u_x^{(2)}(x)]$$
$$\cdot F(x) + [u^{(1)}(x)-u^{(2)}(x)] \int_x^{x_0} dy\, [u^{(1)}(y)-u^{(2)}(y)]F(y) \Big\}$$
$$+ \Big\{ \psi^{(1)}(x)\psi^{(2)}(x)[2k^2 F(x)+F_{xx}(x)-[u^{(1)}(x)+u^{(2)}(x)]F(x)]$$
$$+2\psi_x^{(1)}(x)\psi_x^{(2)}(x)F(x) - [\psi^{(1)}(x)\psi^{(2)}(x)]_x F(x)$$
$$- [\psi_x^{(1)}(x)\psi^{(2)}(x) - \psi^{(1)}(x)\psi_x^{(2)}(x)]$$
$$\cdot \int_x^{x_0} dy\, [u^{(1)}(y)-u^{(2)}(y)]F(y) \Big\} \Big|_{x=x_1}^{x=x_2}.$$

Here $\psi^{(1)}(x)$ and $\psi^{(2)}(x)$ are solutions of the Schroedinger equation with potentials $u^{(1)}(x)$ and $u^{(2)}(x)$,

(2) $$\psi_{xx}^{(j)}(x) + k^2 \psi^{(j)}(x) = u^{(j)}(x)\psi^{(j)}(x), \quad j=1,2,$$

k^2 is a constant (not necessarily positive nor, for that matter, even real); x_0, x_1 and x_2 are arbitrary fixed values of x; and $F(x)$ is an arbitrary function of x, only restricted (together with $u^{(1)}(x)$ and $u^{(2)}(x)$) by the requirement that the various integrations by parts involved in the derivation of (1) (using (2); see below) be permissible, and of course by the (related) requirement that all integrals in (1) be well defined (thus, it is sufficient that $F(x)$ be thrice differentiable).

The proof of (1) is not amusing; we outline here one procedure that the diligent reader may easily follow to arrive at (1).

Clearly (2) implies

(3) $$2k^2 \int_{x_1}^{x_2} dx\, \psi^{(1)}(x)\psi^{(2)}(x)F_x(x)$$
$$\doteq -\int_{x_1}^{x_2} dx\, [\psi_{xx}^{(1)}(x)\psi^{(2)}(x)+\psi^{(1)}(x)\psi_{xx}^{(2)}(x)]F_x(x).$$

Here we have introduced the symbol $\doteq$ to denote equality up to the addition of terms of the form $\int_{x_1}^{x_2} dx\, \psi^{(1)}(x)\psi^{(2)}(x)G(x)$ (with $G(x)$ independent of k^2) or of contributions evaluated at the extremes, x_1 and x_2, of the integration range (such contributions arise from the various integration by parts that are performed below). The reader will have no difficulty in keeping track of these additional contributions, that eventually yield the r.h.s. of (1).

The next step is an integration by parts, yielding

(4) $\quad -\int_{x_1}^{x_2} dx\, [\psi_{xx}^{(1)}(x)\psi^{(2)}(x) + \psi^{(1)}(x)\psi_{xx}^{(2)}(x)] F_x(x)$

$\doteq \int_{x_1}^{x_2} dx\, \{2\psi_x^{(1)}(x)\psi_x^{(2)}(x) F_x(x) + [\psi^{(1)}(x)\psi^{(2)}(x)]_x F_{xx}(x)\}.$

The second term is taken care of by just one more partial integration. As for the first term, a partial integration in reverse order, and the use once more of (2), yields

(5) $\quad 2\int_{x_1}^{x_2} dx\, \psi_x^{(1)}(x)\psi_x^{(2)}(x) F_x(x) \doteq 2k^2 \int_{x_1}^{x_2} dx\, [\psi^{(1)}(x)\psi^{(2)}(x)]_x F(x)$

$- 2\int_{x_1}^{x_2} dx\, [u^{(1)}(x)\psi^{(1)}(x)\psi_x^{(2)}(x) + u^{(2)}(x)\psi_x^{(1)}(x)\psi^{(2)}(x)] F(x).$

The first term in the r.h.s. can now be integrated by parts once more, and it yields precisely the same term as in the l.h.s. of (3), but with the opposite sign. Thus from (3), (4) and (5) there follows

(6) $\quad 4k^2 \int_{x_1}^{x_2} dx\, \psi^{(1)}(x)\psi^{(2)}(x) F_x(x)$

$\doteq -2\int_{x_1}^{x_2} dx\, [u^{(1)}(x)\psi^{(1)}(x)\psi_x^{(2)}(x)$

$+ u^{(2)}(x)\psi_x^{(1)}(x)\psi^{(2)}(x)] F(x).$

It is now convenient to use the identity

(7) $\quad 2[u^{(1)}(x)\psi^{(1)}(x)\psi_x^{(2)}(x) + u^{(2)}(x)\psi_x^{(1)}(x)\psi^{(2)}(x)]$

$= [\psi^{(1)}(x)\psi^{(2)}(x)]_x [u^{(1)}(x) + u^{(2)}(x)]$

$- [\psi_x^{(1)}(x)\psi^{(2)}(x) - \psi^{(1)}(x)\psi_x^{(2)}(x)][u^{(1)}(x) - u^{(2)}(x)].$

This, together with (6) (and one integration by parts to eliminate the first term in the r.h.s.) yields

(8) $\quad 4k^2 \int_{x_1}^{x_2} dx\, \psi^{(1)}(x)\psi^{(2)}(x)F_x(x)$

$\doteq -\int_{x_1}^{x_2} dx\,\bigl[\psi_x^{(1)}(x)\psi^{(2)}(x)-\psi^{(1)}(x)\psi_x^{(2)}(x)\bigr]$

$\cdot \Bigl[\int_x^{x_0} dy\,[u^{(1)}(y)-u^{(2)}(y)]F(y)\Bigr]_x ,$

which has been written so as to suggest what the next integration by parts should be. And indeed this last integration by parts, together with (2), completes the job, yielding finally (1).

Once (1) has been established, it is easy to get all the formulae of subsection 2.4.1.

Note first of all that the arbitrariness of x_0 in (1) implies the well-known formula

(9) $\quad \bigl[\psi_x^{(1)}(x)\psi^{(2)}(x)-\psi^{(1)}(x)\psi_x^{(2)}(x)\bigr]\Big|_{x=x_1}^{x=x_2}$

$= \int_{x_1}^{x_2} dx\, \psi^{(1)}(x)\psi^{(2)}(x)\bigl[u^{(1)}(x)-u^{(2)}(x)\bigr],$

that could of course have been established very simply from the wronskian theorem.

Assume then that k^2 is real and positive, and that

(10) $\quad \psi^{(j)}(x)=\psi^{(j)}(x,k),\quad j=1,2,$

where $\psi^{(j)}(x,k)$ are the solutions of (2) characterized by the boundary conditions (see (2.4.1.-1a, 1b)

(11a) $\quad \psi^{(j)}(x,k)\to T^{(j)}(k)\exp(-ikx),\quad x\to -\infty$

(11b) $\quad \psi^{(j)}(x,k)\to \exp(-ikx)+R^{(j)}(k)\exp(ikx),\quad x\to +\infty.$

Then, for $x_1=-\infty$ and $x_2=+\infty$, (9) yields

(12) $\quad 2ik\bigl[R^{(1)}(k)-R^{(2)}(k)\bigr]$

$= \int_{-\infty}^{+\infty} dx\, \psi^{(1)}(x,k)\psi^{(2)}(x,k)\bigl[u^{(1)}(x)-u^{(2)}(x)\bigr],$

namely (2.4.1-3).

Let instead

(13) $\quad \psi^{(1)}(x) = \psi^{(1)}(x, k), \qquad \psi^{(2)}(x) = \psi^{(2)}(x, -k),$

again with $\psi^{(j)}(x, k)$ characterized by the boundary conditions (11). Then, again for $x_1 = -\infty$ and $x_2 = +\infty$, (9) yields

(14) $\quad 2ik\left[R^{(1)}(k)R^{(2)}(-k) + T^{(1)}(k)T^{(2)}(-k) - 1\right]$
$$= \int_{-\infty}^{+\infty} dx \, \psi^{(1)}(x, k) \psi^{(2)}(x, -k) \left[u^{(1)}(x) - u^{(2)}(x)\right],$$

namely (2.4.1.-21).

Return now to (1), and let $F(x) = 1$ and $x_0 = +\infty$. This yields

(15) $\quad \left\{ \psi^{(1)}(x)\psi^{(2)}(x)\left[u^{(1)}(x) + u^{(2)}(x) - 2k^2\right] - 2\psi_x^{(1)}(x)\psi_x^{(2)}(x) \right.$

$\qquad + \left[\psi_x^{(1)}(x)\psi^{(2)}(x) - \psi^{(1)}(x)\psi_x^{(2)}(x)\right]$

$\qquad \left. \cdot \int_x^{+\infty} dy \left[u^{(1)}(y) - u^{(2)}(y)\right] \right\} \Big|_{x=x_1}^{x=x_2} = \int_{x_1}^{x_2} dx \, \psi^{(1)}(x)\psi^{(2)}(x) \Gamma \cdot 1$

where Γ is the integrodifferential operator defined by the formula

(16) $\quad \Gamma \cdot f(x) = \left[u_x^{(1)}(x) + u_x^{(2)}(x)\right] f(x) + \left[u^{(1)}(x) - u^{(2)}(x)\right]$
$$\cdot \int_x^{+\infty} dy \left[u^{(1)}(y) - u^{(2)}(y)\right] f(y)$$

(see (2.4.1.-5)).

Consider then, always for $k^2 > 0$, the limiting case $x_1 = -\infty$, $x_2 = +\infty$, for the two choices (10) and (13). There obtain

(17) $\quad -4k^2\left[R^{(1)}(k) + R^{(2)}(k)\right] = \int_{-\infty}^{+\infty} dx \, \psi^{(1)}(x, k)\psi^{(2)}(x, k)\Gamma \cdot 1,$

(18) $\quad -4k^2\left[R^{(1)}(k)R^{(2)}(-k) - T^{(1)}(k)T^{(2)}(-k) + 1\right]$
$$+ 2ikT^{(1)}(k)T^{(2)}(-k) \int_{-\infty}^{+\infty} dx \left[u^{(1)}(x) - u^{(2)}(x)\right]$$
$$= \int_{-\infty}^{+\infty} dx \, \psi^{(1)}(x, k)\psi^{(2)}(x, -k)\Gamma \cdot 1,$$

namely (2.4.1.-4a) and (2.4.1.-22).

A.6 Wronskian integral relations: proofs

Return again to (1), and set

(19) $\quad F(x) = \int_{x}^{+\infty} dy \, f(y)$

with

(20) $\quad f(\pm\infty) = f_x(\pm\infty) = 0.$

Let moreover $x_0 = +\infty$, $x_1 = -\infty$ and $x_2 = +\infty$, always assuming k^2 to be real and positive so that the $\psi^{(j)}(x)$'s do not diverge asymptotically. Then (1) reads

(21) $\quad -4k^2 \int_{-\infty}^{+\infty} dx \, \psi^{(1)}(x) \psi^{(2)}(x) f(x) = \int_{-\infty}^{+\infty} dx \, \psi^{(1)}(x) \psi^{(2)}(x) \Lambda f(x)$

$\quad -2 \left\{ \lim_{x \to -\infty} \left[k^2 \psi^{(1)}(x) \psi^{(2)}(x) + \psi_x^{(1)}(x) \psi_x^{(2)}(x) \right] \right\}$

$\quad \cdot \left[\int_{-\infty}^{+\infty} dx \, f(x) \right] + \left\{ \lim_{x \to -\infty} \left[\psi_x^{(1)}(x) \psi^{(2)}(x) - \psi^{(1)}(x) \psi_x^{(2)}(x) \right] \right\}$

$\quad \cdot \int_{-\infty}^{+\infty} dx \left\{ \left[u^{(1)}(x) - u^{(2)}(x) \right] \int_x^{+\infty} dy \, f(y) \right\}.$

The operator Λ in this equation is defined by the formula

(22) $\quad \Lambda f(x) = f_{xx}(x) - 2 \left[u^{(1)}(x) + u^{(2)}(x) \right] f(x) + \Gamma \cdot \int_x^{+\infty} dy \, f(y)$

(see (2.4.1.-6)).

For the choices (10) and (13) the formula (21) yields respectively

(23) $\quad -4k^2 \int_{-\infty}^{+\infty} dx \, \psi^{(1)}(x, k) \psi^{(2)}(x, k) f(x)$

$\quad = \int_{-\infty}^{+\infty} dx \, \psi^{(1)}(x, k) \psi^{(2)}(x, k) \Lambda \cdot f(x),$

(24) $\quad -4k^2 \int_{-\infty}^{+\infty} dx \, \psi^{(1)}(x, k) \psi^{(2)}(x, -k) f(x)$

$\quad = \int_{-\infty}^{+\infty} dx \, \psi^{(1)}(x, k) \psi^{(2)}(x, -k) \Lambda \cdot f(x)$

$\quad - 4k^2 T^{(1)}(k) T^{(2)}(-k) \int_{-\infty}^{+\infty} dx \, f(x) - 2ik T^{(1)}(k) T^{(2)}(-k)$

$\quad \cdot \int_{-\infty}^{+\infty} dy \left[u^{(1)}(y) - u^{(2)}(y) \right] \int_y^{+\infty} dx \, f(x).$

The first of these formulae reproduces (2.4.1.-7); as indicated in section 2.4.1 (see also below), it then yields (2.4.1.-9) and, together with (12) and (17), (2.4.1.-10):

(25) $\quad 2ikg(-4k^2)[R^{(1)}(k)-R^{(2)}(k)]$
$$= \int_{-\infty}^{+\infty} dx\, \psi^{(1)}(x,k)\psi^{(2)}(x,k)g(\Lambda)[u^{(1)}(x)-u^{(2)}(x)],$$

(26) $\quad (2ik)^2 g(-4k^2)[R^{(1)}(k)+R^{(2)}(k)]$
$$= \int_{-\infty}^{+\infty} dx\, \psi^{(1)}(x,k)\psi^{(2)}(x,k)g(\Lambda)\cdot\Gamma\cdot 1.$$

As for equation (24), it is instrumental in the derivation of (2.4.1.-18) and (2.4.1.-19) from (14) and (18); but this requires a bit of additional work. We now indicate how this is done.

Let us introduce the short-hand notation

(27) $\quad \displaystyle\int_{-\infty}^{+\infty} dx\, \psi^{(1)}(x,k)\psi^{(2)}(x,-k)\Lambda^m f(x) = A_m,$

(28) $\quad 4k^2 T^{(1)}(k)T^{(2)}(-k)\displaystyle\int_{-\infty}^{+\infty} dx\, \Lambda^m f(x) + 2ikT^{(1)}(k)T^{(2)}(-k)$
$\quad \cdot \displaystyle\int_{-\infty}^{+\infty} dy\, [u^{(1)}(y)-u^{(2)}(y)] \int_{y}^{+\infty} dx\, \Lambda^m f(x) = B_m.$

Then clearly (24) implies

(29) $\quad A_{m+1} = -4k^2 A_m + B_m.$

This recursion relation is easily solved:

(30) $\quad A_m = (-4k^2)^m A_0 + \displaystyle\sum_{j=0}^{m-1} (-4k^2)^{m-j-1} B_j$

Moreover the sum in the r.h.s. can be formally performed using (28) and the geometric sum formula. There thus obtains, using (27), the equation

(31) $\quad \displaystyle\int_{-\infty}^{+\infty} dx\, \psi^{(1)}(x,k)\psi^{(2)}(x,-k)\Lambda^m f(x) = (-4k^2)^m$
$\quad \cdot \displaystyle\int_{-\infty}^{+\infty} dx\, \psi^{(1)}(x,k)\psi^{(2)}(x,-k)f(x) + 4k^2 T^{(1)}(k)T^{(2)}(-k)$
$\quad \cdot \displaystyle\int_{-\infty}^{+\infty} dx\, \{[(-4k^2)^m - \Lambda^m](-4k^2-\Lambda)^{-1}\}f(x)$
$\quad + 2ikT^{(1)}(k)T^{(2)}(-k) \displaystyle\int_{-\infty}^{+\infty} dy\, [u^{(1)}(y)-u^{(2)}(y)]$
$\quad \cdot \displaystyle\int_{y}^{+\infty} dx\, \{[(-4k^2)^m - \Lambda^m](-4k^2-\Lambda)^{-1}\}f(x),$

and from this, more generally,

(32) $$\int_{-\infty}^{+\infty} dx\, \psi^{(1)}(x,k)\psi^{(2)}(x,-k)g(\Lambda)f(x)$$
$$= g(-4k^2)\int_{-\infty}^{+\infty} dx\, \psi^{(1)}(x,k)\psi^{(2)}(x,-k)f(x)$$
$$+ 4k^2 T^{(1)}(k)T^{(2)}(-k)\int_{-\infty}^{+\infty} dx\, g^{\#}(-4k^2,\Lambda)f(x)$$
$$+ 2ikT^{(1)}(k)T^{(2)}(-k)\int_{-\infty}^{+\infty} dy\,[u^{(1)}(y)-u^{(2)}(y)]$$
$$\cdot \int_{y}^{+\infty} dx\, g^{\#}(-4k^2,\Lambda)f(x).$$

Here $g(z)$ is an arbitrary entire function, and

(33) $$g^{\#}(z_1,z_2) \equiv [g(z_1)-g(z_2)](z_1-z_2)^{-1}$$

(see (2.4.1.-20)).

Clearly (32) is the analog of the (simpler) formula (see (2.4.1.-9))

(34) $$\int_{-\infty}^{+\infty} dx\, \psi^{(1)}(x,k)\psi^{(2)}(x,k)g(\Lambda)f(x)$$
$$= g(-4k^2)\int_{-\infty}^{+\infty} dx\, \psi^{(1)}(x,k)\psi^{(2)}(x,k)f(x),$$

that is implied by (23). Of course both (32) and (34) require, for their validity, not only that $f(x)$ satisfy (20), but indeed that also all $\Lambda^m \cdot f(x)$ satisfy this formula, for any integral positive value of m. A sufficient condition for this, and therefore for the validity of (32) and (34), is that $f(x)$ be infinitely differentiable for all real values of x and that $f(x)$ vanish asymptotically ($x \to \pm\infty$) together with all its derivatives.

Thus, for the class of *bona fide* potentials to which our consideration is generally restricted, these equations are applicable. Then, just as (34) together with (12) and (17) has yielded (25) and (26), so (32) together with (14) and (18) yields (2.4.1.-18) and (2.4.1.-19), which are not rewritten here.

The remaining equations of subsection 2.4.1 for the $k^2 > 0$ case require no proof, since they follow easily from those given above, as explained there.

For the equations for the discrete spectrum, we forsake any detailed treatment. Indeed some of them are proved by techniques so analogous to those used above, that any detailed discussion of their derivation would merely be repetitive. Note in this connection that, as we already mentioned, the basic equation (1) remains valid also for negative $k^2 = -p^2$; although of course special care must now be taken whenever the asymptotic limits $x_1 = -\infty, x_2 = +\infty$ are considered in view of the possible (exponential) divergence of the solutions of the Schroedinger equations (2). The only equations whose derivation involves some novelties are (2.4.1.-57,58,59), as indicated by the appearance of derivatives with respect to the spectral parameter p. But we believe their very structure is sufficiently suggestive of the technique of derivation, not to require an explicit report here of it, that would, in our judgement, be overly cumbersome, also in view of the limited usefulness of these formulae.

A.7. Spectral integral relations: proofs

In this appendix we indicate how the results of subsection 2.4.2 are proved.

Let us begin from the (rather trivial) derivation of (2.4.2.-1b). The second half of it, corresponding to the negative superscript,

(1) $\quad f^{(-)}(x,k) \to [T(k)]^{-1} \exp(-ikx) + [R(k)/T(k)] \exp(ikx),$

$$x \to +\infty,$$

is an immediate consequence of (2.1.-5),

(2a) $\quad T(k) f^{(-)}(x,k) = f^{(+)}(x,-k) + R(k) f^{(+)}(x,k),$

and of the first half (with positive superscript) of (2.4.2.-1a). Now one rewrites (2a) with $k \leftrightarrow -k$,

(2b) $\quad T(-k) f^{(-)}(x,-k) = f^{(+)}(x,k) + R(-k) f^{(+)}(x,-k).$

Then from (2a) and (2b) there immediately follows

(3) $\quad T(k)T(-k) f^{(+)}(x,k) = T(-k) f^{(-)}(x,-k)$
$\quad\quad\quad - R(-k) T(k) f^{(-)}(x,k),$

where we have also used the unitarity relation (2.1.-18), and from this and (the second half of) (2.4.2.-1a) there immediately follows (the first half of) (2.4.2.-1b),

(4) $\quad f^{(+)}(x,k) \to [T(k)]^{-1} \exp(ikx) - [R(-k)/T(-k)] \exp(-ikx),$

$$x \to -\infty.$$

A.7 Spectral integral relations: proofs

We next proof the first basic formula, (2.4.2.-4). This follows directly (indeed see (A.5.-77)) from (A.5.-66),

$$(5) \quad K(x,y) = -(2\pi)^{-1} \int_{-\infty}^{+\infty} dk\, f^{(+)(1)}(x,k) f^{(+)(2)}(y,k)$$

$$\cdot [R^{(1)}(k) - R^{(2)}(k)] - \sum_{n=1}^{N^{(1)}} f^{(+)(1)}(x, i p_n^{(1)}) f^{(+)(2)}(y, i p_n^{(1)}) \rho_n^{(1)}$$

$$+ \sum_{n=1}^{N^{(2)}} f^{(+)(1)}(x, i p_n^{(2)}) f^{(+)(2)}(y, i p_n^{(2)}) \rho_n^{(2)},$$

and the relation (A.5.-15) (with (A.5.-25)),

$$(6) \quad 2K(x,x) = w^{(1)}(x) - w^{(2)}(x).$$

There are various ways to derive the second basic formula, (2.4.2.-5). The simplest is perhaps the following. Replace in (2.4.2.-4) $w^{(1)}(x)$ by $w^{(1)}(x+\varepsilon)$, $w^{(2)}(x)$ by $w^{(2)}(x-\varepsilon)$ and, in the limit $\varepsilon \to 0$, obtain (using the relevant formulae (2.1.1.-2))

$$(7) \quad w_x^{(1)}(x) + w_x^{(2)}(x) = 4 \sum_{n=1}^{N^{(1)}} f^{(+)(1)}(x, i p_n^{(1)}) f^{(+)(2)}(x, i p_n^{(1)}) p_n^{(1)} \rho_n^{(1)}$$

$$+ 4 \sum_{n=1}^{N^{(2)}} f^{(+)(1)}(x, i p_n^{(2)}) f^{(+)(2)}(x, i p_n^{(2)}) p_n^{(2)} \rho_n^{(2)}$$

$$- \pi^{-1} \int_{-\infty}^{+\infty} dk\, f^{(+)(1)}(x,k) f^{(+)(2)}(x,k) 2ik$$

$$\cdot [R^{(1)}(k) + R^{(2)}(k)] + \mathcal{B}(x)$$

where

$$(8) \quad \mathcal{B}(x) = -2 \sum_{n=1}^{N^{(1)}} \rho_n^{(1)} W[f^{(+)(1)}(x, i p_n^{(1)}), f^{(+)(2)}(x, i p_n^{(1)})]$$

$$+ 2 \sum_{n=1}^{N^{(2)}} \rho_n^{(2)} W[f^{(+)(1)}(x, i p_n^{(2)}), f^{(+)(2)}(x, i p_n^{(2)})]$$

$$- \pi^{-1} \int_{-\infty}^{+\infty} dk\, W[f^{(+)(1)}(x,k), f^{(+)(2)}(x,k)]$$

$$\cdot [R^{(1)}(k) - R^{(2)}(k)].$$

In the last equation we are using the wronskian notation,

(9) $\quad W[f^{(1)}(x), f^{(2)}(x)] \equiv f^{(1)}(x) f_x^{(2)}(x) - f_x^{(1)}(x) f^{(2)}(x).$

But the standard wronskian theorem, applied to the Schroedinger equations (2.4.2.-1), implies

(10) $\quad W[f^{(+)(1)}(x, k), f^{(+)(2)}(x, k)] = \int_x^\infty dy\, f^{(+)(1)}(y, k) f^{(+)(2)}(y, k)$
$\quad\quad\quad\quad\quad\quad\quad\quad\quad\quad\quad\quad\quad \cdot [u^{(1)}(y) - u^{(2)}(y)].$

To write this formula, that holds (at least) for Im $k \geq 0$, one has also used the boundary conditions (2.4.2.-1a) (see also (2.4.2.-6a)), that imply the vanishing of the l.h.s. of (10) as $x \to +\infty$.

Now insert (10) in (8) to eliminate all the wronskians, invert the priority of the y- and k-integrations (and of the y-integrations and the sums), and use again the first basic formula (2.4.2.-4). There thus obtains

(11) $\quad \mathcal{B}(x) = -\int_x^\infty dy\, [u^{(1)}(y) - u^{(2)}(y)][w^{(1)}(y) - w^{(2)}(y)],$

and this immediately yields (see (2.4.2.-2b))

(12) $\quad \mathcal{B}(x) = -\tfrac{1}{2}[w^{(1)}(x) - w^{(2)}(x)]^2,$

which together with (7), yields (2.4.2.-5).

The last formulae that must be proved are (2.4.2.-9, 10). This is quite straightforward. Let

(13) $\quad F(x) \equiv f^{(+)(1)}(x, k) f^{(+)(2)}(x, k),$

where $f^{(+)(j)}(x, k)$ is the solution of the Schroedinger equation (2.4.2.-1) characterized by the boundary condition (2.4.2.-1a). Then

(14) $\quad F_x(x) = f_x^{(+)(1)}(x, k) f^{(+)(2)}(x, k) + f^{(+)(1)}(x, k) f_x^{(+)(2)}(x, k)$

and, using the Schroedinger equation (2.4.2.-1),

(15) $\quad F_{xx}(x) = [u^{(1)}(x) + u^{(2)}(x) - 2k^2] F(x) + 2 f_x^{(+)(1)}(x, k) f_x^{(+)(2)}(x, k),$

(16) $\quad F_{xxx}(x) = [u_x^{(1)}(x) + u_x^{(2)}(x)] F(x)$
$\quad\quad\quad\quad\quad + 2[u^{(1)}(x) + u^{(2)}(x) - 2k^2] F_x(x)$
$\quad\quad\quad\quad\quad + [u^{(1)}(x) - u^{(2)}(x)] \int_x^\infty dy\, [u^{(1)}(y) - u^{(2)}(y)] F(y).$

To get (16) from (15), we have used, after differentiation, (14), (10) and (13), in addition to (2.4.2.-1). Integration of (16) from x to ∞ yields precisely (2.4.2.-9) with (2.4.2.-10), which are thereby proved. Note moreover that, if $u^{(1)}(x)+u^{(2)}(x)=0$, (16) is an ordinary third order differential equation for $F(x)$, while if $[u^{(1)}(x)-u^{(2)}(x)]\neq 0$, division by this quantity and one further differentiation yields an ordinary differential equation of fourth order for $F(x)$. The formula generalizing (2.4.2.-9) to the product of any two solutions of the Schroedinger equations (2.4.1.-1) is given in appendix A.9 (see (A.9.-40)).

A.8. A formula for the variation of the coefficients of the asymptotic expansion of the phase of the transmission coefficient

In this appendix we prove the formula

(1) $$\delta\theta_m=(-1)^{m+1}\int_{-\infty}^{+\infty} dx\, L^m \delta u(x), \quad m=0,1,2,\ldots,$$

that relates the (infinitesimal) change of the coefficients of the asymptotic expansion of the transmission coefficient $T(k)$,

(2) $$T(k)=|T(k)|\exp[i\theta(k)], \quad \operatorname{Im} k=0,$$

(3) $$\theta(k)=\sum_{m=0}^{M}\theta_m(2k)^{-2m-1}+O(k^{-2M-3}), \quad \operatorname{Im} k=0,$$

to an (infinitesimal) variation of the corresponding function $u(x)$ (within the class of *bona fide* potentials). The operator L in (1) is of course defined by the formula

(4) $$Lf(x)=f_{xx}(x)-4u(x)f(x)+2u_x(x)\int_x^{+\infty} dy\, f(y),$$

that specifies its action on a generic function $f(x)$ (vanishing as $x\to+\infty$).

Another proof of (1) is provided in appendix A.9.

In writing (2), and some other formulae below, we are implicitly assuming $u(x)$ and $\delta u(x)$ to be real. It is easy to verify that the result (1) remains valid in the more general case of complex $u(x)$ and $\delta u(x)$, provided the definition (2) of the phase $\theta(k)$ is replaced by the more general formula

(2a) $$T(k)=T(-k)\exp[2i\theta(k)], \quad \operatorname{Im} k=0.$$

The starting point for the proof of (1) that is presented here is the formula (2.4.1.-25) with $g(z)=z^m$ (m being an arbitrary positive integer), that, using

(2.1.-12, 13, 14, 15, 17) and (2.4.1.-20), can be recast into the form

(5) $\quad (2k)^{2m+1}\delta\theta(k) - \sum_{j=1}^{m-1}(-1)^j(2k)^{2(m+1-j)}\int_{-\infty}^{+\infty}dx\, L^j\delta u(x)$

$$= |R(k)|^2 \Big\{ (2k)^{2m+1}[\delta\theta(k) - \delta\chi(k)]$$

$$- \sum_{j=1}^{m-1}(-1)^j(2k)^{2(m+1-j)}\int_{-\infty}^{+\infty}dx\, L^j\delta u(x)\Big\}$$

$$+ \int_{-\infty}^{+\infty}dx\,|\psi(x,k)|^2 L^m\delta u(x).$$

Take now the limit of this formula as $k \to \pm\infty$, Im $k=0$. Then the r.h.s. is finite (indeed, the first term vanishes and in the second $|\psi(x,k)|^2$ may, in the limit, be replaced by unity; see (2.1.-39) or (A.3.−1), and (A.3.-5)), while the l.h.s. apparently diverges polynomially in k. Thus the coefficients of this polynomial must all vanish; and by equating them to zero there follows precisely (1). Q.E.D.

For infinitesimal translations and scale transformations (see (2.1.1.-2, 2c) and (2.1.1.-4, 4c)) (1) yields the two formulae

(6) $\quad \int_{-\infty}^{+\infty} dx\, L^m u_x(x) = 0, \quad m = 0, 1, 2, \ldots,$

(7) $\quad \theta_m = (-1)^{m+1}(2m+1)^{-1}\int_{-\infty}^{+\infty}dx\, L^m[xu_x(x) + 2u(x)],$

$$m = 0, 1, 2, \ldots,$$

and a comparison of (an infinitesimal variation of) (7) with (1) yields the formula

(8) $\quad \delta\int_{-\infty}^{+\infty}dx\, L^m[xu_x(x) + 2u(x)] = (2m+1)\int_{-\infty}^{+\infty}dx\, L^m\delta u(x),$

$$m = 0, 1, 2, \ldots$$

Of course the symbol δ in the l.h.s. of the last formula indicates the variation of $u(x)$ wherever it appears, including within the operator L (see (4)).

The expression (7) of θ_m has been obtained (essentially in the same manner, from (1)) also in appendix A.3. The formulae (6) and (8) are clearly properties of the integrodifferential operator L, with no reference to the spectral problem (that was, however, instrumental in the above derivation);

these formulae are also obtained, by a somewhat different technique (also based on the spectral problem) in appendix A.9, and they are further discussed there. (Incidentally, (6) is merely the special case of (8) corresponding to the infinitesimal translation $x \to x + \varepsilon$: use (2.1.1.-2) and note that the variation in the l.h.s. is simply $-\varepsilon \int_{-\infty}^{+\infty} dx\, L^m u_x(x)$).

Let us finally mention that the same method of asymptotic expansion applied here to (2.4.1.-25), can also be applied to the more general formulae (2.4.1.-18, 19). This is done in appendix A.9, where some other expressions for the asymptotic expansion at large k of the transmission coefficient $T(k)$ are also exhibited.

A.9. Properties of the operators $\Lambda, \tilde{\Lambda}, L, \tilde{L}$, and other formulae

In this appendix we collect (and prove) a number of properties of the operators $\Lambda, \tilde{\Lambda}, L, \tilde{L}$. Some of these properties constitute intriguing nonlinear operator identities, which are proven here by roundabout routes using the spectral problem, although their validity is of course quite independent of it (but a direct proof by algebraic and analytic techniques does not appear trivial). For completeness, we also report some properties of the operator M and several formulae for the asymptotic expansion of the transmission coefficient $T(k)$ at large k, that are instrumental to derive the operational identities mentioned above, but are also of interest in their own right.

It is convenient to introduce the multiplicative operators s and m, the differential operator D and the integral operator I by the following formulae, that specify their action on a generic function $f(x)$:

(1) $sf(x) = [u^{(1)}(x) + u^{(2)}(x)] f(x)$,

(2) $mf(x) = [u^{(1)}(x) - u^{(2)}(x)] f(x)$,

(3) $Df(x) = f_x(x)$,

(4) $If(x) = \int_x^{+\infty} dy\, f(y)$.

Clearly D is only applicable to functions $f(x)$ that are differentiable, and I only to functions $f(x)$ that are integrable (so that the r.h.s. of (4) make good sense; thus, in particular, $f(x)$ must vanish sufficiently fast as $x \to +\infty$). In the following such conditions are always understood to hold. Indeed, to be on the safe side, it will be assumed that the functions $f(x)$ on which these operators act are infinitely differentiable for all (real) values of x and vanish with all their derivatives (faster than $1/x$) as $x \to \pm \infty$. Since we assume the two (given) functions $u^{(1)}(x)$ and $u^{(2)}(x)$ to belong to the class of *bona fide* potentials (i.e. to be themselves infinitely differentiable for all real values of x and to vanish with all their derivatives, faster than $1/x$, as $x \to \pm \infty$), it is clear that the four operators defined above, when applied to a function $f(x)$ belonging to the class defined above, not only are well-defined, but yield a function that belongs to the same class (this would not have been the case if the functions $f(x)$ were themselves required to be *bona fide* potentials, implying the requirement to vanish also as $x \to -\infty$, since the fact that $f(x)$ vanishes as $x \to \pm \infty$ does not imply that $If(x)$ also vanishes as $x \to -\infty$; see (4)).

Clearly the definitions (3) and (4) imply the formula

(5) $\quad DI = ID = -1$.

Let us call attention to the apparent contradiction of this formula with the equation

(6) $\quad Dc = 0$,

that clearly follows from (3) (c indicates an arbitrary constant). The point is of course that a constant does not belong to the class of functions $f(x)$ defined above, since it does not vanish as $x \to +\infty$. However occasionally, in the following, we will apply the operators also to constants, since some formulae can be written more compactly using such a notation; we trust any reader to be clever enough to exploit such notational simplifications without incurring into contradictions such as that implied by the last two formulae written above.

The operators whose properties are discussed in this appendix are defined as follows:

(7) $\quad \Gamma = Ds - sD + mIm$,

(8a) $\quad \Lambda = D^2 - 2s + \Gamma I$,

(8b) $\quad \Lambda = D^2 - s + DsI + mImI$,

(9a) $\quad \tilde{\Lambda} = D^2 + 2IsD + I\Gamma$,

(9b) $\quad \tilde{\Lambda} = D^2 - s + IsD + ImIm$.

As for the operators L resp. $\tilde{L}$, they are the special cases of the operators Λ

resp. $\tilde{\Lambda}$, corresponding to

(10) $\quad u^{(1)}(x) = u^{(2)}(x) = u(x)$.

Thus

(11) $\quad L = D^2 - 2u + 2DuI$,

(12) $\quad \tilde{L} = D^2 - 2u + 2IuD$,

where of course the operator u is defined by

(13) $\quad uf(x) = u(x)f(x)$

(so that, when (10) holds, $s = 2u$, $m = 0$).

It is easily seen that the definitions of the operators Γ, Λ, L, $\tilde{\Lambda}$ and $\tilde{L}$ given here coincide with the definitions given elsewhere in this book (see e.g. (2.4.1.-5), (2.4.1.-6), (2.4.1.-13), (2.4.2.-10) and (2.4.2.-18)).

Note that the operators Γ, Λ and L, when applied to *bona fide* potentials, yield *bona fide* potentials, whereas $\tilde{\Lambda}$ and $\tilde{L}$ yield functions that need not vanish as $x \to -\infty$.

Clearly, if the operators s and m were self-adjoint, and the operators D and I were skew-adjoint,

(14a) $\quad s^T = s, \quad m^T = m, \quad D^T = -D$,

(14b) $\quad I^T = -I$,

then the operators Λ and $\tilde{\Lambda}$ (as well as L and $\tilde{L}$) would be the adjoint of each other:

(15) $\quad \Lambda = \tilde{\Lambda}^T, \quad \tilde{\Lambda} = \Lambda^T, \quad L = \tilde{L}^T, \quad \tilde{L} = L^T$.

Here we are of course using the superscript ^{T} to denote the adjoint ("transpose"; see below). However, in the functional space of *bona fide* potentials, equipped with the natural scalar product

(16) $\quad (g, f) \equiv \int_{-\infty}^{+\infty} dx\, g(x) f(x)$,

while (14a) obviously hold,

(17a) $\quad (g, sf) = (sg, f), \quad (g, mf) = (mg, f), \quad (g, Df) = -(Dg, f)$,

(14b) generally does not hold:

(17b) $\quad (g, If) = -(Ig, f) + \left[\int_{-\infty}^{+\infty} dx\, g(x) \right] \left[\int_{-\infty}^{+\infty} dx\, f(x) \right].$

(Note incidentally that, since the definition (16) makes sense even if only one of the two functions $g(x)$ and $f(x)$ vanishes as $x \to \pm\infty$, scalar products like (g, If) or $(g, \tilde{A}f)$ are well-defined, whenever $g(x)$ and $f(x)$ are *bona fide* potentials, in spite of the fact that the functions $If(x)$ and $\tilde{A}f(x)$ are themselves generally not *bona fide* potentials, since they need not vanish as $x \to -\infty$).

Of course hereafter we employ the notation used throughout this book. Let us rewrite here, for convenience, some of the basic formulae:

(18) $\quad -f_{xx}^{(\pm)}(x, k) + u(x) f^{(\pm)}(x, k) = k^2 f^{(\pm)}(x, k),$

(19a) $\quad f^{(\pm)}(x, k) \to \exp(\pm ikx), \quad x \to \pm\infty,$

(19b) $\quad f^{(\pm)}(x, k) \to [T(k)]^{-1} \exp(\pm ikx)$

$\qquad \mp [R(\mp k)/T(\mp k)] \exp(\mp ikx), \quad x \to \mp\infty,$

(20) $\quad \psi(x, k) = T(k) f^{(-)}(x, k) = f^{(+)}(x, -k) + R(k) f^{(+)}(x, k),$

(21) $\quad \varphi_n(x) = c_n^{(+)} f^{(+)}(x, ip_n) = c_n^{(-)} f^{(-)}(x, ip_n), \quad n = 1, 2, \ldots, N;$

$\qquad\qquad\qquad\qquad\qquad\qquad\qquad\qquad\qquad\qquad p_n > 0,$

(22) $\quad \int_{-\infty}^{+\infty} dx\, [\varphi_n(x)]^2 = 1, \quad n = 1, 2, \ldots, N,$

(23) $\quad \lim_{k \to ip_n} [(k - ip_n) T(k)] = i c_n^{(+)} c_n^{(-)}, \quad n = 1, 2, \ldots, N; \quad p_n > 0,$

(24a) $\quad \exp[2i\theta(k)] \equiv T(k)/T(-k), \quad \operatorname{Im} k = 0,$

(24b) $\quad \theta(k) = -\theta(-k), \quad \theta(\pm\infty) = 0.$

The formula (24a) reports merely the definition (mod π) of the phase $\theta(k)$, while (24b) lifts the (mod π) ambiguity (to verify the consistency of (24b) with (24a), rewrite (24a) with k replaced by $-k$ and use again (24a)). Obviously analogous equations hold with all the quantities being identified by a superscript j, $j = 1, 2$.

Let us now display a number of properties of the operators defined above, including some that are reported here merely for completeness.

A.9 Properties of the operators $\Lambda, \tilde{\Lambda}, L, \tilde{L}$

Comments and proofs follow.

(25) $\quad \Lambda^m D = D\tilde{\Lambda}^m, \quad \Lambda^m = -D\tilde{\Lambda}^m I, \quad \tilde{\Lambda}^m = -I\Lambda^m D, \qquad m = 0, 1, 2, \ldots,$

(26) $\quad \{\{g(x), f(x)\}\} \equiv g(x) f_{xx}(x) + f(x) g_{xx}(x) - g_x(x) f_x(x)$
$\qquad - 2s(x) g(x) f(x) + [Img(x)][Imf(x)] - f(x)\tilde{\Lambda}g(x),$

(27) $\quad g(x)\Lambda^m f_x(x) - f_x(x)\tilde{\Lambda}^m g(x)$
$$= \left[\sum_{n=0}^{m-1} \{\{\tilde{\Lambda}^n g(x), \tilde{\Lambda}^{m-n-1} f(x)\}\} \right]_x, \quad m = 1, 2, 3, \ldots,$$

(28) $\quad g(x)\Lambda^m f_x(x) + f(x)\Lambda^m g_x(x)$
$$= \left[\sum_{n=0}^{m-1} \{\{\tilde{\Lambda}^n g(x), \tilde{\Lambda}^{m-n-1} f(x)\}\} + f(x)\tilde{\Lambda}^m g(x) \right]_x,$$
$$m = 1, 2, 3, \ldots,$$

(29) $\quad g_x(x)\tilde{\Lambda}^m f(x) + f_x(x)\tilde{\Lambda}^m g(x)$
$$= \left[-\sum_{n=0}^{m-1} \{\{\tilde{\Lambda}^n g(x), \tilde{\Lambda}^{m-n-1} f(x)\}\} + g(x)\tilde{\Lambda}^m f(x) \right]_x,$$
$$m = 1, 2, 3, \ldots,$$

(30) $\quad L^m D = D\tilde{L}^m, \quad L^m = -D\tilde{L}^m I, \quad \tilde{L}^m = -IL^m D, \qquad m = 0, 1, 2, \ldots,$

(31) $\quad [[g(x), f(x)]] \equiv g(x) f_{xx}(x) + f(x) g_{xx}(x) - g_x(x) f_x(x)$
$\qquad - 4u(x) g(x) f(x) - f(x)\tilde{L}g(x),$

(32) $\quad g(x) L^m f_x(x) - f_x(x)\tilde{L}^m g(x) = \left\{ \sum_{n=0}^{m-1} [[\tilde{L}^n g(x), \tilde{L}^{m-n-1} f(x)]] \right\}_x,$
$$m = 1, 2, 3, \ldots,$$

(33) $\quad g(x) L^m f_x(x) + f(x) L^m g_x(x) = \left\{ \sum_{n=0}^{m-1} [[\tilde{L}^n g(x), \tilde{L}^{m-n-1} f(x)]] \right.$
$$\left. + f(x)\tilde{L}^m g(x) \right\}_x, \quad m = 1, 2, 3, \ldots,$$

(34) $\quad g_x(x)\tilde{L}^m f(x) + f_x(x)\tilde{L}^m g(x) = \left\{ -\sum_{n=0}^{m-1} [[\tilde{L}^n g(x), \tilde{L}^{m-n-1} f(x)]] \right.$
$$\left. + g(x)\tilde{L}^m f(x) \right\}_x, \quad m = 1, 2, 3, \ldots,$$

(35) $$\int_{-\infty}^{+\infty} dx\, \psi^{(1)}(x,k)\psi^{(2)}(x,k)\boldsymbol{m}\cdot 1 = 2ik\left[R^{(1)}(k) - R^{(2)}(k)\right],$$

(36) $$\int_{-\infty}^{+\infty} dx\, \psi^{(1)}(x,k)\psi^{(2)}(x,k)\boldsymbol{\Gamma}\cdot 1 = (2ik)^2\left[R^{(1)}(k) + R^{(2)}(k)\right],$$

(37) $$\int_{-\infty}^{+\infty} dx\, \psi^{(1)}(x,k)\psi^{(2)}(x,k)\Lambda f(x)$$
$$= (2ik)^2 \int_{-\infty}^{+\infty} dx\, \psi^{(1)}(x,k)\psi^{(2)}(x,k)f(x),$$

(38) $$\int_{-\infty}^{+\infty} dx\, [\psi(x,k)]^2 Lf(x) = (2ik)^2 \int_{-\infty}^{+\infty} dx\, [\psi(x,k)]^2 f(x),$$

(39) $$f^{(j)}(x,k) \equiv a^{(j)} f^{(+)(j)}(x,k) + b^{(j)} T^{(j)}(k) f^{(-)(j)}(x,k), \quad j=1,2,$$

(40) $$\tilde{\Lambda}\left[f^{(1)}(x,k)f^{(2)}(x,k)\right] = (2ik)^2\left[f^{(1)}(x,k)f^{(2)}(x,k)\right]$$
$$+ 2ik\left\{a^{(1)}b^{(2)} - a^{(2)}b^{(1)} + b^{(1)}b^{(2)}\left[R^{(1)}(k) - R^{(2)}(k)\right]\right\}\boldsymbol{Im}\cdot 1$$
$$- (2ik)^2\left\{a^{(1)}b^{(2)} + a^{(2)}b^{(1)} + b^{(1)}b^{(2)}\left[R^{(1)}(k) + R^{(2)}(k)\right]\right\},$$

(41) $$\tilde{\Lambda}\left[f^{(+)(1)}(x,k)f^{(+)(2)}(x,k)\right] = (2ik)^2\left[f^{(+)(1)}(x,k)f^{(+)(2)}(x,k)\right],$$

(42) $$\tilde{\Lambda}\left[\psi^{(1)}(x,k)\psi^{(2)}(x,k)\right] = (2ik)^2\left[\psi^{(1)}(x,k)\psi^{(2)}(x,k)\right]$$
$$+ 2ik\left[R^{(1)}(k) - R^{(2)}(k)\right]\boldsymbol{Im}\cdot 1 - (2ik)^2\left[R^{(1)}(k) + R^{(2)}(k)\right],$$

(43) $$\tilde{L}[f(x,k)]^2 = (2ik)^2[f(x,k)]^2 - 2b(2ik)^2[a + bR(k)],$$

(44) $$\tilde{L}[f^{(+)}(x,k)]^2 = (2ik)^2[f^{(+)}(x,k)]^2,$$

(45) $$\tilde{L}[\psi(x,k)]^2 = (2ik)^2[\psi(x,k)]^2 - 2(2ik)^2 R(k),$$

(46a) $$\int_{-\infty}^{+\infty} dx\, [G(\Lambda)\boldsymbol{m}\cdot 1 + H(\Lambda)\boldsymbol{\Gamma}\cdot 1]$$
$$= \pi^{-1}\int_{-\infty}^{+\infty} dk\,\left\{G(-4k^2)\left[T^{(2)}(k)/T^{(1)}(k) - T^{(1)}(k)/T^{(2)}(k)\right]\right.$$
$$\left. - 2ikH(-4k^2)\left[T^{(2)}(k)/T^{(1)}(k) + T^{(1)}(k)/T^{(2)}(k)\right]\right\}$$
$$+ 2\sum_{n=1}^{N^{(2)}} \left[c_n^{(+)(2)}c_n^{(-)(2)}/T^{(1)}(ip_n^{(2)})\right]\left[G\left(4p_n^{(2)^2}\right) + 2p_n^{(2)}H\left(4p_n^{(2)^2}\right)\right]$$
$$- 2\sum_{n=1}^{N^{(1)}} \left[c_n^{(+)(1)}c_n^{(-)(1)}/T^{(2)}(ip_n^{(1)})\right]\left[G\left(4p_n^{(1)^2}\right) - 2p_n^{(1)}H\left(4p_n^{(1)^2}\right)\right],$$

(46b) $$G(z) \equiv \sum_{m=0}^{M} G_m z^m, \quad H(z) \equiv \sum_{m=0}^{M'} H_m z^m, \quad 0 \le M, M' < \infty,$$

(47a) $\int_{-\infty}^{+\infty} dx \left[g(\Lambda)\boldsymbol{m}\cdot 1 + h(\Lambda)\boldsymbol{\Gamma}\cdot 1 \right] \left\{ \boldsymbol{I}\left[G(\Lambda)\boldsymbol{m}\cdot 1 + H(\Lambda)\boldsymbol{\Gamma}\cdot 1 \right] \right\}$

$= -\pi^{-1} \int_{-\infty}^{+\infty} dk\, 2ik \left\{ \left[B_e^{(-)}(k) G(-4k^2) - 2ik B_o^{(-)}(k) \right. \right.$

$\left. \cdot H(-4k^2) \right] \left[T^{(2)}(k)/T^{(1)}(k) + T^{(1)}(k)/T^{(2)}(k) \right]$

$+ \left[B_o^{(-)}(k) G(-4k^2) - 2ik B_e^{(-)}(k) H(-4k^2) \right]$

$\left. \cdot \left[T^{(2)}(k)/T^{(1)}(k) - T^{(1)}(k)/T^{(2)}(k) \right] \right\}$

$+ 4 \sum_{n=1}^{N^{(2)}} \left[p_n^{(2)} c_n^{(+)(2)} c_n^{(-)(2)} / T^{(1)}(ip_n^{(2)}) \right] B^{(-)}(ip_n^{(2)})$

$\cdot \left[G(4p_n^{(2)^2}) + 2 p_n^{(2)} H(4p_n^{(2)^2}) \right]$

$+ 4 \sum_{n=1}^{N^{(1)}} \left[p_n^{(1)} c_n^{(+)(1)} c_n^{(-)(1)} / T^{(2)}(ip_n^{(1)}) \right] B^{(-)}(ip_n^{(1)})$

$\cdot \left[G(4p_n^{(1)^2}) - 2 p_n^{(1)} H(4p_n^{(1)^2}) \right],$

(47b) $g(z) \equiv \sum_{m=0}^{M} g_m z^m, \quad h(z) \equiv \sum_{m=0}^{M'} h_m z^m, \quad 0 \leq M, M' < \infty,$

(47c) $B^{(-)}(k) = B_e^{(-)}(k) + B_o^{(-)}(k),$

(47d) $B_e^{(-)}(k) = g(-4k^2) + h(-4k^2) \int_{-\infty}^{+\infty} dx\, \boldsymbol{m}\cdot 1 + \int_{-\infty}^{+\infty} dx\, \boldsymbol{mI}$
$\cdot \lfloor g^\#(-4k^2, \Lambda)\boldsymbol{m}\cdot 1 + h^\#(-4k^2, \Lambda)\boldsymbol{\Gamma}\cdot 1 \rfloor,$

(47e) $B_o^{(-)}(k) = 2ik \left\{ h(-4k^2) + \int_{-\infty}^{+\infty} dx \left[g^\#(-4k^2, \Lambda)\boldsymbol{m}\cdot 1 \right. \right.$
$\left. \left. + h^\#(-4k^2, \Lambda)\boldsymbol{\Gamma}\cdot 1 \right] \right\},$

(47f) $g^\#(z_1, z_2) \equiv [g(z_1) - g(z_2)]/(z_1 - z_2),$
$h^\#(z_1, z_2) \equiv [h(z_1) - h(z_2)]/(z_1 - z_2),$

(48a) $A_m \equiv \int_{-\infty}^{+\infty} dx\, \Lambda^m \boldsymbol{m}\cdot 1, \quad m = 0, 1, 2, \ldots,$

(48b) $A_m = \int_{-\infty}^{+\infty} dx\, \boldsymbol{mI}\Lambda^{m-1}\boldsymbol{\Gamma}\cdot 1, \quad m = 1, 2, 3, \ldots,$

(48c) $A_m = \pi^{-1} \int_{-\infty}^{+\infty} dk\, (2ik)^{2m} \left[T^{(2)}(k)/T^{(1)}(k) - T^{(1)}(k)/T^{(2)}(k) \right]$

$+ 2 \sum_{n=1}^{N^{(2)}} (2p_n^{(2)})^{2m} \left[c_n^{(+)(2)} c_n^{(-)(2)} / T^{(1)}(ip_n^{(2)}) \right] - 2 \sum_{n=1}^{N^{(1)}} (2p_n^{(1)})^{2m}$

$\cdot \left[c_n^{(+)(1)} c_n^{(-)(1)} / T^{(2)}(ip_n^{(1)}) \right], \quad m = 0, 1, 2, \ldots,$

(48d) $T^{(2)}(k)/T^{(1)}(k) - T^{(1)}(k)/T^{(2)}(k)$

$$= -2 \sum_{m=0}^{M} A_m (2ik)^{-2m-1} + O(k^{-2M-3}), \text{Im } k \geq 0,$$

(48e) $\sin[\theta^{(1)}(k) - \theta^{(2)}(k)]$

$$= \sum_{m=0}^{M} (-1)^{m+1} A_m (2k)^{-2m-1} + O(k^{-2M-3}), \quad \text{Im } k = 0,$$

(49a) $B_m \equiv \int_{-\infty}^{+\infty} dx\, \Lambda^m \Gamma \cdot 1, \quad m = 0, 1, 2, \ldots,$

(49b) $B_m = \int_{-\infty}^{+\infty} dx\, mI\Lambda^m \boldsymbol{m} \cdot 1, \quad m = 0, 1, 2, \ldots,$

(49c) $B_m = -\pi^{-1} \int_{-\infty}^{+\infty} dk\, (2ik)^{2m+1} [T^{(2)}(k)/T^{(1)}(k) + T^{(1)}(k)/T^{(2)}(k)]$

$$+ 2 \sum_{n=1}^{N^{(2)}} (2p_n^{(2)})^{2m+1} \left[c_n^{(+)(2)} c_n^{(-)(2)} / T^{(1)}(ip_n^{(2)}) \right]$$

$$+ 2 \sum_{n=1}^{N^{(2)}} (2p_n^{(1)})^{2m+1} \left[c_n^{(+)(1)} c_n^{(-)(1)} / T^{(2)}(ip_n^{(1)}) \right], \quad m = 0, 1, 2, \ldots$$

(49d) $T^{(2)}(k)/T^{(1)}(k) + T^{(1)}(k)/T^{(2)}(k)$

$$= 2 - 2 \sum_{m=0}^{M} (-1)^m B_m (2k)^{-2m-2} + O(k^{-2M-4}), \quad \text{Im } k \geq 0,$$

(49e) $\cos[\theta^{(1)}(k) - \theta^{(2)}(k)]$

$$= 1 - \sum_{m=0}^{M} (-1)^m B_m (2k)^{-2m-2} + O(k^{-2M-4}), \quad \text{Im } k = 0,$$

(50) $\int_{-\infty}^{+\infty} dx\, (\boldsymbol{\Gamma} \cdot 1)(I\Lambda^m \boldsymbol{m} \cdot 1) = B_m \int_{-\infty}^{+\infty} dx\, \boldsymbol{m} \cdot 1 - A_{m+1}, \quad m = 0, 1, 2, \ldots,$

(51a) $\Lambda^{(0)} \equiv D^2 - u + DuI + uIuI,$

(51b) $\Lambda^{(0)} f(x) = f_{xx}(x) - 2u(x)f(x) + u_x(x) \int_x^{+\infty} dy\, f(y)$

$$+ u(x) \int_x^{+\infty} dy\, u(y) \int_y^{+\infty} dz\, f(z),$$

(52a) $A_m^{(0)} \equiv \int_{-\infty}^{+\infty} dx\, (\Lambda^{(0)})^m u(x), \quad m = 0, 1, 2, \ldots,$

(52b) $A_m^{(0)} = \int_{-\infty}^{+\infty} dx\, u(x) I (\Lambda^{(0)})^{m-1} \left[u_x(x) + u(x) \int_x^{+\infty} dy\, u(y) \right],$

$$m = 1, 2, 3, \ldots,$$

A.9   *Properties of the operators* $\Lambda, \tilde{\Lambda}, L, \tilde{L}$

(52c) $\quad A_m^{(0)} = \pi^{-1} \int_{-\infty}^{+\infty} dk\, (2ik)^{2m} [1/T(k) - T(k)]$

$$-2 \sum_{n=1}^{N} (2p_n)^{2m} c_n^{(+)} c_n^{(-)}, \quad m=0,1,2,\ldots,$$

(52d) $\quad \sin\theta(k) = \sum_{m=0}^{M} (-1)^{m+1} A_m^{(0)} (2k)^{-2m-1} + O(k^{-2M-3}), \quad \text{Im } k = 0,$

(53a) $\quad B_m^{(0)} \equiv \int_{-\infty}^{+\infty} dx\, (\Lambda^{(0)})^m \left[u_x(x) + u(x) \int_x^{+\infty} dy\, u(y) \right],$

$$m=0,1,2,\ldots,$$

(53b) $\quad B_m^{(0)} = \int_{-\infty}^{+\infty} dx\, u(x) I(\Lambda^{(0)})^m u(x), \quad m=0,1,2,\ldots,$

(53c) $\quad B_m^{(0)} = -\pi^{-1} \int_{-\infty}^{+\infty} dk\, (2ik)^{2m+1} [1/T(k) + T(k)]$

$$+2 \sum_{n=1}^{N} (2p_n)^{2m+1} c_n^{(+)} c_n^{(-)}, \quad m=0,1,2,\ldots,$$

(53d) $\quad \cos\theta(k) = 1 - \sum_{m=0}^{M} (-)^m B_m^{(0)} (2k)^{-2m-2} + O(k^{-2M-4}), \quad \text{Im } k = 0,$

(54a) $\quad \int_{-\infty}^{\infty} dx\, L^m u_x(x) = 0, \quad m=0,1,2,\ldots,$

(54b) $\quad \int_{-\infty}^{+\infty} dx\, [L^n u_x(x)] I[L^m u_x(x)] = 0, \quad n,m=0,1,2,\ldots,$

(54c) $\quad \int_{-\infty}^{+\infty} dx\, [\tilde{L}^n u(x)] D[\tilde{L}^m u(x)] = 0, \quad n,m=0,1,2,\ldots,$

(55a) $\quad \int_{-\infty}^{+\infty} dx\, L^{m+1} \delta u(x) = 2 \int_{-\infty}^{+\infty} dx\, \delta u(x) I L^m u_x(x), \quad m=0,1,2,\ldots,$

(55b) $\quad \int_{-\infty}^{+\infty} dx\, L^m \delta u_x(x) = 2 \sum_{n=1}^{m} \int_{-\infty}^{+\infty} dx\, L^{n-1} [\delta u - D(\delta u) I]$

$$\cdot L^{m-n} u_x(x), \quad m=1,2,3,\ldots,$$

(55c) $\quad \int_{-\infty}^{+\infty} dx\, L^m \delta u_x(x) = 2 \sum_{n=1}^{m} \int_{-\infty}^{+\infty} dx\, L^{n-1} [(\delta u) D + D(\delta u)]$

$$\cdot \tilde{L}^{m-n} u(x), \quad m=1,2,3,\ldots,$$

(55d) $\quad \int_{-\infty}^{+\infty} dx\, L^m \delta u(x) = (-1)^{m+1} \pi^{-1} \int_{-\infty}^{+\infty} dk\, (2k)^{2m} \delta$

$$\cdot \ln[T(k) T(-k)] - 4 \sum_{n=1}^{N} (2p_n)^{2m} \delta p_n, \quad m=0,1,2,\ldots,$$

$$\text{(56a)} \quad \theta(k) = \sum_{m=0}^{M} \theta_m (2k)^{-2m-1} + O(k^{-2M-3}), \quad \operatorname{Im} k = 0,$$

$$\text{(56b)} \quad \theta_m = -C_m, \quad m = 0, 1, 2, \ldots,$$

$$\text{(57a)} \quad C_m \equiv (-1)^m (2m+1)^{-1} \int_{-\infty}^{+\infty} dx \, L^m [xu_x(x) + 2u(x)],$$
$$m = 0, 1, 2, \ldots,$$

$$\text{(57b)} \quad C_m = (-1)^m (2m+1)^{-1} \int_{-\infty}^{+\infty} dx \, \tilde{L}^m u(x), \quad m = 0, 1, 2, \ldots,$$

$$\text{(57c)} \quad C_m = -\pi^{-1} \int_{-\infty}^{+\infty} dk \, (2k)^{2m} \ln[T(k)T(-k)]$$
$$+ 2(-1)^{m+1}(2m+1)^{-1} \sum_{n=1}^{N} (2p_n)^{2m+1}, \quad m = 0, 1, 2, \ldots,$$

$$\text{(57d)} \quad C_m = (-1)^m \int_0^1 dy \int_{-\infty}^{+\infty} dx \, [L(yu)]^m u(x), \quad m = 0, 1, 2, \ldots,$$

$$\text{(58a)} \quad C_0 = \int_{-\infty}^{+\infty} dx \, u(x),$$

$$\text{(58b)} \quad C_1 = \int_{-\infty}^{+\infty} dx \, [u(x)]^2,$$

$$\text{(58c)} \quad C_2 = \int_{-\infty}^{+\infty} dx \, \{2[u(x)]^3 + [u_x(x)]^2\},$$

$$\text{(59)} \quad \sin\left\{\sum_{m=0}^{\infty} \left[C_m^{(1)} - C_m^{(2)}\right] z^{2m+1}\right\} = \sum_{m=0}^{\infty} (-1)^m A_m z^{2m+1},$$

$$\text{(60)} \quad \cos\left\{\sum_{m=0}^{\infty} \left[C_m^{(1)} - C_m^{(2)}\right] z^{2m+1}\right\} = 1 - \sum_{m=0}^{\infty} (-1)^m B_m z^{2m+2},$$

$$\text{(61)} \quad \sin\left[\sum_{m=0}^{\infty} C_m z^{2m+1}\right] = \sum_{m=0}^{\infty} (-1)^m A_m^{(0)} z^{2m+1},$$

$$\text{(62)} \quad \cos\left[\sum_{m=0}^{\infty} C_m z^{2m+1}\right] = 1 - \sum_{m=0}^{\infty} (-1)^m B_m^{(0)} z^{2m+2},$$

$$\text{(63a)} \quad \delta C_m = (-1)^m \int_{-\infty}^{+\infty} dx \, L^m \delta u(x), \quad m = 0, 1, 2, \ldots,$$

$$\text{(63b)} \quad \delta \int_{-\infty}^{+\infty} dx \, L^m [xu_x(x) + 2u(x)] = (2m+1) \int_{-\infty}^{+\infty} dx \, L^m \delta u(x),$$

$$\text{(64a)} \quad Mf(x) \equiv f_x(x) - \int_{-\infty}^{x} dy \, u(y) f(y),$$

$$\text{(64b)} \quad Mf(x) = (D + Iu) f(x) - \int_{-\infty}^{+\infty} dy \, u(y) f(y),$$

(64c) $(M^T)^2 = \Lambda^{(0)}$,

(65) $c_m \equiv (-1)^m \int_{-\infty}^{+\infty} dx\, u(x) M^{2m} \cdot 1, \quad m=0,1,2,\ldots,$

(66) $\sin\left[\sum_{m=0} C_m z^{2m+1}\right] = \sum_{m=0} c_m z^{2m+1},$

(67) $c_m = (-1)^m A_m^{(0)}, \quad m=0,1,2,\ldots.$

The formulae (25) are immediate consequences of the definitions (8) and (9), and of the property (5).

The definition (26) is introduced only to express in a compact way the r.h.s. of (27), (28) and (29). These display useful properties of the operators Λ and $\tilde{\Lambda}$, that are instrumental in deriving other results below. Note that the r.h.s. of these equations has been written as the derivative of an explicit expression (that however need not vanish as $x \to -\infty$). In particular (27) holds whenever the two functions $f(x)$ and $g(x)$ belong to that class of functions for which the property (25) is valid (so, for instance, (27) does not apply with $f(x) = 1$ since $D\tilde{\Lambda} \cdot 1 \neq 0 = \Lambda D \cdot 1$); in fact, (27) expresses the failure of (15) to hold in this functional space. Indeed, that $\Lambda^T \neq \tilde{\Lambda}$ is easily seen by integrating both sides of (27), with $m = 1$, thereby obtaining

(68) $(g, \Lambda f_x) - (\tilde{\Lambda} g, f_x) = f(-\infty)[(s \cdot 1, Dg) + (m \cdot 1, Img)]$
$\qquad - (m \cdot 1, g)(m \cdot 1, f),$

where the scalar product is defined by (16). The proof of (27), by induction, is left as an easy exercise.

The formula (28) is easily derived from (27) by adding to both sides of this equation the expression $[f(x)\tilde{\Lambda}^m g(x)]_x$, and using (25). Similarly (29) obtains from (27) by subtracting from both sides of (27) the expression $[g(x)\tilde{\Lambda}^m f(x)]_x$. For future reference we point out here that the resulting equation (29) remains valid for a class of functions that is larger than that considered in the previous cases. In fact (29) is an identity involving only the operator $\tilde{\Lambda}$, and therefore its validity is guaranteed by the weaker condition that IsD and $ImIm$ be well defined operators (see (9b)). Therefore if, for instance, the two potentials $u^{(1)}(x)$ and $u^{(2)}(x)$ (see (1) and (2)) vanish exponentially as $x \to +\infty$, then (29) holds even if $f(x)$ and $g(x)$ are polynomials.

The formulae (30) are merely the subcases of (25) with (10). Analogously, (31), (32), (33) and (34) obtain respectively from (26), (27), (28) and (29) in the special case (10). Of course, the remarks made above on these equations apply as well to these subcases. In particular, the equation

(69) $\quad (g, Lf_x) - (\tilde{L}g, f_x) = 2f(-\infty)(u \cdot 1, Dg)$

shows that L and $\tilde{L}$ are not the adjoint of each other (compare with the comments following (15)).

The formulae (35), (36), (37) and (38) are reported from section 2.4.1 (see also appendix A.6).

The definition (39) introduces the generic solution of the Schroedinger equation with the potential $u^{(j)}(x)$ as the linear combination of the Jost solutions (see (18) and (19)), $a^{(j)}$ and $b^{(j)}$ being arbitrary constants. The validity of (40) can be verified by an elementary, if tedious, computation, using (18) and (19); this is left as an exercise for the diligent reader. The formulae (41) and (42) are subcases of the previous one; in fact, (41) obtains from it with $a^{(j)}=1$, $b^{(j)}=0$, while (42) obtains from it with $a^{(j)}=0$, $b^{(j)}=1$ (see (20)). The formulae (43), (44) and (45) obtain from the preceding ones when the special case (10) holds (i.e. merely by dropping the index j wherever it appears in (39), (40), (41) and (42)).

The formulae (46) and (47), together with their specializations (48) and (49), are the main results of this appendix. Before proving them, the following remarks are in order.

Note first of all that, using the unitarity relation,

(70) $\quad R^{(j)}(k)R^{(j)}(-k) + T^{(j)}(k)T^{(j)}(-k) = 1, \quad j=1,2,$

and the parity of the integrand, the terms $[T^{(2)}(k)/T^{(1)}(k) \mp T^{(1)}(k)/T^{(2)}(k)]$ which appear in the r.h.s. of (46a) and (47a) can both be replaced by the expression

(71) $\quad [R^{(1)}(k)R^{(1)}(-k) - R^{(2)}(k)R^{(2)}(-k)]/[T^{(1)}(k)T^{(2)}(-k)].$

This incidentally confirms the convergence of the integrals, since (see appendix A.3)

(72a) $\quad \lim_{k \to \pm \infty} [T(k)] = 1, \quad \text{Im } k = 0,$

(72b) $\quad \lim_{k \to \pm \infty} [k^m R(k)] = 0, \quad m = 0, 1, 2, \ldots, \quad \text{Im } k = 0.$

(The reader puzzled by the apparent contradiction between (72a) and the convergence of the integral in the r.h.s. of (46a) and (47a) should note that only the part of $[T^{(2)}(k)/T^{(1)}(k) + T^{(1)}(k)/T^{(2)}(k)]$ that is odd in k does in fact contribute to the integral; and this part indeed vanishes faster than any inverse power of k, see (71) and (72b)).

Next observe that (46a) and (47a) have been written for the (generic) case of two potentials $u^{(1)}(x)$ and $u^{(2)}(x)$ having *different* discrete eigenvalues; but they remain valid if some of the eigenvalues coincide, with the simplification implied by the formula (see (23))

(73) $$\lim_{k \to i p_n^{(j)}} \left[T^{(j)}(k) \right]^{-1} = 0,$$

that eliminates the relevant terms in the sums.

Let us now proceed to prove the formulae (46a) and (47a). The starting point are the formulae (2.4.2.-11) and (2.4.2.-12), that we conveniently rewrite here in the form:

(74) $$\int_x^{+\infty} dy \left[G(\Lambda) m \cdot 1 + H(\Lambda) \Gamma \cdot 1 \right]$$

$$= -\pi^{-1} \int_{-\infty}^{+\infty} dk \, f^{(+)(1)}(x,k) f^{(+)(2)}(x,k)$$

$$\cdot \left\{ G(-4k^2) \left[R^{(1)}(k) - R^{(2)}(k) \right] \right.$$

$$\left. + 2ikH(-4k^2) \left[R^{(1)}(k) + R^{(2)}(k) \right] \right\} + 2 \sum_{n=1}^{N^{(2)}} f^{(+)(1)}(x, i p_n^{(2)})$$

$$\cdot f^{(+)(2)}(x, i p_n^{(2)}) \rho_n^{(2)} \left[G\left(4 p_n^{(2)^2}\right) + 2 p_n^{(2)} H\left(4 p_n^{(2)^2}\right) \right]$$

$$- 2 \sum_{n=1}^{N^{(1)}} f^{(+)(1)}(x, i p_n^{(1)}) f^{(+)(2)}(x, i p_n^{(1)}) \rho_n^{(1)}$$

$$\cdot \left[G\left(4 p_n^{(1)^2}\right) - 2 p_n^{(1)} H\left(4 p_n^{(1)^2}\right) \right].$$

This formula obtains setting $g(z) = G(z)$ in (2.4.2.-11), $g(z) = H(z)$ in (2.4.2.-12), summing up the resulting equations and using (25). Of course, as shown by (46b), $G(z)$ and $H(z)$ are arbitrary polynomials.

It is then easily seen that (46a) is merely the limiting case of this formula as $x \to -\infty$. To understand how this result obtains, let us consider firstly the limit of this equation as $x \to +\infty$. Then clearly the l.h.s. vanishes, as well as the sums in the r.h.s. (see (2.4.2.-6a) and (2.4.2.-8a), or (19a) with k on the positive imaginary axis). One therefore concludes that

(75) $$\lim_{x \to +\infty} \left\{ \int_{-\infty}^{+\infty} dk \, f^{(+)(1)}(x,k) f^{(+)(2)}(x,k) \right.$$

$$\cdot \left\{ \left[G(-4k^2) + 2ikH(-4k^2) \right] R^{(1)}(k) \right.$$

$$\left. \left. - \left[G(-4k^2) - 2ikH(-4k^2) \right] R^{(2)}(k) \right\} \right\} = 0.$$

But this is merely a consequence (see (72b)) of (19a) and the standard formula

(76) $\qquad \lim_{x \to \infty} \left[\int_{-\infty}^{+\infty} dk \exp(\pm 2ikx) F(k) \right] = 0,$

that applies provided $F(k)$ is regular (differentiable) for real k and vanishes as $k \to \pm \infty$.

The use of this same property, together with (2.4.2.-6b), (2.4.2.-8b), (2.1.-29), (2.1.-25), and (2.4.2.-1b) (see also (19b) and (21)), allows to perform the limit as $x \to -\infty$ of (74). Using also the first remark given above (see (71)), as well as (72), there obtains precisely (46a), that is thereby proved.

The additional ingredient required to prove (47a) is the following generalized wronskian expression

(77) $\qquad \int_{-\infty}^{+\infty} dx\, f^{(+)(1)}(x,k) f^{(+)(2)}(x,k) [g(\Lambda)\boldsymbol{m} \cdot 1 + h(\Lambda) \boldsymbol{\Gamma} \cdot 1]$

$\qquad = 2ik \{ B^{(-)}(k) R^{(2)}(-k) / [T^{(1)}(k) T^{(2)}(-k)]$

$\qquad \quad - B^{(-)}(-k) R^{(1)}(-k) / [T^{(1)}(-k) T^{(2)}(k)] \},$

that holds (at least) for Im $k \geq 0$. Here the function $B^{(-)}(k)$ is given by (47c) with (47d) and (47e) (clearly representing its even and odd parts). One way, among others, to derive (77) is by integrating on the whole x-axis both sides of (A.20.-28) with $\psi^{(j)}(x) = f^{(+)(j)}(x,k)$, and using (A.20.-24), (19a), (19b), (A.20.-30), (A.20.-31) and (A.20.-32). The two functions $g(z)$ and $h(z)$ entering in (77) are arbitrary polynomials (see (47b)). It remains now to multiply both sides of (74) by the function $[g(\Lambda)\boldsymbol{m} \cdot 1 + h(\Lambda) \boldsymbol{\Gamma} \cdot 1]$ and subsequently integrate over the whole x-axis (exchanging the order of the integrations in the x and k variables in the r.h.s. and using (77)). The resulting expression can be easily made to coincide with (47a) by using again the unitary relation (70), and the parity properties of the integrand. Of course the preceding remarks on (46a) apply as well to (47a).

It should be pointed out that the occurrence in (47a) of four arbitrary polynomials (i.e. $g(z)$, $h(z)$, $G(z)$ and $H(z)$) gives the possibility to obtain, as special cases, a number of important formulae, including operator identities as discussed below.

The formulae (48a) and (49a) are merely definitions. Then (48c) resp. (49c) are subcases of (46a) with $G(z) = z^m$ and $H(z) = 0$ resp. $G(z) = 0$ and $H(z) = z^m$.

It should be noticed that, since the transmission coefficient can be explicitly expressed in terms of the spectral transform via the dispersion relation (see appendix A.4)

(78a) $$T(k) = \prod_{n=1}^{N} \left[1 + (p_n/k)^2\right]^{-1} \exp\left[\pi^{-1} \int_{-\infty}^{+\infty} dq\, (q-k)^{-1} \theta(q)\right],$$
$\operatorname{Im} k > 0,$

(78b) $$\theta(k) = -(2\pi)^{-1} P \int_{-\infty}^{+\infty} dq\, (q-k)^{-1} \ln\left[1 - R(q)R(-q)\right]$$
$$+ 2 \sum_{n=1}^{N} \operatorname{Arctg}(p_n/k), \quad \operatorname{Im} k = 0,$$

the formulae (48c) and (49c) (or, more generally, (46a) and (47a)) provide explicit expressions, in terms of the spectral transforms, of nonlinear functionals of the corresponding potentials.

Consider next (48b) and (49b); these are alternative expressions for the quantities (48a) and (49a), respectively. Clearly the equality of the r.h.s. of these equations has nothing to do with the spectral problem, but it merely displays certain properties of the integrodifferential operators defined by (1), (2), (4), (7) and (8) (with $u^{(j)}(x)$, $j=1,2$, belonging to the class of *bona fide* potentials). The proof of (48b) originates from the remark that the r.h.s. of (47a) with $g(z)=1$, $h(z)=0$, $G(z)=0$, $H(z)=z^{m-1}$ coincides with the r.h.s. of (48c); then (48b) follows from the equality of the corresponding l.h.s.. Similarly the proof of (49b) obtains setting in (47a) $g(z)=1$, $h(z)=0$, $G(z)=z^m$, $H(z)=0$, and noting that the r.h.s. of the resulting equation coincides with the r.h.s. of (49c).

Additional (more intricate) operator identities can obviously be derived from (46a) (with (48c) and (49c)) and (47a). Just to give an example, we have displayed the equation (50) that obtains setting in (47a) $g(z)=0$, $h(z)=1$, $G(z)=z^m$ and $H(z)=0$, and comparing the resulting r.h.s. expression with (48c) and (49c).

Let us now prove (48d), assuming for simplicity that discrete eigenvalues are associated neither with $u^{(1)}(x)$ nor with $u^{(2)}(x)$ ($N^{(1)}=N^{(2)}=0$). Then the two sums in the r.h.s. of (48c) are missing. Moreover the function

(79) $$F(k) = T^{(2)}(k)/T^{(1)}(k) - T^{(1)}(k)/T^{(2)}(k),$$

that appears in the l.h.s. of (48d), and in the integrand of (48c), has the following properties:

(i) $F(k)$ is holomorphic in the half-plane $\operatorname{Im} k \geq 0$. Indeed (see section 2.1) the function $T(k)$, in the half-plane $\operatorname{Im} k \geq 0$, has no zeros and is holomorphic if there are no discrete eigenvalues, as we are now assuming.

(ii) The even part of $F(k)$ vanishes faster than any power of k as $k \to \pm \infty$, this following from the property (see appendix A.3)

(80) $$\lim_{k \to \pm \infty} \{|k|^m [1 - T(k)T(-k)]\} = 0, \quad m = 0, 1, 2, \ldots, \quad \text{Im } k = 0.$$

(iii) The odd part of $F(k)$ (and therefore $F(k)$ itself; see (ii)) has an asymptotic expansion in inverse (odd) powers of k, as implied by (24a) and (56a),

(81) $$F(k) = \sum_{m=0}^{M} F_m (2ik)^{-2m-1} + O(k^{-2M-3}), \quad \text{Im } k = 0.$$

(iv) $F(k)$ vanishes as $\text{Im } k \to +\infty$.

Therefore one can write for $F(k)$ the Cauchy representation (see, e.g., appendix A.4),

(82) $$F(k) = (\pi i)^{-1} P \int_{-\infty}^{+\infty} dq\, F(q)/(q - k), \quad \text{Im } k = 0,$$

and expand (the odd part of) this formula in inverse powers of k, thus obtaining for the coefficient F_m of the asymptotic expansion (81) the formula

(83) $$F_m = -2\pi^{-1} \int_{-\infty}^{+\infty} dq\, (2iq)^{2m} F(q), \quad m = 0, 1, 2, \ldots;$$

note that the integral in the r.h.s. is convergent, since only the even part of $F(q)$ contributes (see (ii) above).

The formula (81), with (79) and (83), reproduces (48d) with (48c). It is left as an exercise to extend the proof to the more general case with discrete eigenvalues; then the function $F(k)$ defined by (79) has (simple) poles in the half-plane $\text{Im } k > 0$, that contribute an additional sum in the Cauchy formula (82); and these additional terms account for the sums in the r.h.s. of (48c), that provide the contribution of the discrete eigenvalues (see (23)).

Since for large k

(84) $$T(k) \approx \exp[i\theta(k)],$$

where the curly equality sign indicates asymptotic equality up to corrections that vanish faster than any inverse power of k, the expression in the l.h.s. of

(48d) can be replaced by $\sin[\theta^{(2)}(k)-\theta^{(1)}(k)]$ (yielding (48e)) or, for instance (see (24a)), by $T^{(2)}(k)T^{(1)}(-k)-T^{(1)}(k)T^{(2)}(-k)$.

The proof of (49d) is quite analogous to the proof of (48d) that we have just given; note however that one is now dealing with a function whose *odd* part vanishes faster than any inverse power of k as $k \to \pm \infty$. This proof, as well as the analogous proof of (49e), are left as an easy exercise for the diligent reader.

It is clear that comparing (48e) with (49e), and using the elementary properties of the circular functions, one can easily express the quantities A_m's in terms of the B_m's, or viceversa: for instance

(85) $\quad B_0 = \tfrac{1}{2} A_0^2, \qquad B_1 = A_0 A_1 - \tfrac{1}{8} A_0^4,$

$\qquad B_2 = A_0 A_2 + \tfrac{1}{2} A_1^2 - \tfrac{1}{2} A_1 A_0^3 + \tfrac{1}{16} A_0^6.$

Of course, because of the definitions (48a) and (49a), these equalities express nontrivial properties of the integro differential operator Λ.

The formulae (52) and (53) are special cases of the previous equations (48) and (49), corresponding to $u^{(1)}(x)=u(x), u^{(2)}(x)=0$. Thus the operator $\Lambda^{(0)}$, explicitly defined by (51), is merely the operator Λ in this special case (see (8)); of course, in this case, $T^{(1)}(k)=T(k), T^{(2)}(k)=1$ and therefore $\theta^{(1)}(k)=\theta(k), \theta^{(2)}(k)=0$, etc.

The formulae (54) are merely the simplest nontrival consequences of (46) and (47), in the limit as $u^{(1)}(x)$ and $u^{(2)}(x)$ coincide. In particular (54a) (that is proved also in appendix A.8) follows straightforwardly setting $u^{(1)}(x)=u^{(2)}(x)=u(x)$ in (49a) and (49b); (54b) follows in the same way from (47a) with $h(z)=z^n$ and $H(z)=z^m$, noting moreover that the r.h.s. integral vanishes because of the parity property of the integrand, while the sums vanish because of (73). Rewriting (54b) by using the relationship (30) between L and $\tilde{L}$ yields (54c).

The next set of formulae, (55), obtain from (48) and (49) by treating the difference $u^{(1)}(x)-u^{(2)}(x)=\delta u(x)$ as an *infinitesimal* variation. Therefore (55a) follows from the equality of (48a) and (48b) with $u^{(1)}(x)=u(x)+\delta u(x)$ and $u^{(2)}(x)=u(x)$; (55b, c) result from the vanishing of the first order term of the expression (49a) of B_m, since this, as clearly shown by (49b), is infinitesimal of second order in $\delta u(x)$. These two equations obviously express properties of the integrodifferential operator L that have nothing to do with the spectral problem. We shall return to this below. The equation (55d) relates instead an infinitesimal variation of the potential to the corresponding variation of the spectral quantities; it obtains from (48a) and

(48c) setting $T^{(1)}(k) = T(k) + \delta T(k)$, $T^{(2)}(k) = T(k)$, etc. (special care must be exercised, using (23), to get the sum in the r.h.s. of (55d)).

The formula (57a) is a definition (consistent with the notation employed elsewhere, see (1.7.3.-5) and (5.-12)). The equation (56b), which shows how simply the integral expressions (57a) are related to the asymptotic expansion (56a) of the phase $\theta(k)$, has been proved in appendix A.8. The explicit expression (57c) of C_m in terms of the spectral transform (see (70)) obtains, via (56b) and (56a), by expanding the r.h.s. of (78b) in inverse powers of k. The examples (58) are reported for completeness.

The fact that the explicit x-dependence in the integrand in (57a) is always integrated away is suggested by the explicit examples (58). That this remarkable property holds for all $m \geq 0$ is implied by the alternative expression (57b). This formula is proved by setting in (34) $f(x) = x$ and $g(x) = 1$, noting that

(86) $\quad \tilde{L} \cdot 1 = -2u(x), \qquad \tilde{L} \cdot x = -2\left[xu(x) - \int_x^{+\infty} dy\, u(y)\right],$

and then by integrating both sides of (34) over the entire x-axis; this yields

(87) $\quad \int_{-\infty}^{+\infty} dx\, \tilde{L}^m u(x) = C_m - 2 \lim_{x \to -\infty} \sum_{n=0}^{m-2} \left[\left[\tilde{L}^n u(x), \tilde{L}^{m-n-2}\right]\right.$

$\left. \cdot \left\{xu(x) - \int_x^{+\infty} dy\, u(y)\right\}\right],$

where, together with the definition (31), we have used the limit

(88) $\quad \lim_{x \to -\infty} \tilde{L}^m u(x) = -\int_{-\infty}^{+\infty} dx\, D\tilde{L}^m u(x)$

$= -\int_{-\infty}^{+\infty} dx\, L^m u_x(x) = 0, \quad m = 0, 1, 2, \ldots,$

that holds because of (12), (30) and (54a). By using again the property (88), one can easily show that the limit in the r.h.s. of (87) vanishes, thereby proving (57b).

To prove (57d), we use (63a) (that is itself proved below, of course without using (57d)); indeed (57d) obtains by replacing, in (63a), $u(x)$ by $yu(x)$, by taking the variation δ to correspond to an infinitesimal variation of y, i.e. replacing $\delta u(x)$ with $u(x)dy$, and by integrating over y from 0 to 1.

The formulae (59) resp. (60) are readily obtained from the asymptotic expansions (48e) resp. (49e), combined with the asymptotic expansion (56a) with (56b), where of course $C_m^{(j)}$ corresponds to the potential $u^{(j)}(x)$ (see (57a) with $u(x)$ replaced by $u^{(j)}(x)$, $j=1,2$). These formulae must be understood by reexpanding the l.h.s. and then equating the coefficients of equal powers of z. As written, they are convenient to express the quantities A_m and B_m as nonlinear combinations of $C_n^{(1)} - C_n^{(2)}$, $0 \leq n \leq m$ (and viceversa, by using the inverse circular functions). The fact that this is possible is a property of the quantities A_m and B_m that is far from evident from their definitions, (48a) and (49a). The first two examples of these formulae read

(89) $\quad A_0 = C_0^{(1)} - C_0^{(2)}, \quad A_1 = \frac{1}{6}[C_0^{(1)} - C_0^{(2)}]^3 - [C_1^{(1)} - C_1^{(2)}],$

(90) $\quad B_0 = \frac{1}{2}[C_0^{(1)} - C_0^{(2)}]^2,$

$\quad B_1 = \frac{1}{24}[C_0^{(1)} - C_0^{(2)}]^4 - [C_0^{(1)} - C_0^{(2)}][C_1^{(1)} - C_1^{(2)}].$

The formulae (61) and (62) are simply the subcases of the preceding expansions obtained by setting $u^{(2)}(x) = 0$ and $u^{(1)}(x) = u(x)$; in this case, they provide nontrivial relations between the operators L (see (57a)) and $\Lambda^{(0)}$ (see (52a)).

The important variational formula (63b), that is merely (63a) with (57a), is proved in appendix A.8. It can be easily reobtained here replacing in (59) the finite difference of the two potentials with an infinitesimal variation (i.e. $u^{(1)}(x) = u(x) + \delta u(x)$, $u^{(2)}(x) = u(x)$, $C_m^{(1)} = C_m + \delta C_m$, $C_m^{(2)} = C_m$, etc.), and inserting for A_m the contribution of first order in $\delta u(x)$ derived from (48a).

The formulae (64) and (65) are merely definitions, reported here for completeness; see appendix A.13, where (66) is also proved by using the spectral problem, although (66) is clearly an operator identity having no reference to the spectral problem. The equality (64c) is easily verified by noticing that

(91) $\quad I^T \cdot f(x) = \int_{-\infty}^{x} dy f(y), \quad M^T = -D - uI,$

where the transposition of an operator is defined with respect to the symmetric scalar product (16), and then comparing with (51a). Finally (67) is easily derived by inspecting (61) and (67); or, directly, rewriting (65) as

(92) $\quad c_m = (-1)^m \int_{-\infty}^{+\infty} dx (M^T)^{2m} u(x),$

and then using (64c).

Let us finally comment on some of the operator identities that have been established. Most interesting for our purposes are the formulae (54). They can be reexpressed in local form, namely via the definitions

(93) $\quad L^m u_x(x) = g_x^{(m)}(x),$

(94) $\quad [L^m u_x(x)] IL^n u_x(x) = G_x^{(n,m)}(x), \quad m, n = 0, 1, 2, \ldots,$

and the properties

(95) $\quad g^{(m)}(\pm\infty) = 0, \quad m = 0, 1, 2, \ldots,$

(96) $\quad G^{(n,m)}(\pm\infty) = 0, \quad m, n = 0, 1, 2, \ldots.$

Of course it is the simultaneous validity of the last two equations at both ends that is nontrivial; this must appear as an automatic consequence of the asymptotic vanishing at both ends of $u(x)$ and all its derivatives.

In fact it is easily seen that the condition $g^{(m)}(+\infty) = 0$, together with the property (30), imply the explicit expression

(97) $\quad g^{(m)}(x) = \tilde{L}^m u(x), \quad m = 0, 1, 2, \ldots,$

for the "currents" $g^{(m)}(x)$, as well as the relation (see also (5))

(98) $\quad G_x^{(m,n)}(x) = -g^{(m)}(x) g_x^{(n)}(x), \quad m, n = 0, 1, 2, \ldots;$

moreover, from this relation it readily follows that

(99a) $\quad G^{(m,n)}(x) = -G^{(n,m)}(x) - g^{(m)}(x) g^{(n)}(x), \quad m, n = 0, 1, 2, \ldots,$

(99b) $\quad G^{(m,m)}(x) = -\tfrac{1}{2}[g^{(m)}(x)]^2, \quad m = 0, 1, 2, \ldots.$

The explicit expressions of the first few "currents" $g^{(m)}(x)$ and $G^{(m,n)}(x)$ read:

(100a) $\quad g^{(0)} = u,$

(100b) $\quad g^{(1)} = u_{xx} - 3u^2,$

(100c) $\quad g^{(2)} = u_{xxxx} - 5u_x^2 - 10 u_{xx} u + 10 u^3,$

(101a) $\quad G^{(0,0)} = -\tfrac{1}{2} u^2,$

(101b) $\quad G^{(1,0)} = -\tfrac{1}{2} u_x^2 + u^3,$

(101c) $\quad G^{(2,0)} = -u_x u_{xxx} + \tfrac{1}{2} u_{xx}^2 + 5 u u_x^2 - \tfrac{5}{2} u^4,$

(101d) $\quad G^{(1,1)} = -\tfrac{1}{2} u_{xx}^2 + 3 u^2 u_{xx} - \tfrac{9}{2} u^4.$

These expressions suggest that both $g^{(m)}$ and $G^{(m,n)}$ are nonlinear (polynomial) combinations of u and its derivatives; a result that guarantees the automatic validity of the properties (95) and (96), but that is itself nontrivial, since the definitions (93) and (94) (or the expressions (97) and (98)) involve the integral operator I. In order to prove that both $g^{(m)}$ and $G^{(m,n)}$ have this property for all nonnegative values of m and n, we start from the recursion relation

(102a) $\quad g^{(m+1)}(x) = \tilde{L} g^{(m)}(x) = Y g^{(m)}(x) + 2 G^{(0,m)}(x), \quad m = 1, 2, 3, \ldots,$

(102b) $\quad g^{(0)}(x) = u(x),$

the operator Y, defined by

(103) $\quad Y \equiv D^2 - 2u,$

being purely differential (i.e., not integral). The recursion equation (102) follows immediately from (97), (12), (98) and (100a), and its formal solution reads

(104) $\quad g^{(m)}(x) = Y^m u(x) + 2 \sum_{n=0}^{m-1} Y^{m-n-1} G^{(0,n)}(x), \quad m = 1, 2, 3, \ldots.$

We then express $G^{(0,n)}$ in terms of the functions $g^{(m)}$ for $0 \leq m \leq n$ by using (33) with $g = f = u$; indeed a simple calculation yields

(105) $\quad G^{(0,n)} = -u g^{(n)} - \sum_{m=0}^{n-1} \left(g^{(m)} Y g^{(n-m-1)} - \tfrac{1}{2} g_x^{(m)} g_x^{(n-m-1)} \right)$
$\qquad + \tfrac{1}{2} \sum_{m=0}^{n} g^{(m)} g^{(n-m)}, \quad n = 1, 2, 3, \ldots.$

Therefore, combining (105) with (104), it follows that if $g^{(n)}(x)$ is a polynomial in u and its derivatives for $0 \leq n \leq m-1$, then also $g^{(m)}$ has this property, that, because of (100) and (101), by induction, holds then for all $m \geq 0$. As for $G^{(m,n)}$, the identity (34), with $g = u$ and $f = g^{(n)}$, yields

(106) $\quad G^{(m,n)} = -G^{(0,m+n)} + \sum_{j=0}^{m-1} \left(g^{(j)} Y g^{(m+n-j-1)} \right.$
$\qquad \left. + g^{(m+n-j-1)} Y g^{(j)} - g_x^{(j)} g_x^{(m+n-j-1)} \right)$
$\qquad - \sum_{j=0}^{m} g^{(j)} g^{(n+m-j)}, \quad m = 1, 2, 3, \ldots, \quad n = 0, 1, 2, \ldots;$

and this, together with (105), provides an explicit expression of $G^{(m,n)}$ in terms of the polynomials $g^{(n)}$ and their derivatives (see (103)), and therefore completes the proof.

Several other operator identities given above can also be recast in local form. For instance the following definitions can be introduced:

(107a) $\quad a_x^{(n)}(x) = (\Lambda^n m - mI\Lambda^{n-1}\Gamma) \cdot 1, \quad n=1,2,3,\ldots,$

(108a) $\quad b_x^{(n)}(x) = (\Lambda^n \Gamma - mI\Lambda^n m) \cdot 1, \quad n=0,1,2,\ldots,$

(109a) $\quad \varepsilon_x^{(n)}(x) = L^{n+1} \delta u(x) - 2[\delta u(x)] IL^n u_x, \quad n=0,1,2,\ldots,$

(110a) $\quad \mu_x^{(n)}(x) = \delta\{L^n[xu_x(x) + 2u(x)]\} - (2n+1)L^n \delta u(x),$
$$n=0,1,2,\ldots,$$

(111a) $\quad q_x^{(n)}(x) = [\Lambda^{(0)}]^n u(x) - u(x) M^{2n} \cdot 1, \quad n=0,1,2,\ldots.$

It is then immediately seen that (48a) and (48b) imply

(107b) $\quad a^{(n)}(\pm\infty) = 0, \quad n=1,2,3,\ldots,$

(49a) and (49b) imply

(108b) $\quad b^{(n)}(\pm\infty) = 0, \quad n=0,1,2,\ldots,$

(55a) implies

(109b) $\quad \varepsilon^{(n)}(\pm\infty) = 0, \quad n=0,1,2,\ldots,$

(63b) implies

(110b) $\quad \mu^{(n)}(\pm\infty) = 0, \quad n=0,1,2,\ldots,$

and (67), (52a) and (65) imply

(111b) $\quad q^{(n)}(\pm\infty) = 0, \quad n=0,1,2,\ldots.$

Of course the nontrivial aspect of these relations is the fact that all these "currents" (that are defined up to an additive constant) vanish at *both* ends, as an automatic consequence of the fact that all the functions that enter in these formulae (i.e. $s(x) = u^{(1)}(x) + u^{(2)}(x)$, $m(x) = u^{(1)}(x) - u^{(2)}(x)$, $u(x)$ and its infinitesimal variation $\delta u(x)$) belong to the class of *bona fide* potentials, namely they vanish at both ends with all their derivatives. To display how this comes about let us exhibit a few examples:

(107c) $\quad a^{(1)}(x) = m_x(x) + s(x) \int_x^\infty dy\, m(y),$

(108c) $\quad b^{(0)}(x) = s(x)$

(108d) $\quad b^{(1)}(x) = s_{xx}(x) - \tfrac{3}{2} s^2(x) - \tfrac{1}{2} m^2(x)$

$\quad\quad\quad + m_x(x) \int_x^\infty dy\, m(y) + s(x) \left[\int_x^\infty dy\, m(y) \right]^2,$

(109c) $\quad \varepsilon^{(0)}(x) = \delta u_x(x) + 2 u(x) \int_x^\infty dy\, \delta u(y),$

(110c) $\quad \mu^{(0)}(x) = x \delta u(x),$

(110d) $\quad \mu^{(1)}(x) = x \delta u_{xx}(x) - 6 x u(x) \delta u(x)$

$\quad\quad\quad + 2 \delta u(x) \int_x^\infty dy\, u(y) - 4 u(x) \int_x^\infty dy\, \delta u(y),$

(111c) $\quad q^{(0)}(x) = 0,$

(111d) $\quad q^{(1)}(x) = u_x(x) + u(x) \int_x^\infty dy\, u(y)$

$\quad\quad\quad + \tfrac{1}{2} \left[\int_{-\infty}^{+\infty} dy\, u(y) \right] \left[\int_x^\infty dy\, u(y) \right] \left[\int_{-\infty}^x dy\, u(y) \right].$

A.10. The two-soliton solution of the KdV and higher KdV equations

In this appendix we display the two-soliton solution of the class of evolution equations

(1) $\quad u_t = \alpha(L) u_x, \quad u \equiv u(x,t),$

where L is the integro-differential operator defined by

(2) $\quad Lf(x) = f_{xx}(x) - 4 u(x,t) f(x) + 2 u_x(x,t) \int_x^\infty dy\, f(y).$

Let us recall that, for $\alpha(z) = -z$, (1) is the KdV equation

(3) $\quad u_t + u_{xxx} - 6 u_x u = 0, \quad u \equiv u(x,t).$

We also discuss some of the properties of this solution. All these results are elementary consequences of the relevant formulae given in the main text.

Let us also recall that it is convenient to associate with $u(x,t)$ its integral,

(4) $\qquad w(x,t)=\int_{x}^{\infty}dy\,u(y,t),\qquad u(x,t)=-w_x(x,t),$

and that in terms of $w(x,t)$ the class (1) can be rewritten in the form

(5) $\qquad w_t=\alpha(\tilde{L})w_x,\qquad w\equiv w(x,t),$

with

(6) $\qquad \tilde{L}F(x)=F_{xx}+4w_x(x,t)F(x)+2\int_{x}^{\infty}dy\,w_{yy}(y,t)F(y).$

The two-soliton solution is more simply written in terms of $w(x,t)$ than $u(x,t)$, and it reads

(7) $\qquad w(x,t)=F^{(2)}\{p_1, p_1[x-\xi_1(t)]; p_2, p_2[x-\xi_2(t)]\}$

where

(8) $\qquad F^{(2)}(p_1, y_1; p_2, y_2)$
$\qquad\qquad =\left[1-\tfrac{1}{4}(p_1+p_2)^{-2}F^{(1)}(p_1,y_1)F^{(1)}(p_2,y_2)\right]^{-1}$
$\qquad\qquad \cdot\left[F^{(1)}(p_1,y_1)+F^{(1)}(p_2,y_2)+(p_1+p_2)^{-1}\right.$
$\qquad\qquad \left.\cdot F^{(1)}(p_1,y_1)F^{(1)}(p_2,y_2)\right],$

(9) $\qquad F^{(1)}(p,y)=-2p(1-\tgh y)=-4p[1+\exp(2y)]^{-1},$

(10) $\qquad \xi_n(t)=\xi_n(0)-\alpha(4p_n^2)t,\qquad n=1,2.$

The corresponding spectral transform is

(11) $\qquad S[u]=\{R(k,t)=0,\ -\infty<k<\infty;\ p_n,\rho_n(t),\ n=1,2\},$

with

(12) $\qquad \rho_n(t)=2p_n\exp[2p_n\xi_n(t)].$

Let us also report the corresponding formulae for the single-soliton solution:

(13) $\quad w(x,t) = F^{(1)}\{p; p[x-\xi(t)]\},$

(14) $\quad \xi(t) = \xi(0) - \alpha(4p^2)t,$

(15) $\quad S[u] = \{R(k,t) = 0, -\infty < k < \infty; p, \rho(t)\},$

(16) $\quad \rho(t) = 2p\exp[2p\xi(t)].$

It is now easy to verify that

(17a) $\quad \lim_{y_2 \to +\infty} \left[F^{(2)}(p_1, y_1; p_2, y_2) \right] = F^{(1)}(p_1, y_1),$

(17b) $\quad \lim_{y_2 \to -\infty} \left[F^{(2)}(p_1, y_1; p_2, y_2) \right] = F^{(1)}(p_1, y_1 + \eta) - 4p_2,$

with

(18) $\quad \exp\eta = (p_1 + p_2)/|p_1 - p_2|.$

Of course in (17) the roles of the subscripts 1 and 2 could be trivially exchanged.

Note that, in writing the last formulae, (17) and (18), we are implicitly assuming that $p_1 \neq p_2$. On the other hand it is easily seen that

(19) $\quad F^{(2)}(p, y_1; p, y_2) = F^{(1)}(p, \tfrac{1}{2}(y_1 + y_2) - \delta)$

with

(20) $\quad \exp(2\delta) = 2\cosh(y_1 - y_2).$

This last formula, together with (7), (10) and (13), show that the two-soliton formula goes into the single-soliton expression, if the parameters p_1 and p_2 of the two solitons coincide, $p_1 = p_2 = p$. Note the consistency of this result with the identity

(21) $\quad \{R(k), -\infty < k < \infty; p_1 = p_2 = p; \rho_1, \rho_2; N = 2\}$
$\quad\quad = \{R(k), -\infty < k < \infty; p, \rho = \rho_1 + \rho_2; N = 1\},$

that is implied by the very definition of the spectral transform, and by the fact that the discrete eigenvalues of the Schroedinger spectral problem are all simple (to check completely this consistency, use (12) and (19)).

On the other hand, from (7), (13) and (17) it is easy to obtain the formula

$$(22) \quad u(x,t) \approx -2 \sum_{n=1}^{2} p_n^2/\cosh^2\{p_n[x-x_n(t)]\},$$

that applies whenever the distance between the positions $x_1(t)$ and $x_2(t)$ of the two solitons is large. A quantitative formulation of this condition obtains from an investigation of the first correction terms to (17). It reads

$$(23) \quad |x_1(t)-x_2(t)| \gg (2/p_2)\ln[(p_1+p_2)/(p_1-p_2)], \quad (p_1>p_2)$$

(this formula is written assuming $p_1>p_2$; otherwise the subscripts 1 and 2 in the r.h.s. must be exchanged).

Note that (22) is written for $u(x,t)$ rather than $w(x,t)$; this has the advantage that constant terms such as that appearing in the r.h.s. of (17b) (whose presence or absence would depend on the signs of $x_1(t)-x_2(t)$, $x-x_1(t)$ and $x-x_2(t)$) can be ignored. But such a dependence is unavoidable if one wishes to relate the soliton positions $x_n(t)$ to the quantities $\xi_n(t)$, see (10) and (12), for indeed there follows from (17) that

$$(24a) \quad x_1(t)=\xi_1(t), \quad x_2(t)=\xi_2(t)-\eta/p_2 \quad \text{if } x_2(t)<x_1(t),$$
$$(24b) \quad x_1(t)=\xi_1(t)-\eta/p_1, \quad x_2(t)=\xi_2(t) \quad \text{if } x_1(t)<x_2(t).$$

These formulae, together with (22) (with (23)), (18) and (10), provide a fairly transparent picture of the two-soliton solution (including its time-evolution; in particular, the reader may verify the results of section 1.6.1 as $t\to\pm\infty$), whenever the two bumps in the r.h.s. of (22) are well separated.

The fact that these formulae imply no inconsistency with the discussion given above of the case when p_1 and p_2 coincide should also be noted; this is due to the divergence of η, see (18), if p_1 and p_2 coincide. Thus the transition of the two-soliton formula to the single-soliton one as p_1 and p_2 coalesce may be associated with the disappearence of one of the two solitons at infinity. Note however that, as p_1 and p_2 come together, also the r.h.s. of (23) diverges. Thus some care must be exercised in discussing this limit. This also applies to the analysis of the time evolution, in particular the limits as $t\to\pm\infty$, to avoid paradoxical results originating from the lack of commutativity of different limits.

Let us emphasize that, whereas for the single-soliton solution,

(25) $$u(x,t) = -2p^2/\cosh^2\{p[x-\xi(t)]\},$$

there is a one-to-one correspondence between the parameters, p and $\xi(t)$, that identify the shape and localization of the soliton in x-space, and the parameters, p and $\rho(t) = 2p\exp[2p\xi(t)]$, that characterize its spectral transform,

(26) $$S[u] = \{R(k,t) = 0; p, \rho(t) = 2p\exp[2p\xi(t)]\},$$

the situation becomes less simple for the two-soliton solution (and a fortiori for the N-soliton solution). Indeed, of the 4 parameters that characterize (for every value of t) the two-soliton solution, only the two constant parameters p_1 and p_2 can be associated in an unambiguous manner to the two solitons, both in the spectral space and in configuration space: in the spectral space, they are associated with the 2 discrete eigenvalues of the spectral transform $\{R(k,t) = 0; p_1, p_2; \rho_1(t), \rho_2(t)\}$; in configuration space, they characterize the shape of the two solitons, whenever these are sufficiently far apart so that the two-soliton solution factors into the sum of two well-separated bumps. Note that this implies the possibility to associate unambiguously, in both spaces, each parameter p_n with the corresponding (say, nth) soliton; such identification being always possible in the spectral space (where each soliton is identified with one discrete eigenvalue), while in configuration space it is of course feasible only if the solitons are well-separated and therefore can be identified as separated entities.

The same sort of one-to-one correspondence does not instead apply to the other parameters, $\rho_n(t)$ or $\xi_n(t)$ (see (12)), that characterize the two-soliton solution. These parameters have a clear significance in the spectral space ($\rho_n(t)$ is the normalization coefficient of the discrete eigenvalue p_n, associated with the nth soliton; see above); in configuration space, they are related to the localization of the solitons (assuming these are well-separated, so that the localization of each soliton is identifiable unambiguously), but not through a one-to-one correspondence, since, say, the position $x_1(t)$ of the soliton associated with p_1 depends not only on $\rho_1(t)$ (and p_1), but also on the parameters of the other soliton (see (22) and (24), with (12) and (18)).

Let us finally recall that an explicit display of the time evolution of the two-soliton solution for the KdV equation is reported and discussed in section 3.2.1.

A.11. Miura and Gardner transformations and related results

In this appendix we treat, in a self-contained if terse manner, some results associated with the Korteweg–de Vries (KdV) equation

(1) $\quad u_t + u_{xxx} - 6u_x u = 0, \quad u \equiv u(x,t).$

These results are important historically, as they have played a key rôle in the discovery of the special nature of the KdV equation, including the spectral transform technique of solution and the existence of an infinite sequence of local conservation laws. Moreover, they yield the main properties of the KdV equation by elementary techniques which are both pleasing and instructive.

The Miura transformation reads

(2) $\quad u(x,t) = v_x(x,t) + v^2(x,t).$

Note that the time variable enters only parametrically.

It is a matter of trivial algebra to verify that, if u and v are related by this formula, then there also holds the relation

(3a) $\quad u_t + u_{xxx} - 6u_x u = \left(v_t + v_{xxx} - 6v_x v^2\right)_x + 2v\left(v_t + v_{xxx} - 6v_x v^2\right),$

(3b) $\quad u_t + u_{xxx} - 6u_x u = (\partial_x + 2v)\left(v_t + v_{xxx} - 6v_x v^2\right).$

It is therefore clear that, if $v(x,t)$ satisfies the modified Korteweg–de Vries (mKdV) equation

(4) $\quad v_t + v_{xxx} - 6v_x v^2 = 0, \quad v \equiv v(x,t),$

then $u(x,t)$, defined by the Miura transformation (2), satisfies the KdV equation (1). Note that the converse statement need not be true; indeed if $u(x,t)$ evolves according to the KdV equation (1) and $v(x,t)$ is related to it by the Miura transformation (2), then there only follows from (3) that $v(x,t)$ evolves in time according to the (nonlinear integrodifferential) equation

(5) $\quad v_t + v_{xxx} - 6v_x v^2 = c(t) \exp\left[-2\int^x dy\, v(y,t)\right].$

The lower limit of the integral in the r.h.s. need not be specified, since its indeterminacy is absorbed by the arbitrariness of the function $c(t)$. As for this quantity, its value is clearly related to the (x-independent) "constant of integration" that, for each value of t, may be arbitrarily chosen when (2) is integrated to evaluate $v(x,t)$ for a given $u(x,t)$.

The main property of the Miura transformation (2) is to associate to every solution $v(x,t)$ of the mKdV equation (4) a solution $u(x,t)$ of the KdV equation (1). Several consequences of this fact are now indicated.

First of all, let us outline the relationship between the Miura transformation and the (linear) Schroedinger equation (note that this has actually been the reasoning that led to the discovery of the spectral transform technique to solve the KdV equation). It is however expedient to consider a generalized version of the Miura transformation, reading

(6) $\quad u(x,t) = v_x(x,k,t) + v^2(x,k,t) + k^2.$

This formula contains the additional parameter k^2, that is going to play the rôle of eigenvalue in the spectral problem, see below. Note incidentally that, as it is easily verified, this formula implies the relation

(7a) $\quad u_t + u_{xxx} - 6u_x u = \left[v_t + v_{xxx} - 6v_x(v^2 + k^2) \right]_x$
$\qquad\qquad + 2v\left[v_t + v_{xxx} - 6v_x(v^2 + k^2) \right];$

(7b) $\quad u_t + u_{xxx} - 6u_x u = (\partial_x + 2v)\left[v_t + v_{xxx} - 6v_x(v^2 + k^2) \right];$

thus, it may again be concluded that, if $v(x,t)$ satisfies the nonlinear evolution equation

(8) $\quad v_t + v_{xxx} - 6v_x(v^2 + k^2) = 0,$

then $u(x,t)$, given by (6), satisfies the KdV equation (1).

Note that, if one assumes, as we actually do, that $u(x,t)$ is independent of the parameter k, then v clearly does depend on this parameter, as we have explicitly indicated in writing (6). Indeed the dependence of v on k may originate not only from the presence of the quantity k^2 in (6) (which, incidentally, would only cause a dependence on k^2, rather than on k), but also via the "integration constant" (that must be independent of x, but may depend on k and t) that enters whenever (6) is integrated to evaluate v.

The (generalized) Miura transformation (6) is a Riccati equation for (the dependent variable) v; it is therefore natural to linearize it by the standard position

(9) $\quad v(x,k,t) = \psi_x(x,k,t)/\psi(x,k,t).$

There thus obtains for the function $\psi(x,k,t)$ the linear second-order ODE

(10) $\quad -\psi_{xx} + u(x,t)\psi = k^2\psi, \quad \psi \equiv \psi(x,k,t),$

which is just the Schroedinger equation.

Moreover, from (8) and (9) there easily obtains the formula

(11) $\quad (\psi_t/\psi)_x = [u_x - 2v(u+2k^2)]_x,$

which immediately yields (integrating and using (9))

(12) $\quad \psi_t = (u_x + a)\psi - 2(u+2k^2)\psi_x.$

This formula contains the "integration constant" a, whose value may of course depend on k and t. If $u(x,t)$ vanishes as $x \to \pm\infty$ and the function $\psi(x,k,t)$ is chosen to satisfy the (standard) boundary conditions

(13a) $\quad \psi(x,k,t) \to \exp(-ikx) + R(k,t)\exp(ikx), \quad x \to +\infty,$
(13b) $\quad \psi(x,k,t) \to T(k,t)\exp(-ikx), \quad x \to -\infty,$

then it is easily seen (by inserting (13) in (12)) that

(14) $\quad a = -4ik^3,$
(15) $\quad R_t(k,t) = -8ik^3 R(k,t),$
(16) $\quad T_t(k,t) = 0.$

The last two equations display the time evolution of the reflection coefficient R and the transmission coefficient T, that correspond to the evolution of $u(x,t)$ according to the KdV equation (1); these formulae, together with the possibility to reconstruct $u(x,t)$ from $R(k,t)$ (up to the contributions of the discrete spectrum) constitute the essence of the spectral transform technique to solve (1).

The relationship of the equations (10) and (12) to the AKNS approach (see appendix A.20) should also be noted.

Let us now proceed to show how, from the (generalized) Miura transformation (6), another important property of the KdV equation (1) can be derived, namely the fact that two solutions of (1) may be related to each other by the (one parameter) Bäcklund transformation (see (4.1.-15))

(17) $\quad w_x^{(1)} + w_x^{(2)} = -\tfrac{1}{2}(w^{(1)} - w^{(2)})(w^{(1)} - w^{(2)} + 4p).$

Here p is a positive parameter, and the two functions $w^{(j)}(x,t), j=1,2$, are related to two solutions $u^{(j)}(x,t)$ of the KdV equation (1) by the standard formulae (see, e.g., (4.1.-16))

(18a) $\quad w^{(j)}(x,t) = \int_{x}^{+\infty} dy\, u^{(j)}(y,t), \quad j=1,2,$

(18b) $\quad u^{(j)}(x,t) = -w_x^{(j)}(x,t), \quad j=1,2.$

Note that, by writing (18a), we are implicitly assuming the functions $u^{(j)}(x,t)$ to vanish (sufficiently fast) as $x \to +\infty$, although the arguments given below are essentially local in character and could be easily reformulated so as to dispense with this asymptotic requirement.

The starting point of the analysis is the observation that the (generalized) mKdV equation (8) is invariant under the transformation $v \to -v$, so that, if v is a solution of (8), $-v$ is also a solution. Thus, if a solution $u^{(2)}$ of the KdV equation (1) is associated with the solution v of the mKdV equation (8) via the Miura transformation (6), a second solution $u^{(1)}$ of the KdV equation (1) is associated with the solution $-v$ of the mKdV equation (8). To obtain the relationship between $u^{(1)}$ and $u^{(2)}$, it is convenient to rewrite (6) with k^2 replaced by $-p^2$, so that the asymptotic vanishing (as $x \to \pm\infty$) of u be consistent with a real v. Thus we write

(19a) $\quad u^{(1)} = -v_x + v^2 - p^2,$

(19b) $\quad u^{(2)} = v_x + v^2 - p^2.$

Note that the asymptotic conditions

(20a) $\quad v(\pm\infty,t) = \pm p, \quad (p>0),$

(20b) $\quad v_x(\pm\infty,t) = 0,$

are consistent with the asymptotic vanishing of $u^{(j)}(x,t)$ (as $x \to \pm\infty$), as well as with (9) and (10), since the latter equation (with k^2 replaced by $-p^2$ and $u(x,t)$ vanishing as $x \to \pm\infty$) implies that the generic solution ψ diverges proportionally to $\exp(\pm px)$ as $x \to \pm\infty$.

Subtraction of (19b) from (19a) yields

(21) $\quad v = \tfrac{1}{2}(w^{(1)} - w^{(2)}) + p,$

where we are of course using the notation (18). To obtain this equation we also used (20a), that, together with (21), yields moreover the condition

(22) $\quad w^{(1)}(-\infty,t) - w^{(2)}(-\infty,t) = -4p.$

On the other hand the sum of (19a) and (19b), together with (21), yields precisely the Bäcklund transformation (17).

These results may be compared with those of section 4.1 (and if need be, of appendix A.12); the fact that (22) corresponds only to one of the two possibilities (4.1.-22) originates from the fact that (20a) corresponds only to the (generic) solution ψ of the Schroedinger equation (10) (with k^2 replaced by $-p^2$), that diverges as $x \to \pm \infty$, while (special) solutions of this equation also exist that vanish exponentially as x diverges to positive or negative infinity. A more detailed discussion of this question is left as a useful exercise; the diligent reader may then compare his findings with the treatments given in section 4.1 and appendix A.12.

Another simple application of the Miura transformation (2) is, to display the connection between the Schroedinger equation (10) underlying the spectral transform used to solve the KdV equation (1) (see above, and section 3.1), and the system of two coupled first order ODEs

(23a) $\quad \psi_x^{(1)} = -ik\psi^{(1)} + v\psi^{(2)}$,

(23b) $\quad \psi_x^{(2)} = ik\psi^{(2)} + v\psi^{(1)}$,

underling the Zakharov–Shabat spectral problem that serves to solve the mKdV equation (4) (see volume II). This connection stems from the equivalence between the Miura transformation (2) and the factorization

(24) $\quad -\partial_x^2 + u = (\partial_x + v)(-\partial_x + v)$

of the Schroedinger operator. Indeed this factorization implies the possibility to rewrite the Schroedinger equation (10) as the coupled system of first order ODEs

(25a) $\quad -\psi_x + v\psi = ik\varphi$,

(25b) $\quad \varphi_x + v\varphi = -ik\psi$;

and it is immediately seen that this goes over into the Zakharov–Shabat system (23) with the positions

(26) $\quad \psi = \psi^{(1)} + \psi^{(2)}, \qquad \varphi = \psi^{(1)} - \psi^{(2)}$.

The Gardner transformation reads

(27a) $\quad u = y + \varepsilon y_x + \varepsilon^2 y^2$,

(27b) $\quad u = (1 + \varepsilon \partial_x + \varepsilon^2 y) y$.

Note that, as in the case of the Miura transformation (2), the time variable enters only parametrically. Also note the presence of the additional (finite) parameter ε, and the fact that the formal position $y=v/\varepsilon$ yields back, in the limit $\varepsilon\to\infty$, the Miura transformation (2). In fact, the Gardner transformation is merely the expression that the generalized version of the Miura transformation (6) takes after the change of variables

(28) $\quad v(x,k,t)=\varepsilon y(x,t,\varepsilon)+(2\varepsilon)^{-1}, \qquad k^2=-(2\varepsilon)^{-2};$

this is convenient to introduce the formal series expansion of y in powers of ε (see below).

It is a matter of trivial algebra to verify that, if u and y are related by (27), then there also holds the relation

(29a) $\quad u_t+u_{xxx}-6u_x u=\varepsilon\left(y_t+y_{xxx}-6y_x y-6\varepsilon^2 y_x y^2\right)_x$
$\qquad\qquad +(1+2\varepsilon^2 y)\left(y_t+y_{xxx}-6y_x y-6\varepsilon^2 y_x y^2\right)$

(29b) $\quad u_t+u_{xxx}-6u_x u=(1+\varepsilon\partial_x+2\varepsilon^2 y)\left(y_t+y_{xxx}-6y_x y-6\varepsilon^2 y_x y^2\right).$

It is therefore clear that, if $y(x,t,\varepsilon)$ satisfies the nonlinear evolution equation

(30a) $\quad y_t+y_{xxx}-6y_x y-6\varepsilon^2 y_x y^2=0,$
(30b) $\quad y_t=\left(-y_{xx}+3y^2+2\varepsilon^2 y^3\right)_x,$

then $u(x,t)$ satisfies the KdV equation (1).

Let us now show how from these results there immediately follows the existence of an infinite sequence of local conservation laws for the KdV equation (1). The idea is to invert the Gardner transformation (27), namely to express $y(x,t,\varepsilon)$ in terms of $u(x,t)$ and its x-derivatives, and then use (30b). We thus write

(31) $\quad y(x,t,\varepsilon)=\sum_{n=0}^{\infty}\varepsilon^n y^{(n)}(x,t).$

(Here, and below, the series need not converge; it is sufficient that they provide an asymptotic expansion). It is then easily seen, by inserting this ansatz in (27) and equating the coefficients of equal powers of ε, that the quantities $y^{(n)}(x,t)$ are determined recursively by the formula

(32) $\quad y^{(n+1)}=-y_x^{(n)}-\sum_{m=0}^{n-1}y^{(m)}y^{(n-m-1)}, \quad n=1,2,3,\dots,$

with

(33a) $\quad y^{(0)} = u(x,t),$

(33b) $\quad y^{(1)} = -u_x(x,t).$

Thus the quantities $y^{(n)}$ are polynomials in u and its x-derivatives:

(33c) $\quad y^{(2)} = u_{xx} - u^2,$

(33d) $\quad y^{(3)} = \left(-u_{xx} + 2u^2\right)_x,$

(33e) $\quad y^{(4)} = \left(u_{xx} - 3u^2\right)_{xx} + u_x^2 + 2u^3,$

and so on.

On the other hand insertion of the ansatz (31) in (30b) yields (again by equating the coefficients of equal powers of ε) the infinite sequence of local conservation laws

(34) $\quad y_t^{(n)} = \eta_x^{(n)}, \quad n = 0, 1, 2, \ldots,$

with

(35a) $\quad \eta^{(0)} = -u_{xx} + 3u^2,$

(35b) $\quad \eta^{(1)} = u_{xxx} - 6u_x u,$

(35c) $\quad \eta^{(n)} = -y_{xx}^{(n)} + 3\sum_{m=0}^{n} y^{(m)} y^{(n-m)} + 2\sum_{m=0}^{n-2} \sum_{j=0}^{n-m-2} y^{(j)} y^{(m)} y^{(n-m-j-2)},$

$$n = 2, 3, 4, \ldots .$$

Note that also the "currents" $\eta^{(n)}$, as well as the "densities" $y^{(n)}$, are polynomials in $u(x,t)$ and its x-derivatives.

If the function $u(x,t)$ vanishes with all its x-derivatives as $x \to \pm\infty$, the local conservation laws (34) yield the conserved quantities

(36) $\quad a_n = \int_{-\infty}^{+\infty} dx\, y^{(n)}(x,t), \quad n = 0, 1, 2, \ldots .$

However, as it is suggested by the explicit expressions (33b) and (33d), for odd n these quantities vanish, since the densities $y^{(2m+1)}$ are exact derivatives. This can be easily proved by setting

(37) $\quad y = y^{(+)} + y^{(-)}$

with

(38a) $$y^{(+)} = \sum_{m=0}^{\infty} \varepsilon^{2m} y^{(2m)},$$

(38b) $$y^{(-)} = \sum_{m=0}^{\infty} \varepsilon^{2m+1} y^{(2m+1)}.$$

Insertion of (37) in (27) then yields (looking at the part of the equation odd in ε)

(39a) $$y^{(-)} + \varepsilon y_x^{(+)} + 2\varepsilon^2 y^{(-)} y^{(+)} = 0;$$

and this equation is immediately solved for $y^{(-)}$, yielding

(39b) $$y^{(-)} = -(2\varepsilon)^{-1} \{\ln[1 + 2\varepsilon^2 y^{(+)}]\}_x.$$

This formula concludes the proof; indeed, be using (38) and reexpanding the r.h.s. of (39b) in powers of ε, explicit expressions of the densities $y^{(2m+1)}$ can be obtained as perfect differentials of combinations of the densities $y^{(2m)}$ (for instance one obtains in this manner $y^{(1)} = -y_x^{(0)}$, $y^{(3)} = -\{y^{(2)} - [y^{(0)}]^2\}_x$, consistently with (33)).

It is on the other hand clear that the densities $y^{(2m)}$ are not pure differentials, as they take the form

(40) $$y^{(2m)} = q_m u^{m+1} + \tilde{y}^{(2m)}, \quad m = 0, 1, 2, \ldots,$$

where $\tilde{y}^{(2m)}$ contains terms at most of order m in u and its x-derivatives. This is suggested by the explicit expressions (33), and it is easily proved by noticing that the (leading) term of $y^{(2m)}$ containing no differentiated terms can be obtained ignoring the differentiated term in (27a) and solving for y. This yields

(41) $$y = \left[-1 + (1 + 4u\varepsilon^2)^{1/2}\right]/(2\varepsilon^2).$$

From this formula, and (38a), there immediately follows the validity of (40), with moreover the explicit expression for the numerical coefficients q_m resulting from the identification

(42) $$(1 + 4z)^{1/2} - 1 = 2 \sum_{m=0}^{\infty} q_m z^{m+1},$$

which implies

(43) $\quad q_m = (-1)^m 2^m (2m-1)!!/(m+1)!, \quad m = 0,1,2,\ldots$.

Since the densities $y^{(n)}$ with n even are not perfect differentials, the corresponding conserved quantities a_{2m} do not vanish identically; they provide an infinite sequence of conserved quantities for the KdV equation (1) (consistently with the results of chapter 5; indeed this treatment is essentially identical to that given after (5.-39a)).

A.12. Bäcklund transformations, Darboux transformations and Bargmann strip

In this appendix we tersely review some relations between the items listed in the title. This treatment is mainly meant to stimulate the more diligent readers to investigate these topics more thoroughly on their own, since such an exercise constitutes the best procedure to acquire and test a sound grasp of these matters.

We focus on the simplest Bäcklund transformation (see (4.1.-15))

(1) $\quad w_x^{(1)} + w_x^{(2)} = -\tfrac{1}{2}(w^{(1)} - w^{(2)})(w^{(1)} - w^{(2)} + 4p)$,

(1a) $\quad u^{(j)}(x) = -w_x^{(j)}(x), \quad w^{(j)}(x) = \int_x^{+\infty} dy\, u^{(j)}(y), \quad j=1,2$.

This is characterized by the real parameter p; and our point of view here is to consider $w^{(2)}$ as given (with $u^{(2)}(x)$ being a *bona fide* potential). Then (1) is a Riccati equation for $w^{(1)}$, that is linearized by the following positions:

(2) $\quad w^{(1)}(x) = w^{(2)}(x) + 2[v(x) - p]$,

(3) $\quad v(x) = f_x(x)/f(x)$,

(4) $\quad f_{xx} + w_x^{(2)} f = p^2 f$.

Thus the problem of finding $w^{(1)}(x)$ is reduced to that of solving the Schroedinger equation (4) with the potential $u^{(2)}$ (see (1a)) and the asymptotic condition

(5) $\quad v(+\infty) = \lim_{x \to +\infty} [f_x(x)/f(x)] = p$.

Recalling the asymptotic condition that characterizes the Jost solutions of the Schroedinger equation ($f^{(\pm)}(x,k) \to \exp(\pm ikx)$ as $x \to \pm\infty$; see (2.1.-4)), it is immediately seen that (5) and (3) imply that, if p is positive, the solution $f(x)$ of (4) can be any linear combination of the Jost solutions, i.e.

(6) $\qquad f(x) = f^{(+)(2)}(x, -ip) + a f^{(+)(2)}(x, ip), \quad p > 0,$

a being an arbitrary (finite) constant, while if p is negative the solution $f(x)$ is fixed (up to a normalization factor, see (4) and (5)) to be

(7) $\qquad f(x) = f^{(+)(2)}(x, -ip), \quad p < 0.$

Therefore the solution $w^{(1)}(x)$ of (1) depends on an arbitrary integration constant a only if the parameter p is positive (via (6),(3),(2)).

However, this is not the whole story: indeed we have only considered the requirement that $w^{(1)}(x)$ vanish as $x \to +\infty$. Another important requirement is that $w^{(1)}(x)$ be regular for all real x. It is easy to see (using arguments closely analogous to those of appendix A.1) that for negative p the condition

(8) $\qquad p \leq -p_N^{(2)},$

where $0 < p_1^{(2)} < p_2^{(2)} < \cdots < p_N^{(2)}$ are the discrete eigenvalues of $u^{(2)}(x)$, is necessary and sufficient to exclude the occurrence of zeros of (7) for real x, and therefore to guarantee the boundedness of $w^{(1)}(x)$. Moreover if p satisfies (8) in the strict form ($p < -p_N^{(2)}$), it follows from (7) and (3) that $v(-\infty) = p$, namely (see (2) and (1))

(9) $\qquad \int_{-\infty}^{+\infty} dx \left[u^{(1)}(x) - u^{(2)}(x) \right] = 0;$

if instead (8) holds with the equality sign, $f(x)$ (see (7)) is the bound state solution of (4), thereby implying that $v(-\infty) = -p$ and, consequently,

(10) $\qquad \int_{-\infty}^{+\infty} dx \left[u^{(1)}(x) - u^{(2)}(x) \right] = -4p.$

For positive p, it is likewise seen that the conditions

(11a) $\qquad p > p_N^{(2)},$
(11b) $\qquad a \geq 0,$

are necessary and sufficient to guarantee that $w^{(1)}(x)$ be regular. The condition (11b) on a corresponds, from the point of view of the spectral transform, to the requirement that the normalization coefficient ρ associated with the discrete eigenvalue p of $u^{(1)}(x)$ be positive (see below). In general, if the two conditions (11) hold, the function (6) diverges (exponentially) as $x \to -\infty$ implying (see (3) and (2)) the relationship (10); the only exception to this occurs for $a=0$ since in this case the function (6) vanishes (exponentially) as $x \to -\infty$, implying instead the validity of (9). This marginal case, in the context of the spectral transform, corresponds to associating with p a vanishing normalization coefficient ρ; in this case, and only in this case (if (11a) holds), the Bäcklund transformation (1) does not produce an additional discrete eigenvalue p. (Recalling the fact that the addition of a discrete eigenvalue corresponds to the addition of a soliton, whose localization depends on the value of the normalization constant—see section 4.1—it may be asserted that in this marginal case the additional soliton is not present because it has escaped to $-\infty$, since it is this value, $\xi = -\infty$, that corresponds to $\rho = 0$ via the formula $\rho = 2p\exp(2p\xi)$ that relates the normalization constant ρ to the soliton position ξ).

These properties of the Bäcklund transformation (1), and the corresponding ones in k-space, can be easily read out of the Darboux transformation. This connects the Jost solutions, $f^{(\pm)(1)}(x,k)$, of the Schroedinger equation with potential $u^{(1)}(x)$, to the Jost solutions, $f^{(\pm)(2)}(x,k)$, corresponding to the potential $u^{(2)}(x)$, where $u^{(1)}(x)$ and $u^{(2)}(x)$ are of course related to each other by the Bäcklund transformation (1).

The Darboux transformation follows from the factorization property

(12a) $\quad (D+v)(D-v) = D^2 + w_x^{(2)} - p^2,$

(12b) $\quad (D-v)(D+v) = D^2 + w_x^{(1)} - p^2,$

that readily imply that the two functions $f^{(1)}(x,k)$ and $f^{(2)}(x,k)$ satisfying the system

(13a) $\quad (D-v)f^{(2)} = i(k+ip)f^{(1)},$

(13b) $\quad (D+v)f^{(1)} = i(k-ip)f^{(2)},$

satisfy also the Schroedinger equations

(14) $\quad f_{xx}^{(j)} = [u^{(j)} - k^2] f^{(j)}, \quad j=1,2.$

Here of course $D = d/dx$ and $v(x)$ is defined by (3), with $f(x) = f^{(2)}(x,k)$.

The first of the connection formulae (13) corresponds to the Darboux transformation mapping a (generic) solution $f^{(2)}(x,k)$ into a (generic) solution $f^{(1)}(x,k)$ of the Schroedinger equations (14).

It is now an easy exercise (that is left to the diligent reader) to derive explicitly the Darboux transformation connecting the Jost solutions $f^{(\pm)(j)}(x,k)$. The resulting formula is, for $p < -p_N^{(2)}$,

$$f^{(\pm)(1)}(x,k) = (p \mp ik)^{-1} \{ [f_x^{(+)(2)}(x,-ip)/f^{(+)(2)}(x,-ip)] \cdot f^{(\pm)(2)}(x,k) - f_x^{(\pm)(2)}(x,k) \}, \quad p < -p_N^{(2)}; \tag{15}$$

this formula provides, in the limits as $x \to \pm\infty$, also the following transformation laws in k-space:

$$R^{(1)}(k) = -R^{(2)}(k)(k+ip)/(k-ip), \tag{16a}$$

$$T^{(1)}(k) = T^{(2)}(k), \quad p < -p_N^{(2)}, \tag{16b}$$

that specify the effect of the Bäcklund transformation (1) on the continuous component (i.e. the reflection coefficient) of the spectral transform, and on the transmission coefficient. As for the discrete component of the spectral transform, the equality of the transmission coefficients (see (16b)) proves that the eigenvalues of $u^{(1)}(x)$ coincide with those of $u^{(2)}(x)$, namely

$$p_n^{(1)} = p_n^{(2)}, \quad n = 1, 2, \ldots, N. \tag{17}$$

Moreover the Darboux transformation for the corresponding eigenfunctions,

$$\varphi_n^{(1)}(x) = (p^2 - p_n^2)^{-1/2} \{ [f_x^{(+)(2)}(x,-ip)/f^{(+)(2)}(x,-ip)] \cdot \varphi_n^{(2)}(x) - \varphi_{nx}^{(2)}(x) \}, \quad n = 1, 2, \ldots, N, \quad p < -p_N^{(2)}, \tag{18}$$

via the normalization condition

$$\int_{-\infty}^{+\infty} dx \, [\varphi_n^{(j)}(x)]^2 = 1, \quad j = 1, 2, \quad n = 1, 2, \ldots, N, \tag{19}$$

readily implies the transformation law

$$\rho_n^{(1)} = \rho_n^{(2)} (p + p_n)/(p - p_n), \quad n = 1, 2, \ldots, N, \quad p < -p_N^{(2)}. \tag{20}$$

Indeed both (18) and (20) follow directly from the formulae:

(21) $\quad [\varphi_n^{(j)}(x)]^2 = \rho_n^{(j)} [f^{(+)(j)}(x, ip_n^{(j)})]^2, \quad j=1,2, \quad n=1,2,\ldots,N,$

(22) $\quad [\varphi_n^{(1)}(x)]^2 = (p+p_n^{(2)})^{-2} (\rho_n^{(1)}/\rho_n^{(2)})$
$\quad \cdot \{(p^2 - p_n^{(2)2})[\varphi_n^{(2)}(x)]^2 + g_x(x)\},$

where

(23) $\quad g(x) \equiv \varphi_n^{(2)}(x) \{\varphi_{nx}^{(2)}(x) - \varphi_n^{(2)}(x)[f_x^{(+)(2)}(x,-ip)$
$\quad /f^{(+)(2)}(x,-ip)]\},$

and from the normalization condition (19) and the asymptotic vanishing of $g(x)$, i.e. $g(\pm\infty)=0$.

Similarly, in the case $p > p_N^{(2)}$, the relevant formulae read

(24) $\quad f^{(\pm)(1)}(x,k)$
$= \pm (p-ik)^{-1} \{ \{[2pT^{(2)}(ip)f_x^{(-)(2)}(x,ip) + \rho f_x^{(+)(2)}(x,ip)]$
$\cdot [2pT^{(2)}(ip)f^{(-)(2)}(x,ip) + \rho f^{(+)(2)}(x,ip)]^{-1}\} f^{(\pm)(2)}(x,k)$
$- f_x^{(\pm)(2)}(x,k)\}, \quad p > p_N^{(2)},$

that obtains from (13), (3) and (6) with

(25) $\quad a = \rho [2pT^{(2)}(ip)]^{-1},$

where, for notational convenience, we have introduced the constant ρ; the Darboux transformation (24) in the limits $x \to \pm\infty$ yields

(26a) $\quad R^{(1)}(k) = -R^{(2)}(k)(k+ip)/(k-ip),$

(26b) $\quad T^{(1)}(k) = T^{(2)}(k)(k+ip)/(k-ip), \quad p > p_N^{(2)}.$

It is thus seen that in this case the transmission coefficient $T^{(1)}(k)$ acquires an additional pole at $k = ip$. Therefore, for the discrete part of the spectral transform of $u^{(1)}(x)$, one concludes that the discrete spectrum has $N+1$ eigenvalues, namely the N eigenvalues of $u^{(2)}(x)$ plus the novel one at p:

(27) $\quad p_n^{(1)} = p_n^{(2)}, \quad n=1,2,\ldots,N, \quad p_{N+1}^{(1)} = p, \quad p > p_N^{(2)};$

the remaining transformation laws are easily found (by using formulae analogous to (22) and (23)) to be

$$(28) \quad \varphi_n^{(1)}(x) = (p^2 - p_n^2)^{-1/2}$$
$$\cdot \left\{ \left\{ \left[2pT^{(2)}(ip)f_x^{(-)(2)}(x,ip) + \rho f_x^{(+)(2)}(x,ip) \right] \right. \right.$$
$$\left. \cdot \left[2pT^{(2)}(ip)f^{(-)(2)}(x,ip) + \rho f^{(+)(2)}(x,ip) \right]^{-1} \right\}$$
$$\left. \cdot \varphi_n^{(2)}(x) - \varphi_{nx}^{(2)}(x) \right\},$$
$$n = 1, 2, \ldots, N, \quad p > p_N, \quad \rho > 0,$$

$$(29a) \quad \int_{-\infty}^{+\infty} dx \left[\varphi_n^{(j)}(x) \right]^2 = 1, \quad n = 1, 2, \ldots, N,$$

$$(29b) \quad \rho_n^{(1)} = \rho_n^{(2)}(p + p_n)/(p - p_n), \quad n = 1, 2, \ldots, N,$$

$$(30a) \quad \varphi_{N+1}^{(1)}(x) = 2p\rho^{1/2} \left[2pT^{(2)}(ip)f^{(-)(2)}(x,ip) + \rho f^{(+)(2)}(x,ip) \right]^{-1},$$

$$(30b) \quad \int_{-\infty}^{+\infty} dx \left[\varphi_{N+1}^{(1)}(x) \right]^2 = 1,$$

$$(30c) \quad \rho_{N+1}^{(1)} = \rho.$$

Note that ρ and a, see (6) and (25), have the same sign (for $p > p_N^{(2)}$, $T^{(2)}(ip)$ is positive; since it is real, it has neither zeros nor poles and $T^{(2)}(i\infty) = 1$).

In the marginal case characterized by the vanishing of the constant a in (6) (or, equivalently, of ρ, see (25)), no additional eigenvalues are introduced by the Bäcklund transformation, and the formulae that therefore apply coincide with (15), (16), (17), (18), (19) and (20).

The consistency of these formulae (see in particular (16), (20) and (26)) with the equations of chapter 4 (see in particular (4.1.-13), (4.1.-18) and (4.1.-21)) should be emphasized.

Let us now investigate the asymptotic behaviour of $u^{(j)}(x)$ as $x \to +\infty$, and the (related) size of the (Bargmann) strip of meromorphy of $R(k)$ (see section 2.1, in particular the discussion after (2.1.-31)). Indeed let us assume that $u^{(2)}(x)$ vanish asymptotically,

$$(31) \quad \lim_{x \to +\infty} \left[w^{(2)}(x) \exp(2\mu x) \right] = b \neq 0,$$

so that $R^{(2)}(k)$ is meromorphic (at least) in the strip $0 \leqslant \text{Im } k < \mu$; and let us recall that, *within this strip*, there is a one-to-one correspondence between the poles of $R^{(2)}(k)$ on the imaginary axis, at $k = ip_n$, and the discrete eigenvalues, the corresponding normalization coefficients $\rho_n^{(2)}$ being moreover related to the residues of $R^{(2)}(k)$ via (2.1.-36).

On the other hand, it is easy to ascertain what the asymptotic behaviour of $w^{(1)}(x)$ is as $x \to +\infty$, since (1) can be linearized (and therefore explicitly solved) in this region by (consistently) neglecting $w^{(1)}(x)$ and $w^{(2)}(x)$ relatively to $4p$ within the last bracket in the r.h.s. In this manner one easily concludes that there is a one-parameter family of solutions having the asymptotic behaviour

$$\text{(32)} \qquad \lim_{x \to +\infty} \left[w^{(1)}(x) \exp(2\mu x) \right] = b(p+\mu)/(p-\mu)$$

(with b defined by (31)), if $p > \mu$ (the parameter that distinguishes the members of the family enters as integration constant of the first order differential equation (1), but does not show up in the asymptotic behaviour (32)). If instead $p < \mu$, then there is only a single solution of (1) having the asymptotic behaviour (32), the remaining one-parameter family of solutions being characterized by the asymptotic behaviour

$$\text{(33)} \qquad \lim_{x \to +\infty} \left[w^{(1)}(x) \exp(2 p x) \right] = \beta,$$

with β an arbitrary (integration) constant.

Let us elaborate on the second case, $p < \mu$. Clearly the single solution $w^{(1)}(x)$ satisfying (32) obtains from (2), (3), (6) and (25) by choosing for the normalization constant ρ the special value obtained by applying (2.1.-36) to the expression (26a) of the reflection coefficient, namely

$$\text{(34)} \qquad \rho = -i \lim_{k \to ip} \left[(k-ip) R^{(1)}(k) \right] = -2 p R^{(2)}(ip).$$

This is consistent with the fact that, in this case, the discrete eigenvalue p falls *inside* the Bargmann strip. All the other solutions $w^{(1)}(x)$ of the Bäcklund transformation (1), being characterized by the asymptotic behaviour (33) rather than (32), have associated a more extended Bargmann strip, so that the discrete eigenvalue p now sits on its boundary rather than inside; thus in this case the normalization coefficient ρ need not satisfy (2.1.-36), and remains therefore as a free parameter, whose value is of course related to the value of the quantity β in the r.h.s. of (33) (the different possible choices of this parameter β, and correspondingly of ρ, correspond to the freedom to change the localization of the additional soliton; see above and section 4.1).

It is left as an exercise for the diligent reader to spell out these points in more detail, as well as to extend some of the arguments given above; for instance, to the case when $u^{(2)}$ vanishes faster than exponentially as $x \to +\infty$;

or to the case when

(35) $\quad p = -p_n$

(see (4.1.-20); and note that, if p is negative, while the solution characterized by the behaviour (32) is still acceptable, there is no acceptable solution associated with the asymptotic behaviour (33); moreover, if $p = -p_n$, (20) yields $\rho_n^{(1)} = 0$, signifying the disappearance of the n-th discrete eigenvalue p_n from the spectrum associated with $u^{(1)}$).

Let us finally note that the analysis outlined above via the Darboux transformation indicates that, if (11) is violated, two effects simultaneously occur: some normalization coefficients $\rho_n^{(1)}$ become negative, and the function $w^{(1)}(x)$ develops some singularities (generally poles resulting in double poles of $u^{(1)}(x)$) for real x. The simplest instance of this phenomenon is provided by the single-soliton solution

(36) $\quad u(x) = -2p^2/\cosh^2[p(x-\xi)],$

that is associated with just one discrete eigenvalue p, whose normalization coefficient ρ is related to ξ by the formula

(37) $\quad \rho = 2p\exp(2p\xi)$

(see 2.2.-9, 10)). For real ξ, $u(x)$ is regular (for all real x) and ρ is positive; but for

(38) $\quad \xi = \xi' + i\pi/2p$

with ξ' real, the function $u(x)$,

(39) $\quad u(x) = 2p^2/\sinh^2[p(x-\xi')],$

although still real, features a double pole at $x = \xi'$, while the "normalization coefficient" ρ,

(40) $\quad \rho = -2p\exp(2p\xi'),$

has turned negative. (For additional discussion of such pathologies, see section 4.2 and especially appendix A.16).

A.13. Asymptotic expansion of $C(k) = 2ik[1 - 1/T(k)]$

The argument given tersely below follows closely the treatment of [C1977]. Define

(1) $\quad C(k) = 2ik[1 - 1/T(k)]$.

There holds then the formula (see (2.4.1.-32))

(2) $\quad C(k) = \int_{-\infty}^{+\infty} dx\, \chi(x, k) u(x)$

where

(3) $\quad \chi(x, k) = \exp(ikx)\, \psi(x, k)/T(k)$.

Thus $\chi(x, k)$ is the solution of the differential equation

(4) $\quad \chi_{xx} - 2ik\chi_x - u\chi = 0$

identified by the boundary condition

(5) $\quad \lim_{x \to -\infty} [\chi(x, k)] = 1$.

Introduce then the asymptotic expansions

(6) $\quad C(k) = \sum_{m=0}^{M} b_m (2ik)^{-m} + O(k^{-M-1})$,

(7) $\quad \chi(x, k) = \sum_{m=0}^{M} \chi^{(m)}(x)(2ik)^{-m} + O(k^{-M-1})$.

Then clearly (2) implies

(8) $\quad b_m = \int_{-\infty}^{+\infty} dx\, u(x) \chi^{(m)}(x), \quad m = 0, 1, 2, \ldots,$

while (4) and (5) yield

(9) $\quad \chi^{(0)}(x) = 1$,

(10) $\quad \chi_x^{(m+1)}(x) = \chi_{xx}^{(m)}(x) - u(x)\chi^{(m)}(x), \quad m = 0, 1, 2, \ldots,$

(11) $\quad \chi^{(m)}(-\infty) = \chi_x^{(m)}(-\infty) = \chi_{xx}^{(m)}(-\infty) = 0, \quad m = 1, 2, 3, \ldots.$

These recursion relations are clearly solved by the formula

(12) $\quad \chi^{(m)}(x) = M^m \cdot 1,$

with the integrodifferential operator M defined by the formula

(13) $\quad M \cdot f(x) = f_x(x) - \int_{-\infty}^{x} dy\, u(y) f(y),$

that specifies its action on the generic function $f(x)$. Thus

(14) $\quad b_m = \int_{-\infty}^{+\infty} dx\, u(x) M^m \cdot 1$

and

(15) $\quad c_m \equiv (-1)^m b_{2m} = (-)^m \int_{-\infty}^{+\infty} dx\, u(x) M^{2m} \cdot 1.$

The reason for introducing the last definition is because the odd-numbered b_m's can in fact be expressed in terms of even-numbered b_m's of lower order (for instance $b_1 = -\tfrac{1}{2} b_0^2$). The last formula is the result quoted and used in chapter 5, see (5.-16).

To prove (5.-18) we note that, for real k,

(16) $\quad T(k) \approx \exp[i\theta(k)].$

Here and below we use the wavy equality sign, $\approx$, to indicate equality up to corrections that vanish faster than any (inverse) power of k as $k \to \pm\infty$. Thus the validity of (16) is implied by (2.1.-15a) and (2.1.-39a); and in its turn this equation, together with the definition of $C(k)$, (1), implies

(17) $\quad \tfrac{1}{2}[C(k) + C(-k)] \approx -2k \sin \theta(k).$

To obtain (5.-18) it is then sufficient to equate the asymptotic expansions of the two sides of this equation, namely (see (6) and (15))

(18) $\quad \tfrac{1}{2}[C(k) + C(-k)] = \sum_{m=0}^{M} c_m (2k)^{-2m} + O(k^{-2M-2}),$

and (see (2.1.-40) and (5.-12))

(19) $\quad \theta(k) = - \sum_{m=0}^{M} C_m (2k)^{-2m-1} + O(k^{-2M-3}).$

A.14. Conserved quantities for generalized KdV equations

In this appendix we report, almost *verbatim* (but with some notational change) the results of [CD1980a]. Results referring to a different kind of "generalized KdV equations" are mentioned at the end.

Let $u(x,t)$ satisfy the evolution equation

(1) $\quad u_t + \alpha(t) u_{xxx} - 6\beta(t) u u_x = 0, \quad u \equiv u(x,t).$

Note that one (but generally not both) of the two functions $\alpha(t), \beta(t)$ could be replaced by a constant by an appropriate redefinition of the time variable; we keep both here for notational convenience (for the same reason (1) has been written with the conventional factor -6). For a numerical study of this equation in an applicative context see [KoKu1978].

We assume u to vanish as $x \to \pm\infty$ and to be regular (for $-\infty < x < +\infty$) so that all integrals written below are convergent.

Define then

(1) $\quad C^{(m)} = \int_{-\infty}^{+\infty} dx \, [u(x,t)]^m, \quad m = 1, 2,$

(2) $\quad X(t) = \int_{-\infty}^{+\infty} dx \, x u(x,t) \bigg/ \int_{-\infty}^{+\infty} dx \, u(x,t).$

Then $C^{(1)}$ and $C^{(2)}$ are time-independent and

(3) $\quad dX(t)/dt = 3(C^{(2)}/C^{(1)}) \beta(t)$

so that

(4) $\quad C^{(3)} = X(t) - 3(C^{(2)}/C^{(1)}) \int^t dt' \, \beta(t')$

is also time-independent. Moreover, if

(5) $\quad \alpha(t) = 1, \quad \beta(t) = a(t - t_0)^{-1/3},$

with a and t_0 arbitrary constants, a fourth constant of the motion is given by the formula

(6) $\quad C^{(4)} = \int_{-\infty}^{+\infty} dx \, \big\{ 6a(t-t_0)^{2/3} [u(x,t)]^3 + 3(t-t_0)[u_x(x,t)]^2 + x[u(x,t)]^2 \big\}.$

If instead

(7) $\quad \alpha(t) = \cosh^2[a(t-t_0)], \quad \beta(t) = \cosh[a(t-t_0)],$

with a and t_0 arbitrary constants, a fourth constant of the motion is given by the formula

(8) $\quad C^{(4)} = \int_{-\infty}^{+\infty} dx \, \{2\cosh^2[a(t-t_0)][u(x,t)]^3$

$\qquad\qquad + \cosh^3[a(t-t_0)][u_x(x,t)]^2$

$\qquad\qquad + ax\sinh[a(t-t_0)][u(x,t)]^2 + \tfrac{1}{6}(ax)^2 u(x,t)\}.$

The proof of these results is elementary (differentiate $C^{(j)}$ with respect to time and use (1)).

The interest of these findings is highlighted by the remark that, by setting

(9) $\quad \alpha(t) = 1, \quad \beta(t) = at^{-p}, \quad v(x,t) = \beta(t)u(x,t),$

with a and p arbitrary constants, (1) goes into

(10) $\quad v_t + v_{xxx} - 6v_x v + (p/t)v = 0, \quad v \equiv v(x,t).$

This nonlinear evolution equation coincides with the "standard", "cylindrical" resp. "spherical" KdV equations for $p=0$, $p=\tfrac{1}{2}$ resp. $p=1$; see section 1.8. The standard ($p=0$) and cylindrical ($p=\tfrac{1}{2}$) KdVs can also be obtained from (1) and (7); the former quite straightforwardly, setting $a=0$; the latter via the position

(11) $\quad u(x,\tau) = (2at)^{1/2} v(x,t), \quad t = 8a\exp(2a\tau),$

taking the limit $a \to \infty$ (with $\tau > 0$). But, of course, in the case of the standard and cylindrical KdVs, the conserved quantities exhibited here are merely (the first) four out of an infinity of conserved quantities; see chapter 5 and section 6.3. For the spherical KdV equation there are instead no other known conserved quantities besides $C^{(1)}$, $C^{(2)}$ and $C^{(3)}$.

A different question is the number N of local conservation laws for the equation

(12) $\quad u_t + g(u)u_x + u_{px} = 0,$

where $u_{px} \equiv \partial^p u / \partial x^p$ (with $p = 2, 3, 4 \ldots$), which may also be considered a "generalized KdV equation". In this case it is known that, except for the "integrable" case characterized by $p=3$ and $g(u)=Au^2+Bu+C$ (with A, B, and C constant; this corresponds essentially to the modified KdV equation, see section 1.8 and volume II; in this case of course $N=\infty$), if p is even $N=1$, while if p is odd $N=3$ unless $g(u)=A(u+a)^{p-1}+B$ (with A, B and a constant), or $g(u)=Au+B$ ($p \geqslant 3$), in which cases $N=4$ (see [FW1981], where the explicit expression of the conserved densities is also exhibited).

A.15. Reflection and transmission coefficients at $k=0$

The main purpose of this appendix is to show that, except for the marginal case when there occurs a "zero-energy bound state" (see below), the reflection and transmission coefficients $R(k)$ and $T(k)$ have the property

(1a) $\quad R(0)=-1$,

(1b) $\quad T(0)=0$.

The exceptional case obtains iff the (*bona fide*) potential $u(x)$ is characterized by the following property (that we actually take as *definition* of the occurence of a "zero-energy bound state" in the Schroedinger spectral problem): the potential $u^{(-)}(x)$ has at least one more discrete eigenvalue than $u^{(+)}(x)$, whenever the inequalities

(2) $\quad u^{(-)}(x) \leqslant u(x) \leqslant u^{(+)}(x), \quad -\infty < x < +\infty,$

hold (strictly at least somewhere!). This phenomenon is, of course, not generic; it occurs only for some special potentials $u(x)$ (it does however happen for all multisoliton potentials; see below. Moreover, the results discussed below and in appendix A.1 imply that there exist such potentials in the immediate neighborhood of any *bona fide* potential; they are obtained by the addition of a very small attractive contribution localized very far away.) Then, instead of the formulae (1), there holds the property

(3a) $\quad R(0)=\cos \alpha,$

(3b) $\quad T(0)=\sin \alpha,$

where α is a real constant (see below).

Note that both the formulae (1) and (3) are consistent with the unitarity equation

(4) $\quad |R(0)|^2+|T(0)|^2=1.$

Throughout this appendix we assume the potential $u(x)$ to vanish asymptotically ($x \to \pm \infty$) faster than any inverse power of x, this condition being sufficient to guarantee that both $R(k)$ and $T(k)$ are analytic in k in the neighborhood of $k=0$. Clearly this condition is sufficient, but not necessary, for the validity of the results described above.

A more detailed version of these results is provided by the following

Theorem. *Let the (real) function $\varphi(x)$ be defined by the (zero-energy Schroedinger) equation*

(5) $\quad \varphi_{xx}(x) = u(x)\varphi(x)$

and by the boundary conditions

(6) $\quad \varphi(-\infty) = 1, \qquad \varphi_x(-\infty) = 0.$

Define the (real) quantities α and β by the formula

(7) $\quad \varphi(x) \underset{x \to +\infty}{\to} \cot(\tfrac{1}{2}\alpha) + \beta x, \quad |\alpha| \leq \pi$

(note that the correction terms to this formula vanish as $u(x)$, namely, by assumption, faster than any inverse power of x). Then, if β does not vanish, the formulae (1) hold, while if β does vanish, the formulae (3) prevail (with α defined by (7)).

The identification of the $\beta = 0$ case with the occurrence of a "zero-energy bound state" is of course consistent with the results of appendix A.1.

To prove the theorem we have just formulated it is convenient to introduce the integral expression

(8) $\quad A(q, k) \equiv \int_{-\infty}^{+\infty} dx \exp(-iqx) f^{(-)}(x, k) u(x)$

(where $f^{(-)}(x, k)$ is the Jost solution, see (2.1.-4), of the Schroedinger equation (2.1.-1)), since this function yields the reflection and transmission coefficients through the formulae

(9a) $\quad R(k) = A(k, k)[2ik - A(-k, k)]^{-1}, \quad \text{Im } k = 0,$

(9b) $\quad T(k) = 2ik[2ik - A(-k, k)]^{-1},$

that follow immediately from the integral relations (2.1.-11) together with (2.1.-5).

The assumed asymptotic behaviour as $x \to \pm\infty$ of the potential $u(x)$ implies the validity of the power expansion

$$(10) \quad A(q,k) = \int_{-\infty}^{+\infty} dx\, f^{(-)}(x,0) u(x) - iq \int_{-\infty}^{+\infty} dx\, x f^{(-)}(x,0) u(x) + k \int_{-\infty}^{+\infty} dx\, f_k^{(-)}(x,0) u(x) + O(k^2, q^2, qk)$$

in the neighborhood of $q=k=0$. The first three coefficients that appear in (10) can be evaluated by taking into account that $f^{(-)}(x,0)$ coincides with the solution $\varphi(x)$ defined by (5) and (6),

$$(11) \quad f^{(-)}(x,0) = \varphi(x),$$

its asymptotic behaviour being therefore characterized by (7), and by introducing the second solution $\chi(x)$ of (5), characterized by the asymptotic behaviour

$$(12) \quad \chi(x) \underset{x \to -\infty}{\to} x,$$

that is readily found to be related to the Jost solution by the formula

$$(13) \quad f_k^{(-)}(x,0) = -i\chi(x).$$

Inserting (11) and (13) in (10) and using (5) (and the analogous equation for $\chi(x)$) together with (7) and (12) there obtains

$$(14) \quad A(q,k) = \beta - iq[1 - \cot g(\tfrac{1}{2}\alpha)] - ik(\gamma - i) + O(k^2, q^2, qk),$$

where the quantity γ is defined by

$$(15) \quad \chi_x(+\infty) = \gamma.$$

The first part of the theorem follows now immediately, if β does not vanish, from (14) and (9). If instead β does vanish, there obtains

$$(16a) \quad R(0) = [\cot g(\tfrac{1}{2}\alpha) - \gamma] / [\cot g(\tfrac{1}{2}\alpha) + \gamma],$$

$$(16b) \quad T(0) = 2 / [\cot g(\tfrac{1}{2}\alpha) + \gamma].$$

The simplest way to obtain (3) (and to complete thereby the proof of the theorem) is then to use the unitarity equation (4) (note that $R(0)$ and $T(0)$ are in fact real), which, together with (16), yields

(17) $\quad \gamma = \text{tg}(\tfrac{1}{2}\alpha)$.

We close this appendix by displaying the effect on the two quantities α and β, of a Bäcklund transformation that adds one novel discrete eigenvalue to the spectral transform. If p is the parameter that characterizes such a Bäcklund transformation (see section 4.1. and appendix A.12), the relevant equations read

(18) $\quad w_x^{(1)} + w_x^{(2)} = -\tfrac{1}{2}(w^{(1)} - w^{(2)})(w^{(1)} - w^{(2)} + 4p),$

$\qquad w^{(j)} \equiv \int_x^{+\infty} dy\, u^{(j)}(y), \quad j = 1, 2,$

(19) $\quad \varphi^{(1)}(x) = p^{-1} \left\{ \varphi_x^{(2)}(x) - \varphi^{(2)}(x) \left(\ln \left[\rho f^{(+)(2)}(x, ip) \right. \right. \right.$

$\qquad \left. \left. \left. + 2pT^{(2)}(ip) f^{(-)(2)}(x, ip) \right] \right)_x \right\},$

(18) being the Bäcklund transformation relating $u^{(1)}$ to $u^{(2)}$, while (19) is the associated Darboux transformation connecting the two functions $\varphi^{(1)}$ and $\varphi^{(2)}$ satisfying (5) (with $u = u^{(1)}$ resp. $u = u^{(2)}$) and (6), while $f^{(\pm)(2)}(x, k)$ are the usual Jost solutions (see (4.1.-30) and (11)). With self-evident notation (19) and (7) yield then the transformation rule

(20a) $\quad \beta^{(1)} = -\beta^{(2)}$

(20b) $\quad \text{cotg}(\tfrac{1}{2}\alpha^{(1)}) = -\text{cotg}(\tfrac{1}{2}\alpha^{(2)}) + \beta^{(2)}/p.$

In particular these results imply that, if $u^{(0)}(x)$ is characterized by the occurrence of a "zero-energy bound state", so that the corresponding $\beta^{(0)}$ vanishes,

(21) $\quad \beta^{(0)} = 0,$

and the potential $u^{(N)}(x)$ is obtained from $u^{(0)}(x)$ by performing N Bäcklund transformations (18) (with parameters $p_1, p_2, \ldots, p_N$), then

(22a) $\quad \beta^{(N)} = 0,$

(22b) $\quad \alpha^{(N)} = (-1)^N \alpha^{(0)}.$

These results are of course consistent with the expressions in k-space corresponding to the Bäcklund transformation (18),

(23a) $\quad R^{(1)}(k) = -R^{(2)}(k)(k+ip)/(k-ip)$,

(23b) $\quad T^{(1)}(k) = T^{(2)}(k)(k+ip)/(k-ip)$,

and with (3). In the special case $u^{(0)}(x)=0$, $u^{(N)}(x)$ is the N-soliton potential; in this case, of course, (22a) still holds, while (22b) reads

(24) $\quad a^{(N)} = (-1)^N \pi/2$.

A.15.N. Notes to appendix A.15

The first indication of the validity of (A.15.-1) for a generic function $u(x)$ may be traced to papers whose main purpose is the investigation of the long-time asymptotic behaviour of solutions of the KdV equation; see e.g. [AS1977a].

The properties displayed in (A.15.-1) and (A.15.-3) are also proved in [DT1979], and in [E1981] (where the treatment is however marred by a misprint, $R(0)=1$ instead of $R(0)=-1$).

A.16. The spectral transform outside of the class of *bona fide* potentials

Several times in this book the question of the spectral transform of functions that are outside the class of *bona fide* potentials has arisen: see sections 2.3, 4.2, 6.2 and appendix A.12. In this appendix we provide a (largely qualitative) discussion of this question.

Let us recall some basic facts. A function $u(x)$ of the real variable x, $-\infty < x < +\infty$, is in the class of *bona fide* potentials, if:

(i) It is infinitely differentiable for all (real) values of x.

(ii) It vanishes asymptotically ($x \to \pm\infty$) with all its derivatives sufficiently fast so that the integrals

(1) $\quad \int_{-\infty}^{+\infty} dx\,(1+|x|)|d^m u(x)/dx^m|, \quad m=0,1,2,\ldots$

are convergent.

(iii) It is real (for real x).

In one-to-one correspondence (via the Schroedinger spectral problem, see e.g. chapter 2) with such a function, there exists a spectral transform

(2) $$S[u] = \{R(k), -\infty < k < +\infty; p_n, \rho_n, n = 1, 2, \ldots, N\}$$

with the following properties:

(a) $R(k)$ vanishes faster than any inverse power of k as $k \to \pm \infty$,

(3) $$\lim_{k \to \pm \infty} [k^m R(k)] = 0, \quad m = 0, 1, 2, \ldots, \quad \operatorname{Im} k = 0.$$

(b) For real (nonvanishing) k, the modulus of $R(k)$ is less than unity,

(4) $$|R(k)| < 1, \quad \operatorname{Im} k = 0, \quad \operatorname{Re} k \neq 0.$$

(c) $R(k)$ satisfies the reflection property

(5) $$R(-k) = R^*(k), \quad \operatorname{Im} k = 0.$$

(d) The number N of discrete eigenvalues is finite (possibly zero),

(6) $$0 \leq N < \infty;$$

(e) The discrete eigenvalues $-p_n^2$ are real and negative, namely the quantities p_n are real (and, by convention, positive), and they are all different:

(7) $$0 < p_n < \infty, \quad p_n \neq p_m$$

(we are ignoring here, for simplicity, the marginal case of a "zero-energy bound state").

(f) The normalization coefficients are all positive,

(8) $$\rho_n > 0.$$

There is moreover a connection between the asymptotic rate of vanishing of $u(x)$ and the structure of (the analytic continuation of) $R(k)$ in the complex k plane; see section 2.1. Furthermore, except for the marginal case of potentials having a "zero-energy bound state", the value of the reflection coefficient at $k = 0$ is also fixed (see appendix A.15):

(9) $$R(0) = -1.$$

Let us now review some of the relationships that exist between the properties of $u(x)$ and those of its spectral transform, with a view to explore to what extent is it possible to go outside of the class of *bona fide* potentials.

Some aspects of this analysis are easy; indeed the definition of *bona fide* potential is considerably more stringent than it is necessary to construct a well-understood theory of the direct and inverse spectral problems. In particular the condition (i) is much more stringent than it is actually necessary to have a well-defined spectral transform; it is indeed sufficient that $u(x)$ be a regular (even discontinuous) function of x for real x. Then however the condition (a) need not hold; the reflection coefficient $R(k)$ will continue to vanish as $k \to \pm \infty$, indeed sufficiently fast so that its (inverse) Fourier transform

$$(10) \quad \hat{M}(x) = (2\pi)^{-1} \int_{-\infty}^{+\infty} dk \exp(ikx) R(k)$$

exist (as a function, not a distribution), but it need not vanish faster than any inverse power of k. In fact, roughly speaking, the relationship between the smoothness (differentiability) of $u(x)$ and the asymptotic behavior as $k \to \pm \infty$ of $R(k)$ reproduces the well-known relation between the smoothness of the function $\hat{M}(x)$ and the asymptotic behavior of its Fourier transform $R(k)$ (see (10)).

Another infringement of the class of *bona fide* potentials whose effects on the spectral transform are well understood is a violation of (iii). If $u(x)$ is complex, then (b), (c), (e) and (f) may all be violated (note incidentally that the validity of (c) implies that $\hat{M}(x)$, defined by (10), is real); but the general framework of the spectral problem remains essentially valid (indeed often we have written equations in a form such as to maintain their validity even for complex potentials; see for instance appendix A.8).

The first infringement that we have just considered, i.e. a breakdown of (i) and (a), can be understood equally well in the framework of the direct and inverse problems; if (i) is violated, then (a) does not hold; if (a) is violated (but (10) exists as a function), then (i) does not hold (but $u(x)$ exists). The second infringement has instead been formulated in the framework of the direct problem; and the fact that its effects are well understood is related to the fact that the Schroedinger spectral problem is well defined even if the function $u(x)$ is complex.

The infringements that we now propose to discuss are instead formulated in the framework of the inverse spectral problem, but they are considerably less understood. Indeed it is perhaps useful to preface this analysis by a brief discussion of an analogous type of problem in the context of the Fourier transform, whose analogy with the spectral transform has always constituted a cornerstone of our treatment.

Let $f(x)$ be a square-integrable function of the real variable x; then its Fourier transform $\hat{f}(k)$ is a square-integrable function of the real variable k; and they are of course related by the integral formulae

(11a) $\quad f(x) = (2\pi)^{-1} \int_{-\infty}^{+\infty} dk\, \hat{f}(k) \exp(ikx),$

(11b) $\quad \hat{f}(k) = \int_{-\infty}^{+\infty} dx\, f(x) \exp(-ikx).$

But suppose one assigns a non-square-integrable function $\hat{f}(k)$, say $\hat{f}(k) = 1$, and asks what is the function $f(x)$ of which this is the Fourier transform. There is no reply to such a question in the framework of ordinary functions; in that framework the question is ruled out of order. There is instead an answer in the wider framework of distributions, or generalized functions; in that framework $f(x)$ is the Dirac delta function, $f(x) = \delta(x)$.

A similar situation may occur for the spectral transform. Suppose a spectral transform (2) is assigned, whose reflection coefficient $R(k)$ does not vanish as $k \to \pm\infty$, but, say, tends to a nonvanishing finite value (a case where precisely such a spectral transform arises is discussed in section 6.2). There is then no possibility to find, within the class of functions for which the direct and inverse spectral problems are well understood, any function $u(x)$ that corresponds to the spectral transform that has been assigned. Is there such a $u(x)$ in the larger framework of distributions or generalized functions? Or is it perhaps necessary to construct an entirely novel theoretical framework, in order to give a consistent definition of $u(x)$? That the latter might possibly be the case is suggested by the nonlinear character of the spectral transform.

(It has been heuristically suggested [Lam1980] that to a spectral transform with $R(k) = -1$ there correspond a potential $u(x)$ proportional to the derivative of the Dirac delta function. But no attempt has been made to justify such a statement in any rigorous sense in the framework of distribution theory. We doubt such an attempt might succeed.)

In conclusion the introduction of a theoretical framework that would play, for the spectral transform, a rôle analogous to that played by the theory of distributions for the Fourier transform, seems to be needed. This task clearly exceeds our present scope. Let us emphasize once more that the nontriviality of such an assignment is presumably associated with the nonlinear character of the mapping that relates bijectively a function to its spectral transform.

An infringement that presumably requires instead a less drastic extension of the theoretical framework is a violation of (f). Indeed the results of section 4.2. and appendix A.12 suggest—to the extent that such statements can be made sense of—that to a spectral transform, that is OK except for the fact that some normalization coefficient ρ_n is negative rather than positive, there corresponds a potential $u(x)$ that has a singularity, and more precisely a double pole, for some real value of x.

Let us end by mentioning another infringement that we also suspect not to require a dramatic extension of the theoretical framework, and yet we can at this stage only put forward as an interesting puzzle. Suppose only (b) is violated (for some real values of k). Then one would suspect that $u(x)$ be complex, since after all (4) is a consequence of the unitarity equation that is clearly closely connected with the reality of $u(x)$. But if (c), (d), (e) and (f) are all satisfied, the kernel M of the GLM equation (2.2.-1) is real; it is then not obvious how the solution $K(x, y)$ of that same equation can go complex, see (2.2.-3,4). Thus in this case, as in the previous one, a singularity must develop in the solution of the GLM integral equation.

A.17. Applications of the wronskian and spectral integral relations to the Schroedinger scattering problem on the whole line

In this appendix we display how some results for the one-dimensional Schroedinger scattering problem can be derived in a straightforward manner from the wronskian integral relations (section 2.4.1) and the spectral integral relations (section 2.4.2). We treat first reflectionless potentials; what we obtain is not new, but the technique of derivation is sufficiently novel to deserve reporting. We then proceed to derive simple formulae relating reflection- and transmission-equivalent potentials, from which explicit instances of such potentials are easily derived. These results are instead, to the best of our knowledge, new; analogous (but less complete) results for the radial Schroedinger equation (corresponding to the multidimensional Schroedinger problem with spherically symmetrical potentials) are given in [C1976].

Reflectionless potentials. Set $g(z) = 2p > 0$ in (2.4.1.-16a) and $g(z) = 1$ in (2.4.1.-16b) and add these two equations. The resulting formula implies that, if the potential $u(x)$ satisfies the nonliner integrodifferential equation

(1) $\quad 2pu(x)+u_x(x)+u(x)\int_x^{+\infty}dy\,u(y)=0,$

the corresponding reflection coefficient vanishes (for all k). But (1) becomes an ordinary differential equation if one changes the dependent variable from $u(x)$ to

(2a) $\quad w(x)=\int_x^{+\infty}dy\,u(y),$

implying of course

(2b) $\quad u(x)=-w_x(x);$

and this differential equation, together with the boundary conditions

(3) $\quad w(+\infty)=w_x(+\infty)=0,$

implied by (2), is easily integrated, yielding

(4a) $\quad w(x)=4p/\{\exp[2p(x-\xi')]-1\},$

namely

(4b) $\quad u(x)=2p^2/\sinh^2[p(x-\xi')].$

In the last two equations ξ' is an arbitrary constant; it need not be real, but if we require $u(x)$ to be real (so that it belongs to the class of functions on which our attention has been heretofore focused), then either ξ' has to be real, or

(5) $\quad \xi'=\xi\pm i\pi/2p$

with ξ real. The first possibility yields however a potential that becomes singular at a real value of x (since the function (4b) is clearly singular at $x=\xi'$); and this is also outside of the class of functions that we have been considering. Thus (5) is the only possibility consistent with our framework. The corresponding potential reads

(6) $\quad u(x)=-2p^2/\cosh^2[p(x-\xi)].$

We have now proved that this potential is reflectionless, namely that the reflection coefficient $R(k)$ corresponding to it vanishes,

(7) $\quad R(k)=0$.

The potential $u(x)$ must therefore possess at least one discrete eigenvalue; call it $E_1 = -p_1^2$, with $p_1 > 0$. But then, setting $g(z) = 2p$ in (2.4.1.-36) and $g(z) = 1$ in (2.4.1.-37) and proceeding as above, one immediately concludes from (1) that $p_1 = p$. Note that this argument also implies that this is the only discrete eigenvalue associated with the potential (6). Also note that these results can be derived as well from the spectral integral relations (set $g(z) = 2p$ in (2.4.2.-21), $g(z) = 1$ in (2.4.2.-22), and add these two equations; the resulting equation with (7) and $N = 1$ with $p_1 = p$, coincides with (1)).

We may therefore conclude that the potential (6) has, associated with it, a vanishing reflection coefficient and precisely one discrete eigenvalue $E = -p^2$. Moreover, the corresponding transmission coefficient can also be explicitly evaluated by applying the same technique used above to (2.4.1.-32) and (2.4.1.-33), getting thereby

(8) $\quad T(k) = (k+ip)/(k-ip)$.

These results reproduce of course those of section 2.2 (see in particular (2.2.-7), (2.2.-10) and (2.2.-12)); note however that they have been obtained here without employing the inverse spectral problem formalism. The approach adopted above, on the other hand, is equivalent to that based on the Bäcklund transformation (4.1.-14); indeed (1) obtains by performing the BT (4.1.-14) on the identically vanishing potential.

Presumably all the sequence of reflectionless potentials with N discrete eigenvalues ($N = 2, 3, \ldots$) could also be obtained by analogous techniques, except for the use of polynomial, rather than constant, $g(z)$.

Reflection- and transmission-equivalent potentials. The starting point of the analysis is the remark that, if two potentials $u^{(1)}(x)$ and $u^{(2)}(x)$ are related by the nonlinear integrodifferential relation

(9) $\quad g(\Lambda)[u^{(1)}(x) - u^{(2)}(x)] + h(\Lambda)\Gamma \cdot 1 = 0$,

the corresponding reflection coefficients $R^{(1)}(k)$ and $R^{(2)}(k)$ are related by the formula

(10a) $\quad g(-4k^2)[R^{(1)}(k) - R^{(2)}(k)]$
$\quad\quad\quad + 2ikh(-4k^2)[R^{(1)}(k) + R^{(2)}(k)] = 0$

or, equivalently,

(10b) $\quad R^{(1)}(k)=R^{(2)}(k)\dfrac{\left[g(-4k^2)-2ikh(-4k^2)\right]}{\left[g(-4k^2)+2ikh(-4k^2)\right]}.$

In this analysis we assume $g(z)$ and $h(z)$ to be real polynomials. The integrodifferential operators Λ and Γ are of course defined by the equations

(11) $\quad \Lambda f(x)=f_{xx}(x)-2\left[u^{(1)}(x)+u^{(2)}(x)\right]f(x)+\Gamma\displaystyle\int_x^{+\infty}dy\,f(y),$

(12) $\quad \Gamma f(x)=\left[u_x^{(1)}(x)+u_x^{(2)}(x)\right]f(x)+\left[u^{(1)}(x)-u^{(2)}(x)\right]$
$\qquad\cdot\displaystyle\int_x^{+\infty}dy\left[u^{(1)}(y)-u^{(2)}(y)\right]f(y).$

These formulae follow directly from the results of section 2.4.1, see in particular (2.4.1.-10). The corresponding formulae for the transmission coefficients $T^{(1)}(k)$ and $T^{(2)}(k)$ follow from (2.4.1.-18), (2.4.1.-19), (2.4.1.-24) and (10):

(13) $\quad T^{(1)}(k)=T^{(2)}(k)\dfrac{\left[g(-4k^2)-2ikh(-4k^2)\right]}{\left[g(-4k^2)-2ikh(-4k^2)+\mathcal{Q}(k)\right]},$

(14) $\quad \mathcal{Q}(k)=-2ik\displaystyle\int_{-\infty}^{+\infty}dx\,g^\#(-4k^2,\Lambda)\left[u^{(1)}(x)-u^{(2)}(x)\right]$
$\qquad+\displaystyle\int_{-\infty}^{+\infty}dy\left[u^{(1)}(y)-u^{(2)}(y)\right]$
$\qquad\cdot\displaystyle\int_y^{+\infty}dx\,g^\#(-4k^2,\Lambda)\left[u^{(1)}(x)-u^{(2)}(x)\right]$
$\qquad+h(-4k^2)\displaystyle\int_{-\infty}^{+\infty}dx\left[u^{(1)}(x)-u^{(2)}(x)\right]$
$\qquad-2ik\displaystyle\int_{-\infty}^{+\infty}dx\,h^\#(-4k^2,\Lambda)\Gamma\cdot 1$
$\qquad+\displaystyle\int_{-\infty}^{+\infty}dy\left[u^{(1)}(y)-u^{(2)}(y)\right]\displaystyle\int_y^{+\infty}dx\,h^\#(-4k^2,\Lambda)\Gamma\cdot 1.$

Of course these formulae hold provided both $u^{(1)}(x)$ and $u^{(2)}(x)$ are *bona fide* potentials (for a different derivation of (9), (10) and (13), see appendix A.20).

We now analyse the implications of these formulae in the simpler cases ($g(z)$ and $h(z)$ polynomials of very low degree). It is convenient to introduce the notation

(15) $$\nu(x)=\int_{x}^{+\infty}dy\left[u^{(1)}(y)-u^{(2)}(y)\right],$$

(16) $$s(x)=u^{(1)}(x)+u^{(2)}(x),$$

implying of course

(17a) $$u^{(1)}(x)=\tfrac{1}{2}\left[s(x)-\nu_x(x)\right],$$
(17b) $$u^{(2)}(x)=\tfrac{1}{2}\left[s(x)+\nu_x(x)\right].$$

Consider first the case

(18) $$g(z)=z-4p^2, \qquad h(z)=0$$

(the justification for the notation will become apparent below). Then (9) reads

(19) $$\nu_{xxx}-2s\nu_x-s_x\nu+\tfrac{1}{2}\nu_x\nu^2-4p^2\nu_x=0.$$

This is easily integrated once, since it becomes a perfect differential upon multiplication by $\nu(x)$. We thus get

(20) $$\nu_{xx}\nu-\tfrac{1}{2}(\nu_x)^2-s\nu^2+\tfrac{1}{8}\nu^4-2p^2\nu^2=0;$$

the integration constant vanishes since $s(x)$, $\nu(x)$ and the derivatives of $\nu(x)$ all vanish as $x\to+\infty$, as implied by (15) and (16).

This equation could in principle be used to determine one of the two reflection-equivalent potentials, say $u^{(1)}(x)$, once the other, say $u^{(2)}(x)$, is given; but generally this is not explicitly feasible. On the other hand (20) can be immediately solved for $s(x)$:

(21) $$s(x)=-2p^2+\tfrac{1}{8}\nu^2(x)+\left\{\nu_{xx}(x)\nu(x)-\tfrac{1}{2}[\nu_x(x)]^2\right\}/\nu^2(x).$$

Thus any choice of a function $\nu(x)$, when inserted in (21) and (17), yields a pair of potentials having the same reflection coefficient,

(22) $$R^{(1)}(k)=R^{(2)}(k)=R(k);$$

of course $\nu(x)$ must be such as to produce *bona fide* potentials, namely both $u^{(1)}(x)$ and $u^{(2)}(x)$ must turn out to be nonsingular and asymptotically vanishing (sufficiently fast). It is easy to find such acceptable $\nu(x)$'s (see below).

A.17 Applications of the wronskian and spectral integral relations

Clearly if the two potentials $u^{(1)}(x)$ and $u^{(2)}(x)$ are different but have the same reflection coefficient, the discrete part of their spectral transform must differ. Let us discuss this briefly here, and then again below. Consider first the wronskian integral relation (2.4.1.-34), with $g(z)$ given by (18), and of course $u^{(1)}(x)$ and $u^{(2)}(x)$ defined by (21) and (17) so that they satisfy (9) (with $h(z)=0$). It then follows that at least one of the two potentials (17), say $u^{(1)}(x)$, has the discrete eigenvalue $E^{(1)}=-p^2$. Actually the implications of (20), (21) and (17) on the spectral transform of $u^{(1)}(x)$ and $u^{(2)}(x)$ are more conveniently ascertained by considering the spectral integral relation (2.4.2.-11) with $g(z)$ given by (18); then it is easily seen that the l.h.s. of this equation vanishes (because of (20) or, equivalently, (9) and (18), and see also (4.-4) and (4.1.-8)), thereby yielding

$$(23) \quad \sum_{n=1}^{N^{(1)}} \left((p_n^{(1)})^2 - p^2 \right) \rho_n^{(1)} f^{(+)(1)}(x, i p_n^{(1)}) f^{(+)(2)}(x, i p_n^{(1)})$$
$$- \sum_{n=1}^{N^{(2)}} \left((p_n^{(2)})^2 - p^2 \right) \rho_n^{(2)} f^{(+)(1)}(x, i p_n^{(2)}) f^{(+)(2)}(x, i p_n^{(2)}) = 0;$$

of course the lack of contribution from the continuous spectrum is due to (22). As the reader can easily verify (for instance, by considering (23) asymptotically as $x \to +\infty$, and taking into account that the parameters $\rho_n^{(j)}$'s are positive for *bona fide* potentials), (23) is satisfied iff each term of the two sums in its l.h.s. vanishes. Therefore only two cases are compatible with (20); either only one of the two potentials (17), say $u^{(1)}(x)$, has the discrete eigenvalue $E_1^{(1)}=-p^2$, and no other discrete eigenvalue, or both potentials have the same eigenvalue $E_1^{(1)}=E_1^{(2)}=-p^2$, and again no other discrete eigenvalue. These two cases are characterized by the expressions of the spectral transforms of $u^{(1)}$ and $u^{(2)}$, that read (see (22))

(24a) $\quad S[u^{(1)}] = \{R(k); p, \rho^{(1)}, N^{(1)}=1\}, \quad S[u^{(2)}] = \{R(k); N^{(2)}=0\}$

in the first case, and

(24b) $\quad S[u^{(1)}] = \{R(k); p, \rho^{(1)}, N^{(1)}=1\},$
$\quad\quad\quad S[u^{(2)}] = \{R(k); p, \rho^{(2)}, N^{(2)}=1\}$

in the second one. Of course the rôle of the two potentials $u^{(1)}(x)$ and $u^{(2)}(x)$ can be exchanged by changing the sign of the function $\nu(x)$ (see (21) and (17)).

Let us now discuss the properties of $\nu(x)$ that are sufficient to insure that the corresponding functions $u^{(1)}(x)$ and $u^{(2)}(x)$ are *bona fide* potentials. Clearly $\nu(x)$ must be regular and nonvanishing for all finite (real) values of x (see (21)),

(25) $\quad \nu(x) \neq 0, \quad -\infty < x < +\infty;$

it must vanish as $x \to +\infty$ (see (15)),

(26) $\quad \nu(x) \to 0 \quad \text{as } x \to +\infty;$

its derivatives must vanish asymptotically (see (21)),

(27) $\quad d^m \nu(x)/dx^m \to 0 \quad \text{as } x \to \pm\infty, \quad m = 1, 2, 3, \ldots,$

and, as $x \to -\infty$, $\nu(x)$ must either (*case a*) tend to the value $-4p$ (the limiting value $+4p$ is also permitted, but since it corresponds merely to the exchange of $u^{(1)}(x)$ with $u^{(2)}(x)$, this case is hereafter ignored), or (*case b*) it must vanish proportionally to $\exp(2px)$:

(28a) $\quad \nu(-\infty) = -4p, \quad (\text{case } a)$

(28b) $\quad \exp(-2px)\,\nu(x) \to \text{finite constant as } x \to -\infty, \, (p>0). \quad (\text{case } b)$

Before giving some examples, let us note that the analog of (22) for the transmission coefficients reads

(29) $\quad T^{(1)}(k) = T^{(2)}(k)(p^2 + k^2)/[p^2 + k^2 + \tfrac{1}{2} i k \nu(-\infty) - \tfrac{1}{8}\nu^2(-\infty)].$

Thus in *case a* one has

(30a) $\quad T^{(1)}(k) = T^{(2)}(k)(k + ip)/(k - ip),$

while in *case b* the simpler conclusion

(30b) $\quad T^{(1)}(k) = T^{(2)}(k)$

holds. Note that both these formulae imply

(31) $\quad |T^{(1)}(k)| = |T^{(2)}(k)|$

consistently with (22) and the unitarity condition (2.1.-13). Incidentally, here the reader may recall from subsection 2.1 that the transmission coefficient $T(k)$ is analytic in the upper half of the complex k-plane, except for simple poles at $k=ip_n$ ($E_n=-p_n^2$ being the eigenvalues). Indeed (24a) and (30a), as well as (24b) and (30b), are certainly consistent with this analyticity property of $T(k)$.

Let us emphasize that, while we have been able to establish simple relationships between the reflection and transmission coefficients associated with the potentials (17), it is generally not possible to compute explicitly these quantities.

Let us return to analyze the discrete part of the spectral transforms of $u^{(1)}(x)$ and $u^{(2)}(x)$. The spectral integral relation (2.4.2.-11) with $g(z)=1$ shows that (24a) corresponds to *case a*, and (24b) to *case b*; in fact, this spectral integral relation reads (here the integral disappears because of (22))

(32a) $\quad \nu(x)=-2\rho^{(1)}f^{(+)(1)}(x,ip)f^{(+)(2)}(x,ip)$

in *case a*, and

(32b) $\quad \nu(x)=-2(\rho^{(1)}-\rho^{(2)})f^{(+)(1)}(x,ip)f^{(+)(2)}(x,ip)$

in *case b*. These relations, together with the definition of the Jost solutions $f^{(\pm)(j)}(x,k)$ (see (2.1.-4)), readily imply that the spectral parameters $\rho^{(j)}$'s are related to the asymptotic behaviour of $\nu(x)$ as $x\to+\infty$, namely

(33a) $\quad \rho^{(1)}=-\tfrac{1}{2}\lim_{x\to+\infty}\left[\exp(2px)\nu(x)\right]$

holds in *case a*, while

(33b) $\quad \rho^{(1)}-\rho^{(2)}=-\tfrac{1}{2}\lim_{x\to+\infty}\left[\exp(2px)\nu(x)\right]$

holds in *case b*. Indeed additional relations obtain by considering also the limit of (32) as $x\to-\infty$; using the notation of section 2.1, and leaving the derivation to the reader, there follows (using (28a))

(34a) $\quad T^{(2)}(ip)=c^{(-)(1)}/\left[2p(\rho^{(1)})^{1/2}\right]$

in *case a*, and

(34b) $\quad \lim_{x\to-\infty}\left[\exp(-2px)\nu(x)\right]=2\left[(\rho^{(2)}/\rho^{(1)})^{1/2}-(\rho^{(1)}/\rho^{(2)})^{1/2}\right]$
$\cdot c^{(-)(1)}c^{(-)(2)}$

in *case b* (note the consistency of this result with (28b)).

Moreover the Jost solution $f^{(\pm)(1)}(x,k)$ is related to $f^{(\pm)(2)}(x,k)$ by the Darboux transformation

(35a) $$f^{(\pm)(1)}(x,k) = [(k-ip)/(k \mp ip)] f^{(\pm)(2)}(x,k)$$
$$+ \{4\nu(x) f_x^{(\pm)(2)}(x,k) - [2\nu_x(x) + \nu^2(x)]$$
$$\cdot f^{(\pm)(2)}(x,k)\} / [8(k+ip)(k \mp ip)]$$

in *case a*, and

(35b) $$f^{(\pm)(1)}(x,k) = f^{(\pm)(2)}(x,k) + \{4\nu(x) f_x^{(\pm)(2)}(x,k)$$
$$- [2\nu_x(x) + \nu^2(x)] f^{(\pm)(2)}(x,k)\} / [8(k^2+p^2)]$$

in *case b*. The Darboux transformation corresponding to the relationship (9) between $u^{(1)}(x)$ and $u^{(2)}(x)$ has been established in appendix A.20; the special transformations (35) obtain therefore from (A.20.-40) by setting (18) in (A.20.-41), (A.20.-31) and (A.20.-32). Remarkably, it is possible to derive, from these equations, the explicit expression (in terms of $\nu(x)$) of the normalized eigenfunctions corresponding to the discrete eigenvalues; in *case a* (see (24a) and the notation of subsection (2.1)) the eigenfunction corresponding to the eigenvalue $E^{(1)} = -p^2$ is

(36a) $$\varphi^{(1)}(x) = [-\nu(x)/2]^{1/2} \exp\left[\int_x^{+\infty} dy\, \nu(y)/4\right],$$

while in *case b* the eigenvalue $E^{(1)} = E^{(2)} = -p^2$ occurs for both potentials $u^{(1)}(x)$ and $u^{(2)}(x)$, and the corresponding eigenfunctions read

(36b) $$\varphi^{(1)}(x) = \{\rho^{(1)} \nu(x) / [2(\rho^{(2)} - \rho^{(1)})]\}^{1/2} \exp\left[\int_x^{+\infty} dy\, \nu(y)/4\right],$$
$$\varphi^{(2)}(x) = \{\rho^{(2)} \nu(x) / [2(\rho^{(2)} - \rho^{(1)})]\}^{1/2} \exp\left[-\int_x^{+\infty} dy\, \nu(y)/4\right].$$

The derivation of (36a) starts by imposing that the r.h.s. of (35a), with the upper sign, be analytic in the upper half of the complex k-plane (this is a well-known property of the Jost solutions, see subsection (2.1.)); this implies that the apparent pole at $k = ip$ of $f^{(+)(1)}(x,k)$ does not occur iff $f^{(+)(2)}(x,ip)$ satisfies the first order differential equation for $y(x)$,

(37a) $$4\nu(x) y_x(x) = [2\nu_x(x) + \nu^2(x)] y(x),$$

with

(37b) $y(x) \to \exp(-px)$ as $x \to +\infty$, $p > 0$;

solving this equation yields the explicit expression of $f^{(+)(2)}(x, ip)$, that, together with

(38) $\varphi^{(1)}(x) = [\rho^{(1)}]^{1/2} f^{(+)(1)}(x, ip)$,

and (32a), imply (36a). The derivation of the expressions (36b) follows by similar arguments (indeed it requires again solving (37)), and is left as an exercise for the diligent reader. Note that in *case b* the normalization (2.1.-21) of the eigenfunctions (36b) implies the additional relation

(39) $\int_{-\infty}^{+\infty} dx\, \nu(x) = 2\ln[\rho^{(2)}/\rho^{(1)}]$.

The simplest example of *case a* results from

(40) $\nu(x) = -2p\{1 - \operatorname{tgh}[p(x-\xi)]\}$,

which yields

(41) $u^{(1)}(x) = -2p^2/\cosh^2[p(x-\xi)]$, $u^{(2)}(x) = 0$.

Thus in this case $u^{(2)}(x)$ vanishes (and therefore $R^{(2)}(k) = 0$, $T^{(2)}(k) = 1$) and $u^{(1)}(x)$ is just the reflectionless potential (6), (note the consistency of $T^{(1)}(k) = (k+ip)/(k-ip)$ as now implied by (30a), with (8)).

A simple example of *case b* obtains if one sets

(42) $\nu(x) = 4c/\cosh[2p(x-\xi)]$,

which yields

(43a) $u^{(1)}(x) = \{c^2 - 3p^2 + 4pc \sinh[2p(x-\xi)]\}/\cosh^2[2p(x-\xi)]$,

(43b) $u^{(2)}(x) = \{c^2 - 3p^2 - 4pc \sinh[2p(x-\xi)]\}/\cosh^2[2p(x-\xi)]$.

In this case the two (equal) reflection, and transmission, coefficients are not known; but we can assert that (for all values of the real constant c) both potentials (43) have only one bound state with energy $E^{(1)} = E^{(2)} = -p^2$. Moreover the corresponding spectral parameters $\rho^{(1)}$ and $\rho^{(2)}$ can be easily computed combining (33b) and (39), thereby obtaining

(44a) $\quad \rho^{(1)} = 2c \exp[(4p^2\xi - \pi c)/(2p)]/\sinh[\pi c/(2p)],$

(44b) $\quad \rho^{(2)} = 2c \exp[(4p^2\xi + \pi c)/(2p)]/\sinh[\pi c/(2p)].$

Also the expressions (36b) of the normalized eigenfunctions can be easily computed, and read

(45a) $\quad \varphi^{(1)}(x) = \dfrac{c^{1/2} \exp\{[\pi c^2 - 4p^2 \operatorname{arctg} \exp[2p(x-\xi)]]/(4pc)\}}{\{\sinh[\pi c/(2p)] \cosh[2p(x-\xi)]\}^{1/2}},$

(45b) $\quad \varphi^{(2)}(x) = \dfrac{c^{1/2} \exp\{[\pi c^2 - 4p^2 \operatorname{arctg} \exp[2p(x-\xi)]]/(-4pc)\}}{\{\sinh[\pi c/(2p)] \cosh[2p(x-\xi)]\}^{1/2}}.$

Note that the value $k = ip$ sits just on the boundary of the (same) Bargmann strip associated with both potentials (43); let us recall that, as implied by the discussion of section 2.1, such a value could not have fallen inside the Bargmann strip, where clearly the analytic continuations of $R^{(1)}(k)$ and $R^{(2)}(k)$ coincide (see also appendix A.12).

Consider next the case

(46) $\quad g(z) = 0, \quad h(z) = 1.$

Then (9) becomes simply

(47) $\quad s(x) = \tfrac{1}{2} \nu^2(x),$

while (10b) and (13) yield

(48) $\quad R^{(1)}(k) = -R^{(2)}(k),$

(49) $\quad T^{(1)}(k) = T^{(2)}(k).$

To obtain the latter formula we have also used the property $\nu(-\infty) = 0$, that is clearly implied by (47).

We therefore conclude that any pair of *bona fide* potentials given by the simple formulae

(50a) $\quad u^{(1)}(x) = \tfrac{1}{4}[\nu^2(x) - 2\nu_x(x)],$

(50b) $\quad u^{(2)}(x) = \tfrac{1}{4}[\nu^2(x) + 2\nu_x(x)],$

have reflection and transmission coefficients related by the simple formulae (48) and (49). Moreover from (9), (46) and (2.4.1.-35a) (with $g(z)=1$) there follows that neither the potential $u^{(1)}(x)$, nor $u^{(2)}(x)$, can support a bound state (discrete eigenvalue), unless this belongs to both of them; a possibility that is however excluded by (9), (46) and (2.4.1.-58) (recall that the normalization constants $c_n^{(j)}$ and $\rho_n^{(j)}$ are positive).

All these results assume both $u^{(1)}(x)$ and $u^{(2)}(x)$ to be *bona fide* potentials. To guarantee this, it is sufficient (see (50)) that $\nu(x)$ be regular for all (real) values of x and that it vanish asymptotically ($x \to \pm\infty$) with its derivatives (fast enough!).

To summarize: given any regular function $\nu(x)$ vanishing asymptotically with its derivatives, the simple formulae (50) yield two potentials, neither of which can support any bound state, and whose reflection and transmission coefficients are related by the simple formulae (48) and (49). This is a remarkably neat, yet far from trivial, result.

It is clearly very easy to exhibit examples of such pairs of potentials. We display only one instance, obtained from

(51) $\quad \nu(x) = 2c/\cosh[\mu(x-\xi)],$

that reads

(52a) $\quad u^{(1)}(x) = c\{c + \mu\sinh[\mu(x-\xi)]\}/\cosh^2[\mu(x-\xi)],$

(52b) $\quad u^{(2)}(x) = c\{c - \mu\sinh[\mu(x-\xi)]\}/\cosh^2[\mu(x-\xi)].$

Here c, μ and ξ are arbitrary real constants. It should perhaps be emphasized that even the mere fact that both of these potentials possess no bound states (for any choice of the real parameters c and μ), let alone the validity of (48) and (49), is far from trivial.

Finally let us work out the implications of (9), (10) and (13) in the case

(53) $\quad g(z) = 2p > 0, \quad h(z) = 1.$

Then

(54) $\quad R^{(1)}(k) = -R^{(2)}(k)(k+ip)/(k-ip),$

(55) $\quad T^{(1)}(k) = T^{(2)}(k)(k+ip)/[k+ip+\tfrac{1}{2}i\nu(-\infty)],$

and

(56) $\quad 2p\nu_x(x) - s_x(x) + \nu(x)\nu_x(x) = 0.$

The last equation can be immediately integrated, and it yields

(57) $\quad s(x) = \frac{1}{2}\nu(x)[\nu(x)+4p]$.

It may therefore be concluded that any pair of *bona fide* potentials given, in terms of a (largely arbitrary; see below) function $\nu(x)$ by the formulae

(58a) $\quad u^{(1)}(x) = \frac{1}{4}\nu^2(x) + p\nu(x) - \frac{1}{2}\nu_x(x)$,
(58b) $\quad u^{(2)}(x) = \frac{1}{4}\nu^2(x) + p\nu(x) + \frac{1}{2}\nu_x(x)$,

have associated reflection and transmission coefficients related by (54) and (55).

The only limitation on the function $\nu(x)$ is that the corresponding functions $u^{(1)}(x)$ and $u^{(2)}(x)$ qualify as *bona fide* potentials. Sufficient conditions for this are that $\nu(x)$ be regular for all (real) x, that it vanish as $x \to +\infty$,

(59) $\quad \nu(x) \to 0 \quad \text{as } x \to +\infty$,

that its derivatives vanish asymptotically,

(60) $\quad d^m \nu(x)/dx^m \to 0 \quad \text{as } x \to \pm\infty, \quad m=1,2,3,\ldots,$

and that, as $x \to -\infty$, either (*case a*) $\nu(x)$ tend to $-4p$ or (*case b*) $\nu(x)$ vanish:

(61a) $\quad \nu(-\infty) = -4p \quad$ (*case a*),
(61b) $\quad \nu(-\infty) = 0 \quad$ (*case b*).

Note that, in *case a*, (55) yields

(62a) $\quad T^{(1)}(k) = T^{(2)}(k)(k+ip)/(k-ip)$,

while of course in *case b*,

(62b) $\quad T^{(1)}(k) = T^{(2)}(k)$.

Moreover, from (9), (53), (2.4.1.-34) and (2.4.1.-35) it is immediately concluded that, if the potential $u^{(1)}(x)$ has a (negative) discrete eigenvalue, say $E_1^{(1)} = -p_1^2$, then either $u^{(2)}(x)$ has the same eigenvalue or $p_1^{(1)} = p$; and in the first case, (9), (53), (2.4.1.-67) and (2.4.1.-68) yield (with obvious notation)

(63) $\quad \rho_1^{(1)} = \rho_1^{(2)}(p+p_1^{(2)})/(p-p_1^{(2)}), \qquad p_1^{(1)} = p_1^{(2)},$

which also implies $p_1^{(2)} < p$. Note that these conclusions can be reached also from the spectral integral relations (2.4.2.-11) and (2.4.2.-12), combined with (9) and (53).

Summarizing, we have shown that the pair of potentials (58), with $\nu(x)$ a regular function, arbitrary except for the conditions (59), (60) and (61), and with $p>0$, has associated reflection and transmission coefficients satisfying (54) and (62); moreover, the only discrete eigenvalue that may be possessed by $u^{(1)}(x)$ and not by $u^{(2)}(x)$ is $E=-p^2$; while any other discrete eigenvalue E' possessed by both $u^{(1)}(x)$ and $u^{(2)}(x)$ must satisfy the inequality $E<E'<0$.

It is easy to construct explicit examples. The choice

(64) $\quad \nu(x) = -2p\{1 - \mathrm{tgh}[p(x-\xi)]\},$

belonging to *case a*, yields $u^{(2)}(x)=0$ and identifies $u^{(1)}(x)$ with the potential (6); $T^{(1)}(k)$ is accordingly identified with (8), consistently with (62a) and $T^{(2)}(k)=1$; while both $R^{(1)}(k)$ and $R^{(2)}(k)$ vanish. The choice

(65) $\quad \nu(x) = 2c/\cosh[\mu(x-\xi)],$

belonging to *case b*, yields

(66a) $\quad u^{(1)}(x) = \dfrac{c\{c + 2p\cosh[\mu(x-\xi)] + \mu\sinh[\mu(x-\xi)]\}}{\cosh^2[\mu(x-\xi)]},$

(66b) $\quad u^{(2)}(x) = \dfrac{c\{c + 2p\cosh[\mu(x-\xi)] - \mu\sinh[\mu(x-\xi)]\}}{\cosh^2[\mu(x-\xi)]}.$

Note finally that in all the examples considered in the second part of this appendix the relationships for the reflection and transmission coefficients preserve the moduli,

(67a) $\quad |R^{(1)}(k)| = |R^{(2)}(k)|, \quad \mathrm{Im}\, k = 0,$

(67b) $\quad |T^{(1)}(k)| = |T^{(2)}(k)|, \quad \mathrm{Im}\, k = 0.$

Indeed this property holds generally for (10b) (provided g and h are real), and therefore also (by unitarity) for (13). This fact justifies the subheading of the second part of this appendix (recall that in a scattering experiment generally only the moduli of the reflection and transmission coefficients are measured).

A.18. On the class of equations $\eta(L)u_t = \alpha(L)u_x$

In this appendix we discuss the class of evolution equations

(1) $\quad \eta(L)u_t(x,t) = \alpha(L)u_x(x,t),$

where η and α are polynomials,

(2a) $\quad \eta(z) = \sum_{m=0}^{M} \eta_m z^m, \quad 0 < M < \infty,$

(2b) $\quad \alpha(z) = \sum_{m=0}^{M'} \alpha_m z^m, \quad 0 \leq M' < \infty,$

and of course L is the usual integro-differential operator,

(3) $\quad Lf(x,t) = f_{xx}(x,t) - 4u(x,t)f(x,t) + 2u_x(x,t)\int_x^{+\infty} dy\, f(y,t).$

We assume for simplicity the (real) functions η and α to be time-independent; the extension of the discussion given below to include an explicit time-dependence in η and α is essentially trivial. Also trivial would be the extension of the treatment to the case when $\eta(z)$ and $\alpha(z)$ are entire functions rather than just polynomials.

Clearly the evolution equation (1) can be rewritten in the (apparently) simpler form

(4) $\quad u_t(x,t) = \bar{\alpha}(L)u_x(x,t)$

by setting formally

(5) $\quad \bar{\alpha}(L) = \alpha(L)/\eta(L).$

Thus the main problem is, whether and how to attribute an unambiguous significance to the operator (5) applied to $u_x(x,t)$. One approach is of course to expand $\bar{\alpha}(L)$ in powers of L, since the operator L^m, $m = 0, 1, 2, \ldots$, has an unambiguous significance; or, more generally, to expand $\bar{\alpha}(L)$ in powers of $L - z$, with z an arbitrary constant. But while this approach yields a formal definition, it does not shed much light on the actual applicability of this procedure, that hinges of course on the convergence of the resulting series.

Much more illuminating is the following approach, that is closely analogous to that adopted, via the Fourier transform, to attribute a meaning to the operator $\bar{a}(D)$, where D is the differential operator,

(6) $\quad Df(x) = f_x(x)$.

Let us recall first how this works. Assume the function $f(x)$ to be Fourier transformable,

(7) $\quad f(x) = (2\pi)^{-1} \int_{-\infty}^{+\infty} dk \exp(ikx) \hat{f}(k)$.

Then formally

(8) $\quad \bar{a}(D) f(x) = (2\pi)^{-1} \int_{-\infty}^{+\infty} dk \exp(ikx) \bar{a}(ik) \hat{f}(k)$.

Thus the application of the operator $\bar{a}(D)$ to $f(x)$ takes a well-defined explicit significance whenever the integral in the r.h.s. of this formula converges; and conditions sufficient to guarantee that this happen are easily ascertained. For instance, if $a(z)$ and $\eta(z)$ are polynomials, see (2) and (5), and if the Fourier transform $\hat{f}(k)$ is regular for real k and vanishes as $k \to \pm\infty$ faster than any inverse power of k (i.e. if $f(x)$ is infinitely differentiable for all real values of x and vanishes with all its derivatives faster than x^{-1} as $x \to \pm\infty$), then a sufficient condition to insure that $\bar{a}(D) f(x)$ be a well-defined function (not a distribution!) is that $\eta(z)$ have no pure imaginary zeros,

(9) $\quad \eta(z) \neq 0 \quad \text{if } \operatorname{Re} z = 0$.

The basic formulae to apply an analogous procedure to the r.h.s. of (4) are

(10) $\quad u(x) = \pi^{-1} \int_{-\infty}^{+\infty} dk \left[f^{(+)}(x, k) \right]^2 2ikR(k) - 4 \sum_{n=1}^{N} \left[f_n(x) \right]^2 p_n \rho_n$,

(11) $\quad L^m D = D \tilde{L}^m$,

(12a) $\quad \tilde{L}\left[f^{(+)}(x, k) \right]^2 = (2ik)^2 \left[f^{(+)}(x, k) \right]^2$,

(12b) $\quad \tilde{L}\left[f_n(x) \right]^2 = 4p_n^2 \left[f_n(x) \right]^2$

(see (1.3.3.-2), or (2.4.2.-16), (A.9.-30) and (A.9.-44); if the reader is unfamiliar with the symbols used here, he may easily retrieve their significance from the context of the formulae we have just quoted). Note that in these formulae (and, wherever convenient, also below) we have eliminated for simplicity the explicit indication of the variable t.

Clearly these formulae imply

(13a) $\quad \bar{a}(L)u_x(x) = D\left\{\pi^{-1}\int_{-\infty}^{+\infty} dk\, [f^{(+)}(x,k)]^2 2ik\bar{a}(-4k^2)R(k)\right.$

$$\left. -4\sum_{n=1}^{N}[f_n(x)]^2 p_n \bar{a}(4p_n^2)\rho_n\right\},$$

(13b) $\quad \bar{a}(L)u_x(x) = (2/\pi)\int_{-\infty}^{+\infty} dk\, f_x^{(+)}(x,k)f^{(+)}(x,k)2ik\bar{a}(-4k^2)$

$$\cdot R(k) - 8\sum_{n=1}^{N} f_{nx}(x)f_n(x)p_n \bar{a}(4p_n^2)\rho_n$$

(indeed this formula coincides essentially with (2.4.2.-16), as it is easily seen using the operator identity (11), namely $\bar{a}(L)D = D\bar{a}(\tilde{L})$). Thus, if $u(x)$ is a *bona fide* potential, a sufficient condition to guarantee that $\bar{a}(L)u_x(x)$ be well-defined (indeed, be itself a *bona fide* potential) is that the polynomial $\eta(z)$ (see (2) and (5)) have no real zeros,

(14) $\quad \eta(z) \neq 0 \quad \text{if Im } z = 0.$

Whenever this condition is met there is hardly any problem in the discussion of (1). Let us indeed recall (see section 3.1) that whenever $u(x, t)$ evolves according to (1), its spectral transform evolves according to the simple formulae

(15a) $\quad R(k, t) = R(k, 0)\exp[2ik\bar{a}(-4k^2)t],$

(15b) $\quad p_n(t) = p_n(0),$

(15c) $\quad \rho_n(t) = \rho_n(0)\exp[-2p_n\bar{a}(4p_n^2)t],$

where we are of course using the notation (5). Note that this implies that, at least as long as the function $u(x, t)$ and its spectral transform

(16) $\quad S[u] = \{R(k, t), -\infty < k < +\infty; p_n, \rho_n(t), n = 1, 2, \ldots, N\}$

are in one-to-one correspondence (which we know certainly to be the case if $u(x, t)$ is a *bona fide* potential), the Cauchy problem (given $u(x, 0)$, to determine $u(x,t)$) is OK. On the other hand, if $u(x,0)$ is a *bona fide* potential, and if the condition (14) holds (with (2) and (5)), then from the formulae (15) themselves one infers that $u(x,t)$ is also a *bona fide* potential. (It is, however, a somewhat peculiar function, since the corresponding reflection coefficient, $R(k,t)$, considered as a function of the complex variable k, generally has, for $t \neq 0$, M essential singularities in the upper k half-plane, and just as many in the lower half-plane; see (2a), (5) and (15a). No explicit example of such a *bona fide* potential is known).

Even if the condition (14) were violated, there clearly might well be a class of initial Cauchy data $u(x,0)$ for which the time-evolution (1) presents no problem. Suppose for instance that (14) holds except for some *positive* zeros,

(17a) $\quad \eta(4q_n^2) = 0,$

(17b) $\quad \eta(z) \neq 0 \quad \text{if Im } z = 0 \text{ and Re } z \neq 4q_n^2.$

Then clearly a condition on the initial datum $u(x,0)$, sufficient to guarantee that the corresponding Cauchy problem be OK, is that none of the discrete eigenvalues p_n (if any) of the spectral transform of $u(x,0)$ coincide with the quantities q_n of (17). And analogously, if there is some negative zero of $\eta(z)$, say $\eta(-4q_n^2) = 0$, it is sufficient that $R(k,0)$ vanish identically in the neighborhoods of $k = \pm q_n$.

An exceptional case, that we mention in spite of its marginal interest, occurs if the polynomials $\eta(z)$ and $\alpha(z)$ have a common *positive* zero, say

(18a) $\quad \eta(z) = (z - 4p^2)\eta'(z),$

(18b) $\quad \alpha(z) = (z - 4p^2)\alpha'(z),$

where $\eta'(z)$ and $\alpha'(z)$ are of course again polynomials (of degree $M-1$ and $M'-1$; see (2)). Then of course the common zero disappears in the definition (5), and therefore plays no role in the formulae (15). But in fact the last of these formulae, (15c), only applies if $p_n \neq p$; if instead $p_n = p$, then the evolution equation (1) implies no restriction on the time-dependence of $\rho_n(t)$, that may in fact be chosen arbitrarily. (The reader who has difficulty to believe this result may for instance easily verify that if

(19) $\quad \eta(4p^2) = \alpha(4p^2) = 0,$

the single-soliton solution

(20) $\quad u(x,t) = -2p^2/\cosh^2\{p[x - \xi(t)]\}$

satisfies (1) for any arbitrary choice of $\xi(t)$, since indeed (see (1.7.1.-18))

(21a) $\quad \eta(L)u_t(x,t) = -\dot{\xi}(t)\eta(4p^2)u_x(x,t) = 0,$

(21b) $\quad \alpha(L)u_x(x,t) = \alpha(4p^2)u_x(x,t) = 0 \;).$

Let us finally discuss the time evolution of the transmission coefficient $T(k)$ associated with a *bona fide* potential $u(x,t)$ that evolves according to (1). The simpler argument to conclude that $T(k)$ is then time-independent follows from the observation that (15a) implies that the modulus of the reflection coefficient is time-independent,

(22) $\quad |R(k,t)| = |R(k,0)|,$

as well as the discrete eigenvalues (see (15b)), while we know (see remark after (2.1.-44)) that these quantities determine $T(k)$ uniquely.

From the time-independence of $T(k)$ there follows (see chapter 5) that also time-independent are the quantities

(23) $\quad C_m = (-1)^m (2m+1)^{-1} \int_{-\infty}^{+\infty} dx\, L^m [xu_x(x,t) + 2u(x,t)],$
$$m = 0,1,2,\ldots,$$

namely that to the class of evolution equations (1) is associated the usual infinite sequence of conserved quantities.

Let us emphasize that, to obtain this result, we have used the fact that $u(x,t)$ remains a *bona-fide* potential for all time (sufficient conditions to guarantee that this happen have been discussed above). If instead one were to use only the evolution equation (1) together with the operator identity (A.8.-8) or (A.9.-63a), as it was done at the end of chapter 5, one would write first

(24) $\quad dC_m/dt = (-1)^m \int_{-\infty}^{+\infty} dx\, L^m u_t(x,t), \quad m = 0,1,2,\ldots$

and then (in order to use (1)) take an appropriate linear combination of these formulae so as to obtain (see (2a))

(25) $\quad \sum_{m=0}^{M} (-1)^{m+n} \eta_m dC_{m+n}/dt = \int_{-\infty}^{+\infty} dx\, L^n \eta(L) u_t(x,t).$

Then (1) and the operator identity (see (A.8.-6) or (A.9.-54a))

(26) $\quad \int_{-\infty}^{+\infty} dx\, L^m u_x(x,t) = 0, \quad m = 0,1,2,\ldots,$

imply that the r.h.s. of (25) vanishes, namely that the quantities

$$Q_n = \sum_{m=0}^{M} (-1)^m \eta_m C_{m+n}, \quad n=0,1,2,\ldots, \tag{27}$$

provide an infinite sequence of conserved quantities. This result is of course consistent with, but a bit weaker than, the time-independence of each of the quantities C_m.

Likewise, an attempt to derive the time independence of the transmission coefficient $T(k)$ by the same procedure used in chapter 5 does not quite work. Indeed in place of (5.-7) and (5.-8) one would now write the formulae

$$2ik\eta(-4k^2)[R_t(k,t)R(-k,t)+T_t(k,t)T(-k,t)] \tag{28}$$
$$+4k^2 T(k,t)T(-k,t)\int_{-\infty}^{+\infty} dx\, \eta^{\#}(-4k^2, \boldsymbol{L})u_t(x,t)$$
$$= \int_{-\infty}^{+\infty} dx\, \psi(x,k,t)\psi(x,-k,t)\eta(\boldsymbol{L})u_t(x,t),$$

$$(2ik)^2 \alpha(-4k^2)R(k,t)R(-k,t) \tag{29}$$
$$= \int_{-\infty}^{+\infty} dx\, \psi(x,k,t)\psi(x,-k,t)\alpha(\boldsymbol{L})u_x(x,t);$$

note in particular the additional term in the l.h.s. of (28), that originates from the second term in the l.h.s. of (2.4.1.-25); here we are of course using the notation (see (2.4.1.-20))

$$g^{\#}(z_1, z_2) \equiv [g(z_1) - g(z_2)]/(z_1 - z_2). \tag{30}$$

From these equations, together with (3.1.-10, 11) (or (1) and (15a)) one obtains the formula

$$\eta(-4k^2)T_t(k,t) = 2ikT(k,t)\int_{-\infty}^{+\infty} dx\, \eta^{\#}(-4k^2, \boldsymbol{L})u_t(x,t), \tag{31}$$

to be contrasted with (5.-9).

Actually it is easily seen that this implies

$$[T(k,t)T(-k,t)]_t = 0, \tag{32a}$$

or equivalently

$$|T(k,t)|_t = 0, \tag{32b}$$

consistently with (15a) and the unitarity relation (2.1.-18) or (2.1.-13), and

(33) $\quad \eta(-4k^2)\theta_t(k,t) = 2k \int_{-\infty}^{+\infty} dx\, \eta^{\#}(-4k^2, L) u_t(x,t),$

where we have introduced as usual the phase of $T(k,t)$ via the formula

(34a) $\quad T(k,t) = T(-k,t) \exp[2i\theta(k,t)],$

or equivalently (for real $u(x,t)$)

(34b) $\quad T(k,t) = |T(k,t)| \exp[i\theta(k,t)], \quad \text{Im}\, k = 0.$

Thus, using (A.3.-2) with (A.3.-3), one cannot immediately conclude that the C_m's, or equivalently the coefficients θ_m's of the asymptotic expansion of θ at large k, are time-independent, since (33) does not appear to imply that θ itself is time-independent. But it is possible, making an asymptotic expansion at large k and using (A.3.-2,3), (30) and (1), to recover from (33) once again the time-independence of the set (27).

On the other hand, having proved (32), and using (15b), one can again conclude from (2.1.-43b), that the phase $\theta(k)$ of the transmission coefficient must be time-independent (and note that to obtain (2.1.-43b) only the analyticity of $T(k)$ in the upper half of the complex k-plane has played a rôle; the analyticity of $R(k)$ has not been used, see appendix A.4).

Of course these results imply that the r.h.s. of (24), (31) and (33) vanish. Indeed such a conclusion might have also been obtained directly using (4) together with the formula

(35) $\quad \int_{-\infty}^{+\infty} dx\, L^m \bar{a}(L) u_x(x,t) = 0, \quad m = 0, 1, 2, \ldots,$

whose validity may be considered to follow from (26), in spite of the fact that $\bar{a}(L)$ is not a polynomial (see (5)), as long as the expression $\bar{a}(L)u_x$ is well-defined. The conditions sufficient to guarantee that this be the case have been discussed above; we also noted that, when they hold, $u(x,t)$ maintains for all time the property to be a *bona fide* potential; and this underscores the general consistency of our treatment (a consistency that is reassuring in view of the essentially heuristic character of this discussion).

A.19. Examples of functions with explicitly known spectral transform

Only a few examples are known of functions $u(x)$ whose spectral transform has an explicit expression. Two approaches may be followed for their determination: either the function $u(x)$ is given and the Schroedinger

A.19 Functions with explicitly known spectral transform

equation

(1) $\quad f_{xx} + [k^2 - u(x)] f = 0, \quad f \equiv f(x,k),$

can be explicitly solved for all values of the (complex) spectral parameter k, or the spectral transform

(2) $\quad S[u] = \{R(k), -\infty < k < +\infty; p_n, \rho_n, n = 1, 2, \ldots, N\}$

is given and the Gel'fand–Levitan–Marchenko equation

(3) $\quad K(x,y) + M(x+y) + \int_x^{+\infty} dz\, K(x,z) M(z+y) = 0, \quad x < y,$

with

(4) $\quad M(x) = \sum_{n=1}^{N} \rho_n \exp(-p_n x) + (2\pi)^{-1} \int_{-\infty}^{+\infty} dk\, R(k) \exp(ikx),$

can be explicitly solved for all (real) values of x.

The aim of this appendix is to discuss some instances of both cases. Of course, we use here the notation introduced in chapter 2, where the direct and inverse problem associated with (1) has been treated (see also appendices A.5, A.12, A.15 and A.17); however, in order to make this appendix largely self-contained, we also report some relevant formulae here.

The (linearly independent) Jost solutions $f^{(\pm)}(x,k)$ of (1) are defined by the boundary condition

(5) $\quad \lim_{x \to \pm\infty} [\exp(\mp ikx) f^{(\pm)}(x,k)] = 1, \quad \text{Im } k \geq 0.$

In the context of the direct problem, these solutions are derived integrating (1) with (5); and from these solutions the transmission and reflection coefficients, $T(k)$ and $R(k)$, are easily obtained via the formula

(6) $\quad T(k) f^{(-)}(x,k) = f^{(+)}(x,-k) + R(k) f^{(+)}(x,k).$

Moreover, if N discrete eigenvalues, $-p_n^2$, occur, they correspond to the (simple) poles of $T(k)$ occurring at $k = i p_n$, $n = 1, 2, \ldots, N$, $p_n > 0$; and the corresponding eigenfunctions $\varphi_n(x)$, satisfying the normalization condition

(7) $\quad \int_{-\infty}^{+\infty} dx\, [\varphi_n(x)]^2 = 1, \quad n = 1, 2, \ldots, N,$

are characterized by the formulae

(8a) $\quad \varphi_n(x) = c_n^{(-)} f^{(-)}(x, i p_n),$

(8b) $\quad \varphi_n(x) = c_n^{(+)} f^{(+)}(x, i p_n),$

(9) $\quad \lim_{k \to i p_n} [(k - i p_n) T(k)] = i c_n^{(-)} c_n^{(+)},$

that also yield the expression of the normalization parameters,

(10) $\quad \rho_n = (c_n^{(+)})^2 = \left\{ \int_{-\infty}^{+\infty} [f^{(+)}(x, i p_n)]^2 \right\}^{-1}, \quad n = 1, 2, \ldots, N.$

In the context of the inverse problem, based on (3) and (4), we recall that the solution of (3) provides not only the function $u(x)$ through the expression

(11) $\quad u(x) = -2 \, dK(x, x+0)/dx,$

but also the Jost solution $f^{(+)}$,

(12) $\quad f^{(+)}(x, k) = \exp(ikx) + \int_x^{+\infty} dy \, K(x, y) \exp(iky), \quad \operatorname{Im} k \geq 0.$

The normalized eigenfunction $\varphi_n(x)$ is then of course given by (8b), (10) and (12) with $k = i p_n$.

Before proceeding to consider those cases in which (1), or (3), is explicitly solvable, we describe below several techniques that can be used to enlarge the class of functions whose spectral transform is explicitly known; in the following we use the term "solvable potential" to indicate a function $u(x)$ such that its associated Schroedinger spectral problem is explicitly solved.

Assume that $u(x)$ is a solvable potential, then also the translated potential

(13a) $\quad u'(x) = u(x + a)$

is solvable, a being a real parameter. In fact, in self-explanatory notation, one easily finds that (see also subsection 2.1.1)

(13b) $\quad f^{(\pm)\prime}(x, k) = \exp(\mp i k a) f^{(\pm)}(x + a, k),$

(13c) $\quad T'(k) = T(k),$

(13d) $\quad R'(k) = \exp(2ika) R(k),$

(13e) $\quad p'_n = p_n, \quad n = 1, 2, \ldots, N,$

(13f) $\varphi'_n(x) = \varphi_n(x+a)$, $n=1,2,\ldots,N$,
(13g) $c_n^{(\pm)'} = \exp(\mp p_n a) c_n^{(\pm)}$, $n=1,2,\ldots,N$,
(13h) $\rho'_n = \exp(-2p_n a) \rho_n$, $n=1,2,\ldots,N$,
(13i) $K'(x,y) = K(x+a, y+a)$, $x<y$.

Also the potential obtained from $u(x)$ by a scale transformation (see subsection 2.1.1), i.e.

(14a) $u'(x) = \mu^2 u(\mu x)$,

where μ is a positive parameter, is solvable. In this case, the relevant formulae read

(14b) $f^{(\pm)'}(x,k) = f^{(\pm)}(\mu x, k/\mu)$,
(14c) $T'(k) = T(k/\mu)$,
(14d) $R'(k) = R(k/\mu)$,
(14e) $p'_n = \mu p_n$, $n=1,2,\ldots,N$,
(14f) $\varphi'_n(x) = \mu^{1/2} \varphi_n(\mu x)$, $n=1,2,\ldots,N$,
(14g) $c_n^{(\pm)'} = \mu^{1/2} c_n^{(\pm)}$, $n=1,2,\ldots,N$,
(14h) $\rho'_n = \mu \rho_n$, $n=1,2,\ldots,N$,
(14i) $K'(x,y) = \mu K(\mu x, \mu y)$, $x<y$.

Consider next the function

(15a) $u'(x) = u(-x)$

that obtains from a solvable potential by reflection with respect to the origin; then also this new potential is solvable, and the reader can easily verify that

(15b) $f^{(\pm)'}(x,k) = f^{(\mp)}(-x, k)$,
(15c) $T'(k) = T(k)$,
(15d) $R'(k) = -R(-k)[T(k)/T(-k)]$,
(15e) $p'_n = p_n$, $n=1,2,\ldots,N$,
(15f) $\varphi'_n(x) = \varepsilon_n \varphi_n(-x)$, $\varepsilon_n^2 = 1$, $n=1,2,\ldots,N$,
(15g) $c_n^{(\pm)'} = \varepsilon_n c_n^{(\mp)}$, $\varepsilon_n^2 = 1$, $n=1,2,\ldots,N$,
(15h) $\rho'_n = c_n^{(-)2}$, $n=1,2,\ldots,N$.

An elementary method of constructing a new solvable potential $u(x)$ from M known solvable potentials $u^{(j)}(x)$, $j=1,2,\ldots,M$, consists in dividing the x-axis in M subintervals by giving $M-1$ real numbers x_j, with the ordering

(16) $\quad -\infty < x_1 < x_2 < \cdots < x_{M-1} < +\infty$,

and setting ($\theta(x)=0$ if $x<0$, $\theta(x)=1$ if $x>0$)

(17a) $\quad u(x) = u^{(1)}(x)\theta(x_1-x) + \sum_{j=1}^{M-2} u^{(j+1)}(x)[\theta(x-x_j)-\theta(x-x_{j+1})]$

$\qquad + u^{(M)}(x)\theta(x-x_{M-1})$

or, equivalently,

(17b) $\quad u(x) = u^{(1)}(x), \quad x<x_1;$

$\qquad u(x) = u^{(j)}(x), \quad x_{j-1} < x < x_j, \quad j=2,3,\ldots,M-1;$

$\qquad u(x) = u^{(M)}(x), \quad x > x_M.$

In fact the Jost solutions $f^{(\pm)}(x,k)$ of (1), with the potential (17), if x is in the subinterval where $u(x)$ coincides with $u^{(j)}(x)$, are obviously given by linear combinations of the two (known) Jost solutions associated with $u^{(j)}(x)$; the coefficients of these linear combinations are then adjusted so that $f^{(\pm)}(x,k)$ satisfy the boundary condition (5), and be continuous, together with their first derivatives, at each point x_j. Once the Jost solutions $f^{(\pm)}(x,k)$ have been obtained, all the relevant spectral quantities can be derived from them via the standard formulae (6), (8), (9) and (10). Note that in the simplest case, $M=2$ and, f.i., $u^{(2)}(x)=0$ (this being indeed a solvable potential!), the new potential obtains by cutting off that part of $u^{(1)}(x)$ that extends from x_1 to $+\infty$.

As an example, consider the case $M=2$ with $x_1=0$ so that

(18a) $\quad u(x) = u^{(1)}(x)\theta(-x) + u^{(2)}(x)\theta(x),$

(18b) $\quad u(x) = u^{(1)}(x), \quad x<0; \quad u(x) = u^{(2)}(x), \quad x>0,$

$u^{(1)}(x)$ and $u^{(2)}(x)$ being two solvable potentials; it is then a trivial exercise to verify that

(18c) $\quad f^{(-)}(x,k) = f^{(-)(1)}(x,k)\theta(-x)$

$\qquad + [f^{(+)(2)}(x,-k) + R(k)f^{(+)(2)}(x,k)]\theta(x)/T(k),$

with

(18d) $\quad T(k)=2\mathrm{i}k\bigl[f^{(-)(1)}(0,k)f_x^{(+)(2)}(0,k)$

$\qquad -f^{(+)(2)}(0,k)f_x^{(-)(1)}(0,k)\bigr]^{-1}$, $\quad \mathrm{Im}\, k\geqslant 0$,

(18e) $\quad R(k)=\bigl[f^{(+)(2)}(0,-k)f_x^{(+)(1)}(0,k)-f^{(+)(1)}(0,k)f_x^{(+)(2)}(0,-k)\bigr]$

$\qquad \cdot \bigl[f^{(-)(1)}(0,k)f_x^{(+)(2)}(0,k)-f^{(+)(2)}(0,k)f_x^{(-)(1)}(0,k)\bigr]^{-1}$,

$\qquad\qquad\qquad\qquad\qquad\qquad\qquad\qquad \mathrm{Im}\, k=0$.

Here, of course, the transmission coefficient (18d), the reflection coefficient (18e) and the Jost solution (18c) correspond to the potential (18a) or (18b), while $f^{(\pm)(j)}(x,k)$ are the Jost solutions corresponding to $u^{(j)}(x)$, $j=1,2$.

A different, and less trivial, method to construct a new solvable potential $u'(x)$ from a known one, say $u(x)$, is provided by the Bäcklund transformation

(19) $\quad u'(x)=u(x)-2\bigl[\ln f(x)\bigr]_{xx}$,

where $f(x)$ is any solution of the Schroedinger equation

(20) $\quad f_{xx}(x)=\bigl[p^2+u(x)\bigr]f(x)$,

p being here a given (real) number. Since this transformation is discussed in great detail in appendix A.12, we limit ourselves here to this brief mention, and refer the reader to that appendix. There the reader can also find the Darboux transformation corresponding to (19), that yields the explicit expression of the Jost solutions associated with the new potential $u'(x)$, together with all the corresponding relevant spectral quantities, including the spectral transform of $u'(x)$.

The techniques we have considered so far to construct a new solvable potential from the knowledge of one, or more, given solvable potentials are all based on the possibility to integrate for the new potential the Schroedinger equation (1). We now consider instead the possibility to construct a new solvable potential from a known one, in the context of the inverse problem; the basic equations are derived in appendix A.5, and are tersely reported below to the extent relevant to the present analysis.

Assume that $u(x)$ is a solvable potential, its spectral transform being given by (2); and let

(21) $\quad S[u']=\{R(k),\ -\infty<k<+\infty;\ p'_n,\ \rho'_n,\ n=1,2,\ldots,N'\}$

be the spectral transform of a second potential $u'(x)$ (note that $u'(x)$ is assumed to have the same reflection coefficient of $u(x)$, while p'_n and ρ'_n are arbitrarily given positive numbers). Then $u'(x)$ can be computed through the following steps: first solve the GLM integral equation (3), with

$$(22) \quad M(x,y) = \sum_{n=1}^{N'} \rho'_n f^{(+)}(x, i p'_n) f^{(+)}(y, i p'_n)$$
$$- \sum_{n=1}^{N} \rho_n f^{(+)}(x, i p_n) f^{(+)}(y, i p_n),$$

and then use the formula

$$(23) \quad u'(x) = u(x) - 2 \, dK(x, x+0)/dx.$$

The feasibility of this construction of $u'(x)$ relies of course on the fact that the kernel (22) of the Fredholm integral equation (3) is separable of rank $N+N'$, and therefore this equation can be reduced to an algebraic system of $N+N'$ linear equations (an explicit example of solution of (3) in a analogous case is given below). Once the solution $K(x,y)$ of (3) is obtained, the Jost solution of "plus" type associated with the new potential $u'(x)$ is given by the formula

$$(24) \quad f^{(+)'}(x,k) = f^{(+)}(x,k) + \int_x^{+\infty} dy\, K(x,y) f^{(+)}(y,k),$$

$$\text{Im } k \geq 0,$$

and from this expression one can then recover also the transmission coefficient $T'(k)$ via the asymptotic formula

$$(25) \quad f^{(+)'}(x,k) \underset{x\to-\infty}{\to} [1/T'(k)] \exp(ikx)$$
$$- [R(-k)/T'(-k)] \exp(-ikx),$$

as well as the normalized eigenfunctions

$$(26) \quad \varphi'_n(x) = (\rho'_n)^{1/2} f^{(+)'}(x, i p'_n), \quad n = 1, 2, \ldots, N'.$$

Leaving a more general treatment of this approach to the reader, we only display here the explicit expressions that obtain in the simple case with

$$(27) \quad p'_n = p_n, \quad n = 1, 2, \ldots, N' = N,$$
$$(28) \quad \rho'_n = \rho_n, \quad n = 2, 3, \ldots, N,$$
$$(29) \quad \rho_1 \equiv \rho, \quad \rho'_1 = \rho + \Delta\rho.$$

(Note that $\Delta\rho$ need not be small). In this case the kernel (22) becomes simply

(30) $\quad M(x, y) = \Delta\rho f^{(+)}(x, ip) f^{(+)}(y, ip),$

and the solution of (3) is easily found:

(31) $\quad K(x, y) = -(\Delta\rho/\rho)\varphi(x)\varphi(y)\left\{1+(\Delta\rho/\rho)\int_x^{+\infty} dy \left[\varphi(y)\right]^2\right\}^{-1},$

where

(32) $\quad \varphi(x) = \varphi_1(x) = \rho^{1/2} f^{(+)}(x, ip)$

is the normalized eigenfunction corresponding to $p_1 = p$. Therefore the new potential reads

(33a) $\quad u'(x) = u(x) - 2\left\{\ln\left[1+(\Delta\rho/\rho)\int_x^{+\infty} dy\left[\varphi(y)\right]^2\right]\right\}_{xx},$

while the expression (24) of the corresponding Jost solution is

(33b) $\quad f^{(+)\prime}(x, k) = f^{(+)}(x, k) - (\Delta\rho/\rho)\varphi(x)\int_x^{+\infty} dy\, \varphi(y) f^{(+)}(y, k)$

$\quad \cdot \left\{1+(\Delta\rho/\rho)\int_x^{+\infty} dy\left[\varphi(y)\right]^2\right\}^{-1}, \quad \text{Im}\, k \geq 0.$

This formula readily shows that

(33c) $\quad T'(k) = T(k),$

a result that was also directly implied by the results of appendix A.4. Moreover, since of course (see (29))

(33d) $\quad c_1^{(+)\prime} = (\rho + \Delta\rho)^{1/2},$

there clearly obtains (see (9) and (33c))

(33e) $\quad c_1^{(-)\prime} = c_1^{(-)}(1+\Delta\rho/\rho)^{-1/2}.$

Finally (33b) yields also the expression of the normalized eigenfunction, that reads

(33f) $\quad \varphi_1'(x) = (1+\Delta\rho/\rho)^{1/2}\varphi(x)\left\{1+(\Delta\rho/\rho)\int_x^{+\infty} dy\left[\varphi(y)\right]^2\right\}^{-1}.$

We proceed now to exhibit a few examples of solvable potentials. We start with the *Dirac distribution*

(34a) $\quad u(x) = 2g\delta(x-x_0),$

where g and x_0 are arbitrary real parameters. Note that the parameter x_0 could be set equal to zero and then reinstated at the end by using the translation (13a) with $a = -x_0$, and also the modulus of g could be replaced by unity and then reinserted via an appropriate scale transformation (14a). In any case, with the potential (34a), (1) can be easily integrated, yielding the Jost solutions (see (5))

(34b) $\quad f^{(\pm)}(x,k) = \big(1 + (ig/k)\{1 - \exp[\pm 2ik(x_0-x)]\}$
$\cdot \theta[\pm(x_0-x)]\big)\exp(\pm ikx);$

here $\theta(x)$ is the step-function, i.e. $\theta(x) = 1$ if $x > 0$, $\theta(x) = 0$ if $x < 0$. This expression readily implies (see (6)) that the reflection and transmission coefficients are

(34c) $\quad R(k) = [g/(ik-g)]\exp(-2ikx_0),$
(34d) $\quad T(k) = ik/(ik-g).$

As for the spectral transform of (34a), the two cases, $g > 0$ and $g < 0$, must be discussed separately. Indeed if $g > 0$ ((34a) being therefore a "repulsive" potential), no discrete eigenvalues occur since $T(k)$ (see (34d)) has no poles in the upper half of the k-plane; in this case then

(34e) $\quad S[u] = \{R(k), -\infty < k < +\infty; N = 0\}$

is the spectral transform of (34a), with $R(k)$ given by (34c).

In the other case, i.e. $g < 0$ ((34a) being then an "attractive" potential), the spectral problem (1) possesses one discrete eigenvalue for $k = ip_1$ with

(34f) $\quad p_1 = -g.$

The corresponding (normalized) eigenfunction is

(34g) $\quad \varphi_1(x) = p_1^{1/2}\{\theta(x-x_0)\exp[-p_1(x-x_0)]$
$+ \theta(x_0-x)\exp[p_1(x-x_0)]\},$

while the asymptotic constants $c_1^{(\pm)}$ (see (8) and (5)) read

(34h) $\quad c_1^{(\pm)} = p_1^{1/2}\exp(\pm p_1 x_0)$

(to derive (34g,h) use (8), (9) and (34d)). In this case ($g<0$), the spectral transform of (34a) has also a discrete component, and its expression is

(34i) $\quad S[u]=\{R(k),\ -\infty<k<+\infty;p_1,\rho_1,N=1\}$,

with (34c,f) and (see (10))

(34j) $\quad \rho_1 = c_1^{(+)^2} = -g\exp(-2gx_0)$.

Our second example is the *"rectangular"* function

(35a) $\quad u(x)=a[\theta(x-x_1)-\theta(x-x_2)], \quad x_1<x_2,$
(35b) $\quad u(x)=a, \quad x_1<x<x_2; \quad u(x)=0, \quad x<x_1 \text{ or } x>x_2,$

a, x_1 and x_2 being real parameters. Also in this case the integration of (1) is an elementary exercise, and we merely report below the relevant formulae. The Jost solutions read

(35c) $\quad f^{(+)}(x,k)=\theta(x-x_2)\exp(ikx)+[\theta(x-x_1)-\theta(x-x_2)]$
$\cdot \exp(ikx_2)\{\cos[q(x-x_2)]+(ik/q)\sin[q(x-x_2)]\}+\theta(x_1-x)$
$\cdot \exp(ikx_2)\sin\chi\{[(2k^2-a)/(2ikq)+\cotg\chi]\exp[ik(x-x_1)]$
$+[a/(2ikq)]\exp[-ik(x-x_1)]\},$

(35d) $\quad f^{(-)}(x,k)=\theta(x_1-x)\exp(-ikx)+[\theta(x-x_1)-\theta(x-x_2)]$
$\cdot \exp(-ikx_1)\{\cos[q(x-x_1)]-(ik/q)\sin[q(x-x_1)]\}$
$+\theta(x-x_2)\exp(-ikx_1)\sin\chi\{[a/(2ikq)]\exp[ik(x-x_2)]$
$+[(2k^2-a)/(2ikq)+\cotg\chi]\exp[-ik(x-x_2)]\},$

where

(36) $\quad q\equiv q(k)=(k^2-a)^{1/2}$,
(37) $\quad \chi\equiv\chi(k)=q(k)(x_2-x_1)$.

From these expressions the reflection and transmission coefficients are easily found:

(35e) $\quad R(k)=[a/(2ikq)]\{(2k^2-a)/(2ikq)+\cotg\chi\}^{-1}\exp(-2ikx_2)$,
(35f) $\quad T(k)=\{[(2k^2-a)/(2ikq)]\sin\chi+\cos\chi\}^{-1}\exp[-ik(x_2-x_1)]$;

the reader may verify that they satisfy the unitarity condition

(38) $R(k)R(-k) + T(k)T(-k) = 1.$

From the analyticity properties of $T(k)$ in the upper half-plane, it is easily seen that no discrete eigenvalues occur for $a > 0$ (this being the case of a repulsive "square barrier" potential). Thus in this case ($a > 0$) the spectral transform of (35a) or (35b) has only its continuous component, namely it is given by (34e) with (35e), (36) and (37).

If instead $a < 0$ (this being the case of an attractive "square well" potential), a discrete spectrum is present, the discrete eigenvalues of (1) being given by the poles of the transmission coefficient (35f) on the positive imaginary axis, i.e. for $k = ip_n$, $p_n > 0$, $n = 1, 2, \ldots, N$, where the numbers p_n are implicitly defined by the equation

(35g) $\cotg \chi_n = (\mu^2 - 2p_n^2)/(2p_n q_n), \quad n = 1, 2, \ldots, N;$

here, consistently with the definitions (36) and (37), we have set

(39) $q_n = (\mu^2 - p_n^2)^{1/2}, \quad \chi_n = q_n(x_2 - x_1), \quad n = 1, 2, \ldots, N,$

with

(40) $a = -\mu^2,$

μ being a (conventionally positive; see below) real number. Indeed, it is easily seen that the poles of $T(k)$ in the upper half-plane can occur only in the interval of the imaginary axis $\mathrm{Re}\, k = 0$, $0 < \mathrm{Im}\, k < \mu$, and that their number N (namely, the number of solutions p_n of (35g)) is fixed by the inequalities

(35h) $(N-1)\pi < \mu(x_2 - x_1) < N\pi.$

(We neglect the marginal case when $\mu(x_2 - x_1)/\pi$ equals an integer, corresponding to the occurrence of a "zero-energy bound state", in the terminology of appendix A.15).

The normalized eigenfunction corresponding to p_n takes the expression

(35i) $\varphi_n(x) = (q_n/\mu)\{2p_n/[2 + p_n(x_2 - x_1)]\}^{1/2}$
$\cdot \{\theta(x_1 - x)\exp[p_n(x - x_1)] + [\theta(x - x_1) - \theta(x - x_2)]$
$\cdot (\cos[q_n(x - x_1)] + (p_n/q_n)\sin[q_n(x - x_1)])$
$+ \theta(x - x_2)\exp[-p(x - x_2)]\},$

from which the asymptotic constants (see (8)) are easily found to be

(35j) $\quad c_n^{(+)} = [\mu \exp(-p_n x_2)]^{-1} q_n \{2 p_n / [2 + p_n(x_2 - x_1)]\}^{1/2},$

(35k) $\quad c_n^{(-)} = [\mu \exp(p_n x_1)]^{-1} q_n \{2 p_n / [2 + p_n(x_2 - x_1)]\}^{1/2}.$

Finally, the spectral transform of (35a) or (35b) with (40) is given by (2) with the reflection coefficient given by (35e), while the N spectral parameters p_n's are defined by (35g) (with (39)); N is fixed by (35h) and the normalization constants are given by the formula

(35l) $\quad \rho_n = \exp(2 p_n x_2)(2 p_n / \mu^2)(\mu^2 - p_n^2) / [2 + p_n(x_2 - x_1)],$

$$n = 1, 2, \ldots, N.$$

The third potential we consider is the *exponential function*

(41a) $\quad u(x) = \gamma \exp(-2|x|) = \gamma [\theta(-x) \exp(2x) + \theta(x) \exp(-2x)].$

Two (real) parameters can of course be added by translating and rescaling the variable x (see (13) and (14)). The way to solve (1) with (41a) is via the change of variables

(42) $\quad z = (-\gamma)^{1/2} \exp(\varepsilon x), \quad \varepsilon = \pm 1,$

(43) $\quad y(z) = f(x, k),$

(44) $\quad k = i\nu;$

in fact, the ODE

(45) $\quad f_{xx} + [k^2 - \gamma \exp(2\varepsilon x)] f = 0,$

that coincides with (1) (with $\varepsilon = +1$ for $x < 0$, $\varepsilon = -1$ for $x > 0$), is thereby transformed into the ODE

(46) $\quad z^2 y_{zz} + z y_z + (z^2 - \nu^2) y = 0,$

whose solutions are the Bessel functions $J_\nu(z)$ and $J_{-\nu}(z)$ (these being linearly independent if $\nu \neq -n$, $n = 1, 2, \ldots$). Thus the general solution of (1) is known for $x > 0$, setting $\varepsilon = -1$ in (42), and for $x < 0$, setting $\varepsilon = +1$ in (42). The Jost solutions are then obtained by requiring them to be continuous, together with their first derivatives, at $x = 0$, and by imposing the boundary conditions (5); this last step is easily performed by noticing that the transformation (42) maps the point at infinity, i.e. $x = \pm \infty$, into the origin, i.e. $z=0$, where the behaviour of the Bessel function is given by the asymptotic formula (see e.g. [AS1965]).

(47) $\quad J_\nu(z)\approx(\tfrac{1}{2}z)^\nu/\Gamma(1+\nu), \quad \nu\neq -n, \quad n=1,2,\ldots$.

Here and below $\Gamma(z)$ is the gamma function. The expression that finally obtains reads

(41b) $\quad f^{(\pm)}(x,k)=(\tfrac{1}{2}c)^{ik}\Gamma(1-ik)$
$\cdot\{\theta(\pm x)J_{-ik}[c\exp(\mp x)]+\{c\pi/[2\sin(i\pi k)]\}$
$\cdot\theta(\mp x)[(J_{ik}(c)J'_{-ik}(c)+J_{-ik}(c)J'_{ik}(c))J_{-ik}[c\exp(\pm x)]$
$-2J_{-ik}(c)J'_{-ik}(c)J_{ik}[c\exp(\pm x)]]\}$.

Here and below the prime indicates differentiation with respect to the argument, and

(48) $\quad c\equiv(-\gamma)^{1/2}$.

Note that the parity transformation (15b), together with the property of the function (41a) to be even, implies that

(49) $\quad f^{(\pm)}(x,k)=f^{(+)}(\pm x,k)$,

that is clearly verified by the expression (41b). The expressions of the transmission and reflection coefficients then follow from (6), and read

(41c) $\quad T(k)=-(\tfrac{1}{2}c)^{-2ik}\sin(i\pi k)[\pi c J_{-ik}(c)J'_{-ik}(c)]^{-1}$
$\cdot[\Gamma(1+ik)/\Gamma(1-ik)]$,

(41d) $\quad R(k)=-(\tfrac{1}{2}c)^{-2ik}\tfrac{1}{2}\{[J_{ik}(c)/J_{-ik}(c)]+[J'_{ik}(c)/J'_{-ik}(c)]\}$
$\cdot[\Gamma(1+ik)/\Gamma(1-ik)]$.

As for the discrete spectrum, we discuss separately the two cases $\gamma>0$ and $\gamma<0$. In the first case, $\gamma>0$ ((41a) being a "repulsive potential"), $T(k)$ has no poles in the upper half-plane; in fact (see (41c)) the poles of $\Gamma(1+ik)$ at $k=in$, $n=1,2,\ldots$, cancel out with the zeros of $\sin(i\pi k)$, and $J_p(c)$ and $J'_p(c)$ vanish, for $p\geq 0$, only for real values of c, while, in the present case, c is purely imaginary (see (48)). We conclude therefore that, if $\gamma>0$, the spectral transform of the function (41a) is (34e), with the reflection coefficient given by (41d). In the other case, $\gamma<0$ ((41a) being then an "attractive potential"), the parameter c is real (see (48)), and $T(k)$ possesses the simple poles $k=ip_n$, $p_n>0$, $n=1,2,\ldots,N$, that come from the vanishing of $J_p(c)$, or of $J'_p(c)$, namely from the solutions of

(41e) $\quad J_{p_n}(c)J'_{p_n}(c)=0, \quad n=1,2,\ldots,N, \qquad p_n>0.$

The expressions of the spectral parameters associated with these eigenvalues can be derived as usual from the formulae (8) and (9), together with (41b), and are

(41f) $\quad c_n^{(+)}=c_n^{(-)}=\left(\tfrac{1}{2}c\right)^{p_n}\left\{-\left[J'_{-p_n}(c)/\tilde{J}'_{p_n}(c)\right]\right.$
$\left.\cdot\left[\Gamma(1-p_n)/\Gamma(1+p_n)\right]\right\}^{1/2} \quad$ if $J'_{p_n}(c)=0,$

(41g) $\quad c_n^{(+)}=-c_n^{(-)}=\left(\tfrac{1}{2}c\right)^{p_n}\left\{-\left[J_{-p_n}(c)/\tilde{J}_{p_n}(c)\right]\right.$
$\left.\cdot\left[\Gamma(1-p_n)/\Gamma(1+p_n)\right]\right\}^{1/2} \quad$ if $J_{p_n}(c)=0;$

here we use the notation

(50) $\quad \tilde{J}_\nu(z)\equiv\partial J_\nu(z)/\partial\nu, \qquad \tilde{J}'_\nu(z)\equiv\partial^2 J_\nu(z)/\partial\nu\,\partial z.$

Finally, the normalized eigenfunctions take the following expressions (see (8) and (41b)):

(41h) $\quad \varphi_n(x)=c_n^{(+)}\left(\tfrac{1}{2}c\right)^{-p_n}\Gamma(1+p_n)$
$\cdot\left\{\theta(x)J_{p_n}[c\exp(-x)]-\left[c\pi J_{p_n}(c)J'_{-p_n}(c)/(2\sin(\pi p_n))\right]\right.$
$\left.\cdot\theta(-x)J_{p_n}[c\exp(x)]\right\} \quad$ if $J'_{p_n}(c)=0,$

with $c_n^{(+)}$ given by (41f), and

(41i) $\quad \varphi_n(x)=c_n^{(+)}\left(\tfrac{1}{2}c\right)^{-p_n}\Gamma(1+p_n)\left\{\theta(x)J_{p_n}[c\exp(-x)]\right.$
$\left.-\left[c\pi J'_{p_n}(c)J_{-p_n}(c)/(2\sin(\pi p_n))\right]\theta(-x)J_{p_n}[c\exp(x)]\right\}$
if $J_{p_n}(c)=0,$

with $c_n^{(+)}$ given now by (41g). Thus, in this case ($\gamma<0$), the spectral transform of (41a) is given by (2) with (41d), with the p_n's (implicitly) defined by (41e) and the ρ_n's given by (10) with (41f,g). Note that the marginal case in which an eigenvalue p_n is a (positive) integer requires a separate treatment, that we leave, as an exercise, to the reader.

The next solvable potential is the $\cosh^{-2}$ *function*,

(51a) $\quad u(x) = h/\cosh^2 x.$

Again, the transformation laws (13) and (14) allow the introduction of two arbitrary parameters in (51a), by translating and rescaling the variable x; the form (51a) is hereafter used, for simplicity. Incidentally, note that for $h = -2$ the function (51a) is (up to rescaling and translating) the well-known one-soliton expression (see (2.2.-10) and also below).

The solvability of (1) with (51a) originates from the possibility to transform it into the hypergeometric ODE

(52) $\quad z(1-z)y_{zz} + [c - (1+a+b)z]y_z - aby = 0,$

satisfied by the hypergeometric function $F(a, b; c; z)$ (see, e.g., [AS1965]). The transformation connecting (1) with (52) reads:

(53) $\quad f(x, k) = (\cosh x)^{ik} y(z),$

(54) $\quad z = \frac{1}{2}(1 - \tgh x) = [1 + \exp(2x)]^{-1},$

(55) $\quad a \equiv a(k) = \frac{1}{2} - ik + r,$

(56) $\quad b \equiv b(k) = \frac{1}{2} - ik - r,$

(57) $\quad r \equiv (\frac{1}{4} - h)^{1/2},$

(58) $\quad c \equiv c(k) = 1 - ik.$

To verify these, and the following formulae, it is useful to notice that

(59) $\quad 1 + a + b = 2c.$

The expression of the Jost solution $f^{(+)}$ is readily obtained by noticing that (54) maps the point $x = +\infty$ into the origin $z = 0$, and that $F(a, b; c; 0) = 1$; therefore (see (5))

(51b) $\quad f^{(+)}(x, k) = [\exp(x) + \exp(-x)]^{ik} F(a, b; c; z),$

of course with a, b, c, and z given by (55), (56), (58) and (54), respectively.

The second Jost solution is more conveniently obtained through the property (49), that certainly holds since the function (51a) is even; the resulting expression is

(51c) $\quad f^{(-)}(x,k) = [\exp(x) + \exp(-x)]^{ik} F(a,b;c;1-z).$

In order to recover from these expressions the transmission and reflection coefficients, we make use of the following property of the hypergeometric function (see e.g. [AS1965]):

(60) $\quad F(a,b;a+b+1-c;1-z)$
$= \{\Gamma(a+b+1-c)\Gamma(1-c)/[\Gamma(b+1-c)\Gamma(a+1-c)]\}$
$\cdot F(a,b;c;z) + z^{1-c}\{\Gamma(a+b+1-c)\Gamma(c-1)/[\Gamma(a)\Gamma(b)]\}$
$\cdot F(b+1-c, a+1-c; 2-c; z), \quad |\arg z| < \pi,$

that implies

(61) $\quad f^{(-)}(x,k) = \{\Gamma(1-ik)\Gamma(ik)/[\Gamma(\tfrac{1}{2}+r)\Gamma(\tfrac{1}{2}-r)]\} f^{(+)}(x,k)$
$+ \{\Gamma(1-ik)\Gamma(-ik)/[\Gamma(\tfrac{1}{2}+r-ik)\Gamma(\tfrac{1}{2}-r-ik)]\}$
$\cdot f^{(+)}(x,-k).$

Comparing this relation with (6) there obtains

(51d) $\quad T(k) = \Gamma(\tfrac{1}{2}+r-ik)\Gamma(\tfrac{1}{2}-r-ik)/[\Gamma(1-ik)\Gamma(-ik)],$
(51e) $\quad R(k) = \pi^{-1}\cos(\pi r)\Gamma(\tfrac{1}{2}+r-ik)\Gamma(\tfrac{1}{2}-r-ik)\Gamma(ik)/\Gamma(-ik);$

note that we have also used the property

(62) $\quad \Gamma(z)\Gamma(1-z) = \pi/\sin(\pi z).$

The poles of the transmission coefficient are the poles of the gamma functions appearing in the numerator of the r.h.s. of (51d), since the function $1/\Gamma(z)$ is entire. It is easy to verify that, if $h > 0$ (so that (51a) is a "repulsive potential"), all poles of $T(k)$ occur in the lower half-plane, implying that there are no discrete eigenvalues; the spectral transform of (51a) reduces then to (34e) with (51e). If instead h is negative (so that (51a) is an "attractive potential"), say

(63) $\quad h = -\mu^2,$

μ being of course real, the poles of $T(k)$ are located at $k = ip_n$ with

(51f) $\quad p_n = \tfrac{1}{2} + (\tfrac{1}{4} + \mu^2)^{1/2} - n, \quad n = 1, 2, \ldots, N;$

therefore the total number N of poles on the upper imaginary axis (corresponding to discrete eigenvalues) is given by the inequalities

(51g) $\quad N(N-1) < \mu^2 < N(N+1)$

(the special values $\mu^2 = m(m+1)$, $m = 1, 2, 3, \ldots$, are considered below).

As for the normalization parameters associated with the discrete eigenvalues, their expression can be derived as usual from (8), (9) and (51b,c); there obtains

(51h) $\quad c_n^{(+)} = (-)^{n+1} c_n^{(-)} = \{\Gamma(2p_n + n)/[\Gamma(n)\Gamma(p_n)\Gamma(1+p_n)]\}^{1/2},$

$$n = 1, 2, \ldots, N,$$

with p_n given by (51f). In the same way one derives for the normalized eigenfunctions the expression

(51i) $\quad \varphi_n(x) = c_n^{(+)} [\exp(x) + \exp(-x)]^{-p_n} F(2p_n + n, 1 - n; 1 + p_n; z),$

$$n = 1, 2, \ldots, N,$$

with $c_n^{(+)}$ given of course by (51h). Note that, using the fact that the hypergeometric function $F(a, 1-m; c; z)$ becomes a polynomial in z if m is a positive integer, (51i) may be rewritten in the more explicit form

(51j) $\quad \varphi_n(x) = c_n^{(+)} [\exp(x) + \exp(-x)]^{-p_n}$

$$\cdot \sum_{j=0}^{n-1} (1-n)_j (2p_n + n)_j / [j!(1+p_n)_j] [1 + \exp(2x)]^{-j},$$

with the conventional notation

(64) $\quad (z)_0 = 1, \quad (z)_n \equiv \Gamma(z+n)/\Gamma(z) = \prod_{j=1}^{n} (z+j-1), \quad n = 1, 2, \ldots .$

Note that $\varphi_n(x)$ satisfies the simple parity property

(65) $\quad \varphi_n(x) = (-1)^{n+1} \varphi_n(-x),$

consistently with the invariance of the potential (51a) under reflection (see

(15f)). In conclusion, the spectral transform of the function (51a) with (63), is given by (2) with the reflection coefficient (51e), the eigenvalues (51f) and the normalization coefficients

(51k) $\quad \rho_n = \Gamma(2p_n+n)/[\Gamma(n)\Gamma(p_n)\Gamma(1+p_n)], \quad n=1,2,\ldots,N.$

We close this discussion pointing out that the k-independent cosine factor in (51e) implies that for the special values

(66) $\quad \mu^2 = \mu_m^2 \equiv m(m+1), \quad m=1,2,\ldots,$

the potential (51a), namely

(67a) $\quad u(x) = -m(m+1)/\cosh^2 x, \quad m=1,2,\ldots,$

is reflectionless,

(67b) $\quad R(k)=0.$

In this case the number of discrete eigenvalues is simply m,

(67c) $\quad N=m,$

and the values of the discrete eigenvalues are given by the neat rule

(67d) $\quad p_n = n, \quad n=1,2,\ldots,m.$

There is, moreover, a "zero-energy bound state" (in the sense of appendix A.15; indeed the rule $R(0) = -1$ is violated, see (67b)). The transmission coefficient takes in this case the expression

(67e) $\quad T(k) = \prod_{n=1}^{m} [(k+in)/(k-in)],$

while the normalization coefficients that enter in the spectral transform of (67a) read

(67f) $\quad \rho_n = (m+n)!/[(n-1)!n!(m-n)!], \quad n=1,2,\ldots,m.$

We leave to the interested reader the simple exercise to specialize to this case the expressions of the Jost solutions and of the normalized eigenfunctions reported above.

The preceding examples of solvable potentials have been discussed in the context of the direct problem, namely by considering cases in which the solutions of the Schroedinger equation (1) can be expressed in terms of known (elementary or special) functions. Now we take the complementary approach, namely we consider cases in which the GLM integral equation (3) can be explicitly solved (for an appropriately given spectral transform (2)).

We consider first the class of *reflectionless potentials*; they are also treated in section 2.2, and the relevant formulae are tersely reported here merely for the sake of completeness. Their spectral transform,

(68a) $\quad S[u] = \{R(k) = 0, -\infty < k < +\infty; p_n, \rho_n, n = 1, 2, \ldots, N\},$

depends on $2N$ positive parameters, p_n and c_n, $n = 1, 2, \ldots, N$, with

(68b) $\quad \rho_n = c_n^2, \quad n = 1, 2, \ldots, N.$

The GLM equation (3) reduces to an algebraic linear system of N equations, whose solution can be expressed as follows. Let $C(x)$ be the $N \times N$ symmetrical matrix whose elements are

(69) $\quad C_{mn}(x) = c_m c_n (p_m + p_n)^{-1} \exp[-(p_m + p_n)x],$

and let I be the $N \times N$ unit matrix; then the function associated with the spectral transform (68a) reads

(68c) $\quad u(x) = u_N(x) \equiv -2 \mathrm{d}^2 \{\ln \det[I + C(x)]\} / \mathrm{d}x^2.$

Indeed, the Schroedinger equation (1) with (68c) possesses N discrete eigenvalues $k = \mathrm{i} p_n$, whose corresponding eigenfunctions (that satisfy the normalization condition (7)) are

(68d) $\quad \varphi_n(x) = \sum_{m=1}^{N} A_{nm}(x) c_m \exp(-p_m x), \quad n = 1, 2, \ldots, N,$

where the functions $A_{nm}(x)$'s are the elements of the (symmetrical $N \times N$) matrix

(70) $\quad A(x) = [I + C(x)]^{-1}.$

The solution of the GLM equation (3) yields moreover, through (12), the

Jost solution $f^{(+)}(x,k)$, that reads

(68e) $$f^{(+)}(x,k)=\exp(ikx)\left\{1-\sum_{m,n=1}^{N}[c_n c_m/(p_n-ik)]\right.$$
$$\left.\cdot\exp[-(p_n+p_m)x]A_{nm}(x)\right\}.$$

While of course the reflection coefficient $R(k)$ corresponding to (68c) vanishes, the transmission coefficient, consistently with the unitarity equation (38), takes the simple expression

(68f) $$T(k)=\prod_{n=1}^{N}[(k+ip_n)/(k-ip_n)].$$

Note that (6), together with this last formula, yields immediately the second Jost solution, i.e.

(68g) $$f^{(-)}(x,k)=f^{(+)}(x,-k)\prod_{n=1}^{N}[(k-ip_n)/(k+ip_n)].$$

Finally the constants $c_n^{(\pm)}$ characterizing the asymptotic behaviour of the eigenfunction (68d) (see (8) and (5)) are given by the formulae

(68h) $$c_n^{(+)}=c_n,$$

(68i) $$c_n^{(-)}=2(p_n/c_n)\prod_{\substack{m=1\\m\neq n}}^{N}[(p_n+p_m)/(p_n-p_m)].$$

As a final remark, we note that the class of reflectionless potentials (68c) has been discovered by Bargmann [B1949] before the theory of the inverse problem was known. In the Bargmann approach the starting point was the ansatz

(71) $$f(x,k)=\exp(ikx)P_N(x,k)$$

for the solution of (1), where $P_N(x,k)$ is a polynomial of degree N in k with x-dependent coefficients; insertion of (71) in (1) yields $N+2$ differential equations for the $N+1$ coefficients of P_N and the unknown function $u(x)$. Once these equations have been integrated (a task that is however not so simple for $N \geq 2$) the Jost solutions (see (5)) are clearly given by the expression

(72) $\quad f^{(\pm)}(x, k) = \exp(\pm ikx) P_N(x, \pm k)/P_N(\pm \infty, \pm k).$

Because of this special dependence of the Jost solutions on k, it is a priori evident that the reflection coefficient $R(k)$ corresponding to such potentials must vanish (see (6)). For instance, for $N=0$, $P_0(x,k)=1$ and the corresponding potential vanishes, $u(x)=0$; for $N=1$

(73) $\quad P_1(x,k) = -2p\,\text{tgh}[p(x-\xi)] + 2ik,$

and, correspondingly,

(74) $\quad u_1(x) = -2p^2/\cosh^2[p(x-\xi)],$

(p and ξ enter in (73) as integration constants). It is clearly seen that (74) is just the potential (68c) for $N=1$ with

(75) $\quad p_1 = p, \quad c_1^2 = 2p\exp(2p\xi).$

In fact, the function (68c) with arbitrary N obtains in the Bargmann approach by using (71), N being here the degree of the polynomial P_N.

Finally we treat tersely the class of *potentials with rational reflection coefficients*, namely we assume that the reflection coefficient $R(k)$ is a rational function of k, and evaluate the corresponding potential $u(x)$. We assume moreover that there are no discrete eigenvalues; this restriction is not essential but it simplifies the presentation (its elimination is left as an instructive exercise).

Let us start by imposing on the reflection coefficient the following conditions:

(76a) $\quad R(\infty) = 0,$
(76b) $\quad |R(k)| < 1 \quad \text{for Im } k = 0,$
(76c) $\quad R(0) = -1$
(76d) $\quad R(k) = R^*(-k^*).$

The first two conditions are necessary in order that a regular $u(x)$ exist (see appendix A.16); the third is assumed in order to make $u(x)$ more generic (see appendix A.15); the last guarantees that $u(x)$ be real. Thus the expression of the reflection coefficient is

(77) $$R(k) = r \prod_{j=1}^{A} (k-a_j) \Big/ \prod_{j=1}^{B} (k-b_j),$$

with

(78) $$A \leq B - 1,$$

(79) $$r \prod_{j=1}^{A} a_j = (-1)^{A+B+1} \prod_{j=1}^{B} b_j,$$

(80) $$r^* \prod_{j=1}^{A} (k+a_j^*) \prod_{j=1}^{B} (k-b_j) = (-1)^{A+B} r \prod_{j=1}^{A} (k-a_j) \prod_{j=1}^{B} (k+b_j^*).$$

These three conditions correspond of course to (76a), (76c) and (76d); as for (76b), it implies the restriction $|r| < \bar{r}$, where $\bar{r}$ is a complicated function of the parameters a_j and b_j, that need not be displayed. Moreover, since $R(k)$ should be regular for real k, its poles must be located away from the real axis, i.e.

(81) $$\operatorname{Im} b_j \neq 0, \quad j = 1, 2, \ldots, B.$$

For future reference, we construct now the transmission coefficient $T(k)$ corresponding to (77). This obtains by combining its analyticity properties with the unitarity equation (38). Indeed (repeating here the argument of appendix A.4), $T(k)$ is analytic in the upper half-plane (as we assume that no discrete eigenvalue occurs), it has no zero there (this being a general property of *bona fide* potentials), and the limit

(82) $$\lim_{k \to \infty} T(k) = 1, \quad \operatorname{Im} k \geq 0,$$

holds; there follows the validity of the following Cauchy integrals,

(83a) $$\ln T(k) = (2\pi i)^{-1} \int_{-\infty}^{+\infty} dq \ln[T(q)] / (q-k), \quad \operatorname{Im} k > 0,$$

(83b) $$0 = (2\pi i)^{-1} \int_{-\infty}^{+\infty} dq \ln[T(q)] / (q+k), \quad \operatorname{Im} k > 0;$$

and, by summing these two equalities, there obtains

(84a) $$\ln T(k) = (2\pi i)^{-1} \int_{-\infty}^{+\infty} dq \ln[1 - R(q)R(-q)] / (q-k),$$

$$\operatorname{Im} k > 0,$$

or, equivalently,

(84b) $\quad \ln T(k) = (2\pi i)^{-1} \int_{-\infty}^{+\infty} dq \ln[1-|R(q)|^2]/(q-k), \quad \operatorname{Im} k > 0.$

Note that, to obtain (84a), the unitarity equation (38) has been used, and to write (84b), the reality condition (76d) has been taken advantage of (incidentally, (84b) also makes apparent the relevance of (76b)).

Let us consider now the function $1 - R(k)R(-k)$; this is of course rational (see (77)), it goes to 1 as $k \to \infty$, and it is even. We conclude that its expression is

(85) $\quad 1 - R(k)R(-k) = \prod_{j=1}^{B} \{[(k-z_j)(k+z_j)]/[(k-s_j)(k+s_j)]\},$

with

(86) $\quad \operatorname{Im} z_j \geq 0, \quad \operatorname{Im} s_j > 0, \quad j = 1, 2, \ldots, B;$

the connection between the zeros z_j's and poles s_j's of (85), and the parameters r, a_j and b_j characterizing the reflection coefficient (77), is given by the equations

(87) $\quad \prod_{j=1}^{B} [(k-s_j)(k+s_j)] = \prod_{j=1}^{B} [(k-b_j)(k+b_j)],$

(88) $\quad \prod_{j=1}^{B} [(k-z_j)(k+z_j)] = \prod_{j=1}^{B} [(k-b_j)(k+b_j)]$

$\quad\quad\quad\quad\quad\quad\quad\quad\quad\quad + (-1)^{A+B+1} r^2 \prod_{j=1}^{A} [(k-a_j)(k+a_j)].$

To compute $T(k)$, it is convenient to eliminate the logarithmic function from (84) by differentiating this equation, writing

(89) $\quad \dfrac{[dT(k)/dk]}{T(k)} = (2\pi i)^{-1} \int_{-\infty}^{+\infty} dq \left\{ \dfrac{d[1-R(q)R(-q)]}{dq} \right\}$

$\quad\quad\quad\quad\quad\quad \cdot \dfrac{[1-R(q)R(-q)]^{-1}}{(q-k)}, \quad \operatorname{Im} k > 0,$

and then to close the integral in the r.h.s. in the lower half-plane. The resulting expression of the transmission coefficient reads

(90) $$T(k)=\prod_{j=1}^{B}\left[(k+z_j)/(k+s_j)\right];$$

furthermore, because of the unitarity condition (38) and of (76c),

(91) $$T(0)=0,$$

so that at least one of the parameters z_j must vanish. Thus (90) reads

(92) $$T(k)=k\prod_{j=2}^{B}(k+z_j)\bigg/\prod_{j=1}^{B}(k+s_j),$$

if we assume conventionally that $z_1=0$.

Let us now turn our attention to the GLM equation (3). Its kernel is readily computed via the Fourier integral appearing in the r.h.s. of (4). By elementary calculations we obtain

(93) $$M(x)=m^{(+)}(x)\theta(x)+m^{(-)}(x)\theta(-x),$$

(94) $$m^{(\pm)}(x)=\sum_{j=1}^{B^{(\pm)}}g_j^{(\pm)}\exp\left(ib_j^{(\pm)}x\right),$$

where $b_j^{(+)}$ ($b_j^{(-)}$) are the $B^{(+)}$ ($B^{(-)}$) poles of $R(k)$ that lie in the upper (lower) half-plane,

(95) $$\pm\operatorname{Im}b_j^{(\pm)}>0,\quad j=1,2,\ldots,B^{(\pm)},\qquad B^{(+)}+B^{(-)}=B,$$

and the constants $g_j^{(\pm)}$ read

(96) $$g_j^{(\pm)}=r\prod_{m=1}^{A}\left(b_j^{(\pm)}-a_m\right)\bigg/\prod_{m=1}^{B}{}'\left(b_j^{(\pm)}-b_m\right),\quad j=1,2,\ldots,B^{(\pm)},$$

where the prime appended to the product symbol in the denominator indicates that the vanishing factor is excluded. Inserting now (93) into (3) it is seen that the GLM equation, for $x\geqslant0$, reduces to

(97) $$K(x,y)+m^{(+)}(x+y)+\int_{x}^{+\infty}dz\,K(x,z)m^{(+)}(z+y)=0,$$

$$y>x\geqslant0,$$

that is separable of rank $B^{(+)}$ (see (94)). On the other hand, as the reader can easily verify, the GLM integral equation in the region $x<0$ is not separable, and we do not attempt to solve it there. As for the solution of (97), this is formally similar to that obtained in the case of the reflectionless potentials discussed above, and it yields eventually the potential

(98) $\quad u(x) = -2\mathrm{d}^2\{\ln\det[I + V^{(+)}(x)]\}/\mathrm{d}x^2, \quad x \geq 0.$

Here the elements of the $B^{(+)} \times B^{(+)}$ matrix $V^{(+)}(x)$ are

(99) $\quad V_{mn}^{(+)}(x) = i\gamma_m^{(+)}\gamma_n^{(+)}\left(b_m^{(+)} + b_n^{(+)}\right)^{-1}\exp\left[i\left(b_m^{(+)} + b_n^{(+)}\right)x\right],$

where we have set

(100) $\quad \gamma_n^{(+)} \equiv \left(g_n^{(+)}\right)^{1/2}.$

Of course, whether the potential so derived is regular for $x \geq 0$ depends on the parameters $\gamma_n^{(+)}$ and $b_n^{(+)}$; further elaboration of this question is left for the diligent reader.

Consider now the problem of constructing the potential corresponding to (77) in the other half of the x-axis, namely for $x<0$. The fact that one is able to solve the GLM equation for $x \geq 0$ whenever the kernel function $M(x)$ has the expression (93) with (94), suggests to perform first a reflection with respect to the origin, thereby introducing the potential

(101) $\quad \bar{u}(x) = u(-x),$

then to compute its corresponding reflection coefficient $\bar{R}(k)$ (that turns out to be also rational, see below), and finally to solve the corresponding GLM equation for $x>0$; this last step will indeed yield the potential $\bar{u}(x)$ for $x>0$, which corresponds, through (101), to $u(x)$ for $x<0$. To start with, we derive, using the transformation law (15d), the new reflection coefficient corresponding to (101),

(102) $\quad \bar{R}(k) = (-)^{A+B+1} r \dfrac{\prod\limits_{j=1}^{A}(k+a_j)}{\prod\limits_{j=1}^{B}(k+s_j)} \prod\limits_{j=2}^{B} \dfrac{(k+z_j)}{(k-z_j)} \prod\limits_{j=1}^{B^{(+)}} \dfrac{\left(k - b_j^{(+)}\right)}{\left(k + b_j^{(+)}\right)};$

it is easily verified that it satisfies the general conditions (76), and that it has only $B-1$ simple poles in the upper half-plane occuring at $k=z_j$, $j=2,3,\ldots,B$. These are indeed the poles that contribute to the kernel of the GLM equation that now remains to be solved. Repeating the previous arguments, the GLM equation now reads

(103) $\quad \overline{K}(x,y)+\overline{m}(x+y)+\int_x^{+\infty}dz\,\overline{K}(x,z)\overline{m}(z+y)=0,\quad y>x>0,$

with

(104) $\quad \overline{m}(x)=\sum\limits_{j=2}^{B}\overline{g}_j\exp(iz_j x),$

(105) $\quad \overline{g}_j=(-)^{A+B+1}r\dfrac{\prod\limits_{n=1}^{A}(z_j+a_n)}{\prod\limits_{n=1}^{B}(z_j+s_n)}\prod\limits_{n=1}^{B^{(+)}}\dfrac{(z_j-b_n^{(+)})}{(z_j+b_n^{(+)})}\dfrac{\prod\limits_{n=2}^{B}(z_j+z_n)}{\prod\limits_{\substack{n=1\\n\neq j}}^{B}(z_j-z_n)},$

and its solution yields

(106) $\quad u(x)=-2d^2\{\ln\det[I+\overline{V}(-x)]\}/dx^2,\quad x<0,$

where $\overline{V}(x)$ is the $(B-1)\times(B-1)$ matrix whose elements are

(107) $\quad \overline{V}_{mn}(x)=i\overline{\gamma}_m\overline{\gamma}_n(z_m+z_n)^{-1}\exp[i(z_m+z_n)x]$

with

(108) $\quad \overline{\gamma}_n\equiv(\overline{g}_n)^{1/2},\quad n=2,3,\ldots,B.$

The task to derive the Jost solutions corresponding to these potentials, and to exhibit some explicit examples, is left as an exercise for the diligent reader. Note however that the simplest case (with $A=0$, $B=1$; see (77) and (78)) has already been treated explicitly above, since the rational reflection coefficient

(109) $\quad R(k)=g/(ik-g)$

corresponds to the Dirac distribution

(110) $\quad u(x)=2g\delta(x)$

(set $x_0=0$ in (34a) and (34c)). This example underscores the need to exercise some care at $x=0$; a point that is already apparent from the above treatment, that yields two separate expressions for the potential $u(x)$ in the regions $x \geqslant 0$ and $x<0$ (see (98) and (106)). A related question is worth a mention. Let us recall that, if the *bona fide* potential $u(x)$ is infinitely differentiable for all (real) values of x, then the corresponding reflection coefficient $R(k)$ vanishes asymptotically ($k \to \pm \infty$) faster than any (inverse) power of k. Now this cannot obviously be the case for a rational $R(k)$. Therefore the corresponding $u(x)$ cannot be infinitely differentiable; and the value of x at which the property of differentiability fails is clearly at $x=0$. It stands to reason to expect that, if $R(k)$ vanishes asymptotically proportionally to k^{-m} (i.e., if $B=A+m$; see (77)), the corresponding $u(x)$ be regular with its derivatives up to the order $m-2$. Indeed for $m=1$, $u(x)$ itself is not regular, having a delta function singularity at $x=0$ (see (110)); while it is easy to verify that, for $m \geqslant 2$, $u(x)$ is (possibly discontinuous but) finite at $x=0$.

A.20. A general approach based on the algebra of differential operators; connections with, and amongst the spectral transform method, the Lax approach and the AKNS technique

In chapter 3 the class of nonlinear evolution equations

(1) $\quad u_t = \alpha(L, t) u_x, \quad u \equiv u(x, t),$

has been identified. Here of course the integrodifferential operator L is defined by the formula

(2) $\quad Lf(x) = f_{xx}(x) - 4uf(x) + 2u_x \int_x^{+\infty} dy f(y),$

where we omit for notational simplicity to display explicitly any dependence on the time variable.

The main property of this class of evolution equations is the simplicity of the corresponding evolution of the spectral transform of $u(x, t)$,

(3) $\quad S[u(x, t)] = \{R(k, t), -\infty < k < \infty; p_n, \rho_n(t), n=1, 2, \ldots, N\},$

as displayed by the equations

(4a) $\quad R_t = 2ik\alpha(-4k^2, t)R, \quad R \equiv R(k, t),$

(4b) $\quad p_n(t) = p_n(0),$

(4c) $\quad \dot{\rho}_n(t) = -2p_n\alpha(4p_n^2, t)\rho_n(t).$

Let us tersely recall that the spectral transform (3) is associated with the spectral problem based on the Schroedinger equation

(5) $\quad \psi_{xx} = [u(x) - k^2]\psi, \quad \psi \equiv \psi(x, k),$

while the main property of the integrodifferential operator L, see (2), is its correspondence, in k-space, merely to multiplication by $-4k^2$, as is for instance apparent from a comparison of (4a) and (4c) with (1).

Many properties of the class of evolution equations (1) are discussed in chapters 3, 4 and 5. The main technique employed there to establish these results is the connection between a function $u(x)$ (belonging to the class of *bona fide* potentials) and its spectral transform, as displayed by the wronskian integral equations and the spectral integral relations of section 2.4. But there also exist other approaches to obtain the same results. Two such methods, that are relatively close to that mainly employed in this book, are associated with the names of Lax and of Ablowitz, Kaup, Newell and Segur (AKNS). Purpose and scope of this appendix is to outline these techniques and to display tersely their relationship to the spectral transform approach. The second part of this program is implemented below by providing a rather general operatorial technique that yields not only the evolution equations (1) and (4), but also the associated Bäcklund and Darboux transformations (see chapter 4 and appendix A.12); this treatment is based on recent work by Bruschi and Ragnisco.

The following developments rely largely on the manipulation of differential and integral operators. Hereafter such operators are denoted by boldface characters. This notation is particularly convenient to distinguish a function, say $u(x)$, belonging to the functional space on which the operators act, from the multiplicative operator, say $\boldsymbol{u}$, whose application to a function $f(x)$ consists merely in the multiplication by $u(x)$, $\boldsymbol{u}f(x) \equiv u(x)f(x)$. Thus for instance, if $\boldsymbol{D}$ indicates the operator of differentiation d/dx, there hold the formulae $\boldsymbol{Du} = \boldsymbol{u}_x + \boldsymbol{u}\boldsymbol{D}$ and $\boldsymbol{D}u(x) = u_x(x)$. Of course an operator may depend on a variable, say x or t, but we generally omit to indicate explicitly such a dependence, except when we intend specifically to draw attention to it.

The Lax approach originates from the observation that the KdV equation, namely (1) with $\alpha(z, t) = -z$ (see (3.1.-33)), corresponds to the operator equation

(6) $\quad H_t = MH - HM$

with

(7a) $\quad H = -D^2 + u,$

(7b) $\quad M = -4D^3 + 6uD + 3u_x,$

where of course D is the differential operator, $D \equiv d/dx$, and the time dependence of the operator H (as well as M) originates from the time dependence of $u \equiv u(x, t)$ (so that $H_t = u_t$, see (7a)). Some connection with the spectral problem based on the Schroedinger equation (5) is of course evident, since H, see (7a), is indeed just the Schroedinger differential operator. The Lax formulation based on (6) clearly focuses attention on the isospectral nature of the time evolution of the Schroedinger operator H (see (4b)). Another aspect of the evolution equation (6) that should be emphasized is the local character of the resulting evolution equation for $u(x, t)$ (see (3.1.-33)), stemming from the pure differential (i.e., not integral) nature of the operators M and H (see (7)).

It is easy (see below) to generalize the result we have just described by showing that every evolution equation of the class (1) can be reproduced by (6) with (7a) and an appropriate choice for the operator M.

The AKNS method originates apparently from a different idea, that is however closely related to the Lax approach. The starting point is the observation that the KdV equation (i.e., (1) with $\alpha(z, t) = -z$; see (3.1.-33)) is just the integrability condition, $\psi_{xxt} = \psi_{txx}$, for the following pair of linear equations satisfied by the function $\psi(x, k, t)$:

(8a) $\quad \psi_{xx} = [u(x, t) - k^2]\psi,$

(8b) $\quad \psi_t = [-u_x(x, t) + 4ik^3]\psi + 2[u(x, t) + 2k^2]\psi_x.$

And also this approach can be easily generalized (see below) so that every evolution equation of the class (1) coincide with the integrability condition

A.20 A general approach based on differential operators

$\psi_{xxt} = \psi_{txx}$ for equation (8a) and for the equation

$$(9) \quad \psi_t = a(x,k,t)\psi + b(x,k,t)\psi_x, \quad \psi \equiv \psi(x,k,t),$$

with an appropriate choice of the two functions $a(x,k,t)$ and $b(x,k,t)$ (of course the integrability condition must hold for all values of k). Indeed the *ansatz* (9) with a and b (low-order) polynomials in k provides perhaps the most elementary and straightforward technique to identify the simplest equations of the class (1).

Let us proceed now to develop the general technique that has been promised above. The starting point is the operator equation

$$(10) \quad H^{(1)}B - BH^{(2)} = A,$$

where of course

$$(11) \quad H^{(j)} = -D^2 + u^{(j)}, \quad j = 1, 2,$$

is the usual Schroedinger operator, see (7a), associated with the potential $u^{(j)}(x)$. The essential feature of the operator formula (10) is, that the operator B be purely differential (i.e. not integral), and the operator A purely multiplicative (i.e., $Af(x) = A(x)f(x)$, with $A(x)$ an appropriate function, see below). Our next task is therefore to show how operators B and A can be constructed that have these properties and satisfy (10) (note incidentally that this equation is linear in A and B).

To realize this task, it is convenient to focus first on a transformation of the form

$$(12a) \quad B' = BT + P,$$

$$(12b) \quad A'(x) = ZA(x) + G(x),$$

that generates a new pair of operators, A' and B', from an assumedly known pair, A and B. Note that (12a) is an operator equation, while (12b) is not, being however adequate to define the purely multiplicative operator A' (we are of course assuming that both pairs of operators satisfy (10) and have the properties detailed above).

It is trivial, if perhaps tedious, to verify that the following formulae define, via (12), a transformation with the required properties:

(13a) $\quad T = H^{(2)}$,

(13b) $\quad Z = -\frac{1}{4}\Lambda$,

(13c) $\quad G(x) = \gamma \Gamma \cdot 1 + cm(x)$,

(13d) $\quad P = ED + F + \gamma(\nu - 2D) + c$,

(13e) $\quad E(x) = \frac{1}{2}IA(x) \equiv \frac{1}{2}\int_x^{+\infty} dy\, A(y)$,

(13f) $\quad F(x) = \frac{1}{4}[A(x) - ImIA(x)]$
$$\equiv \frac{1}{4}\left\{ A(x) - \int_x^{+\infty} dy\, [u^{(1)}(y) - u^{(2)}(y)] \int_y^{+\infty} dz\, A(z) \right\}$$

(13g) $\quad \nu(x) = Im(x) = \int_x^{+\infty} dy\, [u^{(1)}(y) - u^{(2)}(y)]$,

(14) $\quad \Gamma = Ds - sD + mIm$,

(15a) $\quad \Lambda = D^2 - 2s + \Gamma I$,

(15b) $\quad \Lambda = D^2 - s + DsI + mImI$,

(16a) $\quad s(x) = u^{(1)}(x) + u^{(2)}(x)$,

(16b) $\quad m(x) = u^{(1)}(x) - u^{(2)}(x)$,

(17a) $\quad D \equiv d/dx, \quad Df(x) \equiv f_x(x)$,

(17b) $\quad I \equiv \int_x^{+\infty} dy, \quad If(x) \equiv \int_x^{+\infty} dy\, f(y)$.

Note that we are employing systematically the notation introduced above, namely if a character in ordinary type identifies a function, the same character in boldface identifies the corresponding purely multiplicative operator. For instance the quantities γ and c indicate arbitrary constants (i.e., x-independent numbers), and the symbols γ and c indicate the corresponding multiplicative operators, f.i. $cf(x) = cf(x)$.

Note that the operator P, see (13d), is a first order differential operator, and that the two functions $E(x)$ and $F(x)$, see (13e) and (13f), are linear in $A(x)$. Note moreover that all the quantities defined above, except for the two arbitrary constants γ and c (and the operators D and I, see (17)) depend on the two potentials $u^{(1)}(x)$ and $u^{(2)}(x)$. Finally let us emphasize that the

A.20 A general approach based on differential operators

definitions (14)–(17) of the operators $\boldsymbol{\Gamma}$, $\boldsymbol{\Lambda}$, $\boldsymbol{s}$, $\boldsymbol{m}$, $\boldsymbol{D}$ and $\boldsymbol{I}$ given here coincide with the definitions given in appendix A.9, see (A.9.-1,2,3,4,7,8); on the other hand the operator $\boldsymbol{P}$ defined here differs from that of chapter 5 and appendix A.21, see (5.-35) and (A.21.-39), and the operators $\boldsymbol{F}$ and $\boldsymbol{A}$ differ from those of appendix A.21, see (A.21.-11, 27).

The presence of the two arbitrary constants γ and c (and of the corresponding operators $\boldsymbol{\gamma}$ and $\boldsymbol{c}$) in the definitions of $G(x)$ and $\boldsymbol{P}$, see (13c) and (13d), causes the transformation (12) not to be homogeneous; in particular the trivial solution of (10), $\boldsymbol{B} = \boldsymbol{A} = 0$, gets transformed into the nontrivial solution

(18a) $\quad \boldsymbol{B}_0^{(+)} = -2\boldsymbol{D} + \boldsymbol{\nu}$

(18b) $\quad A_0^{(+)}(x) = \boldsymbol{\Gamma} \cdot 1 = u_x^{(1)}(x) + u_x^{(2)}(x)$
$$+ \left[u^{(1)}(x) - u^{(2)}(x) \right] \int_x^{+\infty} dy \left[u^{(1)}(y) - u^{(2)}(y) \right],$$

if one takes $\gamma = 1, c = 0$, or in the solution

(19a) $\quad B_0^{(-)}(x) = 1,$

(19b) $\quad A_0^{(-)}(x) = m(x) = u^{(1)}(x) - u^{(2)}(x),$

if one takes instead $\gamma = 0, c = 1$. Note that $\boldsymbol{B}_0^{(+)}$ is a first-order differential operator, while $\boldsymbol{B}_0^{(-)}$ is purely multiplicative. The motivation for attributing a subscript zero to the quantities defined by (18) and (19) will be immediately apparent.

Two infinite sequences of solutions of (10) can now be obtained by applying the transformation (12) (with (13)–(17) and, without loss of generality, $\gamma = c = 0$) to the two basic solutions (18) and (19). They read:

(20a) $\quad \boldsymbol{B}_n^{(\pm)} = \boldsymbol{B}_0^{(\pm)} [\boldsymbol{H}^{(2)}]^n + \sum_{m=0}^{n-1} \left[\boldsymbol{E}_m^{(\pm)} \boldsymbol{D} + \boldsymbol{F}_m^{(\pm)} \right] [\boldsymbol{H}^{(2)}]^{n-m-1},$

$$n = 1, 2, 3, \ldots,$$

(20b) $\quad A_n^{(\pm)}(x) = \left(-\tfrac{1}{4}\boldsymbol{\Lambda}\right)^n A_0^{(\pm)}(x), \quad n = 0, 1, 2, \ldots,$

(21) $\quad E_n^{(\pm)}(x) = \tfrac{1}{2}\boldsymbol{I}\left(-\tfrac{1}{4}\boldsymbol{\Lambda}\right)^n A_0^{(\pm)}(x), \quad n = 0, 1, 2, \ldots,$

(22) $\quad F_n^{(\pm)}(x) = \tfrac{1}{4}(1 - \boldsymbol{ImI})\left(-\tfrac{1}{4}\boldsymbol{\Lambda}\right)^n A_0^{(\pm)}(x), \quad n = 0, 1, 2, \ldots.$

An arbitrary linear combination of these solutions yields finally the general solution of (10), that is conveniently written in the form

$$(23) \quad B = g(-4H^{(2)}) + \sum_{n=1}^{n-1}(-4)^n g_n \sum_{m=0}^{n-1}\left[E_m^{(-)}D+F_m^{(-)}\right]\left[H^{(2)}\right]^{n-m-1}$$

$$+ [\nu - 2D] h(-4H^{(2)})$$

$$+ \sum_{n=1}^{n-1}(-4)^n h_n \sum_{m=0}^{n-1}\left[E_m^{(+)}D+F_m^{(+)}\right]\left[H^{(2)}\right]^{n-m-1},$$

$$(24) \quad A(x) = g(\Lambda)\left[u^{(1)}(x) - u^{(2)}(x)\right] + h(\Lambda)\Gamma \cdot 1,$$

where $g(z)$ and $h(z)$ are two arbitrary polynomials,

$$(25a) \quad g(z) = \sum_{n=0} g_n z^n,$$

$$(25b) \quad h(z) = \sum_{n=0} h_n z^n.$$

Of course the coefficients g_n and h_n of these polynomials must be x-independent, but are otherwise arbitrary. Let us also note that the two sums in the r.h.s. of (23) are suggestive of the notation $f^{\#}(z_1, z_2) = [f(z_1) - f(z_2)]/(z_1 - z_2)$ (see below), yet they cannot be easily rewritten in more compact form using this notation, due to the noncommutativity of operators.

To conclude this derivation, let us reemphasize that the virtue of the (pure) differential operator B, defined by (23) together with (11), (21), (22), (18), (19), (13g) and (25), is that, when inserted in the l.h.s. of (10), it yields a purely multiplicative operator, see (24) (i.e., all derivative operators cancel out when (23) is inserted in the l.h.s. of (10)).

The formulae (10), (23) and (24) are the basic formulae from which we tersely derive below the results that display the connections between the spectral transform, Lax and AKNS techniques. We restrict attention throughout to functions $u^{(1)}(x)$ and $u^{(2)}(x)$ belonging to the class of *bona fide* potentials (see section 2.3). Note that this assumption guarantees that the function $A(x)$, see (24), vanishes asymptotically,

$$(26) \quad \lim_{x \to \pm \infty} [A(x)] = 0,$$

for any choice of the polynomials $g(z)$ and $h(z)$.

Let us display first of all the connection between these developments and the spectral transform technique that constitutes the main tool of analysis

A.20 A general approach based on differential operators

throughout this book. To this end, we now introduce the eigenfunctions $\psi^{(j)}(x)$ of the Schroedinger operators (11) corresponding to the eigenvalue k^2:

(27) $\quad H^{(j)}\psi^{(j)}(x) = k^2 \psi^{(j)}(x), \quad j=1,2.$

Let us then apply both sides of (10) to the product $\psi^{(1)}\psi^{(2)}$. Using (27) and (11) one obtains easily

(28) $\quad \{\psi_x^{(1)}[\boldsymbol{B}\psi^{(2)}] - \psi^{(1)}[\boldsymbol{B}\psi^{(2)}]_x\}_x = \psi^{(1)}\psi^{(2)}A(x).$

This (local) formula holds clearly for any solutions $\psi^{(j)}$ of the differential equations (27). Let us now focus attention on those eigenfunctions $\psi^{(j)}(x,k)$ characterized, for real k, by the asymptotic boundary conditions (see (2.1.-3))

(29a) $\quad \psi^{(j)}(x,k) \to T^{(j)}(k)\exp(-ikx), \quad x\to -\infty, \quad j=1,2,$

(29b) $\quad \psi^{(j)}(x,k) \to \exp(-ikx) + R^{(j)}(k)\exp(ikx), \quad x\to +\infty, \quad j=1,2,$

and let us integrate (28) from $-\infty$ to $+\infty$. Noting that (23) implies that

(30a) $\quad \boldsymbol{B}\exp(ikx) \to B^{(+)}(k)\exp(ikx), \quad x\to +\infty, \quad \text{Im } k=0,$

(30b) $\quad \boldsymbol{B}\exp(-ikx) \to B^{(-)}(k)\exp(-ikx), \quad x\to -\infty, \quad \text{Im } k=0,$

with

(31) $\quad B^{(+)}(k) = g(-4k^2) - 2ikh(-4k^2),$

(32) $\quad B^{(-)}(k) = g(-4k^2) + 2ik \int_{-\infty}^{+\infty} dx\, g^{\#}(-4k^2, \Lambda)[u^{(1)}(x) - u^{(2)}(x)]$

$\qquad + \int_{-\infty}^{+\infty} dy\, [u^{(1)}(y) - u^{(2)}(y)] \int_{y}^{+\infty} dx\, g^{\#}(-4k^2, \Lambda)$

$\qquad\qquad\qquad\qquad\qquad\qquad \cdot [u^{(1)}(x) - u^{(2)}(x)]$

$\qquad + 2ikh(-4k^2) + h(-4k^2) \int_{-\infty}^{+\infty} dx\, [u^{(1)}(x) - u^{(2)}(x)]$

$\qquad + 2ik \int_{-\infty}^{+\infty} dx\, h^{\#}(-4k^2, \Lambda)\Gamma \cdot 1$

$\qquad + \int_{-\infty}^{+\infty} dy\, [u^{(1)}(y) - u^{(2)}(y)] \int_{y}^{+\infty} dx\, h^{\#}(-4k^2, \Lambda)\Gamma \cdot 1,$

where we use the notation

(33) $$g^{\#}(z_1, z_2) = [g(z_1) - g(z_2)]/(z_1 - z_2),$$
$$h^{\#}(z_1, z_2) = [h(z_1) - h(z_2)]/(z_1 - z_2),$$

one obtains the formula

(34) $$2ik\left[B^{(+)}(-k)R^{(1)}(k) - B^{(+)}(k)R^{(2)}(k)\right]$$
$$= \int_{-\infty}^{+\infty} dx\, \psi^{(1)}(x,k)\psi^{(2)}(x,k)A(x).$$

The arbitrariness of the polynomials $g(z)$ and $h(z)$, together with the expressions (31) and (24), imply that this formula coincides with the wronskian integral relations (2.4.1.-10).

In a similar manner one gets the formula

(35) $$2ik\left[B^{(+)}(-k)R^{(1)}(k)R^{(2)}(-k) + B^{(-)}(-k)T^{(1)}(k)T^{(2)}(-k)\right.$$
$$\left. - B^{(+)}(k)\right] = \int_{-\infty}^{+\infty} dx\, \psi^{(1)}(x,k)\psi^{(2)}(x,-k)A(x)$$

that is easily seen to coincide with the formulae (2.4.1.-18) and (2.4.1.-19) (see also appendix A.6). Moreover, the formulae that obtain when one of (or both) the eigenfunctions $\psi^{(j)}(x)$ correspond to a discrete eigenvalue (if any) are just those discussed in subsection 2.4.1.

It is thus clear that all those results that have been derived in this book from the wronskian integral relations corresponding to (34) and (35) can also be derived from the basic equation (10) with (23) and (24). In particular, two functions $u^{(1)}(x)$ and $u^{(2)}(x)$ are said to be related by a Bäcklund transformation if the corresponding expression (24) vanishes, namely if

(36) $$g(\Lambda)\left[u^{(1)}(x) - u^{(2)}(x)\right] + h(\Lambda)\Gamma \cdot 1 = 0$$

(see (2.4.1.-10c) and (4.1.-1)). Note that, as a consequence of (10), this equation implies the operator formula

(37) $$\boldsymbol{H}^{(1)}\boldsymbol{B} = \boldsymbol{B}\boldsymbol{H}^{(2)}$$

relating, via (23), the Schroedinger operators (11). Clearly this last equation implies that the operator $\boldsymbol{B}$ of equation (23), corresponding to the Bäcklund transformation (36), transforms eigenfunctions of $H^{(2)}$ into eigenfunctions of $H^{(1)}$. In particular, the transformation that maps the Jost solutions $f^{(\pm)(2)}(x,k)$ into the Jost solutions $f^{(\pm)(1)}(x,k)$ is easily found to read

$$(38) \qquad f^{(\pm)(1)}(x,k) = \left[B^{(\pm)}(k)\right]^{-1} B f^{(\pm)(2)}(x,k);$$

indeed this formula, with $B^{(\pm)}(k)$ defined of course by (31) and (32), follows from the definition of the Jost solutions,

$$(39a) \qquad H^{(j)} f^{(\pm)(j)}(x,k) = k^2 f^{(\pm)(j)}(x,k), \quad j=1,2,$$

$$(39b) \qquad \lim_{x \to \pm\infty} \left[\exp(\mp ikx) f^{(\pm)(j)}(x,k)\right] = 1, \quad j=1,2,$$

(see (2.1.-4)), and from the asymptotic expression (30). In the simplest case characterized by constant g and h (namely, when $g(z)$ and $h(z)$ are polynomials of degree zero), by setting $g = 2ph$ with p a real constant, (38) goes over into the Darboux transformation (see (4.1.-30) and appendix A.12). In the following we refer to (38) as a Darboux transformation also in the general case.

The (general) Darboux transformation (38) can actually be written more explicitly noting that in the expression (23) of B the operator $H^{(2)}$, and its powers, enter in all terms as right factors, and can therefore be replaced by the eigenvalue k^2 when B acts on $f^{(\pm)(2)}(x,k)$. Thus in place of (38) one can write

$$(40) \qquad f^{(\pm)(1)}(x,k) = a^{(\pm)}(x,k) f^{(\pm)(2)}(x,k) + b^{(\pm)}(x,k) f^{(\pm)(2)}_x(x,k)$$

with

$$(41a) \qquad a^{(\pm)}(x,k) = \left[B^{(\pm)}(k)\right]^{-1} \{g(-4k^2) + h(-4k^2) \mathit{Im}(x)$$
$$+ \mathit{Im}\left[g^{\#}(-4k^2, \Lambda) m(x) + h^{\#}(-4k^2, \Lambda) \Gamma \cdot 1\right]\},$$

$$(41b) \qquad b^{(\pm)}(x,k) = -2\left[B^{(\pm)}(k)\right]^{-1} \{h(-4k^2) + I\left[g^{\#}(-4k^2, \Lambda) m(x)\right.$$
$$\left. + h^{\#}(-4k^2, \Lambda) \Gamma \cdot 1\right]\},$$

where we are using the definitions (31), (32), (33), (14), (15), (16) and (17).

As for the transformation in k-space that corresponds to the Bäcklund transformation (36), there is little to add to the results of chapter 4. Let us, however, outline how the relevant results can be derived in the present context, limiting our consideration to the continuous part of the spectrum, i.e. to real values of k (for a discussion including also the discrete spectrum see section 4.1 and appendix A.12). The effect of the Bäcklund transformation (36) on the reflection and transmission coefficients $R(k)$ and $T(k)$ can be easily obtained inserting the asymptotic behaviours

(42a) $$f^{(+)(j)}(x,k) \to -\left[R^{(j)}(-k)/T^{(j)}(-k)\right]\exp(-ikx)$$
$$+\left[T^{(j)}(k)\right]^{-1}\exp(ikx), \quad x \to -\infty,$$

(42b) $$f^{(-)(j)}(x,k) \to \left[T^{(j)}(k)\right]^{-1}\exp(-ikx)$$
$$+\left[R^{(j)}(k)/T^{(j)}(k)\right]\exp(ikx), \quad x \to +\infty,$$

in the Darboux transformation (40). One obtains

(43) $$R^{(1)}(k) = R^{(2)}(k)\left[B^{(+)}(k)/B^{(+)}(-k)\right],$$

(44) $$T^{(1)}(k) = T^{(2)}(k)\left[B^{(+)}(k)/B^{(-)}(-k)\right],$$

with $B^{(\pm)}(k)$ defined by (31) and (32). The first of these formulae, (43), coincides of course with (4.1.-2). Note that the requirement of consistency with the unitarity equation,

(45) $$R^{(j)}(k)R^{(j)}(-k) + T^{(j)}(k)T^{(j)}(-k) = 1, \quad j=1,2,$$

yields the formula

(46) $$B^{(+)}(k)B^{(+)}(-k) = B^{(-)}(k)B^{(-)}(-k),$$

that implies, via (31) and (32), a set of intriguing integral identities relating $u^{(1)}$ and $u^{(2)}$ (it is easily seen that, in the special case $g=2p, h=1$, this coincides with (4.1.-22)).

Let us proceed now to investigate problems involving the evolution in time. To this end we consider the one parameter family of Schroedinger differential operators

(47) $$H(t) = -D^2 + u(x,t),$$

associated with a *bona fide* potential u that depends, in addition to the variable x, on the parameter t ("time"). We then set in (10), (23) and (24)

(48a) $\quad u^{(2)}(x) = u(x, t),$

(48b) $\quad u^{(1)}(x) = u(x, t + \Delta t) = u(x, t) + u_t(x, t) \Delta t + O(\Delta t^2),$

(49a) $\quad g(z) = -2\eta(z, t)/\Delta t,$

(49b) $\quad h(z) = \alpha(z, t),$

and let $\Delta t \to 0$. There obtain the following formulae:

(50a) $\quad H_t \eta(-4H, t) - [M, H] = f$

(50b) $\quad f(x, t) = \eta(L, t) u_t - \alpha(L, t) u_x,$

(51) $\quad M = D\alpha(-4H, t) + \sum_{n=1} \left[\alpha_n(t) \sum_{m=0}^{n-1} \left(2\sigma^{(m)} D - \sigma_x^{(m)} \right) \right.$

$\qquad \left. - \eta_n(t) \sum_{m=0}^{n-1} \left(2\mu^{(m)} D - \mu_x^{(m)} \right) \right] (-4H)^{n-m-1},$

with

(52a) $\quad \mu^{(m)}(x, t) = IL^m u_t(x, t),$

(52b) $\quad \sigma^{(m)}(x, t) = IL^m u_x(x, t).$

In these equations the integrodifferential operator L is of course that defined by (2), while the coefficients $\eta_n(t)$ and $\alpha_n(t)$ are those of the polynomials $\eta(z, t)$ and $\alpha(z, t)$:

(53a) $\quad \eta(z, t) = \sum_{n=0} \eta_n(t) z^n,$

(53b) $\quad \alpha(z, t) = \sum_{n=0} \alpha_n(t) z^n.$

These formulae obtain from the expressions of B and $A(x)$, that follow, via (48) and (49), from (23) and (24):

(54) $\quad B = -2[\eta(-4H, t)/\Delta t + M] + O(\Delta t),$

(55) $\quad A(x) = -2[\eta(L, t) u_t(x, t) - \alpha(L, t) u_x(x, t)] + O(\Delta t).$

Let us also report, for completeness, the expressions that follow, via (48) and (49) with $\Delta t \to 0$, from (34) and (35):

(56) $$2ik\bigl[\eta(-4k^2,t)R_t(k,t)-2ik\alpha(-4k^2,t)R(k,t)\bigr]$$
$$=\int_{-\infty}^{+\infty}dx\,[\psi(x,k,t)]^2\bigl[\eta(L,t)u_t(x,t)-\alpha(L,t)u_x(x,t)\bigr],$$

(57) $$2ik\Bigl\{R(-k,t)\bigl[\eta(-4k^2,t)R_t(k,t)-2ik\alpha(-4k^2,t)R(k,t)\bigr]$$
$$+T(-k,t)\Bigl[\eta(-4k^2,t)T_t(k,t)-2ikT(k,t)\int_{-\infty}^{+\infty}dx$$
$$\cdot\bigl[\eta^{\#}(-4k^2,L;t)u_t(x,t)-\alpha^{\#}(-4k^2,L;t)u_x(x,t)\bigr]\Bigr]\Bigr\}$$
$$=\int_{-\infty}^{+\infty}dx\,\psi(x,k,t)\psi(x,-k,t)$$
$$\cdot\bigl[\eta(L,t)u_t(x,t)-\alpha(L,t)u_x(x,t)\bigr].$$

Here of course, consistently with (48), we have set

(58a) $R^{(2)}(k)=R(k,t), \quad T^{(2)}(k)=T(k,t),$
$\psi^{(2)}(x,k)=\psi(x,k,t),$

(58b) $R^{(1)}(k)=R(k,t+\Delta t)=R(k,t)+R_t(k,t)\Delta t+O(\Delta t^2),$

(58c) $T^{(1)}(k)=T(k,t+\Delta t)=T(k,t)+T_t(k,t)\Delta t+O(\Delta t^2),$

(58d) $\psi^{(1)}(x,k)=\psi(x,k,t+\Delta t)=\psi(x,k,t)+O(\Delta t)$

while of course

(59a) $\eta^{\#}(z_1,z_2;t)=[\eta(z_1,t)-\eta(z_2,t)]/(z_1-z_2),$

(59b) $\alpha^{\#}(z_1,z_2;t)=[\alpha(z_1,t)-\alpha(z_2,t)]/(z_1-z_2).$

Note that (57) has been obtained using the unitarity condition (45). Of course (56) coincides with (3.1.-7) and (3.1.-8), while for (57) the reader may refer to the results of appendix A.18.

Among the results that are straightforward consequences of (56) and (57) let us recall the important operator identity

(60) $$\int_{-\infty}^{+\infty}dx\,L^n u_x(x)=0, \quad n=0,1,2,\ldots,$$

(see appendices A.8 and A.9). This follows from (57) with $\eta(z,t)=0$, namely from

(61) $$(2ik)^2 \alpha(-4k^2) R(k) R(-k)$$
$$- (2ik)^2 T(k) T(-k) \int_{-\infty}^{+\infty} dx \left[\alpha^{\#}(-4k^2, L) u_x(x) \right]$$
$$= \int_{-\infty}^{+\infty} dx \, \psi(x, k) \psi(x, -k) \left[\alpha(L) u_x(x) \right],$$

(where we have omitted for simplicity to indicate the variable t, that plays no rôle). Indeed the asymptotic formulae (see appendix A.3)

(62a) $\lim_{k \to \pm \infty} \left[k^M R(k) \right] = 0, \quad \lim_{k \to \pm \infty} \left[T(k) T(-k) \right] = 1, \quad \text{Im } k = 0,$

(62b) $\lim_{k \to \pm \infty} \left[\psi(x, k) \psi(x, -k) \right] = 1, \quad \text{Im } k = 0,$

together with (61), imply

(63) $$\int_{-\infty}^{+\infty} dx \, \alpha^{\#}(-4k^2, L) u_x(x) = 0;$$

and the validity of this formula (see (59b), and recall that $\alpha(z)$ is an *arbitrary* polynomial) clearly implies (60).

All the basic material needed to investigate time-evolution equations is now ready. We actually consider, for completeness, the evolution equation

(64) $\eta(L, t) u_t = \alpha(L, t) u_x, \quad u \equiv u(x, t),$

that is more general than (1), reducing to it if $\eta(z, t) = 1$ (thus the results given below encompass not only those of chapter 3, but also those of appendix A.18; but we avoid here any discussion of the consistency of the extended class of evolution equations (64), for which the reader is referred to appendix A.18).

The evolution equation for the Schroedinger operator (47) that corresponds to (64) reads

(65) $H_t \eta(-4H, t) = MH - HM,$

with M defined by (51). The corresponding equations for the spectral transform follow (for the continuous spectrum part) from (56) and (57) (using (60)). They read

(66) $\eta(-4k^2, t) R_t(k, t) = 2ik \alpha(-4k^2, t) R(k, t),$

(67) $\eta(-4k^2, t) T_t(k, t)$
$$= 2ik T(k, t) \int_{-\infty}^{+\infty} dx \left[\eta^{\#}(-4k^2, L; t) u_t(x, t) \right],$$

(consistently, of course, with (A.18.-15a) and (A.18.-31)). Moreover, from (38), (49), (54) and the formulae (see (48) and (58))

(68a) $\quad f^{(\pm)(2)}(x,k) = f^{(\pm)}(x,k,t),$

(68b) $\quad f^{(\pm)(1)}(x,k) = f^{(\pm)}(x,k,t+\Delta t)$
$$= f^{(\pm)}(x,k,t) + f_t^{(\pm)}(x,k,t)\Delta t + O(\Delta t^2),$$

there follow the evolution equations for the Jost solutions,

(69) $\quad \eta(-4k^2, t) f_t^{(+)} = Mf^{(+)} - ik\alpha(-4k^2, t) f^{(+)},$
$$f^{(+)} \equiv f^{(+)}(x,k,t),$$

(70) $\quad \eta(-4k^2, t) f_t^{(-)} = Mf^{(-)} + ik\left\{\alpha(-4k^2, t) - 2\int_{-\infty}^{+\infty} dx\right.$
$$\left. \cdot \left[\eta^\#(-4k^2, L; t) u_t(x,t)\right]\right\} f^{(-)}, \quad f^{(-)} \equiv f^{(-)}(x,k,t).$$

For the sake of completeness, let us also display the evolution equations satisfied by the eigenfunction (see (29) and (2.1.-5))

(71) $\quad \psi(x,k,t) = T(k,t) f^{(-)}(x,k,t),$

which follows from (70) and (67):

(72) $\quad \eta(-4k^2, t) \psi_t = M\psi + ik\alpha(-4k^2, t)\psi, \quad \psi \equiv \psi(x,k,t).$

Note that in the r.h.s. of this equation, as well as in the r.h.s. of (69) and (70), the operator M, although always defined by (51), can in fact be given an equivalent expression by replacing everywhere the operator H by its eigenvalue k^2:

(73) $\quad M = \left\{\alpha(-4k^2, t) - 2\int_x^{+\infty} dy \left[\eta^\#(-4k^2, L; t) u_t(y,t)\right.\right.$
$$\left.\left. - \alpha^\#(-4k^2, L; t) u_y(y,t)\right]\right\} D$$
$$- \left[\eta^\#(-4k^2, L; t) u_t(x,t) - \alpha^\#(-4k^2, L; t) u_x(x,t)\right].$$

Here of course we use the notation (59) and (2), with an obvious modification (replace x by y) when L appears under the integral; moreover L is *not* meant to act on ψ when (73) is inserted in (72) (see (51)).

A.20 A general approach based on differential operators

As for the time evolution of the part of the spectral transform corresponding to the discrete spectrum, it is easy to reproduce the standard results,

(74a) $\quad \dot{p}_n(t) = 0, \quad n = 1, 2, \ldots, N,$

(74b) $\quad \eta(4p_n^2, t) \dot{\rho}_n(t) = -2 p_n \alpha(4p_n^2, t) \rho_n(t), \quad n = 1, 2, \ldots, N,$

on the assumption (see appendix A.18) that

(75) $\quad \eta(4p_n^2, t) \neq 0.$

Let us now discuss tersely the relationship between the Bäcklund transformation (36) and the time evolution (64). Of course, as regards k-space, one can use the standard argument, namely infer, from the fact that $R^{(2)}(k, t)$ satisfies the evolution equation (66), that $R^{(1)}(k, t)$ satisfies the same evolution equation, provided $g(z)$ and $h(z)$ are time-independent, and therefore, see (31), also time-independent is $B^{(+)}(k)$,

(76) $\quad B_t^{(+)} = 0.$

To conclude from this that the Bäcklund transformation (36) does actually transform a solution $u^{(2)}(x, t)$ of (64) into a new solution $u^{(1)}(x, t)$ of the same equation, it should be moreover ascertained that also the discrete part of the spectral transform of $u^{(1)}(x, t)$ satisfies the corresponding equations (74). For a discussion of this point the reader is referred to chapter 4 and to appendix A.12. Here we report instead two evolution equations that are implied by (65) and (37).

The first equation reads

(77) $\quad B_t \eta(-4H^{(2)}) = M^{(1)} B - B M^{(2)},$

where B is defined by (23) (with arbitrary, but time-independent, $g(z)$ and $h(z)$) and $M^{(j)}$ is the operator defined by (51) and (52) with $u = u^{(j)}$ and $H = H^{(j)}$, (see (11). This formula follows from the remark that, if $H^{(2)}$ is a solution of (65) and (36) is a Bäcklund transformation that maps a solution $u^{(2)}(x, t)$ of (64) into a new solution $u^{(1)}(x, t)$ of the same evolution equation, then also $H^{(1)}$ satisfies (65).

The second equation reads

(78) $$\eta(-4k^2,t)B_t^{(-)}(k,t) = 2ikB^{(-)}(k,t)\int_{-\infty}^{+\infty} dx$$
$$\cdot\left[\eta^\#(-4k^2,\boldsymbol{L}^{(1)};t)u_t^{(1)}(x,t) - \eta^\#(-4k^2,\boldsymbol{L}^{(2)};t)u_t^{(2)}(x,t)\right],$$

where of course $\boldsymbol{L}^{(j)}$ is the operator (2) with $u(x)$ replaced by $u^{(j)}(x,t)$, $j=1,2$ (for the other notation see (59a) and (32)). This equation follows from the evolution equation (70) for the Jost solution $f^{(-)}(x,k,t)$, together with (77) and the Darboux transformation (38). Note that, for the particular, but important, case in which $\eta(z,t)=1$ (so that (64) becomes (1)), the r.h.s. of this equation vanishes (see (59a)); thus in this case $B^{(-)}$ is time-independent (see (32); and note that this conclusion holds provided the polynomials $g(z)$ and $h(z)$ are time-independent, but otherwise arbitrary). Indeed the results of appendix A.18 imply that this conclusion, as well as the simplifications in the formulae written above caused by the vanishing of integrals such as those in the r.h.s. of (78) and (70), hold more generally. Note moreover that $B^{(-)}$, see (32), is a polynomial in k whose coefficients can be expressed in terms of the C_m's (see (A.9.-48a, 48b, 49a, 49b, 59, 60)); thus this conclusion provides one more proof of the time-independence of the "constants of motion" C_m.

Let us finally proceed to discuss directly the connections with the Lax and AKNS approaches.

First of all it is now clear that each evolution equation of the class (1) (with $\alpha(z,t)$ polynomial in z) is equivalent to the Lax formula (6) with the differential operator M given explicitly by (51) (with $\eta(z,t)=1$; (see (50)). We have already emphasized that the time evolution of the Schroedinger operator $H(t)$, see (47), is isospectral, namely the discrete eigenvalues $-p_n^2$ (if any) of $H(t)$ are time-independent. It is easily seen that this property holds as well for the class (65) of operator equations, that is larger than the class originally introduced by Lax, and that corresponds to the class of evolution equations (64); see (74a) (this result could be proved directly from (65), introducing the eigenfunctions of $H(t)$ corresponding to the discrete eigenvalue $-p_n^2$, and of course assuming (75); we leave the explicit derivation as an easy exercise).

Moreover the evolution equation for the eigenfunction $\psi(x,k,t)$, corresponding to the (positive) eigenvalue k^2, of the Schroedinger operator $H(t)$ whose time evolution is specified by the Lax equation (6), obtains immediately by setting $\eta(z,t)=1$ in (72) (and corresponding equations for the Jost functions $f^{(\pm)}(x,k,t)$ obtain similarly from (69) and (70); note that in all these equations the operator M is defined by (73)).

Let us finally proceed to the AKNS technique. All that remains to be shown is, that each evolution equation of the class (1) (or, more generally, of the class (64)) can be expressed as the condition of integrability (i.e., of compatibility) of (8a) and (9). But this follows immediately from the remark that (72) with (73) has indeed the form (9), with the definitions

(79a) $\quad a(x,k,t) = [ik\alpha(-4k^2,t) - \eta^{\#}(-4k^2,L;t)u_t(x,t)$
$\quad\quad\quad + \alpha^{\#}(-4k^2,L;t)u_x(x,t)]/\eta(-4k^2,t),$

(79b) $\quad b(x,k,t) = \left\{\alpha(-4k^2,t) - 2\int_x^{+\infty} dy\left[\eta^{\#}(-4k^2,L;t)u_t(y,t)\right.\right.$
$\quad\quad\quad\left.\left. - \alpha^{\#}(-4k^2,L;t)u_y(y,t)\right]\right\}/\eta(-4k^2,t),$

where we are of course using the notations (59) and (2). Note that a and b, as defined by these formulae, are generally rational functions of k; it is moreover clear that they become polynomials in k in the special, but important, case characterized by $\eta(z,t) = 1$, that corresponds to the class of evolution equations (1).

In conclusion let us note that, while in the (generalized) Lax approach and in the AKNS method the basic equations that introduce the time evolution are respectively (65) and (9) with (79), the resulting evolution equations for $u(x,t)$ are of course the same that also obtain by the spectral transform method, see (64), and also identical are the evolution equations for the spectral data, see (66), (67), (74a) and (74b). In the Lax and AKNS context these last equations obtain from the evolution equations satisfied by the eigenfunctions $\psi(x,k,t)$ or $f^{(\pm)}(x,k,t)$, see (69), (70) and (72) (all with (73)) and the relationship between the asymptotic behaviour $(x \to \pm \infty)$ of these eigenfunctions and the spectral data (see e.g. (29)).

A.20.N. Notes to appendix A.20.

The Lax approach to nonlinear evolution equations, based on the isospectral change with time of a linear differential operator (the Schroedinger operator (A.20.-7a) in this case), is due to [L1968]. A formulation focussing on the compatibility between the equations, (A.20.-8a) and (A.20.-9), satisfied by the corresponding eigenfunctions, is due to [AKNS1974]. These approaches deal only with the infinitesimal change (time-derivative) of a differential operator (in the Lax case), or of its eigenfunctions (in the AKNS case); here these techniques are also extended to the more general case of finite differences of operators, and of their eigenfunctions. This extension, due to [BR1980c], besides reproducing the original results of Lax and

AKNS (in the limit of infinitesimal differences), allows a unified treatment of Bäcklund transformations (see chapter 4), expressed by (A.20.-36) or by the corresponding differential operator equation (A.20.-37), and of the corresponding (Darboux) transformations, (A.20.-38), involving the eigenfunctions. For references to these transformations, see section 4.N.

A.21. Local conservation laws: proofs

In chapter 5 (see (5.-25) and (5.-24)) we have reported the two sequences of local conservation laws

(1) $\quad g_t^{(m)} = \gamma_x^{(m)}, \quad m = 0, 1, 2, \ldots,$

(2) $\quad f_t^{(m)} = \varphi_x^{(m)}, \quad m = 0, 1, 2, \ldots,$

associated with the nonlinear evolution equation

(3) $\quad u_t(x, t) = \alpha(L, t) u_x(x, t),$

where $\alpha(z, t)$ is an arbitrary polynomial in z, i.e.

(4) $\quad \alpha(z, t) = \sum_{n=0}^{M} \alpha_n(t) z^n,$

with time-dependent coefficients. The definition of the integrodifferential operator L, together with its properties, are given in appendix A.9 (see (A.9.-11)), and the solution $u(x, t)$ of (3) is assumed to be a *bona fide* potential as function of x (i.e. to be infinitely differentiable for all real values of x and to vanish with all its derivatives, faster than $1/x$, as $x \to \pm \infty$).

The "densities" $g^{(m)}$ resp. $f^{(m)}$ in the l.h.s. of (1) and (2) are defined by the expressions

(5) $\quad g^{(m)}(x, t) = \tilde{L}^m u(x, t), \quad m = 0, 1, 2, \ldots,$

(6) $\quad f^{(m)}(x, t) = L^m [x u_x(x, t) + 2 u(x, t)], \quad m = 0, 1, 2, \ldots$

in terms of the solution $u(x, t)$ of (3) (the definition, and properties, of the operator $\tilde{L}$ are given in appendix A.9, see (A.9.-12)). These definitions originate from the two following expressions for the constants of motion C_m

(see (A.9.-57a,b), or (5.-12) and (5.-20)):

(7a) $\quad C_m = (-1)^m (2m+1)^{-1} \int_{-\infty}^{+\infty} dx\, g^{(m)}(x,t), \quad m=0,1,2,\ldots,$

(7b) $\quad C_m = (-1)^m (2m+1)^{-1} \int_{-\infty}^{+\infty} dx\, f^{(m)}(x,t), \quad m=0,1,2,\ldots\,.$

Note that the definitions (5) and (6) of the densities $g^{(m)}(x,t)$ and $f^{(m)}(x,t)$ are independent of $\alpha(z,t)$, namely they are the same for all the evolution equations of the class (3). Moreover, the two (generally *different*) densities $g^{(m)}$ and $f^{(m)}$, see (5) and (6), yield the *same* conserved quantities C_m, see (7).

The main purpose of this appendix is to present a terse derivation of the explicit expressions of the "currents" $\gamma^{(m)}$ and $\varphi^{(m)}$ appearing in the r.h.s. of (1) and (2), in terms of the solution $u(x,t)$ of (3). Clearly these currents are defined up to an additive (x-independent) constant, that is hereafter fixed by the asymptotic conditions

(8a) $\quad \gamma^{(m)}(\pm\infty, t) = 0, \quad m=0,1,2,\ldots,$

(8b) $\quad \varphi^{(m)}(\pm\infty, t) = 0, \quad m=0,1,2,\ldots\,.$

Actually these conditions need to be imposed only at one end, say as $x \to +\infty$; then the time-independence of the quantities C_m, see (7), implies, via (1) and (2), that the currents must also vanish at the other end, i.e. as $x \to -\infty$.

Note that, in contrast with the densities, the currents depend on the particular evolution equation (3), namely on the polynomial $\alpha(z,t)$; see below.

The approach adopted here leans on the algebra of differential operators, in analogy to the treatment of appendix A.20, to which the reader is referred for the notation used below. Moreover, in addition to the *integrodifferential* operators (see appendix A.9)

(9) $\quad L = D^2 - 2u + 2 Du I,$

(10) $\quad \tilde{L} = D^2 - 2u + 2 I u D,$

it is convenient to introduce the *purely differential* operator (see (A.9.-5))

(11a) $\quad F = LD = D\tilde{L},$

(11b) $\quad F = D^3 - 2(uD + Du).$

We start from the remark that if $-p^2$ is a discrete eigenvalue of the Schroedinger operator

(12) $\quad H = -D^2 + u,$

namely if

(13) $\quad H\varphi(x,t) = -p^2 \varphi(x,t),$

then $4p^2$ is a discrete eigenvalue of the operator $\tilde{L}$, i.e.

(14) $\quad \tilde{L}[\varphi(x,t)]^2 = 4p^2 [\varphi(x,t)]^2$

(set $k = ip$ in (A.9.-44) and use (A.9.-21), dropping the index n). On the other hand, if $u(x,t)$ evolves in time according to (3), the eigenvalue p is time-independent, a property that is evidently related to the possibility to recast (3) in the equivalent operator ("Lax") form

(15) $\quad H_t = [M, H];$

here H is given by (12), and M by (A.20.-51) (with $\eta(z,t) = 1$). Similarly it may be inferred that, if $u(x,t)$ satisfies (3), or equivalently, H satisfies (15), the operator $\tilde{L}$ undergoes an isospectral evolution, that may be recast in the ("Lax") operator form

(16) $\quad \tilde{L}_t = [\tilde{N}, \tilde{L}];$

where of course the operator $\tilde{N}$ (as indeed M in (15)) depends on the polynomial (4) characterizing the evolution equation (3). Indeed we proceed now to construct explicitly such an operator $\tilde{N}$. But before doing this, let us note that, if (16) holds, then also the operator L satisfies an analogous Lax equation,

(17) $\quad L_t = [N, L],$

with the operator N related to $\tilde{N}$ by the formula (see (11a))

(18) $\quad ND = D\tilde{N}.$

It is convenient to satisfy this relation by formally setting

(19) $\quad \tilde{N} = PD,$

(20) $\quad N = DP;$

the operator P so defined has then, as shown below, the remarkable property to be *purely differential* (i.e. its expression does not contain the integral operator I).

Consider now the evolution equation

(21) $\quad F_t = DPF - FPD,$

that follows from (11a), (17), (18), (19) and (20), and combine it with the time-derivative of the expression (11b), thereby obtaining the operator equation

(22) $\quad DPF - FPD = Da + aD,$

where we have introduced the multiplicative operator a defined by the formulae

(23) $\quad af(x) = a(x, t) f(x)$
(24) $\quad a(x, t) = -2\alpha(L, t) u_x(x, t).$

Note that the equation (22) has nothing to do with the time evolution, as the time enters only parametrically; indeed, we are now left with the problem of determining the pair of operators P and a that satisfy (22) with the condition that a be multiplicative (see (23) and (24)), and F be given by (11b).

The general solution of this problem can be obtained by first noting that, if P and a satisfy (22), then also P' and a', given by the transformation

(25) $\quad P' = PL + c - A + aI,$
(26) $\quad a'(x, t) = La(x, t) - 2cu_x(x, t),$

satisfy the same equation (22); the reader may easily verify that, in these formulae, c is an arbitrary constant (i.e. x-independent, but possibly dependent on t; and c is of course the corresponding multiplicative operator), while A is the multiplicative operator defined by the expressions

(27) $\quad Af(x) = A(x, t) f(x),$
(28) $\quad A(x, t) \equiv Ia(x, t) = \int_x^{+\infty} dy\, a(y, t)$

$\qquad = -2 \int_x^{+\infty} dy\, \alpha(L, t) u_y(y, t).$

Since the transformation equations (25) and (26) are not homogeneous, one can now easily produce a sequence of solutions of (22) by repeatedly applying (25) and (26) (with $c=0$ without loss of generality) to the first nontrivial solution

(29a) $\quad P^{(0)}=1$

(29b) $\quad a^{(0)}(x,t)=-2u_x(x,t),$

that itself obtains, from the trivial solution $P=a=0$, through (25) and (26) with $c=1$. It is easily found that after n iterations the solution thus obtained reads

$$(30) \quad P^{(n)}=L^n-2\sum_{j=0}^{n-1}\left(g^{(n-1-j)}+g_x^{(n-1-j)}I\right)L^j, \quad n=1,2,3,\ldots,$$

(31a) $\quad a^{(n)}(x,t)=-2L^n u_x(x,t), \quad n=1,2,3,\ldots,$

(31b) $\quad a^{(n)}(x,t)=-2g_x^{(n)}(x,t), \quad n=1,2,3,\ldots,$

where, of course, $g^{(m)}$ is the operator that multiplies by the function $g^{(m)}(x,t)$ defined by (5), while (31b) follows from (31a) by using (11a) and (5).

To prove that the operator $P^{(n)}$ is purely differential, it is convenient to consider the recursion relation

$$(32) \quad P^{(n+1)}=P^{(n)}J-2\left(g^{(n)}+Q^{(n)}\right), \quad n=0,1,2,\ldots,$$

that obtains by setting in (25), consistently with (30), $P=P^{(n)}$, $P'=P^{(n+1)}$, $c=0$, $a=a^{(n)}=-2g_x^{(n)}$, $A=2g^{(n)}$ and by rewriting the integrodifferential operator L in the convenient form

(33) $\quad L=J+2u_x I,$

with (see (9))

(34) $\quad J\equiv D^2-4u;$

moreover the operator $Q^{(n)}$ is defined by the operator equation

(35) $\quad P^{(n)}u_x=g_x^{(n)}+Q^{(n)}D.$

The proof proceeds now by recursion. Clearly $P^{(0)}$ is a purely differential operator (in fact, purely multiplicative; see (29a)). Assume then that $P^{(n)}$ is purely differential. Since

(36) $\quad P^{(n)}u_x(x,t) = g_x^{(n)}(x,t), \quad n = 0, 1, 2, \ldots,$

(as implied by the explicit expression (30) with (31)), also $Q^{(n)}$ is then purely differential (see (35) and (36); and note that the first of these two formulae, but not the second, is an operator equation). Hence, $P^{(n+1)}$ is also a purely differential operator (see (32) and (34)). Thus, by recursion, $P^{(n)}$ is, for all nonnegative integral values of n, a purely differential operator. Q.E.D.

In fact, combining (35) with the formal solution of (32),

(37) $\quad P^{(n)} = J^n - 2\sum_{j=0}^{n-1}\left(g^{(n-1-j)} + Q^{(n-1-j)}\right)J^j, \quad n = 1, 2, 3, \ldots,$

and starting from (29a) and $Q^{(0)} = 0$, there obtains a technique of computation by iteration, that is alternative to the explicit expression (30). For instance, the first two operators of the sequence (37) are

(38a) $\quad P^{(1)} = D^2 - 6u,$

(38b) $\quad P^{(2)} = D^4 - 10uD^2 - 10u_x D - 10u_{xx} + 30u^2.$

Since the equation (22) is linear, it is finally clear that its solution P corresponding to (24), i.e. to the evolution equation (3), obtains by taking the linear combination

(39) $\quad P = \sum_{n=0}^{M} \alpha_n(t) P^{(n)},$

where the coefficients $\alpha_n(t)$ are defined by (4) and $P^{(n)}$ is given by (30), or, alternatively, by (32) and (35). This expression, together with (19) ((20)), yields the operator $\tilde{N}(N)$ that enters into the "Lax" evolution equation (16) ((17)) for the operator $\tilde{L}(L)$.

Let us proceed now to derive the set (1) of local conservation laws. By differentiating (5) with respect to time, and by using (16) together with

(40) $\quad u_t(x,t) = \tilde{N}u(x,t),$

(that follows from (39), (36) and (3)), it is immediately found that all the densities $g^{(m)}$ satisfy the same equation

(41) $\quad g_t^{(m)}(x,t) = \tilde{N}g^{(m)}(x,t), \quad m = 0, 1, 2, \ldots .$

Comparing this equation with (1) there then follows that the currents $\gamma^{(m)}$ are defined by the formula

(42) $\quad \tilde{N}g^{(m)}(x,t) = \gamma_x^{(m)}(x,t), \quad m = 0, 1, 2, \ldots$.

This equation is readily seen to be satisfied by the expression

(43) $\quad \gamma^{(m)}(x,t) = \alpha(\tilde{L},t)g^{(m)}(x,t) + 2 \sum_{n=1}^{M} \alpha_n(t)$
$\cdot \sum_{j=0}^{n-1} \left[G^{(n-j-1, m+j)}(x,t) - G^{(m+j, n-j-1)}(x,t) \right], \quad m = 0, 1, 2, \ldots,$

where the quantities $G^{(m,n)}(x,t)$ are defined by

(44a) $\quad G_x^{(m,n)}(x,t) = -g^{(m)}(x,t)g_x^{(n)}(x,t), \quad m, n = 0, 1, 2, \ldots,$

with

(44b) $\quad G^{(m,n)}(+\infty, t) = 0, \quad m, n = 0, 1, 2, \ldots;$

indeed, this formula is easily derived using (19), (39), (30), (11a), (5).

It is remarkable that the currents $\gamma^{(m)}$ are nonlinear polynomial combination of u and its derivatives, this being a consequence of the expression (43) and of the fact, proved in appendix A.9, that $g^{(m)}$ and $G^{(m,n)}$ also have this same property. This result, moreover, automatically guarantees that the asymptotic condition (8a) is satisfied.

As for the second set (2) of conservation laws, the expressions of the currents $\varphi^{(m)}$ may be derived directly by differentiating with respect to time the expression (6) of the corresponding density $f^{(m)}$. In this manner, and using (17), one obtains

(45) $\quad f_t^{(m)}(x,t) = Nf^{(m)}(x,t) + L^m[f_t^{(0)}(x,t) - Nf^{(0)}(x,t)],$
$\qquad\qquad\qquad\qquad\qquad\qquad\qquad\qquad m = 0, 1, 2, \ldots$

But (6), (3), (11a) and (20) imply

(46) $\quad f_t^{(0)}(x,t) - Nf^{(0)}(x,t) = \{[x\alpha(\tilde{L},t)u(x,t)]_x$
$\qquad\qquad\qquad\qquad\qquad - P[xu_x(x,t) + 2u(x,t)]\}_x.$

There immediately follows, from (45) and (2), that

(47) $\quad \varphi^{(m)}(x,t) = Pf^{(m)}(x,t) + \tilde{L}^m\{[x\alpha(\tilde{L},t)u(x,t)]_x$
$\qquad\qquad\qquad - P[xu_x(x,t) + 2u(x,t)]\}, \quad m = 0, 1, 2, \ldots,$

or, equivalently,

(48) $\quad \varphi^{(m)}(x,t) = (PL^m - \tilde{L}^m P)[xu_x(x,t) + 2u(x,t)]$
$\quad\quad\quad + \tilde{L}^m[x\alpha(\tilde{L},t)u(x,t)]_x, \quad m = 0,1,2,\ldots .$

Of course, in these expressions, the operator P is given by the formula (39), that explicitly shows its dependence on the polynomial $\alpha(z,t)$. Note that the asymptotic condition (8b) is obviously satisfied by the expression (48).

It should be finally pointed out that neither the densities $f^{(m)}$ nor the currents $\varphi^{(m)}$ have the locality property possessed by $g^{(m)}$ and $\gamma^{(m)}$, namely the property to be polynomials in u and its derivatives (see below).

We conclude this appendix by displaying the explicit expressions of the first 3 densities (of both types):

(49a) $\quad g^{(0)} = u$

(49b) $\quad g^{(1)} = u_{xx} - 3u^2,$

(49c) $\quad g^{(2)} = u_{xxxx} - 10u_{xx}u - 5u_x^2 + 10u^3,$

(50a) $\quad f^{(0)} = (xu)_x + u,$

(50b) $\quad f^{(1)} = \left[x(u_{xx} - 3u^2) + 3u_x + 2u\int_x^{+\infty} dy\, u(y,t) \right]_x - 3u^2,$

(50c) $\quad f^{(2)} = \left[(f^{(1)} - 6u^2)_x + x(-4u_{xx}u + u_x^2 + 10u^3) \right.$
$\quad\quad\quad \left. - 6u^2 \int_x^{+\infty} dy\, u(y,t) - 6u \int_x^{+\infty} dy\, u^2(y,t) \right]_x + 5(u_x^2 + 2u^3).$

We also report the corresponding currents, for the KdV case ($\alpha(z,t) = -z$, $u_t + u_{xxx} - 6u_x u = 0$):

(51a) $\quad \gamma^{(0)} = -u_{xx} + 3u^2,$

(51b) $\quad \gamma^{(1)} = -u_{xxxx} + 3u_x^2 + 12u_{xx}u - 12u^3,$

(51c) $\quad \gamma^{(2)} = -u_{xxxxxx} + 16uu_{xxxx} + 24u_x u_{xxx} + 23u_{xx}^2$
$\quad\quad\quad - 90u^2 u_{xx} - 60uu_x^2 + 45u^4,$

(52a) $\quad \varphi^{(0)} = -[x(u_{xx} - 3u^2)]_x,$

(52b) $\quad \varphi^{(1)} = -x(u_{xxxx} - 12u_{xx}u - 3u_x^2 + 12u^3)_x - 2(u_{xx} - 3u^2)_x$
$\quad\quad\quad \cdot \int_x^{+\infty} dy\, u(y,t) - 3u_{xxxx} + 26uu_{xx} + 15u_x^2 - 18u^3,$

(52c) $\varphi^{(2)} = \varphi_{xx}^{(1)} + x\left(4uu_{xxxx} - 6u_x u_{xxx} + 5u_{xx}^2 - 54u^2 u_{xx} + 12uu_x^2 + 30u^4\right)_x$
$\quad + 6(u_{xxx} - 6uu_x)\left[2u\int_x^{+\infty} dy\, u(y,t) - \int_x^{+\infty} dy\, u^2(y,t)\right]$
$\quad + 12uu_{xxxx} + 2u_x u_{xxx} + 5u_{xx}^2 - 120u^2 u_{xx} - 78uu_x^2 + 87u^4.$

A.22. "Variable phase approach" to the Schroedinger scattering problem on the whole line

The (three-dimensional) nonrelativistic quantal scattering problem of a beam of particles off a spherically symmetrical potential can be reduced to the calculation of the scattering phase shifts, each of which represents the effect of the potential on one angular momentum component of the (incoming and outgoing) beam. The values of these quantities are related to the asymptotic behavior, as the radial coordinate tends to infinity, of an appropriate solution of the radial (stationary) Schroedinger equation. Thus the evaluation of the scattering phase shift requires the solution of the radial Schroedinger equation, namely of a second order ODE, in the interval from $r=0$ to $r=\infty$. For instance, to compute the S-wave scattering phase shift $\delta(k)$ one solves the S-wave radial Schroedinger equation

(1) $\quad -\psi_{rr}(r,k) + v(r)\psi(r,k) = k^2 \psi(r,k),$

with boundary conditions

(2) $\quad \psi(0,k) = 0, \quad \psi_r(0,k) = 1,$

and then extracts $\delta(k)$ from the asymptotic behavior of ψ via the formula

(3) $\quad \psi(r,k) \to c(k)\sin[kr + \delta(k)], \quad r \to \infty.$

Here $c(k)$ is a quantity whose value is irrelevant to the scattering process, and $v(r)$ in (1) is the scattering potential ($v(r)$ is of course assumed to vanish as $r \to \infty$, consistently with the asymptotic behavior (3)).

An alternative approach to the computation of the scattering phase shift $\delta(k)$ is based on the introduction of the "variable phase function" $\delta(k,r)$. This function is defined by the first-order nonlinear ODE

(4) $\quad \delta_r(k,r) = -k^{-1} v(r) \sin^2[kr + \delta(k,r)],$

with boundary condition

(5) $\quad \delta(k,0)=0.$

It can be shown that $\delta(k,\rho)$ is then precisely the (S-wave) scattering phase shift produced by the potential $v(r)\theta(\rho-r)$ (where θ is the usual step function, $\theta(x)=1$ if $x>0$, $\theta(x)=0$ if $x<0$), namely by the potential $v(r)$ amputated of its part extending beyond ρ. This is clearly consistent with the boundary condition (5) (if the potential is completely eliminated, there is no scattering, and therefore the scattering phase shift vanishes), and it implies obviously that the scattering phase shift produced by $v(r)$ coincides with the asymptotic value of $\delta(k,\rho)$ as ρ tends to infinity,

(6) $\quad \delta(k)=\delta(k,\infty).$

This approach to potential scattering theory presents a number of advantages [C1967] [B1968]. Purpose and scope of this appendix is to indicate how an analogous approach can be developed for the (one-dimensional) Schroedinger scattering problem on the whole line, and in particular for the evaluation of the reflection coefficient $R(k)$, that plays such a crucial rôle in the spectral transform. Our treatment is quite terse, and limited to display a few formulae that exhibit rather transparently the relationship between the reflection coefficient and the potential $u(x)$ that underlies it, this being the main merit of the "variable phase approach." The reader interested in a more complete study of this method and of its implications in the context of the one dimensional Schroedinger scattering problem should have no difficulty to pursue the matter on his own, perhaps relying for guidance on the analogous treatment in the context of the radial Schroedinger equation [C1967] [B1968].

Let us then consider the one-dimensional Schroedinger scattering problem of chapter 2. We report a few relevant formulae:

(7) $\quad -\psi_{xx}(x,k,\xi)+u(x,\xi)\psi(x,k,\xi)=k^2\psi(x,k,\xi),$

(8) $\quad \psi(x,k,\xi)\to\exp(-ikx)+R(k,\xi)\exp(ikx), \quad x\to\infty,$

(9) $\quad 2ikR_\xi(k,\xi)=\int_{-\infty}^{+\infty}dx\left[\psi(x,k,\xi)\right]^2 u_\xi(x,\xi).$

Here we have introduced an explicit dependence on a parameter ξ; apart from this insertion, the first two formulae, (7) and (8), reproduce (2.1.-1) and (2.1.-3b), while (9) coincides with the special case of (2.4.1.-12a) corresponding to $g(z)=1$ and to the attribution of the variation to an (infinitesimal) change of the parameter ξ (so that $\delta R=R_\xi d\xi$, $\delta u=u_\xi d\xi$).

Let us now set

(10) $\quad u(x,\xi)=u(x)\theta(\xi-x);$

thus $u(x,\xi)$ coincides with $u(x)$ for $x<\xi$ and vanishes identically for $x>\xi$. Then (8) implies

(11) $\quad \psi(x,k,\xi)=\exp(-ikx)+R(k,\xi)\exp(ikx) \quad$ for $x\geqslant\xi$,

while (10) implies

(12) $\quad u_\xi(x,\xi)=u(\xi)\delta(x-\xi)$

and therefore (9) yields

(13) $\quad 2ikR_\xi(k,\xi)=u(\xi)[\psi(\xi,k,\xi)]^2.$

This last equation, together with (11), yields the basic equation of the "variable phase method," namely

(14) $\quad R_x(k,x)=(2ik)^{-1}u(x)[\exp(-ikx)+R(k,x)\exp(ikx)]^2.$

It is obvious that this first order ODE (having the form of a Riccati equation) is supplemented by the boundary condition

(15) $\quad R(k,-\infty)=0,$

and that the function $R(k,x)$, uniquely defined by (14) and (15), converges asymptotically to the reflection coefficient $R(k)$:

(16) $\quad R(k,\infty)=R(k).$

Indeed by definition (see (10)) $R(k,x)$ is just the reflection coefficient produced by the potential u amputated of its part extending beyond (i.e., to the right of) x.

A natural *ansatz* suggested by (14) reads

(17) $\quad R(k,x)=-\exp[2i\eta(k,x)],$

since this transforms (14) into the (real) equation

(18) $\quad \eta_x(k,x)=-k^{-1}u(x)\sin^2[kx+\eta(k,x)],$

whose resemblance (say, coincidence) with (4) is quite remarkable.

But the *ansatz* (17) is inconvenient when it comes to satisfying the boundary condition (15); thus this approach does not provide a satisfactory procedure to evaluate $R(k, x)$, and therefore also $R(k)$, see (16). An amusing way to bypass this difficulty, that might indeed provide the most convenient way for the actual (numerical) computation of the reflection coefficient $R(k)$, can be based on the general property of the Riccati equation, according to which any 4 different solutions $R^{(j)}(k, x)$ of (14) are related by the formula

$$(19) \quad (R^{(1)} - R^{(4)})(R^{(2)} - R^{(3)}) / [(R^{(1)} - R^{(3)})(R^{(2)} - R^{(4)})] = C$$

where C is a constant (i.e., C does not depend on x; it might of course depend on k). One may then introduce 3 (real) solutions $\eta^{(j)}(k, x)$ of (18), identified by the 3 (different) boundary conditions

$$(20) \quad \eta^{(j)}(k, -\infty) = \eta_-^{(j)}(k), \quad j=1,2,3$$

(a convenient simple choice might be $\eta_-^{(1)} = 0$, $\eta_-^{(2)} = \frac{1}{4}\pi$, $\eta_-^{(3)} = \frac{1}{2}\pi$; these three constants must of course differ mod (π), see (17)). The quantities $\eta^{(j)}(k, x)$ are then easily computed by integrating numerically (18) with (20) (the 3 integrations can be conveniently carried out in parallel, starting of course from a negative value of x sufficiently large so that $u(x)$ be negligibly small; the use of a variable integration step, roughly proportional to $|u(x)|^{-1}$, might moreover be profitable). The quantity $R(k, x)$ is then provided by the explicit formula

$$(21) \quad R(k,x) = -\{\exp[i\sigma^{(1)}]\sin[\delta^{(2)} - \delta^{(3)}] + \exp[i\sigma^{(2)}]\sin[\delta^{(3)} - \delta^{(1)}]$$
$$+ \exp[i\sigma^{(3)}]\sin[\delta^{(1)} - \delta^{(2)}]\} / \{\exp[-i\delta^{(1)}]\sin[\sigma^{(2)} - \sigma^{(3)}]$$
$$+ \exp[-i\delta^{(2)}]\sin[\sigma^{(3)} - \sigma^{(1)}] + \exp[i\delta^{(3)}]\sin[\sigma^{(1)} - \sigma^{(2)}]\}$$

where

$$(22a) \quad \delta^{(j)} \equiv \delta^{(j)}(k, x) = \eta^{(j)}(k, x) - \eta_-^{(j)}(k), \quad j=1,2,3,$$
$$(22b) \quad \sigma^{(j)} \equiv \sigma^{(j)}(k, x) = \eta^{(j)}(k, x) + \eta_-^{(j)}(k), \quad j=1,2,3.$$

Note that this formula is obviously consistent with (15), and it yields the reflection coefficient $R(k)$ via (16) (it is of course sufficient to consider an adequately large but finite value of x, since $R(k, \infty) \approx R(k, x_\infty)$ provided $u(x)$ is sufficiently small for $x \geq x_\infty$; see (14)).

An alternative *ansatz* that might appear appropriate to solve (14) and to accomodate the boundary condition (15) reads

(23) $\quad R(k,x) = \text{tgh}[\gamma(k,x)] \exp[i\chi(k,x)],$

with γ and χ real and

(24) $\quad \gamma(k, -\infty) = 0.$

Indeed this yields for $\gamma(k, x)$ the simple equation

(25) $\quad \gamma_x(k,x) = -(2k)^{-1} u(x) \sin[2kx + \chi(k,x)]$

that, together with (24), implies

(26) $\quad \gamma(k,x) = -(2k)^{-1} \int_{-\infty}^{x} dy\, u(y) \sin[2ky + \chi(k,y)],$

while the equation for $\chi(k, x)$ is also rather neat, reading

(27) $\quad \chi_x(k,x) = -k^{-1} u(x) \{1 + \cotgh[2\gamma(k,x)] \cos[2kx + \chi(k,x)]\}.$

But the disadvantage of the *ansatz* (23) is that, wherever R, and therefore γ, vanish, the definition of the phase χ becomes ambiguous. In fact, at the finite values x_n where $\gamma(k, x)$ vanishes, it follows from (27) that χ must coincide (mod 2π) with $-2kx_n + \pi$. But this prescription is not sufficient to identify the value of $\chi(k, x)$ as $x \to -\infty$, that is on the other hand needed in order to evaluate $\chi(k, x)$ using (27) with (26). Note moreover that the zeros x_n of $\gamma(k, x)$ are themselves known only after $\gamma(k, x)$ is known, and this in its turn requires knowledge of $\chi(k, y)$ for $-\infty < y < x$ (see (26)).

A more pedestrian approach to the numerical solution of (14) with (15) may be based on the standard separation of $R(k, x)$ into its real and imaginary parts:

(28) $\quad R(k,x) = A(k,x) + iB(k,x).$

Then (14) yields the two coupled real equations

(29a) $\quad A_x = (2k)^{-1} u(x) [2B + 2AB \cos(2kx) + (A^2 - B^2 - 1) \sin(2kx)],$

(29b) $\quad B_x = (2k)^{-1} u(x)$
$\quad\quad\quad \cdot [-2A + 2AB \sin(2kx) - (A^2 - B^2 + 1) \cos(2kx)],$

while (15) reads of course

(30a) $A(k, -\infty) = 0$,

(30b) $B(k, -\infty) = 0$.

Qualitative features and approximate expressions of the reflection coefficient are readily obtained in the framework of this approach, by relying on the analogous results obtained in the framework of the radial Schroedinger problem [C1967], [B1968]. Here we outline only the results relevant to the $k \to 0$ limit, that may be compared with the findings reported in appendices A.15 and A.1.

It is clear from (14) that, for very small k, $R(k, x)$ changes very rapidly (as a function of x), unless $R(k, x) \approx -1$. This suggests the expansion

(31) $$R(k, x) = -1 + \sum_{m=1}^{M} (2ik)^m r^{(m)}(x) + O(k^{M+1}).$$

Insertion of this expansion in (14) yields the set of (real) equations

(32) $$r_x^{(m)}(x) = u(x) \left\{ [1 - (-1)^m] x^{m+1}/(m+1)! \right.$$
$$+ \sum_{n=1}^{m} r^{(n)}(x) r^{(m+1-n)}(x) + \sum_{n=1}^{m} \left[x^{m+1-n}/(m+1-n)! \right]$$
$$\left. \cdot \sum_{j=0}^{n} r^{(j)}(x) r^{(n-j)}(x) \right\}, \quad m = 1, 2, 3, \ldots .$$

Note that, while the first equation of this set,

(33) $r_x^{(1)}(x) = u(x) [r^{(1)}(x) - x]^2$

is nonlinear, all the others are linear ODEs:

(34a) $r_x^{(2)} = u(x) [2 r^{(2)} + x r^{(1)}] [r^{(1)} - x],$

(34b) $r_x^{(3)} = u(x) \{ 2 r^{(3)} [r^{(1)} - x] + r^{(2)} [r^{(2)} + 2x r^{(1)} - x^2]$
$\qquad + \frac{1}{6} x^2 r^{(1)} [3 r^{(1)} - 2x] + \frac{1}{12} x^4 \},$

and so on.

It is convenient (see below) to replace the function $r^{(1)}(x)$ by the function $\mu(x)$ via the position

(35) $\quad r^{(1)}(x) = y \cotg[\mu(x)] + z$

where y and z are arbitrary real constants (with the restriction $y>0$). Then (33) reads

(36) $\quad \mu_x(x) = -yu(x)\{[(x-z)/y]\sin\mu(x) - \cos\mu(x)\}^2.$

The coincidence of this equation with (A.1.-12) has motivated the position (35); indeed the coincidence of the function $\mu(x)$ introduced here via (35), with the function $\mu(x)$ introduced via (A.1.-12) and (A.1.-13) (see (36), and (37) below), provides (in close analogy to the treatment of chapter 22 of [C1967]) a result that may be considered an analog, valid in the context of the Schroedinger scattering problem on the whole line, of the "Levinson's theorem" valid for the radial Schroedinger problem. (Let us recall in this connection that there exists another result, referring to the value at $k=0$ of the phase $\theta(k)$ of the transmission coefficient $T(k)$, that may also be considered an analog of Levinson's theorem; see (2.1.-44) and appendix A.4).

It is easily seen that the boundary condition that supplements (36) reads

(37) $\quad \mu(-\infty) = 0;$

indeed in this manner the apparent contradiction between (15) and (31) is bypassed, since (37) and (35) imply that $r^{(1)}(x)$ is singular at $x=-\infty$. This indicates a breakdown of the expansion (31) at $x=-\infty$, as it is of course implied by (15). An analogous breakdown of the expansion (31) occurs at all the other values x_n of x (if any) such that

(38) $\quad \mu(x_n) = 0 \pmod{\pi};$

this is consistent with the findings of appendix A.15, since the results of appendix A.1 imply that the potential $u(x)\theta(x_n-x)$ is precisely one possessing a "zero-energy bound state" (as defined in appendix A.15), and therefore $R(0, x_n) \neq -1$. Note incidentally that this argument applies as well to the value of $R(k, x)$ at $x=-\infty$ ($R(k, -\infty) \neq -1$; see (15)), since the potential $u(x)\theta(-\infty-x)$, namely the identically vanishing potential $u(x)=0$, is indeed in the class of the potentials possessing a zero energy bound state (as defined in appendix A.15; a potential $u(x)$ that does not vanish identically and is nowhere positive, $u(x) \leq 0$, always possesses at least one bound state, while of course a potential cannot support any bound state if it is nowhere negative).

A.23. KdV and higher KdV equations as hamiltonian flows: an outline

An important contribution to "soliton theory" comes from the remarkable fact that the nonlinear evolution equations, that are solvable by the spectral transform technique, can also be written as Hamiltonian flows. This appendix provides an elementary introduction to this topic, restricted to a terse presentation of the basic ideas and results.

Let us recall some basic definitions concerning a Hamiltonian system with N degrees of freedom. As usual, let us denote by $q \equiv (q_1, \ldots, q_N)$, respectively by $p \equiv (p_1, \ldots, p_N)$, the N configuration, respectively momentum, coordinates characterizing the state of the dynamical system. The evolution in time is then described by the Hamiltonian function $H(q, p)$ of $2N$ variables via Hamilton's equations,

(1) $\quad q_t = \nabla_p H,$

(2) $\quad p_t = -\nabla_q H.$

Here

(3) $\quad q_t = (dq_1/dt, \ldots, dq_N/dt),$

(4) $\quad p_t = (dp_1/dt, \ldots, dp_N/dt),$

and

(5) $\quad \nabla_q H = (\partial H/\partial q_1, \ldots, \partial H/\partial q_N),$

(6) $\quad \nabla_p H = (\partial H/\partial p_1, \ldots, \partial H/\partial p_N).$

The expression of the time-derivative of any differentiable function of the coordinates q and p, say $A(q, p)$, reads (see (1) and (2))

(7) $\quad dA/dt = \nabla_q A \cdot \nabla_p H - \nabla_p A \cdot \nabla_q H = \sum_{j=1}^{N} (A_{q_j} H_{p_j} - A_{p_j} H_{q_j});$

this suggests to introduce the operation

(8) $\quad \{A, B\} \equiv \nabla_q A \cdot \nabla_p B - \nabla_p A \cdot \nabla_q B = \sum_{j=1}^{N} (A_{q_j} B_{p_j} - A_{p_j} B_{q_j}),$

that associates with the pair of functions $A(q, p)$ and $B(q, p)$, the function $\{A, B\}$, called the *Poisson bracket* of A and B. The set of (infinitely differentiable) functions defined on the $2N$-dimensional phase space (whose points have coordinates q and p) is thereby equipped with the bilinear product operation (8), that is antisymmetric,

(9a) $\quad \{A, B\} = -\{B, A\},$

and satisfies the Jacobi identity

(9b) $\quad \{A, \{B, C\}\} + \{B, \{C, A\}\} + \{C, \{A, B\}\} = 0.$

In fact this set of functions is said to have a Poisson structure, to indicate that also the Leibnitz rule

(9c) $\quad \{A, BC\} = B\{A, C\} + C\{A, B\}$

holds.

Thus (7) reads

(10) $\quad \mathrm{d}A/\mathrm{d}t = \{A, H\}.$

A constant of motion (or integral) C of (4) is therefore a function of q and p, whose Poisson bracket with the Hamiltonian function vanishes, i.e.

(11) $\quad \{C, H\} = 0.$

An obvious consequence of the Jacobi identity (9b) is that the Poisson bracket of two constants of motion is itself a constant of motion. The following definitions are useful in the present context: two constants of motion C_1 and C_2 are said to be *in involution* if their Poisson bracket vanishes,

(12) $\quad \{C_1, C_2\} = 0;$

two constants of motion C_1 and C_2 are said to be mutually *independent* if (a_1 and a_2 being constants)

(13a) $\quad a_1 \nabla_q C_1 + a_2 \nabla_q C_2 = 0,$
(13b) $\quad a_1 \nabla_p C_1 + a_2 \nabla_p C_2 = 0$

imply $a_1 = a_2 = 0$. Moreover a dynamical system with N degrees of freedom is said to be *completely integrable* if there exist N independent constants of motion in involution. Indeed in this case a classical theorem by Liouville guarantees that, given the values of these N constants, the Hamilton's equations, (1) and (2), can be solved by quadratures.

In order to apply these definitions to a broad class of evolution equations, it is convenient to rewrite Hamilton's equations, (1) and (2), in the form

(14) $\quad u_t = \Omega \nabla H,$

where u is the set of $2N$ coordinates in phase-space, i.e.

(15) $\quad u \equiv (q_1, \ldots, q_N, p_1, \ldots, p_N),$

u_t is the (tangent) vector

(16) $\quad u_t = (dq_1/dt, \ldots, dq_N/dt, dp_1/dt, \ldots, dp_N/dt),$

and ∇H is the gradient (or cotangent vector, or one-form)

(17) $\quad \nabla H \equiv (\partial H/\partial q_1, \ldots, \partial H/\partial q_N, \partial H/\partial p_1, \ldots, \partial H/\partial p_N).$

Comparing (14) with (4) shows that Ω is the $2N \times 2N$ antisymmetric matrix

(18) $\quad \Omega \equiv \begin{pmatrix} 0 & 1 \\ -1 & 0 \end{pmatrix},$

where of course the off-diagonal entries are the unit $N \times N$ matrix (with opposite sign). In this notation, with the standard meaning of the symbol $(\cdot, \cdot)$, the Poisson bracket (8) takes the expression

(19) $\quad \{A, B\} = (\nabla A, \Omega \nabla B);$

and the equations (9), (10), (11) and (12) do not change, while (13) should be replaced by $a_1 \nabla C_1 + a_2 \nabla C_2 = 0$.

This notation suggests the following generalization. A dynamical system is Hamiltonian if its time-evolution equation takes the form (14), with $H = H(u)$ a function defined in phase space, and Ω an antisymmetrical linear operator taking gradients (i.e. one-forms) into (tangent) vectors, and such that the expression (19) defines a Poisson structure in the set of (smooth) functions defined in phase-space, i.e. it satisfies the conditions (9). Note that in general the gradient ∇A of $A(u)$ is defined through the formula for an infinitesimal variation,

(20) $\quad \delta A \equiv (\delta u, \nabla A).$

Thus, in a given phase space, a Hamiltonian system is defined by the pair (Ω, H) formed by the ("symplectic") operator Ω and the (Hamiltonian) function $H(u)$. Note that with this generalization, the number of degrees of freedom need not be finite.

Let us now identify the phase space, the operator Ω and the Hamiltonian $H(u)$ for the KdV equation

(21) $\quad u_t = -u_{xxx} + 6u_x u, \quad u \equiv u(x,t),$

within the class of spectral transformable solutions. The phase space is the set of *bona fide* potentials (see subsection 2.3); the state of the system is characterized by a *bona fide* potential $u(x)$, and its evolution by $u(x,t)$. This dynamical system has clearly infinitely many degrees of freedom. If $A(u)$ is a functional defined on the set of *bona fide* potentials $u(x)$, the formula (20) reads

(22) $\quad \delta A = \int_{-\infty}^{+\infty} dx \, \delta u(x) \nabla A;$

thus the gradient of $A(u)$ is the usual functional derivative,

(23) $\quad \nabla A = \delta A / \delta u(x).$

As for the other identifications, it is convenient to rewrite the KdV equation (21) in the operator form (set $\alpha(z) = -z$ in (5.1))

(24) $\quad u_t = -Lu_x = -D\tilde{L}u,$

where the second equality follows from the operator identity (A.9.-30); here D is the differential operator,

(25) $\quad D \equiv d/dx,$

while L and $\tilde{L}$ are the integrodifferential operators (see (A.9.-11, 12))

(26a) $\quad L = D^2 - 2u + 2DuI,$
(26b) $\quad \tilde{L} = D^2 - 2u + 2IuD,$

with

(27) $\quad I \equiv \int_{x}^{+\infty} dy$

(for their properties see appendix A.9). The form (24) of the KdV equation, by comparison with the r.h.s. of (14), suggests the following expressions:

(28) $\quad \Omega = D$

(29) $\quad H(u) = \int_{-\infty}^{+\infty} dx \, (\tfrac{1}{2} u_x^2 + u^3) = \tfrac{1}{2} C_2,$

the last equality referring to the notation introduced in subsection 1.7.3, or chapter 5 or appendix A.9 (see (1.7.3.-12c) or (A.9.-58c)). Indeed the Poisson bracket (19), that, as implied by (23) and (25), reads

(30) $\quad \{A, B\} = \int_{-\infty}^{+\infty} dx \, [\delta A/\delta u(x)] [\delta B/\delta u(x)]_x,$

satisfies the conditions (9); moreover, it is easily verified that, with the definition (29), there holds the formula

(31) $\quad \delta H/\delta u(x) = -\tilde{L} u(x),$

and this implies that (24), and therefore also the KdV equation (21), coincides with the Hamilton's equations (14) (see (23) and (28)),

(32) $\quad u_t(x, t) = \Omega \delta H/\delta u(x, t).$

In this context, the time-independence of the quantities (see (A.9.-57), or (5.-20))

(33) $\quad C_m = (-1)^m (2m+1)^{-1} \int_{-\infty}^{+\infty} dx \, \tilde{L}^m u(x, t), \quad m = 0, 1, 2, \ldots,$

when $u(x, t)$ evolves according to the KdV equation (21), follows from their property of being (obviously independent functionals) in involution, i.e.

(34) $\quad \{C_m, C_n\} = 0, \quad m, n = 0, 1, 2, \ldots,$

and from the fact that the Hamiltonian functional (29), corresponding to the KdV equation, is proportional to C_2 (see (29)). The validity of the important formula (34) is a consequence of (30), and of the identity

(35) $\quad \delta C_m/\delta u(x) = (-1)^{m+1} 2 \tilde{L}^{m-1} u(x), \quad m = 1, 2, \ldots,$

that follows from (A.9.-63a, 55a, 30), (22) and (23); in fact the resulting expression,

(36) $\{C_m, C_n\} = (-1)^{m+n} 4 \int_{-\infty}^{+\infty} dx [\tilde{L}^{m-1} u(x)] D[\tilde{L}^{n-1} u(x)],$

$$m, n = 1, 2, 3, \ldots,$$

together with the operator identity (A.9.-54c), imply precisely (34) (note that $\{A, C_0\} = 0$ is trivially satisfied for any functional $A(u)$ because $\delta C_0 / \delta u(x)$ = constant, see (33) with $m = 0$).

From the involutory property (34) of the infinite set of functionals C_m with respect to the Poisson bracket (30), it is clear that the C_m's are constants of the motion not only for the KdV equation, but also for the evolution equation (32) obtained by choosing as Hamiltonian any linear combination of the functionals C_m. Thus the quantities C_m are constants of motion for any higher KdV equation,

(37) $u_t = (-1)^n D \tilde{L}^n u, \quad n = 0, 1, 2, 3, \ldots,$

since this can be recast in the Hamiltonian form (32) with

(38) $H(u) = \frac{1}{2} C_{n+1}$

(see (28) and (35)).

In fact any equation of the class

(39) $u_t(x, t) = \alpha(L, t) u_x(x, t),$

with (see (5.-1, 2))

(40) $\alpha(z, t) = \sum_{m=0}^{M} \alpha_m(t) z^m,$

describes an Hamiltonian system, the corresponding Hamiltonian functional being

(41) $H = \frac{1}{2} \sum_{m=0}^{M} (-1)^m \alpha_m(t) C_{m+1};$

note that now a time-dependence of H may (trivially) originate from the time-dependence of the coefficients $\alpha_m(t)$. However the quantities (33) are still constants of the motion, i.e.

(42) $\{C_m, H\} = 0, \quad m = 0, 1, 2, \ldots .$

As a simple application of the Hamiltonian formalism, we prove that the evolution of the Schroedinger operator $-d^2/dx^2 + u(x,t)$ is isospectral for any flow associated with (38) (and therefore also with (41)), namely that

(43) $\quad dE/dt = 0,$

if $u(x,t)$ evolves according to (38) and the eigenvalue E is defined as follows:

(44a) $\quad [-d^2/dx^2 + u(x,t)] \varphi(x,t) = E\varphi(x,t),$

(44b) $\quad \int_{-\infty}^{+\infty} dx [\varphi(x,t)]^2 = 1.$

To prove this, it is sufficient to show that

(45) $\quad \{E, C_m\} = 0, \quad m = 0, 1, 2, \ldots;$

but this relation is implied by (30), (35), (2.4.1.-51), that reads

(46) $\quad \delta E/\delta u(x,t) = [\varphi(x,t)]^2,$

and by the wronskian integral relation (2.4.1.-56) (setting $g(z) = z^{m-1}$ and using (A.9.-30)).

As a second example, let us compute the time evolution, again corresponding to the flow with the hamiltonian (38), of the reflection coefficient $R(k,t)$ corresponding to $u(x,t)$ via the formulae (see (2.1.-1) and (2.1.-3)):

(47a) $\quad [-d^2/dx^2 + u(x,t)] \psi = k^2 \psi, \quad \psi \equiv \psi(x,k,t),$

(47b) $\quad \psi(x,k,t) \to T(k) \exp(-ikx), \quad x \to -\infty,$

(47c) $\quad \psi(x,k,t) \to \exp(-ikx) + R(k,t) \exp(ikx), \quad x \to +\infty.$

Its time-derivative, given by (10), i.e. (see (30))

(48) $\quad R_t(k,t) = \{R(k,t), H\} = \int_{-\infty}^{+\infty} dx [\delta R(k,t)/\delta u(x,t)]$
$\quad \cdot [\delta H/\delta u(x,t)]_x,$

where H is the Hamiltonian (38), takes the expression (see also (A.9.-30))

(49) $\quad R_t(k,t) = (-1)^n (2ik)^{-1} \int_{-\infty}^{+\infty} dx [\psi(x,k,t)]^2 L^n u_x(x,t),$

where we have taken into account (35) and the formula

(50) $\quad \delta R(k,t)/\delta u(x,t) = (2ik)^{-1}[\psi(x,k,t)]^2,$

that follows from the wronskian integral relation (2.4.1.-12a) (with $g(z)=1$). Finally, by using the wronskian integral relation (2.4.1.-12b) with $g(z)=z^n$, one obtains

(51) $\quad R_t(k,t) = (-1)^n (2ik)^{2n+1} R(k,t),$

which is precisely the evolution equation corresponding to the higher KdV equation (37) (cf. (3.1.-42a) with $\alpha(z,t)=(-1)^n z^n$).

A remarkable property of the evolution equation (39) is that it does not identify uniquely the pair (Ω, H) through which it can be written in the Hamiltonian form (32). This originates from the fact that any one of the operators

(52) $\quad \Omega_n \equiv (-1)^n D(\mathcal{L}^T)^n = (-1)^n \mathcal{L}^n D$

can be equally well chosen to define, via the expression (see (19) and (23))

(53) $\quad \{A, B\}_n = \int_{-\infty}^{+\infty} dx [\delta A/\delta u(x)] \Omega_n [\delta B/\delta u(x)], \quad n=0,1,2,\ldots,$

a Poisson bracket satisfying the condition (9). In the definition (52) we have introduced the operator

(54) $\quad \mathcal{L} \equiv D^2 - 2u - 2Du\mathcal{I},$

that differs from the operator L, see (26a), due to the replacement of the integral operator I, see (27), by the operator $-\mathcal{I}$, where

(55) $\quad \mathcal{I} = \tfrac{1}{2}(I^T - I) = \tfrac{1}{2}\left(\int_{-\infty}^{x} dy - \int_{x}^{+\infty} dy\right).$

It is easily seen that the equations

(56a) $\quad \mathcal{I}^T = -\mathcal{I},$

(56b) $\quad \mathcal{I}D = D\mathcal{I} = 1$

hold for the class of functions $f(x)$ such that $f(+\infty) + f(-\infty) = 0$; moreover the second equality in (52) follows from the identity

(57) $\quad \mathcal{L} D = D \mathcal{L}^T$,

and from the property of the operator $\mathcal{L}^T$,

(58) $\quad \mathcal{L}^T = D^2 - 2u - 2\mathcal{I} uD$,

to map a function $f(x)$ satisfying the condition $f(+\infty)+f(-\infty)=0$ into a function, $g(x)=\mathcal{L}^T f(x)$, that satisfies the same condition, $g(+\infty)+g(-\infty)=0$ (as noted above, this condition is necessary for (56a) to hold). Thus the operator Ω_n defined by (52) is clearly antisymmetrical, as it is required by the property (9a) of the Poisson bracket (53); however, while the property (9c) is trivially satisfied by (53), the proof of the Jacobi identity (9b) is not straightforward for $n>0$ (and is therefore omitted), due to the dependence of Ω_n on u (see (52) and (54)).

As for the evolution equations, it should be noted that the class (39) can also be rewritten as

(59) $\quad u_t(x,t) = \alpha(\mathcal{L}, t) u_x(x, t)$,

with (40) and (54); indeed the equality

(60) $\quad \mathcal{L}^n u_x(x, t) = L^n u_x(x, t), \quad n = 0, 1, 2, \ldots$,

easily follows by recursion from the formula

(61) $\quad Lf(x) = \mathcal{L} f(x) + u_x(x, t) \int_{-\infty}^{+\infty} dy f(y)$,

and from the identity (A.9.-54a),

(62) $\quad \int_{-\infty}^{+\infty} dx \, L^n u_x(x, t) = 0, \quad n = 0, 1, 2, \ldots$.

Moreover it is easily seen that the functionals C_m, see (33), are in involution also with respect to the Poisson bracket (53), i.e.

(63) $\quad \{C_m, C_j\}_n = 0, \quad m, j, n = 0, 1, 2, \ldots$.

In fact, because of the equality

(64) $\quad (\mathcal{L}^T)^n u(x, t) = \tilde{L}^n u(x, t), \quad n = 0, 1, 2, \ldots$,

(which itself follows from the formula

(65) $\quad \tilde{L}f(x) = \pounds^T f(x) + \int_{-\infty}^{+\infty} dy\, u(y,t) f_y(y),$

and from the identity (see (A.9.-54c))

(66) $\quad \int_{-\infty}^{+\infty} dx\, u(x,t) \boldsymbol{D} \tilde{L}^n u(x,t) = 0, \quad n = 0,1,2,\ldots),$

one can replace the operator $\tilde{L}$ by $\pounds^T$ in (33) and (35), thus obtaining the alternative expressions

(67) $\quad C_m = (-1)^m (2m+1)^{-1} \int_{-\infty}^{+\infty} dx\, (\pounds^T)^m u(x,t), \quad m = 0,1,2,\ldots,$

(68) $\quad \delta C_m / \delta u(x) = (-1)^{m+1} 2 (\pounds^T)^{m-1} u(x), \quad m = 0,1,2,\ldots\,.$

Therefore, the involutory property (63) follows from the expression

(69) $\quad \{C_m, C_j\}_n = (-1)^{m+j+n} 4 \int_{-\infty}^{+\infty} dx \left[(\pounds^T)^{m-1} u(x) \right] \boldsymbol{D} \left[(\pounds^T)^{n+j-1} u(x) \right],$

together with (64) and the operator identity (A.9.-54c). Note that the previous equations, (28), (30) and (34), correspond merely to the simplest choice, $n=0$, in (52), (53) and (63). Of course, once the symplectic operator has been chosen, the Hamiltonian corresponding to a given flow is uniquely fixed. Thus, for instance, the KdV equation (21), or equivalently (24), in addition to the Hamiltonian form (see (52), (31) and (29))

(70) $\quad u_t = \tfrac{1}{2} \Omega_0\, \delta C_2 / \delta u,$

admits the second Hamiltonian structure

(71) $\quad u_t = \tfrac{1}{2} \Omega_1\, \delta C_1 / \delta u.$

More generally, there easily follows that the higher KdV equation (37) admits $n+1$ different Hamiltonian structures, namely (37) can be written as

(72) $\quad u_t = \tfrac{1}{2} \Omega_m\, \delta C_{n-m+1} / \delta u, \quad m = 0,1,\ldots,n.$

A.23.N. Notes to appendix A.23

The Hamiltonian structure of the KdV equation was first pointed out (in the class of its periodic solutions) in [G1971]. The remarkable connection between the Hamiltonian formalism and the spectral transform method is given in [ZF1971]. There it is shown that the action-angle variables have a simple expression in terms of the spectral transform, and therefore that the mapping associating with a *bona fide* potential $u(x)$ its spectral transform $S[u]$ (see chapter 2) is essentially the canonical transformation yielding the action-angle variables. An elementary introduction to the Hamiltonian formalism as applied to the KdV equation is given in [L1977]; see also [FN1974]. A good textbook on the mathematical properties of Hamiltonian systems is, for instance, [Ar1974].

For a general algebraic method to construct the functionals C_m (see (A.23.-33), and to prove their involutory property (A.23.-34), see [GD1975] and [Wil1979].

That the KdV equation (A.23.-21) possesses a second Hamiltonian structure (see (A.23.-71)) was first noted in [Mag1978]; further expansions of this point are given in [Mag1979]; see also [D1980] and [KuW1981].

An interesting relationship between group theory and the Hamiltonian structure of the KdV equation has been investigated in [LM1978], and in [A1979a]. A group theoretical approach to the KdV equation has been also considered in [BePe1980]. An interpretation of the KdV equation as describing the motion of a set of constrained harmonic oscillators has been recently pointed out in [Mo1978]; see also [DLT1980] and [DLT1980a].

For an up-dated general mathematical treatment of infinite-dimensional Hamiltonian systems see [KS1981] and the references quoted there.

REFERENCES

The list of references is in alphabetical order of the acronyms. The number(s) in parentheses after each reference indicate the page(s) where that reference is mentioned.

[A1976]   V.A. Andreev, Application of the inverse scattering method to the equation $\sigma_{xt} = \exp(\sigma)$, *Theoret. Math. Phys.* **29** (1976) 1027–1032 [*Teoret. Mat. Fiz.* **29** (1976) 213–220]. (67)

[A1977]   M. Adler, Some finite dimensional integrable systems, in: [FML1978], 237–243. (177)

[A1977a]   M. Adler, Some finite dimensional integrable systems and their scattering behaviour, *Commun. Math. Phys.* **55** (1977) 195–230. (177)

[A1979]   M. Adler, Completely integrable systems and symplectic actions, *J. Math. Phys.* **20** (1979) 60–67. (5)

[A1979a]   M. Adler, On a trace functional for formal pseudodifferential operators and symplectic structure of the Korteweg–de Vries type equations, *Inventiones Math.* **50** (1979) 219–248. (5, 487)

[Ab1977]   M.J. Ablowitz, Nonlinear evolution equations—continuous and discrete, *Siam Review* **19** (1977) 663–684. (5)

[Ab1977a]   M.J. Ablowitz, The inverse scattering transform—continuous and discrete, and its relationship with Painlevé transcendents, in: [C1978], 9–32. (5, 178)

[Ab1978]   M.J. Ablowitz, Lectures on the inverse scattering transform, *Stud. Appl. Math.* **58** (1978) 17–94. (4, 5)

[Ab1979]   M.J. Ablowitz, Remarks on nonlinear evolution equations and ordinary differential equations of Painlevé type, in: [MZ1979], 129–141. (178)

[ABCOP1979]   S. Ahmed, M. Bruschi, F. Calogero, M.A. Olshanetsky and A.M. Perelomov, Properties of the zeros of the classical polynomials and of the Bessel functions, *Nuovo Cimento* **49B** (1979) 173–199. (177)

[ABM1976]   Kh.O. Abdulloev, I.L. Bogolubsky and V.G. Makhankov, One more example of inelastic soliton interaction, *Phys. Lett.* **56A** (1976) 427–428. (176)

[AC1979]   M.J. Ablowitz and H. Cornille, On solutions of the Korteweg–de Vries equation, *Phys. Lett.* **72A** (1979) 277–280. (297)

[ADHM1978]   M.F. Atiyah, V.G. Drinfeld, N.J. Hitchin and Yu.I. Manin, Construction of instantons, *Phys. Lett.* **65A** (1978) 185–137. (67)

[AF1978]   D. Anker and N.C. Freeman, On the soliton solutions of the Davey–Stewartson equation for long waves, *Proc. Roy. Soc. London* **A 360** (1978) 529–540. (66)

[AKNS1973]   M.J. Ablowitz, D.J. Kaup, A.C. Newell and H. Segur, Method for solving the Sine–Gordon equation, *Phys. Rev. Lett.* **30** (1973) 1262–1264. (67)

[AKNS1974]   M.J. Ablowitz, D.J. Kaup, A.C. Newell and H. Segur, The inverse scattering transform—Fourier analysis for nonlinear problems, *Stud. Appl. Math.* **53** (1974) 249–315. (4, 63, 118, 119, 175, 461)

[AKS1979]	M.J. Ablowitz, M. Kruskal and H. Segur, A note on Miura's transformation, *J. Math. Phys.* **20** (1979) 999–1003. (65)
[AM1963]	Z.S. Agranovitch and V.A. Marchenko, *The Inverse Problem of Scattering Theory*. Gordon and Breach, New York, 1963 (translated from Russian). (118)
[AM1978]	M. Adler and J. Moser, On a class of polynomials connected with the Korteweg–de Vries equation, *Commun. Math. Phys.* **61** (1978) 1–30. (177)
[AMM1977]	H. Airault, H.P. McKean and J. Moser, Rational and elliptic solutions of the Korteweg–de Vries equation and a related many-body problem, *Comm. Pure Appl. Math.* **30** (1977) 95–148. (177)
[AN1973]	M.J. Ablowitz and A.C. Newell, The decay of the continuous spectrum for solutions of the Korteweg–de Vries equation, *J. Math. Phys.* **14** (1973) 1277–1284. (177)
[Ar1974]	V. Arnold, *Mathematical Methods of Classical Mechanics*, Nauka, Moscow, 1974 (in Russian). [French translation: *Les Méthodes Mathématiques de la Mécanique Classique*, Mir, Moscow, 1976; English translation: *Mathematical Methods of Classical Mechanics*, Graduate Texts in Mathematics **60**, Springer, New York, 1978]. (4, 487)
[ARS1978]	M.J. Ablowitz, A. Ramani and H. Segur, Nonlinear evolution equations and ordinary differential equations of Painlevé type, *Lett. Nuovo Cimento* **23** (1978) 333–338. (178)
[ARS1980a]	M.J. Ablowitz, A. Ramani and H. Segur, A connection between nonlinear evolution equations and ordinary differential equations of P-type. I, *J. Math. Phys.* **21** (1980) 715–721. (178)
[ARS1980b]	M.J. Ablowitz, A. Ramani and H. Segur, A connection between nonlinear evolution equations and ordinary differential equations of P-type. II, *J. Math. Phys.* **21** (1980) 1006–1015. (178)
[AS1965]	M. Abramowitz and I.A. Stegun. (editors), *Handbook of Mathematical Functions*, Dover, New York, 1965. (430, 432, 433)
[AS1977a]	M.J. Ablowitz and H. Segur, Asymptotic solutions of the Korteweg–de Vries equation, *Studies Appl. Math.* **57** (1977) 13–44. (177, 394)
[AS1977b]	M.J. Ablowitz and H. Segur, Exact linearization of a Painlevé transcendent, *Phys. Rev. Lett.* **38** (1977) 1103–1106. (178)
[AS1981]	M.J. Ablowitz and H. Segur, *Solitons and the Inverse Scattering Transform*, SIAM, Philadelphia, 1981. (4)
[Au1976]	S. Aubry, A unified approach to the interpretation of displacive and order-disorder systems. II: Displacive systems, *J. Chem. Phys.* **64** (1976) 3392–3402. (67)
[B1871]	M.J. Boussinesq, Théorie de l'intumescence liquide appellé onde solitaire ou de translation se propageant dans un canal rectangulaire, *Comptes Rendus* **72** (1871) 755–759. (65)
[B1872]	M.J. Boussinesq, Théorie des ondes et des remous qui se propagent le long d'un canal horizontal, en communiquant au liquide contenu dans ce canal des vitesses sensiblement pareilles de la surface au fond, *J. Math. Pures Appl.* **7** (1872) 55–108. (65)
[B1903]	L. Bianchi, *Lezioni di Geometria Differenziale*, Vol. 2, 2nd Ed. Spoerri, Pisa, 1903 (see sect. 352). (67)
[B1949]	V. Bargmann, On the connection between phase shifts and scattering potential, *Rev. Modern Phys.* **21** (1949) 488–493. (437)
[B1953]	A. Sh. Bloch, On the determination of a differential equation by its spectral matrix functions, *Dokl. Akad. Nauk SSSR* **92** (1953) 209–212 (in Russian). (330)

[B1968] V.V. Babikov, *The Method of Phase Functions in Quantum Mechanics*, Nauka, Moscow, 1968 (in Russian). (471, 475)

[B1974] T.B. Benjamin, Lectures on nonlinear wave motion, in: [N1974], 3–47. (64)

[B1978] F.A. Berezin, Models of Gross–Neveu type are quantization of a classical mechanics with nonlinear phase space, *Commun. Math. Phys.* **63** (1978) 131–153. (5)

[Ba1978] A.O. Barut (editor), *Nonlinear Equations in Physics and Mathematics*, Reidel, Dordrecht, 1978. (4, 491, 492, 502)

[BB1980] C. Bardos and D. Bessis (editors), *Bifurcation Phenomena in Mathematical Physics and Related Topics*, Reidel, Dordrecht, 1980. (4, 492)

[BBM1972] T.B. Benjamin, J.L. Bona and J.J. Mahony, Model equations for long waves in nonlinear dispersive systems, *Phil. Trans. Roy. Soc. London* **A272** (1972) 47–78. (64, 176)

[BC1980] R.K. Bullough and P.J. Caudrey (editors), *Solitons*, Topics in Current Physics, **17**, Springer, Berlin, 1980. (4, 490, 494, 502, 505, 507, 508)

[BC1981] M. Bruschi and F. Calogero, Finite-dimensional matrix representations of the operator of differentiation through the algebra of raising and lowering operators: general properties and explicit examples, *Nuovo Cimento* **62B** (1981) 337–351. (177)

[BCG1980] R.K. Bullough, P.J. Caudrey and H.M. Gibbs, The double Sine–Gordon equations: a physically applicable system of equations, in: [BC1980], 107–141. (67)

[BEMS1971] A. Barone, F. Esposito, C.J. Magee and A.C. Scott, Theory and applications of the Sine–Gordon equation, *Riv. Nuovo Cimento* **1** (1971) 227–267. (67)

[Ben1967] T. Benjamin, Internal waves of permanent form in fluids of great depth, *J. Fluid Mech.* **29** (1967) 559–592. (65)

[BePe1980] F.A. Berezin and A.M. Perelomov, Group-theoretical interpretation of the Korteweg–de Vries type equations, *Commun. Math. Phys.* **74** (1980) 129–140. (5, 487)

[BeZ1978] A.A. Belavin and V.E. Zakharov, Yang–Mills equations as inverse scattering problem, *Phys. Lett.* **73B** (1978) 53–57. (67)

[Bi1978] A.R. Bishop, Solitons and physical perturbations, in: [LS1978], 61–87. (67)

[BJKS1979] R.K. Bullough, P.M. Jack, P.W. Kitchenside and R. Saunders, Solitons in laser physics, in: [W1979], 364–381. (66)

[BK1979] T.L. Bock and M.D. Kruskal, A two-parameter Miura transformation of the Benjamin–Ono equation, *Phys. Lett.* **74A** (1979) 173–176. (65)

[BLOPR1982] M. Bruschi, D. Levi, M.A. Olshanetsky, A.M. Perelomov and O. Ragnisco, The quantum Toda lattice, *Phys. Lett. A* **88A** (1982) 7–12. (4)

[BLR1978] M. Bruschi, D. Levi and O. Ragnisco, Evolution equations associated to the triangular-matrix Schroedinger problem solvable by the inverse spectral transform, *Nuovo Cimento* **45A** (1978) 225–237. (4)

[BLR1978a] M. Bruschi, D. Levi and O. Ragnisco, Nonlinear evolution equations solvable by the inverse spectral transform associated to the matrix Schroedinger equation of rank 4, *Nuovo Cimento* **43B** (1978) 251–270. (4)

[BLR1978b] M. Bruschi, D. Levi and O. Ragnisco, Discrete version of the modified Korteweg–de Vries equation with x-dependent coefficients, *Nuovo Cimento* **48A** (1978) 213–226. (4)

[BLR1979] M. Bruschi, D. Levi and O. Ragnisco, Discrete version of the nonlinear Schroedinger equation with x-dependent coefficients, *Nuovo Cimento* **53A** (1979) 21–30. (4)

[BLR1980] M. Bruschi, D. Levi and O. Ragnisco, Toda lattice and generalized wronskian technique, *J. Phys. A* **13** (1980) 2531–2533. (4)

References

[BLR1982] M. Bruschi, D. Levi and O. Ragnisco, The chiral field hierarchy, *Phys. Lett. A* (in press). (4)

[BLR1982a] M. Bruschi, D. Levi and O. Ragnisco, Nonlinear partial differential equations and Bäcklund transformations related to the 4-dimensional self-dual Yang–Mills equations, *Lett. Nuovo Cimento* **33** (1982) 263–266. (4)

[BLR1982b] M. Bruschi, D. Levi and O. Ragnisco, The discrete chiral field hierarchy, *Lett. Nuovo Cimento* **33** (1982) 284–288. (4)

[BM1979] L.A. Bordag and V.B. Matveev, Explicit analytic solutions of the KdV equation and their application to cylindrical KdV equation, Preprint, Section Matematik, Karl Marx Universität, Leipzig, DDR. (257, 297)

[BMRL1980] M. Bruschi, S.V. Manakov, O. Ragnisco and D. Levi, The non-abelian Toda lattice (discrete analogue of the matrix Schroedinger spectral problem), *J. Math. Phys.* **21** (1980) 2749–2753. (4)

[BP1979] M. Boiti and F. Pempinelli, Similarity solutions of the Korteweg–de Vries equation, *Nuovo Cimento* **51B** (1979) 70–78. (178)

[BP1980] M. Boiti and F. Pempinelli, Similarity solutions and Bäcklund transformations of the Boussinesq equation, *Nuovo Cimento* **56B** (1980) 148–156. (65, 178)

[BPS1980] M. Boiti, F. Pempinelli and G. Soliani (editors), *Nonlinear Evolution Equations and Dynamical Systems*, Proceedings of a meeting in Lecce, June 1979. Lecture Notes in Physics **120**, Springer, Berlin, 1980. (4, 492, 495, 503, 506)

[BR1980a] M. Bruschi and O. Ragnisco, Existence of a Lax pair for any member of the class of nonlinear evolution equations associated to the matrix Schroedinger spectral problem, *Lett. Nuovo Cimento* **29** (1980) 321–326. (4)

[BR1980b] M. Bruschi and O. Ragnisco, Extension of the Lax method to solve the class of nonlinear evolution equations with x-dependent coefficients associated to the matrix Schroedinger spectral problem, *Lett. Nuovo Cimento* **29** (1980) 327–330. (4, 297)

[BR1980c] M. Bruschi and O. Ragnisco, Bäcklund transformations and Lax technique, *Lett. Nuovo Cimento* **29** (1980) 331–334. (4, 207, 461)

[BR1981a] M. Bruschi and O. Ragnisco, Nonlinear differential-difference equations, Bäcklund transformations and Lax technique, *J. Phys. A: Math. Gen.* **14** (1981) 1075–1081. (4)

[BR1981b] M. Bruschi and O. Ragnisco, Nonlinear differential-difference matrix equations with n-dependent coefficients, *Lett. Nuovo Cimento* **31** (1981) 492–496. (4)

[BR1981c] M. Bruschi and O. Ragnisco, Nonlinear evolution equations associated to the 3rd order scalar differential operator, Rome preprint, No. 254 (1981). (4)

[BRL1981] M. Bruschi, O. Ragnisco and D. Levi, Evolution equations associated to the discrete analogue of the matrix Schroedinger spectral problem solvable by the inverse spectral transform, *J. Math. Phys.* **22** (1981) 2463–2471. (4)

[BS1978] A.R. Bishop and T. Schneider (editors), *Solitons and Condensed Matter Physics*, Springer Series in Solid-State Sciences, **8**, Springer, Berlin, 1978. (4)

[Bu1974] J.M. Burgers, *The Nonlinear Diffusion Equation*, Reidel, Dordrecht, 1974. (65)

[Bu1977] R.K. Bullough, Solitons, in: *Interaction of Radiation with Condensed Matter*, IAEA, Vienna, 1977, Vol. I, 381–469. (4, 66, 67)

[Bu1977a] R.K. Bullough, Solitons in physics, in: [Ba1978], 99–141. (4, 66, 67)

[Bu1978] R.K. Bullough, Solitons, *Physics Bulletin* (February 1978) 78–82. (4)

[BZ1978] V.A. Belinsky and V.E. Zakharov, Integration of the Einstein equations by means of the inverse scattering problem technique and construction of exact soliton solutions, *Sov. Phys. JETP* **48** (1978) 985–994 [*Zh. Eksp. Teor. Fiz.* **75** (1978) 1955–1971]. (67)

[BZ1979] V.A. Belinsky and V.E. Zakharov, Stationary gravitational solitons with axial symmetry, *Sov. Phys. JETP* **50** (1979) 1–9 [*Zh. Eksp. Teor. Fiz.* **77** (1979) 3–19]. (67)

[C1967] F. Calogero, *Variable Phase Approach to Potential Scattering*, Academic Press, New York, 1967. (118, 299, 302, 471, 475, 476)

[C1971] F. Calogero, Solution of the one-dimensional N-body problem with quadratic and/or inversely quadratic pair potentials, *J. Math. Phys.* **12** (1971) 419–436. (5, 177)

[C1975] F. Calogero, A method to generate solvable nonlinear evolution equations, *Lett. Nuovo Cimento* **14** (1975) 443–448. (4, 118, 175, 297)

[C1975a] F. Calogero, Bäcklund transformations and functional relation for solutions of nonlinear partial differential equations solvable via the inverse-scattering method, *Lett. Nuovo Cimento* **14** (1975) 537–543. (4, 206, 207)

[C1976] F. Calogero, Generalized wronskian relations: a novel approach to Bargmann-equivalent and phase-equivalent potentials, in: E.H. Lieb, B. Simon and A.S. Wightman (editors), *Studies in Mathematical Physics (Essays in honor of Valentine Bargmann)*, Princeton University Press, Princeton, NJ, 1976. (4, 398)

[C1976a] F. Calogero, Generalized wronskian relations, one-dimensional Schroedinger equation and nonlinear partial differential equations solvable by the inverse-scattering method, *Nuovo Cimento* **31B** (1976) 229–249. (4, 118)

[C1977] F. Calogero, Nonlinear evolution equations solvable by the inverse spectral transform, in: [DDJ1978], 235–269. (xvii, 224, 386)

[C1977a] F. Calogero, Integrable many-body problems, in: [Ba1978], 3–53. (5, 177)

[C1978] F. Calogero (editor), *Nonlinear Evolution Equations Solvable by the Spectral Transform*, Research Notes in Mathematics **26**, Pitman, London, 1978. (4, 488, 495, 496, 499, 504, 505, 508)

[C1978a] F. Calogero, Motion of poles and zeros of special solutions of nonlinear and linear partial differential equations, and related "solvable" many-body problems, *Nuovo Cimento* **43B** (1978) 177–241. (4, 177)

[C1978b] F. Calogero, Spectral transform and nonlinear evolution equations, in: [R1979], 29–34. (xvii, 4)

[C1979] F. Calogero, Spectral transform and solitons: tools to solve and investigate nonlinear evolution equations, Mimeographed notes of lectures given at the Institute of Theoretical Physics in Groningen, April–June, 1979. (xiii, xvii)

[C1979a] F. Calogero, Nonlinear evolution equations solvable by the spectral transform: some recent results, in: [BPS1980], 1–14. (xvii, 4)

[C1979b] F. Calogero, Solvable many-body problems and related mathematical findings (and conjectures), in: [BB1980], 371–384. (4, 5, 177)

[C1980a] F. Calogero, Isospectral matrices and polynomials, *Nuovo Cimento* **58B** (1980) 169–180. (177)

[C1980b] F. Calogero, Integrable many-body problems and related mathematical results, in: [Co1980], 151–164. (5, 177)

[C1980c] F. Calogero, Finite transformations of certain isospectral matrices, *Lett. Nuovo Cimento* **28** (1980) 502–504. (177)

[C1980d] F. Calogero, Spectral transform and solitons: an introduction to a novel technique to solve (certain classes of) nonlinear evolution equations, in: [Co1980], 143–150. (xvii, 4)

[C1981a] F. Calogero, Matrices, differential operators and polynomials, *J. Math. Phys.* **22** (1981) 919–932. (177)

References

[C1981b] F. Calogero, Additional identities for certain isospectral matrices, *Lett. Nuovo Cimento* **30** (1981) 342–344. (177)

[C1982] F. Calogero, Isospectral matrices and classical polynomials, *Linear Alg. Appl.* (in press). (177)

[Ca1980] P.J. Caudrey, The inverse problem for the third order equation $u_{xxx} + q(x)u_x + r(x)u = -i\zeta^3 u$, *Phys. Lett.* **79A** (1980) 264–268. (65)

[Ca1982] P.J. Caudrey, The inverse problem for a general $N \times N$ spectral equation, *Physica D* (to be published). (65)

[CaC1977] K.M. Case and S.C. Chiu, Some remarks on the wronskian technique and the inverse scattering transform, *J. Math. Phys.* **18** (1977) 2044–2052. (4, 118, 207)

[Cas1978] K.M. Case, The N-soliton solution of the Benjamin–Ono equation, *Proc. Nat. Acad. Sci. USA* **75** (1978) 3562–3563. (65)

[Cas1979a] K.M. Case, Properties of the Benjamin–Ono equation, *J. Math. Phys.* **20** (1979) 972–977. (65)

[Cas1979b] K.M. Case, Benjamin–Ono-related equations and their solutions, *Proc. Nat. Acad. Sci. USA* **76** (1979) 1–3. (65)

[Cas1979c] K.M. Case, Meromorphic solutions of the Benjamin–Ono equation, *Physica* **96A** (1979) 173–182. (65)

[CB1981] D.K. Campbell and A.R. Bishop, Soliton excitations in polyacetylene and relativistic field theory models. Preprint LA-UR-81-2114, Los Alamos Scientific Lab. (67)

[CC1977] D.V. Choodnovsky and G.V. Choodnovsky, Pole expansion of nonlinear differential equations, *Nuovo Cimento* **40B** (1977) 339–353. (177)

[CC1978] D.V. Chudnovsky and G.V. Chudnovsky, Seminaire sur les equations non linaires. I, Mimeographed lecture notes, Centre de mathématiques, Ecole polytechnique, Paris, 1977–78. (4, 5)

[CD1976a] F. Calogero and A. Degasperis, Nonlinear evolution equations solvable by the inverse spectral transform. I, *Nuovo Cimento* **32B** (1976) 201–242. (4, 118, 207)

[CD1976b] F. Calogero and A. Degasperis, Nonlinear evolution equations solvable by the inverse spectral transform associated with the multichannel Schroedinger problem, and properties of their solutions, *Lett. Nuovo Cimento* **15** (1976) 65–69. (4)

[CD1976c] F. Calogero and A. Degasperis, Transformations between solutions of different nonlinear evolution equations solvable via the same inverse spectral transform, generalized resolvent formulas and nonlinear operator identities, *Lett. Nuovo Cimento* **16** (1976) 181–186. (4)

[CD1976d] F. Calogero and A. Degasperis, Coupled nonlinear evolution equations solvable via the inverse spectral transform and solitons that come back: the boomeron, *Lett. Nuovo Cimento* **16** (1976) 425–433. (4, 66)

[CD1976e] F. Calogero and A. Degasperis, Bäcklund transformations, nonlinear superposition principle, multisoliton solutions and conserved quantities for the "boomeron" nonlinear evolution equation, *Lett. Nuovo Cimento* **16** (1976) 434–438. (4, 66)

[CD1977] F. Calogero and A. Degasperis, Spectral transform and nonlinear evolution equations, in: [Sab1978], 274–295. (xvii)

[CD1977a] F. Calogero and A. Degasperis, Nonlinear evolution equations solvable by the inverse spectral transform. II, *Nuovo Cimento* **39B** (1977) 1–54. (4, 118)

[CD1977b] F. Calogero and A. Degasperis, Special solutions of coupled nonlinear evolution equations with bumps that behave as interacting particles, *Lett.*

Nuovo Cimento **19** (1977) 525–533. (4)

[CD1978a] F. Calogero and A. Degasperis, Extension of the spectral transform method for solving nonlinear evolution equations, *Lett. Nuovo Cimento* **22** (1978) 131–137. (4, 297)

[CD1978b] F. Calogero and A. Degasperis, Exact solution via the spectral transform of a nonlinear evolution equation with linearly x-dependent coefficients, *Lett. Nuovo Cimento* **22** (1978) 138–141. (4, 297)

[CD1978c] F. Calogero and A. Degasperis, Extension of the spectral transform method for solving nonlinear evolution equations. II, *Lett. Nuovo Cimento* **22** (1978) 263–269. (4, 297)

[CD1978d] F. Calogero and A. Degasperis, Exact solution via the spectral transform of a generalization with linearly x-dependent coefficients of the modified Korteweg–de Vries equation, *Lett. Nuovo Cimento* **22** (1978) 270–273. (4, 297)

[CD1978e] F. Calogero and A. Degasperis, Exact solution via the spectral transform of a generalization with linearly x-dependent coefficients of the nonlinear Schroedinger equation, *Lett. Nuovo Cimento* **22** (1978) 420–424. (4, 297)

[CD1978f] F. Calogero and A. Degasperis, Conservation laws for classes of nonlinear evolution equations solvable by the spectral transform, *Commun. Math. Phys.* **63** (1978) 155–176. (4, 118, 224, 297)

[CD1978g] F. Calogero and A. Degasperis, Inverse spectral problem for the one-dimensional Schroedinger equation with an additional linear potential, *Lett. Nuovo Cimento* **23** (1978) 143–149. (4, 297)

[CD1978h] F. Calogero and A. Degasperis, Solution by the spectral transform method of a nonlinear evolution equation including as a special case the cylindrical KdV equation, *Lett. Nuovo Cimento* **23** (1978) 150–154. (4, 65, 297)

[CD1978i] F. Calogero and A. Degasperis, Conservation laws for a nonlinear evolution equation that includes as a special case the cylindrical KdV equation, *Lett. Nuovo Cimento* **23** (1978) 155–160. (4, 297)

[CD1980] F. Calogero and A. Degasperis, Nonlinear evolution equations solvable by the inverse spectral transform associated with the matrix Schroedinger equation, in: [BC1980], 301–323. (xvii, 4)

[CD1980a] F. Calogero and A. Degasperis, Conserved quantities for generalized KdV equations, *Lett. Nuovo Cimento* **28** (1980) 12–14. (4, 388)

[CD1981] F. Calogero and A. Degasperis, Reduction technique for matrix nonlinear evolution equations solvable by the spectral transform, *J. Math. Phys.* **22** (1981) 23–31. (4, 65)

[Ce1977] C. Cercignani, Solitons, theory and applications, *Riv. Nuovo Cimento* **7** (1977) 429–469. (4, 67)

[CE1977] P.J. Caudrey and J.C. Eilbeck, Numerical evidence for breakdown of soliton behaviour in solutions of the Maxwell–Bloch equations, *Phys. Lett.* **62A** (1977) 65–66. (66)

[Ch1974] H.-H. Chen, General derivation of Bäcklund transformations from inverse scattering problems, *Phys. Rev. Lett.* **33** (1974) 925–928. (206)

[Ch1975] H.-H. Chen, Relation between Bäcklund transformations and inverse scattering problems, in: [Mi1976], 241–252. (65)

[ChD1979] D.V. Chudnovsky, One and multidimensional completely integrable systems arising from the isospectral deformation, in: *Complex analysis, microlocal calculus and relativistic quantum theory*, edited by D. Iagolnitzer, Lecture Notes in Physics **126**, Springer, Berlin, 1980, 352–416. (4, 5)

[CK1978] M. Chaichian and P.P. Kulish, On the method of inverse scattering problem

	and Bäcklund transformations for supersymmetric equations, *Phys. Lett.* **78B** (1978) 413–416. (67)
[CK1980]	H.-H. Chen and D.J. Kaup, Conservation laws of the Benjamin–Ono equation, *J. Math. Phys.* **21** (1980) 19–20. (65)
[CL1977]	S.C. Chu and J.F. Ladik, Generating exactly soluble nonlinear discrete evolution equations by a generalized wronskian technique, *J. Math. Phys.* **18** (1977) 690–700. (4)
[CL1979]	H.-H. Chen and Y.C. Lee, Internal wave solitons of fluids with finite depth, *Phys. Rev. Lett.* **43** (1979) 264–266. (65)
[CLP1978]	H.-H. Chen, Y.C. Lee and N.R. Pereira, Algebraic internal wave solitons and the integrable Calogero–Moser–Sutherland n-body problem, *Phys. Fluids* **22** (1979) 187–188. (65, 177)
[Co1950]	J.D. Cole, On a quasilinear parabolic equation occurring in aerodynamics, *Q. Appl. Math.* **9** (1950) 225–236. (65)
[Co1980]	E.G.D. Cohen (editor), *Fundamental Problems in Statistical Mechanics V*, Proceedings of the 1980 Enschede Summer School, North-Holland, Amsterdam, 1980. (492)
[COP1979]	F. Calogero, M.A. Olshanetsky and A.M. Perelomov, Rational solutions of the KdV equation with damping, *Lett. Nuovo Cimento* **24** (1979) 97–100. (4, 177)
[Cor1979]	H. Cornille, Solutions of the nonlinear three-wave equations in three spatial dimensions, *J. Math. Phys.* **20** (1979) 1653–1666. (66)
[Cos1980]	C.M. Cosgrove, Relationships between the group-theoretic and soliton-theoretic techniques for generating stationary axisymmetric gravitational solutions, *J. Math. Phys.* **21** (1980) 2417–2447. (67)
[Cra1978]	M.G. Crandall (editor), *Nonlinear Evolution Equations*, Proceedings of a Symposium at the Mathematics Research Center, University of Wisconsin–Madison, October 17–19, 1977, Academic Press, New York, 1978. (4, 501)
[CS1977]	K. Chadan and P.C. Sabatier, *Inverse Problems in Quantum Scattering Theory*, Springer, Berlin, 1977. (4, 64, 118)
[D1975]	B.A. Dubrovin, The periodic problem for the Korteweg–de Vries equation in a class of finite zone potentials, *Func. Anal. Appl.* **9** (1975) 215–223 [*Funk. Anal. Priloz.* **9** (1975) 41–52]. (176)
[D1977]	A. Degasperis, Solitons, boomerons, trappons, in: [C1978], 97–126. (xvii, 66)
[D1978]	A. Degasperis, Spectral transform and solvability of nonlinear evolution equations, in: [R1979], 35–90. (xvii, 118)
[D1979]	A. Degasperis, Reduction technique for matrix nonlinear evolution equations, in: [BPS1980], 16–34. (xvii, 4)
[D1979a]	A. Degasperis, Solutions of the Korteweg–de Vries equation and their spectral transform, in: [Sab1980], 189–222. (xvii, 4, 297, 298)
[D1980]	A. Degasperis, The spectral transform method to integrate nonlinear dynamical systems, in: [L1980], 5–27. (xvii, 4, 487)
[D1982]	A. Degasperis, On the conservation laws associated with Lax equations, *Lett. Nuovo Cimento* **33** (1982) 425–432. (4, 224)
[DAR1965]	V. De Alfaro and T. Regge, *Potential Scattering*, Wiley, New York, 1965. (118)
[DB1979]	R.K. Dodd and R.K. Bullough, The generalized Marchenko equation and the canonical structure of the A.K.N.S.–Z.S. inverse method, *Physica Scripta* **20** (1979) 514–530. (119)
[DD1979]	H. Dijkhuis and J.K. Drohm, Some investigations on a new class of nonlinear partial differential equations solvable by the inverse spectral

[DDJ1978] transform technique, Institute of Theoretical Physics, Groningen, Intern Rapport 148, June 1979 (unpublished). (298)
G. Dell'Antonio, S. Doplicher and G. Jona-Lasinio, *Mathematical Problems in Theoretical Physics*, Lecture Notes in Physics **80**, Springer, Berlin, 1978. (4, 492)

[DEGM1982] R.K. Dodd, J.C. Eilbeck, J.D. Gibbon and H.C. Morris, *Solitons and Nonlinear Waves*, Academic Press, New York, 1982. (4)

[DHN1974] R.F. Dashen, B. Hasslacher and A. Neveu, Nonperturbative methods and extended-hadron models in field theory. II: Two-dimensional models and extended hadrons, *Phys. Rev.* **D10** (1974) 4130–4138. (67)

[DLT1980] P. Deift, F. Lund and E. Trubowitz, Nonlinear wave equations and constrained harmonic motion, *Proc. Natl. Acad. Sci. USA* **77** (1980) 716–719. (487)

[DLT1980a] P. Deift, F. Lund and E. Trubowitz, Nonlinear wave equations and constrained harmonic motion, *Commun. Math. Phys.* **74** (1980) 141–188. (487)

[DMN1976] B.A. Dubrovin, V.B. Matveev and S.P. Novikov, Nonlinear equations of Korteweg–de Vries type, finite-zone linear operators and abelian varieties, *Russian Math. Surveys* **31**(1) (1976) 59–146 [*Usp. Mat. Nauk* **31**(1) (1976) 55–136]. (4, 5, 176, 177)

[DN1974] B.A. Dubrovin and S.P. Novikov, Periodic problem for the Korteweg–de Vries and the Sturm–Liouville equations. Their connection with algebraic geometry, *Dokl. Akad. Nauk. SSSR* **219** (1974) 19–22 (in Russian). (176)

[DOP1980] A. Degasperis, M.A. Olshanetsky and A.M. Perelomov, Group-theoretical approach to a class of Lax equations, including those solvable by the spectral transform, *Nuovo Cimento* **59A** (1980) 245–262. (4)

[Dr1974] V.S. Dryuma, Analytic solution of the two-dimensional Korteweg–de Vries (KdV) equation, *Soviet JETP Lett.* **19** (1974) 387–388 [*Zh. ETF Pis. Red.* **19** (1974) 753–755]. (65)

[Dr1976] V.S. Dryuma, An analytical solution of the axial symmetric KdV equation, *Isv. Akad. Nauk Mold. SSSR* **3** (1976) 89 (in Russian). (297)

[DS1974] A. Davey and K. Stewartson, On three-dimensional packets of surface waves, *Proc. Roy. Soc. London* **A338** (1974) 101–110. (66)

[DT1979] P. Deift and E. Trubowitz, Inverse scattering on the line, *Comm. Pure Appl. Mat.* **32** (1979) 121–251. (118, 394)

[E1968] F.J. Ernst, New formulation of the axially symmetric gravitational field problem, *Phys. Rev.* **167** (1968) 1175–1178. (67)

[E1977] J.C. Eilbeck, Boomerons. A computer-produced film, Mathematics Dept., Heriot–Watt University, Edinburgh, 1977. (4, 66)

[E1977a] J.C. Eilbeck, Zoomerons. A computer-produced film, Mathematics Dept., Heriot–Watt University, Edinburgh, 1977. (4, 66)

[E1981] G. Eilenberger, *Solitons. Mathematical Method for Physicists*, Springer Series in Solid-State Science **19**, Springer, Berlin, 1981. (4, 394)

[EGCB1973] J.C. Eilbeck, J.D. Gibbon, P.J. Caudrey and R.K. Bullough, Solitons in nonlinear optics. I: A more accurate description of the 2π pulse in self-induced transparency, *J. Phys. A: Math. Gen.* **6** (1973) 1337–1347. (66)

[Eis1960] L.P. Eisenhart, *A treatise on the differential geometry of curves and surfaces*, Dover, New York, 1960 (see p. 280). (67)

[EMG1975] J.C. Eilbeck and G.R. McGuire, Numerical study of the regularized long-wave equation. I: Numerical methods, *J. Compt. Phys.* **19** (1975) 43–57. (176)

[EMG1977] J.C. Eilbeck and G.R. McGuire, Numerical study of the regularized long-wave equation. II: Interaction of solitary waves, *J. Compt. Phys.* **23** (1977) 63–73. (176)

[ES1982] W. Eckhaus and P. Schuur, The emergence of solitons of the Korteweg–de Vries equation from arbitrary initial conditions, *Math. Methods Appl. Sci.* (in press). (177)

[EVH1981] W. Eckhaus and A. Van Harten, *The Inverse Scattering Transformation and the Theory of Solitons. An Introduction*, North-Holland Mathematics Studies **50**, North-Holland, Amsterdam, 1981. (4, 5, 177)

[EW1977] F.B. Estabrook and H.D. Wahlquist, Prolongation structures, connection theory and Bäcklund transformation, in: [C1978], 64–83. (5)

[F1958] L.D. Faddeev, On the relation between S-matrix and potential for the one-dimensional Schroedinger operator, *Dokl. Akad. Nauk SSSR* **121** (1958) 63–66 (in Russian). (64, 330)

[F1959] L.D. Faddeev, The inverse problem in the quantum theory of scattering, *J. Math. Phys.* **4** (1963) 72–104 [*Uspekhi Matem. Nauk* **14** (1959) 57]. (64)

[F1964] L.D. Faddeev, On the relation between the S-matrix and potential for the one-dimensional Schroedinger operator, *Trudi Mat. Inst. Steklov* **73** (1964) 314–336 (in Russian). (64, 118)

[F1974] L.D. Faddeev, The inverse problem of quantum scattering theory. II, *Soviet J. Math.* **5** (1976) [*Sovrem. Probl. Mat.* **3**, VINITI, Moscow (1974) 93–181]. (64)

[F1974a] H. Flaschka, The Toda lattice. I: Existence of integrals, *Phys. Rev.* **B9** (1974) 1924–1925. (5)

[F1974b] H. Flaschka, On the Toda lattice. II: Inverse scattering solution, *Prog. Theor. Phys.* **51** (1974) 703–716. (5)

[FA1982] A.S. Fokas and M.J. Ablowitz, On a unified approach to transformations and elementary solutions of Painlevé equations, *J. Math. Phys.* **23** (July 1982). (178)

[FG1980] A.P. Fordy and J. Gibbons, A class of integrable nonlinear Klein–Gordon equations in many dependent variables, *Commun. Math. Phys.* **77** (1980) 21–30. (67)

[FML1974] H. Flaschka and D.W. McLaughlin, Some comments on Bäcklund transformations, canonical transformations, and the inverse scattering method, in: [Mi1976], 253–295. (206)

[FML1978] H. Flaschka and D.W. McLaughlin (editors), Proceedings of a conference on the theory and applications of solitons, *Rocky Mountain J. Math.* **8** (1,2) (1978). (4, 488, 499, 503, 506)

[FN1974] H. Flaschka and A.C. Newell, Integrable systems of nonlinear evolution equations, in: [Mo1975], 355–440. (5, 487)

[FN1979] H. Flaschka and A.C. Newell, Multiphase similarity solutions of integral evolution equations, in: [MZ1979], 203–221. (178)

[Fok1979] A.S. Fokas, Generalized symmetries and constants of motion of evolution equations, *Lett. Math. Phys.* **3** (1979) 467–473. (224)

[Fok1980] A.S. Fokas, A symmetry approach to exactly solvable evolution equations, *J. Math. Phys.* **21** (1980) 1318–1325. (225)

[Fou1822] J.B. Fourier, *Théorie Analytique de la Chaleur*, Firmin Didot, Paris, 1822. (3)

[FPU1955] E. Fermi, J.R. Pasta and S.M. Ulam, Studies of nonlinear problems. Los Alamos Sci. Lab. Rep. LA-1940, 1955. Reprinted in *Collected works of Enrico Fermi*, University of Chicago Press, Chicago, 1965, Vol. II, 978; and also in: [N1974], 143–156. (1)

[FW1981] A. Fujimoto and Y. Watanabe, Conserved densities of certain nonlinear evolution equations, *Math. Japonica* **26** (1981) 203–221. (390)

[FY1982] A.S. Fokas and Y.C. Yortos, On the exactly solvable equation $S_t = [(\beta S + \gamma)^{-2} S_x]_x + \alpha(\beta S + \gamma)^{-2} S_x$ occurring in two-phase flow in porous media, *SIAM J. Appl. Math.* **42** (1982) 318–332. (65)

[G1971] C.S. Gardner, Korteweg–de Vries equation and generalizations. IV: The Korteweg–de Vries equation as a Hamiltonian system, *J. Math. Phys.* **12** (1971) 1548–1551. (3, 5, 175, 486)

[G1976] B.S. Getmanov, Bound states of soliton in the ϕ_2^4 field-theory model, *JETP Lett.* **24** (1976) 291–294 [*Pis'ma Zh. Eksp. Teor. Fiz.* **24** (1976) 323–327]. (67)

[GCBE1973] J.D. Gibbon, P.J. Caudrey, R.K. Bullough and J.C. Eilbeck, An N-soliton solution of a nonlinear optics equation derived by a general inverse method, *Lett. Nuovo Cimento* **8** (1973) 775–779. (66)

[GD1975] I.M. Gel'fand and L.A. Dikii, Asymptotic behaviour of the resolvent of Sturm-Liouville equations and the algebra of the Korteweg–de Vries equations, *Russian Math. Surveys* **30**(5) (1975) 77–113 [*Usp. Mat. Nauk.* **30**(5) 67–100 (1975)]. (5, 64, 225, 487)

[GD1976] I.M. Gel'fand and L.A. Dikii, Fractional powers of operators and Hamiltonian systems, *Func. Anal. Appl.* **10** (1976) 259–273. [*Funk. Anal. Priloz.* **10**(4) 13–29 (1976)]. (5)

[GD1977] I.M. Gel'fand and L.A. Dikii, The resolvent and hamiltonian systems, *Func. Anal. Appl.* **11** (1977) 93–105 [*Funk. Anal. Priloz.* **11**(2) (1977) 11–27]. (5)

[GD1978a] I.M. Gel'fand and L.A. Dikii, A family of hamiltonian structures connected with integrable, nonlinear differential equations, Preprint Inst. Prikl. Mat. Akad. Nauk SSSR. No. 136 (1978). (5)

[GD1978b] I.M. Gel'fand and L.A. Dikii, The calculus of jets and nonlinear hamiltonian systems, *Func. Anal. Appl.* **12** (1978) 81–94 [*Funk. Anal. Priloz.* **12**(2) (1978) 8–23]. (5)

[GD1979] I.M. Gel'fand and L.A. Dikii, Integrable nonlinear equations and the Liouville theorem, *Func. Anal. Appl.* **13** (1979) 6–15 [*Funk. Anal. Priloz.* **13**(1) 8–20 (1979)]. (5, 64)

[GDo1979] I.M. Gel'fand and I.Ya. Dorfman, Hamiltonian operators and algebraic structures related to them, *Func. Anal. Appl.* **13** (1979) 248–262 [*Funk. Anal. Priloz.* **13** (1979) 13–30]. (5)

[GE1979] W. Guttinger and H. Eikemeier (editors), *Structural Stability in Physics*, Springer Series in Synergetics **4**, Springer, Berlin, 1979. (4)

[GF1980] K. Goda and Y. Fukui, Numerical studies of the regularized long wave equation, *J. Phys. Soc. Japan* **48** (1980) 623–630. (176)

[GFJ1978] J.D. Gibbon, N.C. Freeman and R.S. Johnson, Correspondence between the classical $\lambda\phi^4$, double and single sine–Gordon equations for three-dimensional solitons, *Phys. Lett.* **65A** (1978) 380–382. (65)

[GGKM1967] C.S. Gardner, J.M. Greene, M.D. Kruskal and R.M. Miura, Method for solving the Korteweg–de Vries equation, *Phys. Rev. Lett.* **19** (1967) 1095–1097. (2, 175)

[GGKM1974] C.S. Gardner, J.M. Greene, M.D. Kruskal and R.M. Miura, Korteweg–de Vries and generalizations. VI: Methods for exact solution, *Comm. Pure Appl. Math.* **27** (1974) 97–133. (3, 175)

[GIK1980] V.S. Gerdjikov, M.I. Ivanov and P.P. Kulish, Quadratic bundle and nonlinear equations, *Theor. Math. Phys.* **44** (1980) 372 [*Teor. Mat. Fiz.* **44** (1980) 342–357]. (66, 67)

[GJ1975] J. Goldstone and R. Jackiw, Quantization of nonlinear waves, *Phys. Rev.* **D11** (1975) 1486–1498. (67)

[GK1980a]	V.S. Gerdjikov and E.Kh. Khristov, On the evolution equations solvable through the inverse scattering method. I: Spectral Theory, *Bulg. J. Phys.* **7** (1980) 28–41 (in Bulgarian). (4, 118, 119)
[GK1980b]	V.S. Gerdjikov and E.Kh. Khristov, On the evolution equations solvable through the inverse scattering method. II: Hamiltonian structure and Bäcklund transformations, *Bulg. J. Phys.* **7** (1980) 119–133 (in Bulgarian). (4, 118, 119)
[GKu1980]	J. Gibbons and B. Kupershmidt, A linear scattering problem for the finite depth equation, *Phys. Lett.* **79A** (1980) 31–32. (65)
[GL1951]	I.M. Gel'fand and B.M. Levitan, On the determination of a differential equation by its spectral function, *Amer. Math. Soc. Transl. Ser.* 2 **1** (1955) 253–304 [*Izv. Akad. Nauk. SSR Ser. Math.* **15** (1951) 309–360]. (330)
[GSV1958]	I.M. Gel'fand et al., *Generalized Functions* (Vols. 1–5), Academic Press, New York, 1964–68 (translated from the Russian edition of 1958). (3)
[GW1964]	M.L. Goldberger and K.M. Watson, *Collision Theory*, New York, Wiley, 1964. (118)
[H1899]	D. Hilbert, *Grundlagen der Geometrie*, Teubner, Leipzig, 1899. (67)
[H1974]	M. Hénon, Integrals of the Toda lattice, *Phys. Rev.* **B9** (1974) 1921–1923. (5)
[H1976]	R. Hermann, *The geometry of nonlinear differential equations, Bäcklund transformations, and solitons. Part. A*, Math. Sci. Press, Brookline, MA, 1976. (4, 5)
[H1977]	R. Hermann, *Geometric theory of nonlinear differential equations, Bäcklund transformations and solitons. Part B*, Math. Sci. Press, Brookline, MA, 1977. (4, 5)
[H1980]	F.S. Henyey, Finite-depth and infinite-depth internal wave solitons, *Phys. Rev.* **A21** (1980) 1054–1056. (65)
[Ha1978]	B.K. Harrison, Bäcklund transformation for the Ernst equation of general relativity, *Phys. Rev. Lett.* **41** (1978) 1197–1200. (67)
[Ha1980]	B.K. Harrison, New large family of vacuum solutions of the equations of general relativity, *Phys. Rev.* **D21** (1980) 1695–1697. (67)
[HE1980]	I. Hauser and F.J. Ernst, A homogeneous Hilbert problem for the Kinnersley–Chitre transformations, *J. Math. Phys.* **21** (1980) 1126–1140. (67)
[Hi1971]	R. Hirota, Exact solution of the Korteweg–de Vries equation for multiple collisions of solitons, *Phys. Rev. Lett.* **27** (1971) 1192–1194. (3, 64)
[Hi1973]	R. Hirota, Exact envelope-soliton solutions of a nonlinear wave equation, *J. Math. Phys.* **14** (1973) 805–809. (66)
[Hi1973a]	R. Hirota, Exact N-soliton solutions of the wave equation of long waves in shallow-water and in nonlinear lattices, *J. Math. Phys.* **14** (1973) 810–814. (65)
[Hi1980]	R. Hirota, Direct methods in soliton theory, in: [BC1980], 157–176. (64)
[Ho1950]	E. Hopf, The partial differential equation $u_t + uu_x = \mu u_{xx}$, *Comm. Pure Appl. Math.* **3** (1950) 201–230. (65)
[HS1978]	R. Hirota and J. Satsuma, A simple structure of superposition formula of the Bäcklund transformation, *J. Phys. Soc. Japan* **45** (1978) 1741–1750. (207)
[J1979]	R.S. Johnson, On the inverse scattering transform, the cylindrical Korteweg–de Vries equation and similarity solutions, *Phys. Lett.* **72A** (1979) 197–199. (55)
[JE1978]	R.I. Joseph and R. Egri, Multisoliton solutions in a finite depth fluid, *J. Phys. A: Math. Gen.* **11** (1978) L97–L102. (65)
[JJ1975]	M. Jaulent and C. Jean, The inverse problem for the one-dimensional

	Schroedinger equation with an energy-dependent potential. I and II, *Ann. Inst. Henri Poincaré* **25** (1975) 105–118 and 119–137. (330)
[Jo1977]	R.I. Joseph, Solitary waves in a finite depth fluid, *J. Phys. A: Math. Gen.* **10** (1977) L225–L227. (65)
[JRT1972]	J.R. Taylor, *Scattering theory: the quantum theory on nonrelativistic collisions*, New York, Wiley, 1972. (118)
[K1974]	M.D. Kruskal, Nonlinear wave equations, in: [Mo1975], 310–354. (33, 64, 176, 298)
[K1974a]	M.D. Kruskal, The Korteweg–de Vries equation and related evolution equations, in: [N1974], 61–83. (64, 128, 176, 177)
[K1975]	D.J. Kaup, Method for solving the sine–Gordon equation in laboratory coordinates, *Studies in Appl. Math.* **54** (1975) 165–179. (67)
[K1976]	D.J. Kaup, Closure of the squared Zakharov–Shabat eigenstates, *J. Math. Anal. Appl.* **54** (1976) 849–864. (119)
[K1976a]	D. J. Kaup, The three-wave interaction: a nondispersive phenomenon, *Stud. Appl. Math.* **55** (1976) 9–44. (66)
[K1977]	D.J. Kaup, Applications of the inverse scattering transform. II: The three-wave resonant interaction, in: [FML1978], 283–308. (66)
[K1978]	M.D. Kruskal, The birth of the soliton, in: [C1978], 1–8. (1)
[K1980a]	D.J. Kaup, A method for solving the separable initial-value problem of the full three-dimensional three-wave interaction, *Stud. Appl. Math.* **62** (1980) 75–83. (66)
[K1980b]	D.J. Kaup, Determining the final profiles from the initial profiles for the full three dimensional three-wave resonant interaction, *Lecture Notes in Physics* **130** (1980) 247–252. (66)
[K1980c]	D.J. Kaup, The inverse scattering solutions for the full three dimensional three-wave resonant interaction, *Physica* **1D** (1980) 45–67. (66)
[K1980d]	D.J. Kaup, On the inverse scattering problem for cubic eigenvalue problems of the class $\psi_{xxx} + 6Q\psi_x + 6R\psi = \lambda\psi$, *Studies in Appl. Math.* **62** (1980) 189–216. (65)
[K1981]	D.J. Kaup, The solution of the general initial value problem for the full three-dimensional three wave resonant interaction, *Physica* **3D** (1981) 374–395. (66)
[KA1981]	Yu. Kodama and M.J. Ablowitz, Perturbations of solitons and solitary waves, *Studies Appl. Math.* **64** (1981) 225–245. (5)
[Kau1980]	D.J. Kaup, The Estabrook–Wahlquist method with examples of applications, *Physica* **1D** (1980) 391–411. (5)
[Kay1955]	I. Kay, The inverse scattering problem, Research Report No. EM.74. NY., N.Y. University, Inst. Math. Sc. Electromagnetic Research. (330)
[Kay1960]	I. Kay, The inverse scattering problem when the reflection coefficient is a rational function, *Comm. Pure Appl. Math.* **16** (1960) 371–393. (330)
[KayMo1956]	I. Kay and H.E. Moses, The determination of the scattering potential from the spectral measure function. III: Calculation of the scattering potential from the scattering operator for the one dimensional Schroedinger equation, *Nuovo Cimento* **3** (1956) 276–304. (330)
[KayMo1957]	I. Kay and H.E. Moses, The determination of the scattering potential from the spectral measure function. IV: "Pathological" scattering problems in one-dimension, *Nuovo Cimento Suppl.* **5** (1957) 230–242. (330)
[KdV1895]	D.J. Korteweg and G. de Vries, On the change of form of long waves advancing in a rectangular canal, and on a new type of long stationary waves, *Phil. Mag.* **39** (1895) 422–443. (1, 29, 49, 128)

[Khr1980] E.Kh. Khristov, On spectral properties of operators generating KdV-type equations, Dubna preprint P5-80-381 (in Russian), to appear in *Differenzialnie Uravnienia*. (119)

[KKS1978] D. Kazhdan, B. Kostant and S. Sternberg, Hamiltonian group actions and dynamical systems of Calogero type, *Comm. Pure Appl. Math.* **31** (1978) 481–508. (117, 177)

[KM1978] V.I. Karpman and E.M. Maslov, Structure of tails produced under the action of perturbations on solitons, *Sov. Phys. JETP* **48** (1978) 252–259 [*Zh. Eksp. Teor. Fiz.* **75** (1978) 504–517]. (5)

[KM1979] F. Kako and N. Mugibayashi, Complete integrability of general nonlinear differential-difference equations solvable by the inverse method. II, *Prog. Theor. Phys.* **61** (1979) 776–790. (4)

[KMGZ1970] M.D. Kruskal, R.M. Miura, C.S. Gardner and N.J. Zabusky, Korteweg–de Vries equation and generalizations. V: Uniqueness and nonexistence of polynomial conservation laws, *J. Math. Phys.* **11** (1970) 952–960. (3, 175, 224)

[KMi1974] E.A. Kuznetsov and A.V. Mikhailov, Stability of stationary waves in nonlinear weakly dispersive media, *Sov. Phys. JETP* **40** (1975) 855–859 [*Zh. Eksp. Teor. Fiz.* **67** (1974) 1717–1727]. (297)

[KMi1977] E.A. Kuznetsov and A.V. Mikhailov, On the complete integrability of the two-dimensional classical Thirring model, *Theoret. Math. Phys.* **30** (1977) 193–200 [*Teoret. Mat. Fiz.* **30** (1977) 303–314]. (67)

[KML1977] J.P. Keener and D.W. McLaughlin, Solitons under perturbations, *Phys. Rev.* **A16** (1977) 777–790. (5)

[KN1978] D.J. Kaup and A.C. Newell, An exact solution for a derivative nonlinear Schroedinger equation, *J. Math. Phys.* **19** (1978) 798–801. (66)

[KN1978a] D.J. Kaup and A.C. Newell, The Goursat and Cauchy problems for the sine–Gordon equation, *SIAM J. Appl. Math.* **34** (1977) 37–54. (67)

[KN1978b] D.J. Kaup and A.C. Newell, Solitons as particles, oscillators, and in slowly changing media: a singular perturbation theory, *Proc. Roy. Soc.* **A361** (1978) 413–446. (5)

[KN1979] D.J. Kaup and A.C. Newell, Evolution equations, singular dispersion relations and moving eigenvalues, *Advances in Math.* **31** (1979) 67–100. (119, 175)

[KN1980] C.J. Knickerbocker and A.C. Newell, Internal solitary waves near a turning point, *Phys. Lett.* **75A** (1980) 326–330. (64)

[KoKu1978] K. Ko and H.H. Kuehl, Korteweg–de Vries soliton in a slowly varying medium, *Phys. Rev. Lett.* **40** (1978) 233–236. (388)

[KP1970] B.B. Kadomtsev and V.I. Petviashvili, On the stability of solitary waves in weakly dispersing media, *Sov. Phys. Dokl.* **15** (1970) 539–541 [*Dokl. Akad. Nauk SSSR* **192** (1970) 753–756]. (65)

[Kri1977] I.M. Krichever, Methods of algebraic geometry in the theory of nonlinear equations, *Russian Math. Surveys* **32**(6) (1977) 185–213 [*Usp. Math. Nauk* **32**(6) (1977) 183–208]. (5)

[Kri1978] I.M. Krichever, Rational solutions of the Kadomtsev–Petviashvili equation and many-body problems, *Func. Anal. Appl.* **12** (1978) 59–61 [*Funk. Anal. Priloz.* **12** (1978) 76–78]. (177)

[Kri1980] I.M. Krichever, Elliptic solutions of the Kadomtsev–Petviashvili equation and integrable systems of particles, *Func. Anal. Appl.* **14** (1980) 282–290 [*Funk. Anal. Priloz.* **14**(4) (1980) 45–54]. (177)

[KS1981] Y. Kosmann–Schwarzbach, Hamiltonian systems on fibered manifolds,

	Lett. Math. Phys. **5** (1981) 229–237. (487)
[Kup1980a]	B.A. Kupershmidt, On the nature of the Gardner transformation, J. Math. Phys. **22** (1980) 449–451. (224)
[KuW1981]	B.A. Kupershmidt and G. Wilson, Modifying Lax equations and the second hamiltonian structure, Invent. Math. **62** (1981) 403–436. (487)
[KuW1981a]	B.A. Kupershmidt and G. Wilson, Conservation laws and symmetries of generalized sine–Gordon equations, Commun. Math. Phys. **81** (1981) 189–202. (67)
[KZ1975]	I.A. Kunin, *Theory of Elastic Media with Microstructures*, Nauka, Moscow, 1975 (in Russian). Chapter 5 (The inverse scattering method) is due to V.E. Zakharov. (4)
[L1853]	J. Liouville, Sur l'equation aux differences partielles $d^2\log\lambda/dudv \pm \lambda/2a^2 = 0$, J. Math. Pures et Appliqueés **18** (1853) 71–72. (67)
[L1968]	P.D. Lax, Integrals of nonlinear equations of evolution and solitary waves, Comm. Pure Appl. Math. **21** (1968) 467–490. (2, 64, 175, 176, 207, 461)
[L1975]	P.D. Lax, Periodic solutions of the KdV equation, Comm. Pure Appl. Math. **28** (1975) 141–188. (5)
[L1977]	P.D. Lax, A hamiltonian approach to the KdV and other equations, in: [Cra1978], 207–224. (5, 487)
[L1980]	U. Lindner (editor), *Proceedings of the 3rd International School on Modern Trends in Solid State Theory*, Reinhardsbrunn, November 1980, Karl-Marx-Universität, Leipzig, 1980. (495)
[Lak1977]	M. Lakshmanan, Continuum spin system as an exactly solvable dynamical system, Phys. Lett. **61A** (1977) 53–54. (66)
[Lam1971]	G.L. Lamb, jr., Analytical descriptions of ultrashort optical pulse propagation in a resonant medium, Rev. Modern Phys. **43** (1971) 99–124. (207)
[Lam1974a]	G.L. Lamb, jr., Bäcklund transformations at the turn of the century, in: [Mi1976], 69–79. (205, 207)
[Lam1974b]	G.L. Lamb, jr., Bäcklund transformations for certain nonlinear evolution equations, J. Math. Phys. **15** (1974) 2157–2165. (207)
[Lam1980]	G.L. Lamb, jr., *Elements of Soliton Theory*, Wiley, New York, 1980. (4, 397)
[LB1980]	D. Levi and R. Benguria, Bäcklund transformations and nonlinear differential-difference equations, Proc. Natl. Acad. Sci. USA **77** (1980) 5025–5027. (4)
[Le1978]	D. Levi, The spectral transform as a tool for solving nonlinear discrete evolution equations, in: [R1979], 91–106. (4, 5)
[Le1981]	D. Levi, Nonlinear differential-difference equations as Bäcklund transformations, J. Phys. A: Math. Gen. **14** (1981) 1083–1098. (4)
[LL1964]	A.A. Lugovzov and B.A. Lugovzov, Investigation of axisymmetrical longwaves in the Korteweg–de Vries approximation, in: *Dynamics of Continuous Media*, Vol. 1, Nauka, Novosibirsk, 1964, 195–206. (in Russian). (177, 297)
[LL1979]	P.D. Lax and C.D. Levermore, The zero dispersion limit for the KdV equation, Proc. Natl. Acad. Sci. USA **76** (1979) 3602–3606. (177)
[LLSM1982]	M. Leo, R.A. Leo, G. Soliani and L. Martina, Prolongation algebra and Bäcklund transformations for the cylindrical Korteweg–de Vries equation, Phys. Rev. D (submitted). (297)
[LM1978]	D.R. Lebedev, Yu.I. Manin, Gel'fand-Dikii hamiltonian operator and coadjoint representation of Volterra group, Preprint ITEF-155, Moscow, 1978. (487)
[LML1980]	G.L. Lamb, jr. and D.W. MacLaughlin, Aspects of soliton physics, in: [BC1980], 65–106. (65)

[LOPR1980] D. Levi, M.A. Olshanetsky, A.M. Perelomov and O. Ragnisco, Group-theoretical approach to nonlinear evolution equations of Lax type. III: The Boussinesq equation, *Phys. Lett.* **77A** (1980) 307–311. (4)

[LPS1981] D. Levi, L. Pilloni and P.M. Santini, Integrable three-dimensional lattices, *J. Phys. A: Math. Gen.* **14** (1981) 1567–1575. (4)

[LR1976] F. Lund and T. Regge, Unified approach to strings and vortices with soliton solutions, *Phys. Rev.* **D14** (1976) 1524–1535. (67)

[LR1978] D. Levi and O. Ragnisco, Extension of the spectral transform method for solving nonlinear differential-difference equations, *Lett. Nuovo Cimento* **22** (1978) 691–696. (4)

[LR1979] D. Levi and O. Ragnisco, Nonlinear differential-difference equations with n-dependent coefficients. I, *J. Phys. A: Math. Gen.* **12** (1979) L157–L162 (4)

[LR1979a] D. Levi and O. Ragnisco, Nonlinear differential-difference equations with n-dependent coefficients. II, *J. Phys. A: Math. Gen.* **12** (1979) L163–L167. (4)

[LR1982] D. Levi and O. Ragnisco, Bäcklund transformations for chiral field equations, *Phys. Lett.* **87A** (1982) 381–384. (4)

[LRB1980] D. Levi, O. Ragnisco and M. Bruschi, Extension of the generalized Zakharov–Shabat inverse method for solving differential-difference and difference-difference equations, *Nuovo Cimento* **58A** (1980) 56–66. (4)

[LRS1982] D. Levi, O. Ragnisco and A. Sym, Bäcklund transformation vs. the dressing method, *Lett. Nuovo Cimento* **33** (1982) 401–406. (4)

[LS1974] S. Leibovich and A.R. Seebass (editors), *Nonlinear Waves*, Cornell University Press, Ithaca, 1974. (503)

[LS1978] K. Lonngren and A. Scott, *Solitons in Action*, Academic Press, New York, 1978. (4, 490, 504)

[LSP1981] D. Levi, P.M. Santini and L. Pilloni, Bäcklund transformations for $(2+1)$-dimensional integrable systems, *Physics Lett.* **A81** (1981) 419–423. (4)

[Lu1977] F. Lund, Solitons and geometry, in: [Ba1978], 143–175. (67)

[LYS1981] Li Yishen, One special inverse problem of the second order differential equation on the whole real axis, *Chin. Ann. Math.* **2** (1981) 147–156. (297)

[LYS1982] Li Yishen, Evolution equations associated with the eigenvalue problem based on the ODE $\psi_{xx}(x) = [u(x) - k^2\rho^2(x)]\psi(x)$, *Nuovo Cimento B* (submitted). (298)

[M1950] V.A. Marchenko, Concerning the theory of a differential operator of the second order, *Dokl. Akad. Nauk SSSR* **72** (1950) 457–460 (in Russian). (330)

[M1979a] Y. Matsuno, Exact multisoliton solution of the Benjamin–Ono equation, *J. Phys. A: Math. Gen.* **12** (1979) 619–621. (65)

[M1979b] Y. Matsuno, N-soliton and N-periodic wave solutions of the higher order Benjamin–Ono equation, *J. Phys. Soc. Japan* **47** (1979) 1745–1746. (65)

[M1979c] Y. Matsuno, Exact multi-soliton solution for nonlinear waves in a stratified fluid of finite depth, *Phys. Lett.* **74A** (1979) 233–235. (65)

[M1980a] Y. Matsuno, Solutions of the higher order Benjamin–Ono equation, *J. Phys. Soc. Japan* **48** (1980) 1024–1028. (65)

[M1980b] Y. Matsuno, N-soliton solution of the higher order wave equation for a fluid of finite depth, *J. Phys. Soc. Japan* **48** (1980) 663–668. (65)

[M1980c] Y. Matsuno, Interaction of the Benjamin–Ono solitons, *J. Phys. A: Math. Gen.* **13** (1980) 1519–1536. (65)

[M1981] S.V. Manakov, The inverse scattering transform for the time-dependent Schroedinger equation and Kadomtsev–Petviashvili equation, in: [MZ1979],

420–427. (65)

[Ma1974] S. V. Manakov, Complete integrability and stochastization of discrete dynamical systems, *Soviet Phys. JETP* **40** (1975) 269–274 [*Zh. Eksp. Teor. Fiz.* **67** (1974) 543–555]. (5)

[Ma1978] D. Maison, Are the stationary, axially symmetric Einstein equations completely integrable?, *Phys. Rev. Lett.* **41** (1978) 521–522. (67)

[Ma1979] D. Maison, On the complete integrability of the stationary, axially symmetric Einstein equations, *J. Math. Phys.* **20** (1979) 871–877. (67)

[Ma1980] L. Martinez Alonso, Gel'fand–Dikii method and nonlinear equations associated to Schroedinger operators with energy-dependent potentials, *Lett. Math. Phys.* **4** (1980) 215–222. (5)

[Mag1978] F. Magri, A simple model of the integrable hamiltonian equation, *J. Math. Phys.* **19** (1978) 1156–1162. (487)

[Mag1979] F. Magri, A geometrical approach to the nonlinear solvable equations, in: [BPS1980], 233–263. (487)

[Mak1978] V.G. Makhankov, Dynamics of classical solitons (in non-integrable systems), *Phys. Rep.* **35C** (1978) 1–128. (4)

[Man1978] Yu.I. Manin, Algebraic aspects of nonlinear differential equations, *J. Sov. Math.* **12** (1979) 1–122 [*Sovrem. Probl. Matem.* **11** (1978) 5–152]. (4, 5, 176)

[Mar1974] V.A. Marchenko, The periodic Korteweg–de Vries problem, *Mat. Sb.* **95** (1974) 331–356 (in Russian). (176)

[Mar1977] V.A. Marchenko, *Sturm–Liouville operators and their applications*, Naukova Dumka, Kiev, 1977 (in Russian). (4, 64, 118)

[Mat1976] V.B. Matveev, Abelian functions and solitons, Institute of Theoretical Physics, Wrocław University preprint 373, 1976. (4, 176)

[Mat1979a] V.B. Matveev, Darboux transformation and explicit solutions of the Kadomtzev–Petviashvili equation, depending on functional parameters, *Lett. Math. Phys.* **3** (1979) 213–216. (207)

[Mat1979b] V.B. Matveev, Darboux transformation and explicit solutions of differential-difference and difference-difference evolution equations. I, *Lett. Math. Phys.* **3** (1979) 217–222. (207)

[Mat1979c] V.B. Matveev, Darboux transformations and nonlinear equations, in: [Sab1980], 247–264. (207)

[Max1976] S. Maxon, Cylindrical and spherical solitons, in: [FML1978], 269–281. (64)

[McG1978] M.J. McGuinness, The conserved densities of the Korteweg–de Vries equation, *J. Math. Phys.* **19** (1978) 2285–2288. (225)

[MD1979] H.C. Morris and R.K. Dodd, The two-component derivative nonlinear Schroedinger equation, in: [W1979], 505–508. (66)

[MD1979a] H.C. Morris and R.K. Dodd, A two-connection and operator bundles for the Ernst equation for axially symmetric gravitational fields, *Phys. Lett.* **75A** (1979) 20–22. (67)

[MGK1968] R.M. Miura, C.S. Gardner and M.D. Kruskal, Korteweg–de Vries equation and generalizations. II: Existence of conservation laws and constants of motion, *J. Math. Phys.* **9** (1968) 1204–1209. (3, 175, 224)

[Mi1968] R.M. Miura, Korteweg–de Vries equation and generalizations. I: A remarkable explicit nonlinear transformation, *J. Math. Phys.* **9** (1968) 1202–1204. (3, 175)

[Mi1974] R.M. Miura, The Korteweg–de Vries equation: a model equation for non-linear dispersive waves, in: [LS1974], 212–234. (64, 65)

[Mi1976] R.M. Miura (editor), *Bäcklund transformations*, Lecture Notes in Mathematics **515**, Springer, Berlin, 1976. (4, 205, 494, 497, 501, 505)

[Mi1976a] R.M. Miura, The Korteweg–de Vries equation: a survey of results, *SIAM*

	Rev. **18** (1976) 412–459. (65, 224)
[Mi1978]	R.M. Miura, An introduction to solitons and the inverse scattering method via the Korteweg–de Vries equation, in: [LS1978], 1–19. (64, 65)
[Mik1976]	A.V. Mikhailov, Integrability of the two-dimensional Thirring model, *JETP Lett.* **23** (1976) 320–323 [*Pis'ma Zh. Eksp. Teor. Fiz.* **23** (1976) 356–358]. (67)
[Mik1979]	A.V. Mikhailov, Integrability of a two-dimensional generalization of the Toda chain, *JETP Letters* **30**(7) (1979) 414–418 [*Pis'ma Zh. Eksp. Teor. Fiz.* **30**(7) (1979) 443–448]. (67)
[Mil1981]	J.W. Miles, The Korteweg–de Vries equation: a historical essay, *J. Fluid Mech.* **106** (1981) 131–147. (64)
[MK 1978]	H.P. McKean, Boussinesq's equation as a hamiltonian system, *Topics in Functional Analysis*, Advances in Mathematics Supplementary Studies, Vol. 3 (Academic Press, New York, 1978), 217–226. (65)
[MKM1975]	H.P. McKean and P. van Moerbeke, The spectrum of Hill's equation, *Invent. Math.* **30** (1975) 217–274. (5)
[MKT1976]	H.P. McKean and E. Trubowitz, Hill's operator and hyperelliptic function theory in the presence of infinitely many branch points, *Comm. Pure Appl. Math.* **29** (1976) 143–226. (5)
[MLS1977]	D.W. McLaughlin and A.C. Scott, Soliton perturbation theory, in [C1978], 225–243. (5)
[Mo1975]	J. Moser (editor), *Dynamical Systems, Theory and Applications*, Lecture Notes in Physics **38**, Springer, Berlin, 1975. (4, 497, 499)
[Mo1975a]	J. Moser, Three integrable hamiltonian systems connected with isospectral deformations, *Adv. in Math.* **16** (1975) 197–220. (5, 177)
[Mo1978]	J. Moser, Various aspects of integrable hamiltonian systems, in: *Dynamical Systems*, Proceedings of the CIME Conference, Bressanone, Italy, June 1978, Liguori, Napoli, 1980, 137–195. (487)
[MOP1981]	A.V. Mikhailov, M.A. Olshanetsky and A.M. Perelomov, Two-dimensional generalized Toda lattice, *Commun. Math. Phys.* **79** (1981) 473–488. (67)
[Mos1956]	H.E. Moses, Calculation of the scattering potential from reflection coefficients, *Phys. Rev.* **102** (1956) 559–567. (330)
[Mos1964]	H.E. Moses, Generalizations of the Jost functions, *J. Math. Phys.* **5** (1964) 833–840. (330)
[Mos1979]	H.E. Moses, Gel'fand–Levitan equations with comparison measures and comparison potentials, *J. Math. Phys.* **20** (1979) 2047–2053. (330)
[MS1979]	V.B. Matveev and M.A. Salle, Differential-difference evolution equations. II: Darboux transformation for the Toda lattice, *Lett. Math. Phys.* **3** (1979) 425–429. (207)
[MS1980]	L. Martina and P.M. Santini, Propagation of ion-acoustic waves in cold inhomogeneous plasmas, *Lett. Nuovo Cimento* **29** (1980) 513–516. (4)
[MST1980]	S.V. Manakov, P.M. Santini and L.A. Takhtajan, Asymptotic behaviour of the solutions of the Kadomtsev–Pyatviashvili equation (two dimensional Korteweg–de Vries equation), *Phys. Lett.* **75A** (1980) 451–454. (4, 65, 177)
[MZ1979]	S.V. Manakov and V.E. Zakharov, *Soliton Theory*, Proceedings of the Soviet-American Symposium on Soliton Theory, Kiev, September 1979, *Physica* **3D** (1,2) (July 1981). (4, 488, 497, 503, 506)
[MZ1981]	S.V. Manakov and V.E. Zakharov, Three-dimensional model of relativistic-invariant field theory, integrable by the inverse scattering transform, *Lett. Math. Phys.* **5** (1981) 247–253. (67)
[MZBIM1977]	S.V. Manakov, V.E. Zakharov, L.A. Bordag, A.R. Its and V.B. Matveev, Two-dimensional solitons of the Kadomtsev-Petviashvili equation and their

interaction, *Phys. Lett.* **63A** (1977) 205–206. (65)

[N1966] R.G. Newton, *Scattering theory of waves and particles*, McGraw Hill, New York, 1966. (64, 118)

[N1974] A.C. Newell (editor), *Nonlinear Wave Motion*, Lect. Appl. Math. **15**, American Mathematical Society, Providence, RI, 1974. (4, 176, 489, 499)

[N1975] A.C. Newell, The interrelation between Bäcklund transformations and the inverse scattering transform, in: [Mi1976], 227–240. (178)

[N1977] A.C. Newell, Near integrable systems, nonlinear tunnelling and solitons in slowly changing media, in: [C1978], 127–179. (5, 297)

[N1978] A.C. Newell, Long waves-short waves; a solvable model, *SIAM J. Appl. Math.* **35** (1978) 650–664. (66)

[N1980] A.C. Newell, The inverse scattering transform, in: [BC1980], 177–242. (65, 297)

[Na1979a] A. Nakamura, Bäcklund transform and conservation laws of the Benjamin–Ono equation, *J. Phys. Soc. Japan* **47** (1979) 1335–1340. (65)

[Na1979b] A. Nakamura, Exact N-soliton solution of the modified finite depth fluid equation, *J. Phys. Soc. Japan* **47** (1979) 2043–2044. (65)

[Na1979c] A. Nakamura, N-periodic wave and N-soliton solutions of the modified Benjamin–Ono equation, *J. Phys. Soc. Japan* **47** (1979) 2045–2046. (65)

[Na1980] A. Nakamura, Bäcklund transformation of the cylindrical KdV equation, *J. Phys. Soc. Japan* **49** (1980) 2380–2386. (297)

[Ne1980] R.G. Newton, Inverse scattering. I: One dimension, *J. Math. Phys.* **21** (1980) 493–505. (118)

[Neu1979] G. Neugebauer, Bäcklund transformations of axially symmetric stationary gravitational fields, *J. Phys. A: Math. Gen.* **12** (1979) L67–L70. (67)

[NN1958] E. Nagel and J.R. Newman, *Gödel's Proof*, New York University Press, New York, 1958. (6)

[No1974] S.P. Novikov, The periodic problem of the Korteweg–de Vries equation. I, *Func. Anal. Appl.* **8** (1974) 236–246 [*Funk. Anal. Priloz.* **8** (1974) 54–66]. (64)

[O1980] S.J. Orfanidis, σ models of nonlinear evolution equations, *Phys. Rev.* **D21** (1980) 1513–1522. (67)

[OB1978] J. Oficjalski and I. Bialynicki-Birula, Collision of gaussons, *Acta Phys. Polon.* **B9** (1978) 759–775. (176)

[OB1980] A.R. Osborne and T.L. Burch, Internal solitons in the Andaman sea, *Science* **208** (1980) 451–460. (64)

[Ol1977] P.J. Olver, Evolution equations possessing infinitely many symmetries, *J. Math. Phys.* **18** (1977) 1212–1215. (225)

[Ono1975] H. Ono, Algebraic solitary waves in stratified fluids, *J. Phys. Soc. Japan* **39** (1975) 1082–1091. (65)

[OP1976] M.A. Olshanetsky and A.M. Perelomov, Explicit solution of the Calogero model in the classical case and geodesic flows on symmetric spaces of zero curvature, *Lett. Nuovo Cimento* **16** (1976) 33–39. (177)

[OP1981] M.A. Olshanetsky and A.M. Perelomov, Classical integrable finite-dimensional systems related to Lie algebras, *Phys. Rep.* **71** (1981) 313–400. (5, 177)

[OW1981] M. Omote and M. Wadati, Bäcklund transformations for the Ernst equation, *J. Math. Phys.* **22** (1981) 961–964. (67)

[OW1981a] M. Omote and M. Wadati, The Bäcklund transformations and the inverse scattering method of the Ernst equation, *Prog. Theor. Phys.* **65** (1981) 1621–1631. (67)

[P1966] D.H. Peregrine, Calculations of the development of an undular bore, *J. Fluid Mech.* **25** (1966) 321–330. (64, 176)

[P1967]	J.C. Portinari, Finite-range solutions to the one-dimensional inverse scattering problem, *Ann. Phys.* **45** (1967) 445–451. (330)
[P1976]	K. Pohlmeyer, Integrable hamiltonian systems and interactions through quadratic constraints, *Commun. Math. Phys.* **46** (1976) 207–221. (67)
[Par1805]	M.A. Parseval–Deschênes, in: *Mémoires par Divers Savants*. I, 1805, 639–648 (as referred to in [WW1902], 182 of fourth edition). (3)
[Pil1980]	L. Pilloni, Extension of the spectral transform method for solving nonlinear evolution equations, and related conservation laws, *Nuovo Cimento* **56B** (1980) 87–109. (4)
[PRS1979]	F.A.E. Pirani, D.C. Robinson and W.F. Shadwick, Local jet bundle formulation of Bäcklund transformations, *Mathematical Physics Studies* **1** (1979) 1–132. (5, 205)
[R1975]	R. Rajamaran, Some non-perturbative semi-classical methods in quantum field theory (a pedagogical review), *Phys. Rep.* **21C** (1975) 227–313. (176)
[R1979]	A.F. Rañada (editor), *Nonlinear Problems in Theoretical Physics*, Lecture Notes in Physics **98**, Springer, Berlin, 1979. (4, 492, 495, 501)
[R1981]	O. Ragnisco, Conservation laws for the whole class of nonlinear evolution equations associated to the matrix Schroedinger spectral problem, *Lett. Nuovo Cimento* **31** (1981) 651–656. (4)
[Re1979]	C. Rebbi, Solitons, *Scientific American* **240**(2) (1979) 76–91. (176)
[S1974]	J. Satsuma, Higher conservation laws for the Korteweg–de Vries equation through Bäcklund transformation, *Prog. Theor. Phys.* **52** (1974) 1396–1397. (224)
[Sa1978]	A.R. Santarelli, Numerical analysis of the regularized long-wave equation: anelastic collision of solitary waves, *Nuovo Cimento* **46B** (1978) 179–188. (176)
[SA1979]	H. Segur and M.J. Ablowitz, Asymptotic solutions of nonlinear evolution equations and a Painlevé transcendent, in: [MZ1979], 165–184. (178)
[Sab1978]	P.C. Sabatier (editor), *Applied inverse problems*, Lecture Notes in Physics **85**, Springer, Berlin, 1978. (4, 493)
[Sab1979a]	P.C. Sabatier, Around the classical string problem, in: [BPS1980], 85–102. (298)
[Sab1979b]	P.C. Sabatier, On some spectral problems and isospectral evolutions connected with the classical string problem. I: Constants of motion, *Lett. Nuovo Cimento* **26** (1979) 477–482. (298)
[Sab1979c]	P.C. Sabatier, On some spectral problems and isospectral evolutions connected with the classical string problem. II: Evolution equations, *Lett. Nuovo Cimento* **26** (1979) 483–486. (293)
[Sab1980]	P.C. Sabatier (editor), *Problèmes Inverses, Evolution Non Linéaire*, CNRS, Paris, 1980. (495, 503, 507)
[SAK1979]	J. Satsuma, M.J. Ablowitz and Y. Kodama, On an internal wave equation describing a stratified fluid with finite depth, *Phys. Lett.* **73A** (1979) 283–286. (65)
[San1978]	P.M. Santini, Asymptotic behaviour (in t) of solutions of the boomeron equation, *Nuovo Cimento* **47B** (1978) 228–243. (4, 177)
[San1979]	P.M. Santini, Asymptotic behavior (in t) of solutions of the cylindrical KdV equation. I, *Nuovo Cimento* **54A** (1979) 241–258. (4, 177)
[San1980]	P.M. Santini, Asymptotic behaviour (in t) of solutions of the cylindrical KdV equation. II, *Nuovo Cimento* **57A** (1980) 387–396. (4)
[San1981]	P.M. Santini, On the evolution of two-dimensional packets of water waves over an uneven bottom, *Lett. Nuovo Cimento* **30** (1981) 236–240. (4)
[Sch1950]	L. Schwartz, *Théorie des distributions*, Vol. I & II, Hermann, Paris, 1950 & 1951. (3)

[SCM1973] A.C. Scott, F.Y.F. Chu and D. McLaughlin, The soliton: a new concept in applied science, *Proc. IEEE* **61** (1973) 1443–1483. (4, 65, 66)

[SCYA1976] H.H. Szu, C.E. Carroll, C.C. Yang and S. Ahn, A new functional equation in the plasma inverse problem and its analytic properties, *J. Math. Phys.* **17** (1976) 1236–1247. (330)

[Seg1976] H. Segur, Solitons as approximate descriptions of physical phenomena, in: [FML1978], 15–24. (64, 65)

[Seg1976a] H. Segur, Asymptotic solutions and conservation laws for the nonlinear Schroedinger equation, II, *J. Math. Phys.* **17** (1976) 714–716. (177)

[SG1969] C.S. Su and C.S. Gardner, Korteweg–de Vries equation and generalizations. III: Derivation of the Korteweg–de Vries equation and Burgers' equation, *J. Math. Phys.* **10** (1969) 536–539. (3, 64, 175)

[SI1979] J. Satsuma and Y. Ishimori, Periodic wave and rational soliton solutions of the Benjamin–Ono equation, *J. Phys. Soc. Japan* **46** (1979) 681–687. (65)

[Sky1958] T.H.R. Skyrme, A nonlinear theory of strong interactions, *Proc. Roy. Soc.* **A247** (1958) 260–278. (66, 176)

[Sky1961] T.H.R. Skyrme, Particle states of a quantized field, *Proc. Roy. Soc.* **A262** (1961) 237–245. (66, 176)

[SR1845] J. Scott-Russell, Report on waves in: *Report of the fourteenth meeting of the British association for the advancement of science*, John Murray, London, 1845, 311–390. (1)

[St1975] H. Steudel, Noether's theorem and the conservation laws of the Korteweg–de Vries equation, *Ann. Physik Leipzig* **32** (1975) 445–455. (225)

[SW1980] T. Shimizu and M. Wadati, A new integrable nonlinear evolution equation, *Prog. Theor. Phys.* **63** (1980) 808–820. (65)

[Sym1978] A. Sym, Formal series solutions and rational solutions to soliton equations, *Lett. Nuovo Cimento* **22** (1978) 142–146. (177)

[T1958] W.E. Thirring, A soluble relativistic field theory, *Ann. Phys.* **3** (1958) 91–112. (67)

[T1967] M. Toda, Vibration of a chain with nonlinear interaction, *J. Phys. Soc. Japan* **22** (1967) 431–436. (5)

[T1971] F.D. Tappert, Nonlinear wave propagation as described by the Korteweg–de Vries equation and its generalizations, Computer-produced film, Bell Laboratories, 1971. (3, 176)

[T1972] S. Tanaka, Analogue of Fourier's method for Korteweg–de Vries equation, *Proc. Japan. Acad.* **48** (1972) 647–650. (63)

[T1975] M. Toda, Studies in a non-linear lattice, *Phys. Rep.* **18C** (1975) 1–125. (5)

[T1977] L.A. Takhtajan, Integration of the continuous Heisenberg spin chain through the inverse scattering method, *Phys. Lett.* **64A** (1977) 235–237. (66)

[T1980] M. Toda, On a nonlinear lattice (the Toda lattice), in: [BC1980], 143–155. (5)

[T1981] M. Toda, *Theory of Nonlinear Lattices*, Springer Series in Solid-State Sciences **20**, Springer, Berlin, 1981. (5)

[Taf1981] E. Taflin, Analytic linearization, hamiltonian formalism and infinite sequences of constants of motion for Burger's equation, *Phys. Rev. Lett.* **47** (1981) 1425–1428. (65)

[TB1979] J. Timonen and R.K. Bullough, Solitons in physics: an application to spin waves in the one-dimensional ferromagnet $CsNiF_3$, in: [Sab1980], 133–187. (4, 67)

[TF1974] L.A. Tahtadžjan and L.D. Faddeev, Essentially nonlinear one-dimensional model of classical field theory, *Theoret. Math. Phys.* **21** (1974) 1046–1057 [*Teoret. Mat. Fiz.* **21** (1974) 160–174]. (67)

[TF1976] L.A. Tahtadžjan and L.D. Faddeev, The hamiltonian system connected with the equation $u_{\xi\eta}+\sin u = 0$, *Proceed. Steklov Inst. Math.* **3** (1979) 277–289 [*Trudi Mat. Inst. Steklov* **142** (1976)]. (67)

[Th1976] W.R. Thikstun, A system of particles equivalent to solitons, *J. Math. Anal. Appl.* **55** (1976) 335–346. (177)

[Ts1980] M. Tsutsumi, On solutions of Liouville's equation, *J. Math. Anal. Appl.* **76** (1980) 116–123. (67)

[TY1969] T. Taniuti and N. Yajima, Perturbation method for a nonlinear wave modulation, I. *J. Math. Phys.* **10** (1969) 1369–1372. (65)

[W1968] S. Weinberg, Nonlinear realizations of chiral symmetry, *Phys. Rev.* **166** (1968) 1568–1577. (67)

[W1972] M. Wadati, The exact solution of the modified Korteweg–de Vries equation, *J. Phys. Soc. Japan* **32** (1972) 1681. (65)

[W1974] G.B. Whitham, *Linear and Nonlinear Waves*, Wiley, New York, 1974. (4, 63)

[W1977] M. Wadati, Infinitesimal transformations and conservation laws; field theoretic approach to the theory of soliton, in: [C1978], 33–63. (224)

[W1978] M. Wadati, Invariances and conservation laws of the Korteweg–de Vries equation, *Stud. Appl. Math.* **59** (1978) 153–186. (224, 225)

[W1979] H. Wilhelmsson (editor), *Solitons in Physics*, Topical issue of Physica Scripta, Royal Swedish Academy of Sciences, *Phys. Scr.* **20** (1979). (4, 490, 503)

[WE1973] H.D. Wahlquist and F.B. Estabrook, Bäcklund transformations for solutions of the Korteweg–de Vries equation, *Phys. Rev. Lett.* **31** (1973) 1386–1390. (205, 207)

[Wil1979] G. Wilson, Commuting flows and conservation laws for Lax equations, *Math. Proc. Camb. Phil. Soc.* **86** (1979) 131–143. (487)

[WKI1979] M. Wadati, K. Konno and Y.H. Ichikawa, New integrable nonlinear evolution equations, *J. Phys. Soc. Japan* **47** (1979) 1698–1700. (65)

[WW1902] E.T. Whittaker and G.N. Watson, *A Course of Modern Analysis*, Cambridge University Press, Cambridge, 1902. (506)

[Z1973] V.E. Zakharov, On stochastization of one-dimensional chains of nonlinear oscillators, *Sov. Phys. JETP* **38** (1974) 108–110 [*Zh. Eksp. Teor. Fiz.* **65** (1973) 219–225]. (65)

[Z1980] V.E. Zakharov, The inverse scattering method, in: [BC1980], 243–285. (65)

[ZDK1968] N.J. Zabusky, G.S. Deem and M.D. Kruskal, Formation, propagation and interaction of solitons, Computer-produced film, Bell Laboratories, 1968. (3, 176)

[ZF1971] V.E. Zakharov and L.D. Faddeev, Korteweg–de Vries equation, a completely integrable hamiltonian system, *Func. Anal. Appls.* **5** (1971) 280–287 [*Funk. Anal. Pril.* **5** (1971) 18–27]. (3, 5, 224, 487)

[ZhS1979] A.V. Zhiber and A.B. Shabat, Klein–Gordon equations with a non-trivial group, *Sov. Phys. Dokl.* **24**(8) (1979) 607–609 [*Dokl. Akad. Nauk SSSR* **247** (1979) 1103–1107]. (67)

[ZK1965] N.J. Zabusky and M.D. Kruskal, Interactions of "solitons" in a collisionless plasma and the recurrence of initial states, *Phys. Rev. Lett.* **15** (1965) 240–243. (1, 32, 176)

[ZM1973] V.E. Zakharov and S.V. Manakov, Resonant interaction of wave packets in non-linear media, *JETP Lett.* **18** (1973) 243–245. (66)

[ZM1975] V.E. Zakharov and S.V. Manakov, The theory of resonant interaction of wave packets in nonlinear media, *Soviet Phys. JETP* **42** (1976) 842–850 [*Zh. Eksp. Teor. Fiz.* **69** (1975) 1654–1673]. (66)

[ZM1976] V.E. Zakharov and S.V. Manakov, Asymptotic behaviour of nonlinear wave systems integrated by the inverse scattering method, *Sov. Phys. JETP* **44** (1976) 106–112 [*Zh. Eksp. Teor. Fiz.* **71** (1976) 203–215]. (177)

[ZM1979] V.E. Zakharov and S.V. Manakov, Soliton theory, *Soviet Scien. Reviews* **A1** (1979) 133–190. (4, 65)

[ZMi1978] V.E. Zakharov and A.V. Mikhailov, Relativistically invariant two-dimensional models of field theory which are integrable by means of the inverse scattering problem method, *Sov. Phys. JETP* **47** (1978) 1017–1027 [*Zh. Eksp. Teor. Fiz.* **74** (1978) 1953–1973]. (67)

[ZMi1980] V.E. Zakharov and A.V. Mikhailov, On the integrability of classical spinor models in two-dimensional space-time, *Commun. Math. Phys.* **74** (1980) 21–40. (67)

[ZMNP1980] V.E. Zakharov, S.V. Manakov, S.P. Novikov and L.P. Pitaievski, *Theory of Solitons. The Inverse Problem Method*, Nauka, Moscow, 1980 (in Russian). (4, 177)

[ZS1971] V.E. Zakharov and A.B. Shabat, Exact theory of two-dimensional self-focusing and one-dimensional self-modulation of waves in nonlinear media, *Soviet Phys. JETP* **34** (1972) 62–69 [*Zh. Eksp. Teor. Fiz.* **61** (1971) 118–134]. (2, 66)

[ZS1974] V.E. Zakharov and A.B. Shabat, A scheme for integrating the non-linear equations of mathematical physics by the method of the inverse scattering problem. I, *Func. Anal. Appl.* **8** (1974) 226–235 [*Funk. Anal. Pril.* **8** (1974) 43–53]. (4, 66, 175)

[ZTF1974] V.E. Zakharov, L.A. Tahtadžjan and L.D. Faddeev, A complete description of the solution of the sine–Gordon equation, *Soviet Phys. Dokl.* **19** (1975) 824–826 [*Dokl. Akad. Nauk SSSR* **219** (1974) 1334–1337]. (67)

SUBJECT INDEX

action-angle variables, 487
Airy functions, 257, 258 ff
AKNS approach, 440 ff
applications, 4, 48 ff, 64 ff, 141 ff
asymptotic (long-time) behavior, 9, 30 ff, 32 ff, 34, 166, 177

background, 32 ff,
Bäcklund transformations, 35 ff, 179 ff, 224, 370 ff, 378 ff, 393, 423, 452 ff
Bargmann approach (to reflectionless potentials), 437
Bargmann strip, 75, 378 ff
BBM equation, *see* regularized long-wave equation
Benjamin–Ono equation, 53, 65, 177
Bessel functions, 429
bona fide potentials, 88, 195 ff, 394 ff
boomeron, 58, 66
boomeron equation, 57, 66
Born approximation, 20
bound states, *see* discrete eigenvalues
Boussinesq equation, 54, 65
Burgers equation, 55, 65

canonical transformations, 487
Cauchy problem, *see* initial value problem
center of mass (motion of), 47, 221 ff, 272, 276, 279, 388 ff
chiral field equation, 61, 67
closure relations, 117, 119, 306 ff
cnoidal waves, 160
collision of one pole and one soliton, 284 ff
collision (of solitary waves), 161 ff, 176
collision (of solitons), 31, 134 ff

commutativity (of Bäcklund transformations), 39, 191 ff
completeness relations, *see* closure relations
conservation laws, 43 ff, 208 ff, 233, 243 ff, 272 ff, 278 ff, 290, 293, 294, 375 ff, 388 ff, 417, 462 ff, 478 ff
continuum part of the spectrum, 16 ff, 69 ff
$\cosh^{-2}$ potential, 432 ff
cylindrical Korteweg–de Vries equation, 50, 64, 171, 177, 227, 255, 268 ff, 272, 276 ff, 297, 389

damping (in KdV equation), 177
Darboux transformations, 179, 189, 205, 207, 378 ff, 423, 453 ff
Davey–Stewartson equations, 57, 66
delta function, *see* Dirac distribution
derivative nonlinear Schroedinger equation, 56, 66
diffusion equation, 55
Dirac distribution, 316, 397, 426
direct method, 64
direct spectral problem, 16 ff, 68 ff, 419 ff
discrete eigenvalues, 16 ff, 123, 182 ff, 230, 236, 299 ff, 320, 400 ff, 418 ff, 459, 483
discrete spectrum, *see* discrete eigenvalues
discretized equations, 2, 5
dispersion (of waves), 9, 32, 153
dispersion relations (for transmission coefficient), 78, 311 ff, 357
double Sine–Gordon equation, 60, 67
Dym, *see* Harry Dym equation
dynamical systems, 2, 5, 168 ff, 177, 477 ff

Einstein equation, 62, 67
elementary particles, 2, 141, 163, 176

511

Ernst equations, 62, 67
exponential potential, 429 ff

Fermi–Pasta–Ulam problem, 1
ferromagnet equation, *see* Heisenberg ferromagnet equation
finite-depth fluid equation, 54, 65
Fourier transform, 2, 8 ff, 20, 21, 22, 26 ff, 40, 63, 396, 413

gamma function, 430
Gardner transformation, 224, 370 ff
Gel'fand–Levitan–Marchenko equation, 18 ff, 81 ff, 314 ff, 398, 419
generalized KdV equation, 52, 65, 388 ff
Goursat problem, 59, 317
Gross–Neveu model, 62, 67
group theoretical approaches, 487
group velocity, 9, 34

hamiltonian dynamics, 2, 168, 177, 224, 477 ff
hamiltonian flow, 477 ff
Hamilton's equations, 168, 477 ff
Harry Dym equation, 53, 292, 298
hash, 33
Heisenberg ferromagnet equation, 56, 66
higher Benjamin–Ono equations, 65
higher finite-depth fluid equations, 65
higher KdV equations, 64, 128, 281, 482 ff
Hilbert operator, 53
Hirota equation, 56, 66
Hirota method, *see* direct method
Hopf–Cole transformation, 55, 65
hypergeometric function, 432

initial value problem, 8 ff, 14 ff, 24 ff, 121 ff, 231 ff, 253, 277, 292, 296, 414 ff
integrability conditions, *see* AKNS approach
integrable dynamical systems, *see* dynamical systems
inverse spectral problem, 18 ff, 81 ff, 118, 263 ff, 297, 313 ff, 330, 436 ff
involution (in), 478, 487
isospectral evolution, 24, 124, 210, 459

Jacobi identity, 478
Jost function, 260 ff, 268 ff
Jost solutions, 69, 108 ff, 189, 306 ff, 314 ff, 346, 348, 419 ff, 453

Kadomtsev–Petviashvili equation, 54, 65
KdV equation, *see* Korteweg–de Vries equation

kink, 60
Korteweg–de Vries equation, 1, 14, 29 ff, 46 ff, 48 ff, 64, 128, 134 ff, 164, 167 ff, 172, 176, 177, 205, 215, 217, 220, 224 ff, 255, 281 ff, 296, 297, 365 ff, 370 ff, 389, 446, 469, 480 ff, 487

lagrangian density (for the KdV equation), 225
Lamb diagram, 40, 207
Lax approach, 2, 444 ff, 464
Leibnitz rule, 478
Levinson's theorem, 78, 312, 476
linear evolution equations, 7 ff, 15, 26 ff, 147, 204
Liouville equation, 60, 67
Liouville theorem, 479

massive Thirring model, 62, 67
matrix Schroedinger equation, 58
Maxwell–Bloch equations, *see* reduced Maxwell–Bloch equations
Miura transformation, 52, 65, 207, 370 ff
modified Benjamin–Ono equation, 65
modified finite-depth fluid equation, 65
modified Korteweg–de Vries equation, 51, 65, 370 ff, 390
multisoliton solution, 30, 64, 133, 198, 207, 212, 221, 248, 257, 365 ff

Nambu–Jona Lasinio–Vaks–Larkin model, 62, 67
Navier–Stokes equations, 55
nonlinear evolution equations, 2, 4, 14 ff, 22 ff, 48 ff, 88, 120 ff, 175 ff, 180, 228 ff, 235 ff, 268 ff, 277, 283, 289 ff, 296 ff, 412 ff, 457 ff, 482
nonlinear operator identities, 201 ff, 207, 342, 343 ff
nonlinear Schroedinger equation, 2, 55, 65
nonlinear superposition, 41 ff, 196, 207
normalization coefficient, 17, 19, 72, 76, 124, 182, 185, 193 ff, 230, 236, 324 ff, 366, 381, 385, 395, 408, 420 ff
Novikov equation, 64
N-soliton solution, *see* multisoliton solution
N-soliton + 1-pole solution, 284, 297
number of discrete eigenvalues, 17, 73, 299 ff

operator identities, 11, *see also* nonlinear operator identities
orthogonality relations, *see* closure relations

Subject Index

Painlevé transcendents, 171 ff, 178
PBBM equation, see regularized long-wave equation
periodic boundary conditions, 2, 160, 171, 176, 486
perturbation theory, 3, 5
phi-four equation, 60, 67
Pohlmeyer–Lund–Regge model, 61, 67
Poisson bracket, 478 ff
Poisson structure, 478
principal chiral field equation, see chiral field equation
prolongation structures, 5, 205

radiation, 33
rational reflection coefficient, 330, 438 ff
rectangular function, 427 ff
reduced Maxwell–Bloch equations, 59, 66
reference potential, 226, 296, 315 ff, 330, 423 ff
reflection- and transmission-equivalent potentials, 400 ff
reflection coefficient, 17, 69 ff, 76, 77, 122, 125, 181, 309 ff, 325 ff, 330, 333, 338, 372, 381, 390 ff, 394, 395, 398 ff, 419 ff, 483
reflectionless potentials, 19, 84 ff, 398 ff, 436 ff
regularized long-wave equation, 49, 64, 160 ff, 176
relativistically-invariant equations, 59, 60, 61, 62, 66, 67
resolvent formula, 11, 199 ff, 204
Riccati equation, 371, 378, 473

scale transformations 80 ff, 342, 421
scattering phase shifts, 470
Schroedinger equation, 16 ff, 68 ff, 226, 257, 258, 283, 299 ff, 314 ff, 330, 331, 346, 372, 380, 418 ff, 445 ff, 470 ff
second Hamiltonian structure, 486
self-induced transparency, 59, 66
self-similar solutions, see similarity solutions
Shimizu–Wadati equation, 53, 65
shock equation, 153
sigma model, 61, 67
similarity solutions, 171 ff, 178
Sine–Gordon equation, 59, 66
solitary wave, 132, 160 ff
soliton, 1, 3, 28 ff, 34 ff, 39, 42, 52, 54, 55, 56, 58, 60, 64, 132 ff, 166, 176, 188, 198, 207, 212, 221, 237, 246 ff, 257, 270, 280, 284 ff, 365 ff, 415
soliton ladder, 42, 207
solitron, 132 ff, 176
solvable potentials, 420 ff
space reflection, 421
space translation, 79 ff, 339, 342, 420
spectral integral relations, 89, 108 ff, 118, 338 ff
spectral transform, 1, 15 ff, 17, 20 ff, 87 ff, 121, 175, 179, 206, 226, 256, 283, 296, 326, 394 ff, 418 ff, 444 ff, 487
spherical Korteweg–de Vries equation, 51, 64, 280, 389
square-well potential, see rectangular function
string equation, 298
Sturm–Liouville problem, 16, 68
superposition principle, 11, see also nonlinear superposition
symmetries, 224
symplectic operator, 480 ff

Thirring model, see massive Thirring model
three-dimensional three-wave resonant interaction equation, 57, 66
three-wave resonant interaction equation, 57, 66
transitional Korteweg–de Vries equation, 50, 64
translation, see space translation
transmission coefficient, 17, 69 ff, 74, 77, 78, 79, 95 ff, 187, 210, 224, 232, 238, 288, 309 ff, 311 ff, 325, 341 ff, 356 ff, 372, 381 ff, 386, 390 ff, 400 ff, 417, 419 ff
trappon, 58, 66
two-dimensional KdV equation, see Kadomtsev–Petviashvili equation

unitarity relation, 71, 97, 323, 354, 390, 398, 428, 454

virial theorem, 297

wronskian, 70, 323
wronskian integral relations, 89 ff, 118, 338 ff

Yang–Mills equations, 62, 67

Zakharov–Shabat spectral problem, 2, 52, 56, 57, 59, 65, 118, 119, 314, 330, 374
zoomeron equation, 58, 66
zoomeron, 58, 66

LIST OF SYMBOLS

To help the casual reader we list below some of the mathematical symbols used in this volume, with an indication of the number(s) of the equation(s) and pages(s) where they are defined. The list includes mainly symbols that appear in more than one section or appendix. The symbol * denotes those symbols that have been used with more than one meaning (of course, in different contexts). The order of presentation is alphabetical (first Latin alphabet, then Greek). Finally let us note that a subscripted *variable* always indicates partial differentiation with respect to that variable; a superimposed dot indicates differentiation with respect to the "time" variable, t.

a_m	(5.-55, 56) 219; (A.11.-36) 376
* A_m	(A.6.-27) 336
* A_m	(A.9.-48) 349
$Ai(y)$	(6.3.1.-4 ff) 258 ff
B	(A.20.-23) 450
* B_m	(A.6.-28) 336
* B_m	(A.9.-49) 350
$B_n^{(\pm)}$	(A.20.-20a) 449
$B^{(+)}(k)$	(A.20.-31) 451
$B^{(-)}(k)$	(A.9.-47c) 349; (A.20.-32) 451
$Bi(y)$	(6.3.1.-5 ff) 258 ff
* c_m	(1.7.3.-8, 10) 44; (1.7.3.-13, 14) 45; (5.-16) 212; (5.-21) 213; (A.9.-65) 353; (A.13.-15) 387
* c_m	(6.3.1.-14, 15) 259
* c_n	(2.1.-24, 26) 72, 73; (2.1.-29) 74
* $c_n^{(1)}, c_n^{(2)}$	99
* $c_n^{(\pm)}$	(2.1.-28) 74
C_m	(1.7.3.-5, 6) 44; (1.7.3.-12) 45; (5.-12) 211; (5.-14) 212; (5.-19, 20) 213; (A.9.-57, 58) 352; (A.21.-7) 463; (A.23.-33) 481
D	(2.4.2.-26) 115; (2.4.2.1.-6) 116; (A.9.-3) 343; (A.20.-17a) 448; (A.23.-25) 480
$Ei(y)$	(6.3.1.-16) 260

List of Symbols

* $f(x)$	generic function
* $f(z)$	(6.3.1.-25) 260; (6.3.1.-29) 261
* f_n	(6.3.1.-33) 262; (6.3.1.-41) 263
* $f_n(x)$	(1.3.1.-1, 4, 5, 6) 16, 17; (2.1.-22, 19, 20, 1) 72, 68
* $f^{(m)}(x,t)$	(1.7.3.-15) 46; (5.-26, 24, 28) 214; (A.21.-6) 462; (A.21.-50) 469
* $f^{(j)}(x,k)$	(2.4.2.-1) 108
* $f^{(+)}(x,k)$	(1.3.3.-4) 21; (2.1.-9) 70; (2.2.-5a) 82
* $f^{(+)(j)}(x,\mathrm{i}p_n^{(j)})$	(2.4.2.-6, 1) 110, 108
* $f^{(\pm)}(x,k)$	(2.1.-4, 1) 69, 68; (2.1.-9) 70
* $f^{(\pm)(j)}(x,k)$	(2.4.2.-1a, 1) 108
$g^{(m)}(x), g^{(m)}(x,t)$	(1.2.-3) 14; (1.7.3.-16) 46; (2.4.1.-27, 28) 97; (3.1.-30, 32) 127; (5.-27, 25) 214; (A.9.-97, 100) 362; (A.9.-104) 363; (A.21.-5) 462; (A.21.-49) 469
$g^{\#}(z_1, z_2)$	(2.4.1.-20) 96; (A.9.-47 f) 349
$G^{(j,k)}(x,t)$	(5.-30) 215; (A.9.-98, 101) 362; (A.9.-106, 105) 363; (A.21.-44) 468
* H	(1.8.-12a) 53
* $H, H(t)$	(A.20.-7a) 446; (A.20.-47) 454
$H^{(j)}$	(A.5.-7) 315; (A.20.-11) 447
* I	(1.3.2.-13) 19; (1.6.1.-9) 30; (2.2.-20) 85; (A.19.-68c) 436
* I	(A.9.-4) 343; (A.20.-176) 448; (A.23.-27) 480
J	(A.21.-34) 466
* $K(x, y)$	(1.3.2.-3) 18; (2.2.-1) 81; (2.2.-5b, 5c) 82; (A.5.-70) 327
* $K(x, y)$	(6.3.1.-44) 264
* $K(x, y)$	(A.5.-66, 67) 326, 327
L	(1.2.-2) 14; (2.4.1.-13) 94; (3.1.-9) 122; (4.-2) 180; (5.-13) 211; (A.9.-11) 345; (A.21.-9) 463; (A.23.-26a) 480
$\tilde{L}$	(1.7.3.-7) 44; (2.4.2.-18, 19) 113; (3.1.-38) 129; (4.-6) 180; (5.-22) 213; (A.9.-12) 345; (A.21.-10) 463; (A.23.-26b) 480
* $\mathcal{L}$	(6.5.-11) 288
* $\mathcal{L}$	(A.23.-54) 484
m	(A.9.-2) 343
$M(x)$	(1.3.2.-2) 18; (2.2.-2) 81; (A.5.-71) 327
* $M(x, y)$	(6.3.1.-43) 264
* $M(x, y)$	(A.5.-68) 327
* M	(1.7.3.-9) 44; (5.-17) 212; (A.9.-64) 352
* M	(6.3.1.-36) 262
* M	(A.20.-7b) 446; (A.20.-51) 455
N	(1.3.1.-4) 17; 299 ff
* N	(6.3.1.-37) 262
* N	(A.21.-20) 464
$\tilde{N}$	(A.21.-19) 464
* P	(A.20-13d) 448
* P	(A.21.-19, 20) 464; (A.21.-39) 467
p_n	(1.3.1.-4) 17; (2.1.-20, 19) 72
* Q_0, Q_1	(6.3.4.-18) 278
* Q_n	(A.18.-27) 417
$R(k)$	(1.3.1.-3b) 17; (2.1.-3b) 69
$R(k,t)$	$22 \div 24$
$R^{(j)}(k)$	(2.4.1.-1b) 90
s	(A.9.-1) 343
$S[u]$	(1.3.1.-7) 17

List of Symbols

Symbol	Reference
$T(k)$	(1.3.1.-3a) 17; (2.1.-3a) 69
$T^{(j)}(k)$	(2.4.1.-1a) 90
$w(x)$	(1.3.2.-5) 18; (2.2.-4) 82
$w(x,t)$	(3.1.-27) 126; (4.-4) 180
$w^{(m)}(x), w^{(j)}(x)$	(1.7.1.-12) 37; (2.4.2.-2) 109
$W[f^{(1)}(x), f^{(2)}(x)]$	(2.1.-7) 70
$\alpha_m(t)$	(3.1.-29) 127; (5.-2) 208
$\gamma^{(m)}(x,t)$	(1.7.3.-18) 46; (5.-29) 215; (A.21.-1) 462; (A.21.-43) 468; (A.21.-51) 469
Γ	(1.7.1.-7) 36; (2.4.1.-5) 90; (4.1.-3) 181; (A.6.-16) 334; (A.9-7) 344; (A.20.-14) 448
* Γ_m	(6.1.-39) 233
* Γ_m	(6.3.3.-24) 275
* $\theta(k)$	(2.1.-15a) 71; (A.3.-4) 309; (A.8.-2) 341
* $\theta(x)$	(A.1.-6, 7) 300
* $\theta(x)$	(A.5.-13) 316; (A.19.-17) 422
* $\theta(t, t_0; \lambda_0)$	(6.2.-43, 44) 242
θ_m	(2.1.-40, 41) 77; (2.1.-45) 78; (5.-12) 211; (A.3.-2, 3) 309; (A.8.-3) 341; (A.8.-7) 342; (A.9.-56) 352
Λ	(1.7.1.-8) 36; (2.4.1.-6) 91; (4.1.-4) 181; (A.6.-22) 335; (A.9.-8) 344; (A.20.-15) 448
$\Lambda^{(0)}$	(1.7.3.-11) 45; (2.4.1.-17) 95; (5.-23) 213; (A.9.-51) 350
$\tilde{\Lambda}$	(2.4.2.-10, 9) 111, 110; (A.9.-9) 344
* $\xi_n(t)$	(3.2.1.-6) 134; (A.10.-12) 366
* $\xi_j(t)$	(3.2.4.1.-3) 167
$\xi^{(j)}(t)$	(4.2.-25) 197
$\xi_n^{(\pm)}$	(1.6.1.-14) 30
* $\rho(z)$	(6.3.1.-42) 263
* $\rho(x)$	(6.5.-1) 287
* $\rho(k)$	(A.5.-28) 319; (A.5.-47) 323
ρ_n	(1.3.1.-6) 17; (2.1.-27) 73; (2.1.-36) 76
$\rho_n(t)$	22÷24
$\rho_n^{(j)}$	103; (2.4.2.-7) 110
$\varphi_n(x)$	(2.1.-21, 19, 20, 1) 72, 68
$\varphi^{(m)}(x,t)$	(1.7.3.-17) 46; (5.-34) 216; (A.21.-2) 462; (A.21.-47, 48) 468, 469; (A.21.-52) 469, 470
* $\chi(k)$	(2.1.-15b) 71
* $\chi(k, x)$	(A.22.-23) 474
* $\chi(t, k_0)$	(6.2.-12, 13) 236
* $\chi(x, k)$	(A.13.-3) 386
$\psi(x, k)$	(1.3.1.-1, 3) 16, 17; (2.1.-1, 3) 68, 69
$\psi^{(j)}(x, k)$	(2.4.1.-1, 1a, 1b) 89, 90
$\psi(x, k, t)$	122
* Ω	(A.5.-7) 315
* Ω	(A.23.-14) 479
∇_p, ∇_q	(A.23.-5, 6) 477
$\{\cdot, \cdot\}$	(A.23.-8) 477
$\{\{\cdot, \cdot\}\}$	(A.9.-26) 347
$[[\cdot, \cdot]]$	(A.9.-31) 347
#	see $g^{\#}(z_1, z_2)$ above